Heat Transport in Micro- and Nanoscale Thin Films

Bekir Sami Yilbas

Mechanical Engineering Department, King Fahd University of Petroleum and Minerals, Dhahran, Saudi Arabia

Saad Bin Mansoor

Mechanical Engineering Department, King Fahd University of Petroleum and Minerals, Dhahran, Saudi Arabia

Haider Ali

Mechanical Engineering Department, King Fahd University of Petroleum and Minerals, Dhahran, Saudi Arabia

ELSEVIER

Elsevier
Radarweg 29, PO Box 211, 1000 AE Amsterdam, Netherlands
The Boulevard, Langford Lane, Kidlington, Oxford OX5 1GB, United Kingdom
50 Hampshire Street, 5th Floor, Cambridge, MA 02139, United States

British Library Cataloguing-in-Publication Data
A catalogue record for this book is available from the British Library

Library of Congress Cataloging-in-Publication Data
A catalog record for this book is available from the Library of Congress

ISBN: 978-0-323-42979-5

Publisher: Matthew Deans
Acquisition Editor: Simon Holt
Editorial Project Manager: Sabrina Webber
Production Project Manager: Poulouse Joseph
Cover Designer: Greg Harris

Typeset by MPS Limited, Chennai, India

Contents

Preface

Dielectric and metallic micro/nanosized films are widely used in electronic, biomedical, sensor, and other industries. The recent developments in science and engineering enable us to reduce device sizes at micro/nanolevels, which becomes vital for the technological progresses in this century. The performance of micro/nanosized devices is highly related to heat generation and dissipation in the device. The fundamental understanding of the device response to heat generation and dissipation is critical for reducing device size and improving device performance. Micro/nanoscale energy transport becomes important in maintaining device performance over long operational periods. For example, tissue engineering is critical for biomedical applications; the micro/nanoscale energy transport remains important for the development of cells in various forms in biological tissues. In general, phonon transport governs the energy transfer in terms of heat in micro/nanoscale devices and thin films. The Fourier heat equation predicts temperature field and heat transfer rates within small-sized devices and thin films. Consequently, when modeling energy transfer in terms of heat, care should be taken to incorporate the wave nature of transport characteristics. Although hyperbolic equations can provide some useful information about the heat diffusion characteristics in micro/nanosized devices, it overlooks the frequency dependence of the transport. Therefore, energy transport in such devices can be modeled using the Boltzmann transport equation, which incorporates the frequency dependence of transport characteristics via dispersion relations of the device materials. Further adoption of the Boltzmann equation toward micro/nanoscale energy transport leads to the development of the equation for phonon radiative transport, which is currently used to tackle phonon transport in micro/nanoscale devices.

In some applications, micro/nanosized devices are composed of layered structures such as thin film pairs or the structures involved with minute-sized gaps such as nanosized pores and voids. The energy transport characteristics across the film pairs or minute-sized gaps are modified through thermal boundary resistance at the interfaces. Thermal boundary resistance lowers phonon transport across the interface; however, the amount of energy transport across the films pair interface depends on the mismatch between cut-off frequencies of phonon corresponding to both films. On the other hand, quantifying the phonon intensity distribution inside the film is challenging in terms of thermodynamic properties, such as temperature, because of the small scale of the film size. Therefore, equivalent equilibrium temperature can be defined anywhere in the thin film as an analog of the thermodynamic temperature as defined in the diffusive limit in any medium. Equivalent equilibrium temperature represents the average energy of all phonons around a local point and it is equivalent to the equilibrium temperature of phonons when they redistribute adiabatically to an equilibrium state. In the case of the presence of a minute-sized vacuum gap at the interface, ballistic phonons play a major role in energy transport across the films despite the fact that near field radiation contributes to energy transport across vacuum gaps. In metallic films, when the size of the vacuum gap becomes greater than that of the mean free path of electrons, energy transport by electron jump ceases across the vacuum gap. The reflected electrons from interfaces cause electron resistance in the close region of interface while influencing the energy state of electrons in this region. This, in turn, complicates energy transport across the vacuum gap between metallic thin films.

An analytical solution to the micro/nanoscale heating problem is possible after considerable simplification of the physical processes involved in micro/nanoscale thermal transport. Although an analytical solution of the heating problem provides a functional relation between the thermal response and the physical parameters of the film, oversimplification of the physical process via assumptions limits the practical importance of the solution. On the other hand, numerical treatment of the heating process can require excessive computational efforts to capture and apprehend the physical insight of micro/nanolevel processes. Nevertheless, the accuracy of analytical and numerical solutions remains in question because of the limitations in model studies featuring the complete physical aspects of the real phenomena. Analytical solution of heating complements numerical results. Formulation of micro/nanoscale heat transfer ends up with integrodifferential equations and an analytical approach for such equations requires significant efforts to obtain closed form solutions. However, once it is achieved, it serves the purpose and eliminates the excessive computations efforts; therefore, such efforts deserve acknowledgment and support for trial.

In this book, some basics of crystallography are introduced in line with energy transport at micro/nanoscale. In addition, relevant aspects of statistical thermodynamics pertinent to energy transfer are presented to form the background of energy transport characteristics in small-scale devices such as thin films. The basics of the Boltzmann equation and the equation for phonon transport are presented in the physical frame associated with the micro/nanoscale transport. Since thermal separation of electron and lattice subsystems takes place in metallic films when subjected to short duration of thermal disturbance, the electron kinetic theory approach is introduced to formulate governing equations of heat transfer. An attempt is made to reduce the heat equations into hyperbolic form, which covers the wave nature of the heating problem. Later, comparison of the heat equation derived from electron kinetic theory and that corresponding to a well-stated two-temperature model is made. An analytical solution of the equation for phonon radiative transfer is presented for the conditions of a steady and one-dimensional micro/nanoscale heating situation. A numerical solution of radiative phonon transport is provided for one-dimensional dielectric and metallic thin films, and later this effort is extended to include two-dimensional thin films. Since electron contribution to the energy transport is significant in metallic thin films, the equation for phonon radiative transport is modified to include electron–phonon coupling, and phonon intensity distribution in an electronic subsystem is, then, obtained accordingly.

Case studies are included to demonstrate the application of the equation for phonon transport in one- and two-dimensional films. These case studies involve laser short-pulse irradiation at film surfaces and temperature disturbance across film edges. Cross-plane transport across thin films pair is also included and some case studies are introduced to understand the phonon transport characteristics. Importance and application of thermal boundary resistance for cross-plane phonon transport is introduced and some case studies are included. The minute-sized vacuum gap is introduced at the films pair interface and its effects on phonon cross-transport characteristics are presented. A case study is incorporated to demonstrate the energy balance, including ballistic phonon contributions and near-field radiation across the vacuum gap, across the interface of films pair. However, other issues related to micro/nanoscale heat transfer are possible and these are not presented in this book due to space limitations. Future treatments will review the other processes.

Acknowledgment

We would like to acknowledge the role of the Deanship of Research of King Fahd University of Petroleum & Minerals in extending strong support, via project BW151002, from beginning to end, facilitating every aspect during the preparation of the book. The author wishes to thank to the colleagues who contributed to the work presented in the book through previous cooperation of the authors. In particular, thanks to Dr. Nasser Al-Aqeeli, Dr. Hussein Al-Qahtani, Dr. Ahmed Al-Dweik, and all the graduate students involved with micro/nanoscale transport research.

Chapter 1

Introduction

1.1 GENERAL CONSIDERATIONS

Thin films of dielectric materials of thickness within the range of 10 nm–100 μm are used extensively in semiconductor electronic devices. This is because of the fact that there has been an increasing focus on miniaturization of devices by researchers. In addition, micro- and nanoscale technologies have become one of the major research areas in academia. In the micro- and nanoscale analysis, some of the physical effects, which can be ignored at the macroscale analysis, become important while resulting in the error in estimations due to the differences in the properties of bulk, and micro- and nanoscale material. In order to design high efficiency micro- and nanoscale devices and to achieve desirable performances from these devices, a deep physical understanding of the transport process becomes necessary. These films form barrier layers between electrically conducting regions, and also have to conduct the heat generated in the devices. Since the performance of an electronic device depends on the device temperature, the study into energy transport characteristics in dielectric thin films becomes crucial [1]. A design methodology based on the Fourier law of heat conduction does not provide desirable accuracy since the transport characteristics do not match with the experimental data. Numerous researchers have concentrated their efforts toward the proper modeling and the solution of energy transport processes at micro- and nanoscale levels for such devices. The researchers face the problems in modeling of micro- and nanoscale energy transport mainly due to the transport properties of the small-size devices. When device sizes are comparable to the mean free path of the heat carriers, conventional methods of modeling, which are based on a continuum hypothesis, may no longer be applicable. Fourier law, which is quite accurate for common engineering situations [2] and based on the continuum hypothesis, becomes invalid for energy transport at micro- and nanoscale. It is now well established that conventional Fourier analysis leads to large erroneous thermal predictions when the mean free path of heat carriers is comparable to or larger than the device size [3].

On the other hand, energy transport at micro- and nanoscale is one of the important aspects of the process; therefore, it is the essential part of the

Heat Transport in Micro- and Nanoscale Thin Films.
DOI: http://dx.doi.org/10.1016/B978-0-323-42979-5.00001-3

area of research. The knowledge of thermal characteristics including energy transport and properties can help in determining the performance of micro- and nanoscale devices; e.g., effective thermal management of heat sources can improve the performance of devices, such as transistors on computer microprocessors. In solar energy technology, for direct conversion of solar energy into electrical energy, the photovoltaic (PV) cells are commercially used. Silicon is commonly used as a photoactive material in the manufacturing of PV cells. Since the absorption depth for solar radiation in silicon is large, a thick layer of material is required for sufficient absorption of solar radiation, making the PV cells expensive. In an effort to make solar technology economically viable, quantum dots are placed on the solar cells. Quantum dots with their high absorption characteristics are a promising solution for significantly reducing the absorber material requirement. In this regard, the physical understanding of energy transport in silicon thin film with the presence of quantum dots is of vital importance. Thermoelectric generators (TEGs) are cost-effective solid-state devices that are used to convert thermal energy into electrical energy. The TEG efficiency depends upon the operating temperature, figure of merit, and design configuration of the device. A material with high electrical conductivity and low thermal conductivity is required in order to increase the figure of merit and efficiency of TEG. In metals, at moderate temperatures, the ratio of the thermal conductivity to that of the electrical conductivity is directly proportional to the temperature, with the proportionality constant being independent of any particular metal. Therefore, improvements in the efficiency of thermoelectric elements using metals are difficult. The thermal conductivities of dielectric thin films are often smaller than those of their corresponding bulk materials; therefore, in microsized semiconductors, it may be possible to reduce the thermal conductivity without reducing the electrical conductivity. This in fact increases the figure of merit and efficiency of a thermoelectric device. In this case, the heat is mainly transported by phonons in a semiconductor whereas the electricity is transported by the electrons. Hence, it is important to have a thorough understanding of the transport processes in thin films.

Dielectric thin films are often in contact with a substrate material that may be a dielectric or a metal. The continuum hypothesis becomes invalid if the mean free path for the heat carrier is comparable or higher than the film and substrate thickness. Thermal energy transport between a dielectric film and a metal film is further complicated due to the nonequilibrium energy transfer between the electron and phonon subsystems in the metallic film, near the interface. As a result of this effect, thermal boundary resistance increases. It is important in such situations to explore a basic understanding of the thermal transport mechanisms. Net heat flux for two solids, which are a small distance apart by vacuum gap, is much greater than that predicted by the Stefan–Boltzmann law between two blackbodies [4]. Since energy transport between surfaces at close vicinity has important applications in nanoscale energy conversion devices and near-field scanning thermal microscopy [5], investigation of energy transport in thin films paired with the presence of minute gap is of importance. Energy transport in dielectric thin films with the combination of metallic/nonmetallic films and quantum dots plays an important role for the device performance. The thermal properties of the thin film mainly depend on the film size and heat source location. Therefore, investigation into the energy transport characteristics due to the film size and heat source location becomes necessary. Since the model study provides useful information on the transport characteristics, research focusing on the model study is essential.

1.2 SCALE OF ENERGY TRANSPORT

There exist two important size parameters that govern the energy transport in a thin film: mean free path (Λ) and characteristic dimension of the material (L). In dielectric materials, thermal energy is transported by means of lattice vibration about their equilibrium position. Among these vibrations, only certain frequencies of waves are responsible for carrying energy due to quantization. These waves can travel from one surface of thin film toward the other surface, which results in the energy transport between surfaces. These lattice vibrations are called phonons. The interaction of the two phonon waves is termed collision and average distance traveled between two successive interactions of phonon wave is known as mean free path of phonon [6]. In the diffusive transport, the characteristic dimension of the materials is much larger than the mean free path, that is ($L \gg \Lambda$); the energy transport is said to be macroscopic. In this case, Fourier law is applicable and the transport is governed by a purely diffusive nature. When the mean free path is much greater than the characteristic dimension of the material, that is ($\Lambda \gg L$), the energy transport is said to be microscopic. Since ($\Lambda \gg L$), the lattice wave reaches another surface without being scattered and is not affected by resistance. This is known as ballistic transport. Temperature gradient could not be established inside the film, and according to the Fourier law, it is impossible to prescribe the thermal conductivity [7]. Ballistic transport is also observed in a specific phonon wave in which the wavelength of phonon (λ) is greater than the distance in the x-direction, that is $\lambda > x$. Therefore in thin film at a specific distance it is possible that some phonon waves are reached by ballistic transport and some by diffusive transport. When the characteristic dimension of the material is comparable with the mean free path, then the transport is known as ballistic diffusive transport. This is a combination of ballistic and diffusive transport. The temperature jump can be observed at the films edges due to the boundary resistance.

However, in 1807, French scientist Joseph Fourier gave the relationship between heat flux q'' and temperature gradient ∇T, which are important thermal quantities in energy transport in solids. For homogeneous and isotropic medium, the Fourier law of heat conduction can be written as:

$$q'' = -k\nabla T \tag{1.1}$$

where k is the thermal conductivity. The Fourier law considers that heat flux q'' and temperature gradient ∇T appears at the same time instant; the thermal wave traveled with infinite speed. It means if the thermal disturbance is applied in the material, its effects can be measured at infinite distance from the thermal disturbance within no time. Although infinite speed of thermal wave is practically imperceptible, but for most common engineering situations Fourier's law is quite accurate [2]. However, with the fast development in science and technology creating situations involving temperature near absolute zero, extreme thermal gradients, high heat flux conduction, and short time behavior, such as laser–material interaction, under such situations the Fourier's law becomes invalid [3]. In order to remedy the short comings of the Fourier law, a single-phase lagging model is introduced [8]. Tzou [8] proposed the single phase lagging model, which considers the lagging behavior in heat flux, given by

$$q''(r, t + \tau_0) = -k\nabla T(r, t) \tag{1.2}$$

where r is the position vector, t is time, and τ_0 represents the time lag. According to this model for a particular position vector r, temperature gradient is established at time t, whereas heat flux appears at some later time $t + \tau_0$. Time lag (τ_0) is the relaxation time that physically represents the time difference needed to establish the heat flux at a position when a temperature gradient is applied. The first order expansion of τ_0 in Eq. (1.2) with respect to t bridges all the physical quantities at the same instant of time. This results in the expression

$$q''(r, t) + \tau_0 \frac{\partial q''}{\partial t}(r, t) \cong -k\nabla T(r, t) \tag{1.3}$$

which is the Cattaneo–Vernotte model provided by Cattaneo [9] and Vernotte [10]. This equation is reduced to Fourier law as τ_0 approaches to zero. Analysis of the single lagging model indicates that higher $\partial q''/\partial t$

results in a large deviation from the Fourier law. On the other hand, many experiments have showed that the single lagging model provides more accurate predications when compared with Fourier law. However, some of the predications by the single lagging model are not in good agreement with the experimental results [2,11,12]. In a detailed study, it has been revealed that the single lagging model does not consider size effects and only accounts for fast transients. To consider the size effect and fast transition the dual phase lag model is presented here [12]:

$$q''(r, t + \tau_0) = -k\nabla T(r, t + \tau_T) \tag{1.4}$$

In this expression, τ_T is the time lag in temperature gradient. It is the time delay due to the size effect and the microstructural interaction such as the phonon–electron interaction or phonon scattering in temperature gradient. τ_0 is the time lag of the heat flux, which incorporates fast transient effects. τ_T and τ_0 are intrinsic thermal properties of the bulk material.

Although single and dual-phase lagging models provide solutions to micro/nanoscale transport in solids, the physical fundamentals are not fully incorporated in the analysis, particularly metallic films. Moreover, the kinetic theory approach can be used for the analysis of nonequilibrium energy transport processes in metallic substrate. One nonequilibrium energy transport process is the heating of a metallic substrate when the heating time approaches the electron–phonon energy relaxation time. In this case, the Fourier heating model fails to predict the correct temperature distribution in the surface vicinity of the substrate [13]. In such a situation, a kinetic theory approach can be used to estimate the energy transport characteristics [14–16]. The electron kinetic theory approach considers the electron–phonon collision mechanism through which the energy exchange between the electrons and lattice site atoms occur. For ultrafast external pulse heating application the electron kinetic theory approach gives [17]

$$\rho C_p \frac{\partial T}{\partial t} = k\frac{\partial^2 T}{\partial x^2} + f\frac{\Lambda_e^2 \partial^2}{\partial x^2}\left(\rho C_p \frac{\partial T}{\partial t}\right) - \rho C_p \tau_s \frac{\partial^2 T}{\partial t^2} + I_0 \delta \exp(-\delta|x|) \tag{1.5}$$

where ρ is density, C_p is the specific heat, x is the spatial coordinate corresponding to x axis, f is the fraction of excess energy exchange, Λ_e is the mean free path of electrons, τ_s is the thermal relaxation time, I_0 is the peak power intensity, and δ is the absorption coefficient. When the second and third terms of the right-hand side in this equation are neglected, the equation results in the Fourier conduction equation with external heating term.

One of the fundamental solutions of micro/nanoscale energy transport is to use the Schrödinger wave equation with the appropriate boundary conditions. The wave particle duality of light, motivated de Broglie [18] to postulate that a material particle may also have wave properties. In continuation to the wave nature of material particles, Schrödinger introduced the equation in the form of the wave functions that govern the motion of small particles, and specify how these waves are altered by external influence. Schrödinger states that the wave function of any material obeys the following equation [19]:

$$-\frac{\hbar}{2m}\nabla^2\psi_t + U\psi_t = i\hbar\frac{\partial\psi_t}{\partial t} \tag{1.6}$$

where m is mass, t is the time, U is the potential energy constraint, and ψ_t is the wave function. In 1926, Born defined the wave function [6]. He stated that the product of wave function and its complex conjugate is the probability density function. Schrödinger's equation can be solved for the analysis of energy transport characteristics for some simple cases [20].

Molecular dynamics (MD) simulations can be used to calculate the energy transport characteristics in nanosize devices. These simulations calculate the motion of atoms in a molecular assembly using Newtonian dynamics to determine the net force and acceleration experienced by each particle. For ith particle with a mass m_i at r_i, MD simulation solved the following equation [6]:

$$m_i\frac{d^2r_i}{dt^2} = \sum_{\substack{j=1 \\ j\neq i}}^{N} F_{ij} \tag{1.7}$$

where N is the number of particles in the system and F_{ij} denotes the force exerted on particle i by particle j. MD simulation predicts the position and momentum of particles in the system, which can be used to investigate the transport characteristics in the system.

The first principle solution of energy transport at micro/nanoscale transport can be formulated using the Boltzmann transport equations (BTEs). The BTE, which is one of the subcontinuum methods, can also be incorporated to examine the nonequilibrium energy transport characteristics in micro- and nanoscale devices. The BTE predicts the nonequilibrium probability distribution, which can be used to determine the energy transport characteristics in nano- and microscale devices. The BTE can be written as [21]

$$\frac{\partial f}{\partial t} + \frac{p}{m}\cdot\nabla_r f + \dot{p}\cdot\nabla_p f = \left(\frac{\partial f}{\partial t}\right)_{\text{collision}} \tag{1.8}$$

where f is the nonequilibrium probability distribution function and p denotes momentum.

There exist the limitations of the models dictated for the micro/nanoscale energy transport. Fourier law, which is commonly used in macroscale energy transport analysis, does not provide the accurate estimation of transport characteristics at micro- and nanoscale level. Although single-phase lagging models overcome the deficiency of Fourier law regarding the infinite speed thermal wave, some predications by the single lagging model are not in good agreement with the experimental results [2,11,12]. The dual-phase lagging model considers the microstructural interaction, which was neglected in the single phase model. This model can only be used for the predication of energy transport in metallic thin film [22]. The kinetic theory approach can be used for the analysis of nonequilibrium energy transport processes in metallic substrate [14–16], but as the solution requires electron and lattice subsystems, it cannot be used for the analysis of dielectric thin films. The Schrödinger equation is used for the estimation of electrical characteristics in thin films [23,24], but for the analysis of energy transport characteristics in thin films it becomes very complicated to obtained proper solution. An MD simulation overcomes the major limitation of BTE methods, which is that they invariably require experiments to determine fitting parameters for the models. Therefore, for new and novel nanoscale materials for which no experiments have been performed or for cases where experiments are too difficult to perform, BTE approaches cannot be effectively used. Despite the fact that MD simulation can be used to determine the energy transport characteristics in thin films, it requires excessive computational power and can become unphysical due to periodic boundary conditions implemented in the solution. Thin films are too large to yield to atomistic-level simulation, and estimation of the energy transport characteristic must rely on solutions of the BTE [25]. Therefore, in this book, the BTE, which is one of the subcontinuum methods, is used to assess the energy transport characteristics in thin films including the ballistic effects.

In statistical mechanics, nonequilibrium processes are described by using the BTE, which is a semiclassical approach. The BTE is a nonclassical approach as it does not consider the matter as a continuum, but is considered to be composed of an extremely large number of particles, atoms, molecules, electrons, and so on. Therefore, this models the microscopic behavior of the material and macroscopic properties are comprehended from microscopic behavior. The BTE is classical in the sense that the particles are assumed to obey the laws of classical mechanics and no quantum mechanical effect is assumed. This may not be entirely true because some results from the quantum theory may be needed to model the scattering terms in the BTE. However, the wave mechanics formulation is never used directly and hence we may assume a classical treatment. Hamiltonian mechanics formulation of classical mechanics is usually used in the derivation of the relevant equations [26].

1.3 SOME ASPECTS OF STATISTICAL APPROACH FOR MICRO/NANOSCALE TRANSPORT

Statistical mechanics is the modern form of the kinetic theory of matter and was developed mainly by the scientists Clausius, Maxwell, Boltzmann, and Gibbs. In this section, a summary of the statistical mechanical approach of modeling is presented as applied to an ideal gas.

Consider a dilute, ideal, monatomic gas. This gas is composed of an extremely large number of exactly identical particles that are traveling in all directions, at tremendous speeds. During their motion, these particles undergo elastic collisions with one another. In each second there are an enormous number of collisions occurring within the gas. These collisions are assumed to be instantaneous and they change the momentum of the particles discontinuously, without changing their spatial positions. Furthermore, it is assumed that only two particles take part in any given collision and three or more particles colliding are extremely rare. In between the collisions, the particles travel in a straight trajectory, obeying the second law of motion or equivalently, obeying the laws of Hamiltonian mechanics. At any given time, different particles have different speeds and traveling directions. It is this microscopic picture of the gas that has to be used to deduce the macroscopic properties of it.

Two points should be noted at this juncture. The first is that to theoretically describe the exact dynamics of such a large number of particles that are also undergoing elastic collisions is virtually impossible. Even by employing modern numerical computations, the problem is without solution. The second thing to note is that the main goal is not to determine the exact dynamics of the particles but to determine the macroscopic manifestations of such dynamics. This comes about in the form of a few macroscopic properties of the gas, such as its total energy, entropy, density, temperature, and such. In fact, it is these properties that we can measure experimentally and not the exact speed of a particular atom or molecule. It is clear therefore that statistical behavior of the collection of a large number of particles having the dynamic characteristics is the main interest. Even in experiments conducted on the gas, it is the time averaged and space averaged quantities that we can measure, since the measurement time and the dimensions of the probe are much larger than the characteristic time of the dynamics of the gas particles and the size of these particles, respectively. Hence, it becomes necessary to apply statistical techniques to the motion of the gas particles, and this is the domain of the subject of statistical mechanics.

A single particle, moving in three dimensions and thus having three degrees of freedom, has associated with it three momentums. If the three coordinates of the particle along with the three momentums are prescribed at any given time then the dynamics of the particle is completely prescribed. If these six quantities are given as functions of time then the complete dynamic history of the particle is known. A Cartesian coordinate system with each single axis is assigned to each of the six quantities. There are three axes for the three position variables and three axes for the three momentum variables. This is called the phase space of the particle. The space consisting of only three position axes is called the configuration space. And the space consisting of only three momentum axes is called the velocity/momentum space. A single point in a phase space completely determines the dynamics of the particle at any given time. A trajectory in such a phase space completely determines the dynamic time history of the particle. This trajectory is obviously parameterized by the time variable. In order to describe the dynamic behavior of a single particle statistically, we may assume that the phase point corresponding to the dynamics of a particle can be anywhere in the phase space, but with a certain probability. That is, there is a certain probability associated with each location in the phase space. The magnitude of this probability tells the likelihood of the particle's phase point to be in that location. The probability can be defined through the probability distribution function; i.e., $f = f(x, y, z, p_x, p_y, p_z, t)$.

Another view of looking at this feature of statistical mechanics is by means of the concept of an ensemble. An ensemble is the name given to the collection of a large number of identical mental copies of a single particle. Each mental copy of a particle differs from the other mental copies in position, momentum, or both. Then, the density of this ensemble in phase space can be defined by the probability distribution function, which was defined earlier. Now, the minimum and maximum values of the position variables that any particle can take are dictated by the size of the container in which the particle is confined. The maximum value of the momentum is fixed by the total energy of the single particle. The probability density function will then give the number of phase points per unit infinitesimal volume of the phase space around any location (x, y, z, p_x, p_y, p_z) (where p_x, p_y, p_z are the momentums) in the phase space and at any given time. The point (x, y, z, p_x, p_y, p_z), for which this density is maximum, is the most probable point to which the dynamics of a single particle is assumed to correspond at that time. This is in turn will give the most probable position and the most probable momentums of a single particle at any given time. It turns out that the function attains sharp peak at the most probable values of the phase variables and therefore the most probable values of the phase variables are also equal to their corresponding expectation values that are calculated with respect to the probability distribution function. These are given by the expressions, $\bar{r}(t) = \int rf(r,p,t)dr\,dp$ and $\bar{p}(t) = \int pf(r,p,t)dr\,dp$ (where r and p are the position and momentum vectors, respectively). Hence, the actual trajectory of a single particle may now be given as the expectation values of the dynamical variables, corresponding to the time-dependent probability distribution function f. For N number of particles, the phase space is actually a $6N$ dimensional space. $3N$ axes correspond to the position variables and the other $3N$ corresponds to the momentum variables. Any point in the phase space completely specifies the dynamics of the N particles. If these N particles are those of our previously described collection of particles, then the dynamics of that collection is also completely prescribed at any given time. The probability density function can be defined as $f = f(x_i, y_i, z_i, p_{xi}, p_{yi}, p_{zi}, t)$ [6], which is now a function of the $6N$ variables as well as that of time t. It can be shown that the function f satisfies a continuity equation in phase space, which is given as [21]

$$\frac{\partial f}{\partial t} + \frac{1}{m}\sum_{i=1}^{N}\frac{\partial(p_{xi}f)}{\partial x_i} + \frac{1}{m}\sum_{i=1}^{N}\frac{\partial(p_{yi}f)}{\partial y_i} + \frac{1}{m}\sum_{i=1}^{N}\frac{\partial(p_{zi}f)}{\partial z_i} + \sum_{j=1}^{N}\frac{\partial(\dot{p}_{xj}f)}{\partial p_{xj}} + \sum_{j=1}^{N}\frac{\partial(\dot{p}_{yj}f)}{\partial p_{yj}} + \sum_{j=1}^{N}\frac{\partial(\dot{p}_{zj}f)}{\partial p_{zj}} = 0 \tag{1.9}$$

In vector notation this equation is written as

$$\frac{\partial f}{\partial t} + \frac{1}{m}\sum_{i=1}^{N}\nabla_{r_i}(p_i f) + \sum_{j=1}^{N}\nabla_{p_j}(\dot{p}_j f) = 0 \tag{1.10}$$

This equation, when expanded using the product rule and simplified using the Hamilton's equations of motion yield the Liouville's equation. These derivations are given as follows.

$$\frac{\partial f}{\partial t} + \frac{1}{m}\sum_{i=1}^{N} f\nabla_{r_i}p_i + \frac{1}{m}\sum_{i=1}^{N} p_i\nabla_{r_i}f + \sum_{j=1}^{N} f\nabla_{p_j}\dot{p}_j + \sum_{j=1}^{N}\dot{p}_j\nabla_{p_j}f = 0 \tag{1.11}$$

Hamilton's equations of motion for the N particles are

$$\dot{x}_i = \frac{\partial H}{\partial p_{x_i}}, \quad \dot{y}_i = \frac{\partial H}{\partial p_{y_i}}, \quad \dot{z}_i = \frac{\partial H}{\partial p_{z_i}} \quad \text{or} \quad \frac{p_i}{m} = \nabla_{p_i}H \tag{1.12a}$$

$$\dot{p}_{x_j} = -\frac{\partial H}{\partial x_j}, \quad \dot{p}_{y_j} = -\frac{\partial H}{\partial y_j}, \quad \dot{p}_{z_j} = -\frac{\partial H}{\partial z_j} \quad \text{or} \quad \dot{p}_j = -\nabla_{r_j}H \tag{1.12b}$$

$$\frac{\partial f}{\partial t} + \frac{1}{m}\sum_{i=1}^{N} p_i \nabla_{r_i} f + \sum_{j=1}^{N} \dot{p}_j \nabla_{p_j} f = 0 \tag{1.13}$$

The previous equation is the Liouville's equation. It describes the time-dependent behavior of the probability distribution function in the $6N$ dimensional phase space. In particular it can describe the dynamics of the previously described collection of N particles.

Due to the assumption that the particles are completely identical, the Liouville's equation can be applied in an extremely simplified form to a collection of particles. For identical particles, the function $f = f(x, y, z, p_x, p_y, p_z, t)$ will be exactly same for each particle; therefore, it is only necessary to consider a single particle and a six-dimensional phase space. However, the consequences of the collision processes between the particles have to be explicitly accounted for. This is done by adding a collision term on the right-hand side of the simplified Liouville's equation. It is this equation that is referred to as the BTE [27]. The three-dimensional form of the BTE is presented here [21].

$$\frac{\partial f}{\partial t} + \frac{p}{m} \cdot \nabla_r f + \dot{p} \cdot \nabla_p f = \left(\frac{\partial f}{\partial t}\right)_{\text{collision}} \tag{1.14}$$

In dielectric material, the term $(\partial f/\partial t)_{\text{collision}}$ describes the phonon scattering (collision) rate. There are two categories of phonon scattering: elastic and inelastic scattering. In elastic scattering, the phonon interacts with the lattice imperfection: crystal defects, dislocation, and the grain boundaries. However, the energy is conserved and the frequency of incident phonon is unchanged during an elastic scattering. Inelastic scattering occurs because of the interaction of three or more phonons. This is also known as intrinsic process [1] since it takes place because of the deviation from periodicity in perfect crystal due to the thermal lattice vibrations [27]. An inelastic event that involves four or more phonons is important only at higher temperatures [28]; therefore, it is neglected in the present study because of low temperature involvement ($300\text{K} \leq T \leq 301\text{K}$). The scattering term solution of the energy transport equation, in its rigorous form, is very complicated because we need to solve the Schrödinger equation for all particles system (i.e., electron and phonons). One way is to use the perturbation method [29] to solve the scattering term, but this also results in the complicated integral-differential equation, which is difficult to solve [6]. The other ways to treat the scattering term is the relaxation time approximation, based on the observation that scattering or collision of electron or

Table 1.1 Equilibrium Probability Distribution

f^0	Application
Maxwell–Boltzmann distribution	Ordinary or ideal gas
Fermi–Dirac distribution	Fermion particles (such as electrons)
Bose–Einstein distribution	Boson particles (such as phonons)

phonon tends to restore the equilibrium. Scattering term is based on the relaxation time approximation, which is given as [30]

$$\left(\frac{\partial f}{\partial t}\right)_{\text{collision}} = \frac{f^0 - f}{\tau} \tag{1.15}$$

where τ is the relaxation time and f^0 and f are the equilibrium and the nonequilibrium probability distribution functions. For different particles the equilibrium probability distribution f^0 is defined by different distribution functions, which are given in Table 1.1.

In order to quantify the relaxation time τ, the understanding of elastic and inelastic phonon scattering processes and their relaxation times are required. The elastic scattering relaxation τ_i time can be obtained from the kinetic theory approach [31]. There are two types of inelastic three-phonon scattering: the N process and the U process. In both processes, the energy conversation takes place with alteration in the frequencies. In the N process, momentum conservation takes place, therefore it does not directly contribute to the resistance in the energy flow [27]. However, due to frequencies alteration during the N process, it indirectly imposes the resistance in the heat flow [1]. In the U process, energy flow encountered the direct resistance as the momentum is not conserved during the scattering. The expressions for U process relaxation time τ_U and N process relaxation time τ_N are available in the open literature [32]. When the scattering processes are independent of each other, the relaxation time τ can be calculated using the Matthiessen's rule as follows:

$$\frac{1}{\tau} = \frac{1}{\tau_i} + \frac{1}{\tau_U} + \frac{1}{\tau_N} \tag{1.16}$$

where τ_i is the elastic scattering relaxation, τ_U is the U process relaxation time, τ_N is the N process relaxation time, and τ is the relaxation time. The assumption of independent scattering processes, in Matthiessen's rule, is not valid for inelastic scattering. But error involved in applying Matthiessen's rule is usually very small [27], and it gives correct results for most situations [6]. Therefore, it can be used in the measurement of relaxation time.

In general, the relaxation time approximation is adopted for the simplification of the scattering term in the BTE. Therefore, after incorporating the relaxation time approximation, the BTE can be written as

$$\frac{\partial f}{\partial t} + \frac{p}{m} \cdot \nabla_r f + \dot{p} \cdot \nabla_p f = \frac{f^0 - f}{\tau} \tag{1.17}$$

In a dielectric material, thermal energy is stored as the kinetic and potential energies of vibrating atoms or molecules. In crystalline materials, these particles are arranged in the shape of a periodic crystal lattice. Long chains of atoms and molecules vibrate with different frequencies, in longitudinal as well as transverse polarizations. Thermal energy is also transported through these lattice vibrations. According to quantum mechanics, the vibrational energy of an oscillator is quantized and is proportional to the frequency of vibration. Hence, only certain frequencies are allowed due to the specified distribution of energy. It is beneficial to regard a single quantum of vibrational energy as a particle called the phonon. This is analogous to the electromagnetic energy quantum, the photon. However, the phonons and photons are not completely equivalent.

According to this picture we may assume that thermal energy is stored and transported by the motion of phonons and since the number of phonons is very large we may conceive of a phonon gas permeating the dielectric material. To this phonon gas the statistical mechanics concepts developed for the particles of an ideal gas can be applicable. To obtain the equilibrium properties, an equilibrium probability distribution function may be used, whereas to compute the transport properties BTE has to be used. In this respect, it is important to realize that the phonons do not obey the Maxwell–Boltzmann distribution function like the gases. This is because the number of phonons is not conserved. Furthermore, they do not obey the Pauli Exclusion principle. Hence, we may consider the phonons as a type of boson, just as the photons are. The correct equilibrium distribution to use in this case is the Bose–Einstein distribution. Based on this distribution and using the BTE, we may compute the thermal conductivity of the dielectric material under a Fourier law assumption or without it. Usually the Fourier law breaks down in that direction along which the material dimension is of the order of the phonon mean free path.

REFERENCES

[1] A. Majumdar, Microscale heat conduction in dielectric thin films, J. Heat Transfer 115 (1) (1993) 7–16.

[2] L. Wang, Generalized Fourier law, Int. J. Heat Mass Transfer 37 (17) (1994) 2627–2634.

[3] G. Atefi, A. Bahrami, M.R. Talaee, Analytical solution of dual phase lagging heat conduction in a hollow sphere with time-dependent heat flux, New Aspects Fluid Mech. Heat Transfer Environ. 1 (2010) 114–126.
[4] J.-P. Mulet, K. Joulain, R. Carminati, J.-J. Greffet, Enhanced radiative heat transfer at nanometric distances, Microscale Thermophys. Eng. 6 (3) (2002) 209–222.
[5] C.J. Fu, Z.M. Zhang, Nanoscale radiation heat transfer for silicon at different doping levels, Int. J. Heat Mass Transfer 49 (9–10) (2006) 1703–1718.
[6] G. Chen, Nanoscale Energy Transport and Conversion, Oxford University Press, Oxford, 2005.
[7] C.V.D.R. Anderson, PhD Thesis, Thermal Heat Transport Characterization for Macroscale, Microscale, and Nanoscale Heat Conduction, The University of Minnesota, 2008.
[8] D.Y. Tzou, Thermal shock phenomena under high-rate response in solids, Annu. Rev. Heat Transfer 4 (1992) 111–185.
[9] C. Cattaneo, A form of heat conduction equation which eliminates the paradox of instantaneous propagation, Compte Rendus 247 (1958) 431–433.
[10] P. Vernotte, Les paradoxes de la theorie continue de l'equation de la chaleur, Compte Rendus 246 (1958) 3154–3155.
[11] D.Y. Tzou, Unified field approach for heat conduction from macro- to micro-scales, J. Heat Transfer 117 (1) (1995) 8–16.
[12] T.-W. Tsai, Y.-M. Lee, Y.-H. Liao, Analysis of microscale heat transfer and ultrafast thermoelasticity in a multi-layered metal film, in Proceedings of the ASME Summer Heat Transfer Conference 2009, HT2009, 2009, vol. 2, pp. 87–95.
[13] B.S. Yilbas, Validity of Fourier theory of radiation heating of metals, Res. Mech. 24 (4) (1988) 377–382.
[14] B.S. Yilbas, Heating of metals at a free surface by laser irradiation an electron kinetic theory approach, Int. J. Eng. Sci. 24 (8) (1986) 1325–1334.
[15] B.S. Yilbas, Electron kinetic theory approach—one- and three-dimensional heating with pulsed laser, Int. J. Heat Mass Transfer 44 (10) (2001) 1925–1936.
[16] B.S. Yilbas, Improved formulation of electron kinetic theory approach for laser ultra-short-pulse heating, Int. J. Heat Mass Transfer 49 (13–14) (2006) 2227–2238.
[17] B.S. Yilbas, Electron kinetic theory approach for picosecond laser pulse heating, Numer. Heat Transfer, Part A 39 (8) (2001) 823–845.
[18] L. De Broglie, Recherches sur la Theorie des Quanta, Ann. Phys. (Paris) 3 (1925).
[19] E. Schrödinger, Quantisation as a problem of characteristic values, Ann. Phys. 79 (1926) 361–376, 489–529.
[20] R.K. Kawakami, E. Rotenberg, H.J. Choi, E.J. Escorcia-Aparicio, M.O. Bowen, J.H. Wolfe, et al., Quantum-well states in copper thin films, Nature. 398 (6723) (1999) 132–134.
[21] R.L. Liboff, Kinetic Theory, second ed., Willey, New York, 1998.
[22] D.Y. Tzou, Lagging behavior in ultrafast laser heating, in American Society of Mechanical Engineers, Heat Transfer Division, 1995, 304, 2, p. 139.
[23] M. Chen, J.M. Nash, C.E. Patton, A numerical study of nonlinear Schrödinger equation solutions for microwave solitons in magnetic thin films, J. Appl. Phys. 73 (8) (1993) 3906.

[24] H. Wong, V.A. Gritsenko, Defects in silicon oxynitride gate dielectric films, Microelectron. Reliab. 42 (4–5) (2002) 597–605.

[25] D.G. Cahill, W.K. Ford, K.E. Goodson, G.D. Mahan, A. Majumdar, H.J. Maris, et al., Nanoscale thermal transport, J. Appl. Phys. 93 (2) (2003) 793.

[26] S. Bin Mansoor, PhD Thesis, Absorption of Light Waves and Energy Transport in Thin Silicon Film in Relation to Photovoltaic Applications, KFUPM, 2010.

[27] J.M. Ziman, Electrons and Phonons: The Theory of Transport Phenomena in Solids, Oxford University Press, London, 1960.

[28] Y. Joshi, M. Tiwari, G. Verma, Role of four-phonon processes in the lattice thermal conductivity of silicon from 300 to 1300°K, Phys. Rev. B 1 (2) (1970) 642–646.

[29] L.D. Landau, E.M. Lifshitz, Quantum Mechanics, Pergamon, 1977.

[30] C. Kittel, Introduction to Solid State Physics, sixth ed., Willey, 1986.

[31] W.G. Vincenti, Introduction to Physical Gas Dynamics, Robert E. Krieger, Florida, 1966.

[32] M.G. Holland, Analysis of lattice thermal conductivity, Phys. Rev. 132 (6) (1963) 2461–2471.

Chapter 2

Crystal Dynamics and Lattice Waves

2.1 INTRODUCTION

Solids may be classified into two types based on their structure: crystalline and amorphous. Metals and dielectric materials have a crystalline structure and this book is only concerned with crystalline materials. In a crystal, the atoms and/or molecules are arranged in a periodic manner, such that the local structure of the solid is the same everywhere. Due to this periodicity, the mathematical analysis of the crystalline structure is considerably simplified. An example of a crystal lattice can be observed in Fig. 2.1.

As we can observe from Fig. 2.1, a crystal lattice can be broken down into a fundamental unit called the unit cell. The crystal lattice is made of a repetition of this unit cell. A unit cell is not unique; a crystal lattice can be constructed from a repetition of various unit cells. A unit cell that consists of a single atom/molecule is called the primitive unit cell, which is unique for that crystal lattice. An example of a primitive unit cell is displayed in Fig. 2.2.

In Fig. 2.2, the lengths a, b, c and the angles α, β, γ can have various values. All primitive unit cells can be divided in to seven groups based on the relations between the lengths a, b, c and the relations between the angles α, β, γ. These relations as well as the names of the primitive unit cell are described in Table 2.1.

The atoms/molecules in a crystal are not free to move as in a liquid. However, they vibrate about their mean positions. Since the atoms/molecules apply a force on their neighbors, the atoms/molecules do not vibrate in isolation but vibrate in unison and therefore, longitudinal as well as transverse waves are produced in the crystal. These waves can only take on specific, allowed frequencies and wavelengths. This is because of the discrete nature of the crystal lattice. In addition, the phase velocity of the waves is not constant but is a function of the wavelength. Consequently, the frequency is also a function of the wavelength. The relation between the frequency and wavelength is called the dispersion relation. A single wavelength must be comprised of at least two atoms/molecules and therefore there is a minimum

Heat Transport in Micro- and Nanoscale Thin Films.
DOI: http://dx.doi.org/10.1016/B978-0-323-42979-5.00002-5

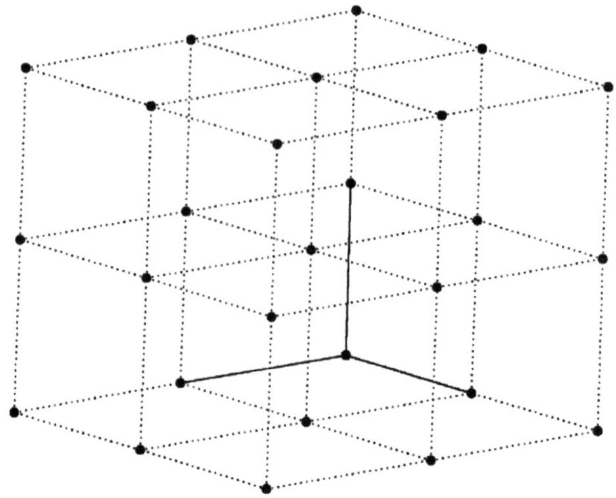

■ **FIGURE 2.1** A crystal lattice.

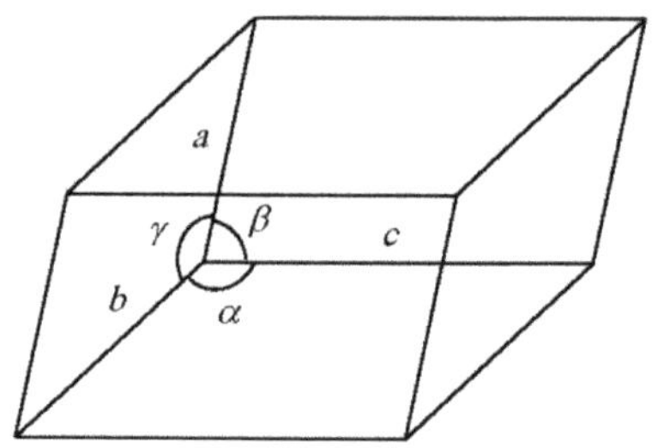

■ **FIGURE 2.2** A unit cell of the crystal lattice.

Table 2.1 The Seven Primitive Cells [1]

Primitive Cell	Lengths	Angles
Triclinic	$a \neq b \neq c$	$\alpha \neq \beta \neq \gamma$
Monoclinic	$a \neq b \neq c$	$\alpha = \beta =$ 90 degree $\neq \gamma$
Orthorhombic	$a \neq b \neq c$	$\alpha = \beta = \gamma =$ 90 degree
Tetragonal	$a = b \neq c$	$\alpha = \beta = \gamma =$ 90 degree
Cubic	$a = b = c$	$\alpha = \beta = \gamma =$ 90 degree
Hexagonal	$a = b \neq c$	$\alpha = \beta =$ 90 degree, $\gamma =$ 120 degree
Trigonal	$a = b = c$	$\alpha = \beta = \gamma \neq$ 90 degree

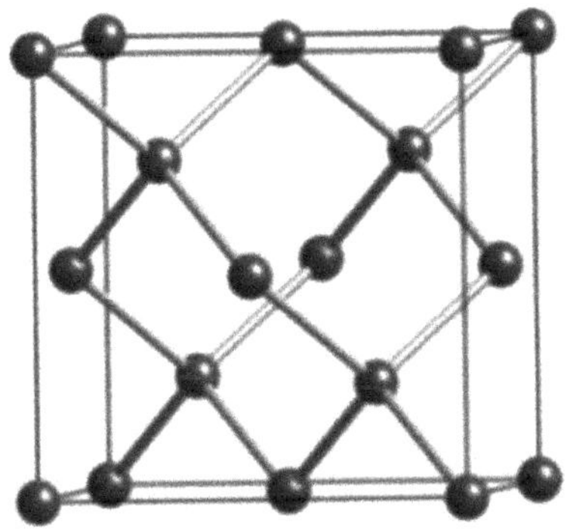

■ **FIGURE 2.3** Crystal structure of diamond.

wavelength called the cut-off wavelength that the crystal can sustain. Likewise, there is a maximum frequency called the cut-off frequency that the crystal can sustain.

2.2 ELEMENTARY CRYSTALLOGRAPHY

Crystallography is the study of atomic and molecular structures in a crystalline material. A crystalline material is one in which the atoms or molecules are arranged in a periodic or repeating array. This periodic arrangement extends to distances that are much larger than the atomic distances. Therefore, there is long range order in such a material. Additionally, due to the repetitive structure the properties of a crystal are also periodic. Under normal conditions, all metals, and some ceramics and polymers, acquire a crystalline structure. Many properties of a crystal depend on its specific crystalline structure. The number of various possible crystal structures is very large. Metals usually have simple crystal structures whereas some ceramic and polymers have very complex crystal structures.

Crystal structures have usually been studied by means of X-ray diffraction techniques. This is because the wavelength of the X-ray radiation is of the same order of magnitude as that of the atomic distances; hence X-rays can resolve the fine details of the crystalline structure. Electron and neutron beams are also used to probe the crystalline structure. These diffraction studies produce diffraction patterns from which the crystalline structure can be deduced.

To analyze the properties of a crystal that depend on its structure, it is necessary to first describe that structure in mathematical terms. In the next section, the mathematical description of a crystal structure is presented.

2.2.1 Structure of Crystal Lattices

To mathematically describe waves in a crystal, it is first necessary to be able to describe the position of each atom/molecule in the crystal relative to some reference location. This is accomplished by defining a position vector $\mathbf{r}$ from the reference location to the point in question occupied by the atom/molecule. The position vector is defined in terms of the basis vectors such as $\mathbf{i}, \mathbf{j}, \mathbf{k}$ in a Cartesian coordinate system. The most fundamental choice for the basis vectors are those vectors with magnitudes a, b, c and having directions along the lengths a, b, c as described in Fig. 2.2. However, as may be observed, a Cartesian coordinate system can only be defined for some types of primitive unit cells. It is therefore necessary to use an oblique coordinate system defined by the basis vectors as previously discussed. The basis vectors will be denoted as $\mathbf{E}_1, \mathbf{E}_2, \mathbf{E}_3$ and have as their magnitudes $|\mathbf{E}_1| = a$, $|\mathbf{E}_2| = b$, $|\mathbf{E}_3| = c$. These are shown in Fig. 2.4.

Using an oblique coordinate system makes it necessary to use the methods of tensor analysis. The basis vectors $\mathbf{E}_1, \mathbf{E}_2, \mathbf{E}_3$ are called the covariant basis vectors. The position of any atom in the crystal lattice can then be written as

$$\mathbf{r} = l_1\mathbf{E}_1 + l_2\mathbf{E}_2 + l_3\mathbf{E}_3 \tag{2.1}$$

In Eq. (2.1) the variables l_1, l_2, l_3 take on integer values; i.e.,

$$l_1, l_2, l_3 = 1, 2, 3, \ldots \tag{2.2}$$

In this manner, the position of any atom is described with reference to the origin. The primitive cell is also called a direct cell. Finally, we note that the covariant basis vectors $\mathbf{E}_1, \mathbf{E}_2, \mathbf{E}_3$ do not change from one lattice point to another. They have the same magnitude as well as direction at each lattice point.

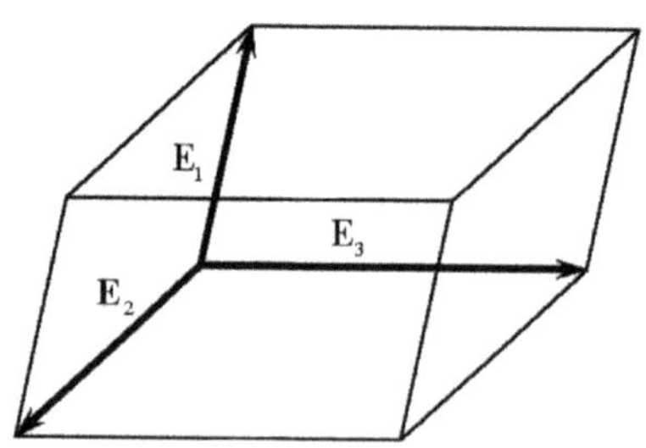

■ **FIGURE 2.4** The oblique coordinate system and basis vectors.

2.2.2 **Reciprocal Lattices**

In addition to the covariant basis vectors $\mathbf{E}_1, \mathbf{E}_2, \mathbf{E}_3$ we can also define the contravariant basis vectors at the origin. These are defined as

$$\mathbf{E}^i \cdot \mathbf{E}_j = \delta^i_j \quad i,j = 1,2,3 \tag{2.3}$$

Another equivalent approach of defining the contravariant basis vectors is as follows. We first describe the volume of the primitive cell V_d, which is defined as

$$V_\mathrm{d} = \mathbf{E}_1 \cdot (\mathbf{E}_2 \times \mathbf{E}_3) \tag{2.4}$$

Then the contravariant basis vectors are defined as

$$\mathbf{E}^1 = \frac{\mathbf{E}_2 \times \mathbf{E}_3}{V_\mathrm{d}} \qquad \mathbf{E}^2 = \frac{\mathbf{E}_3 \times \mathbf{E}_1}{V_\mathrm{d}} \qquad \mathbf{E}^3 = \frac{\mathbf{E}_1 \times \mathbf{E}_2}{V_\mathrm{d}} \tag{2.5}$$

A new position vector can be defined by using the contravariant basis vectors,

$$\mathbf{a} = h_1\mathbf{E}^1 + h_2\mathbf{E}^2 + h_3\mathbf{E}^3 \tag{2.6}$$

In Eq. (2.6) the variables h_1, h_2, h_3 take on integer values; i.e.,

$$h_1, h_2, h_3 = 1, 2, 3, \ldots \tag{2.7}$$

In this manner the position vector in Eq. (2.6) will define an imaginary point in space, which we may designate as a lattice point in a new lattice called the reciprocal lattice. From these imaginary lattice points we may construct a primitive cell in the reciprocal space. The volume of a primitive cell in the reciprocal space is given as

$$V_\mathrm{r} = \mathbf{E}^1 \cdot (\mathbf{E}^2 \times \mathbf{E}^3) \tag{2.8}$$

It can be shown that

$$V_\mathrm{d} V_\mathrm{r} = 1 \tag{2.9}$$

Substitute the definition of $\mathbf{E}^2$ and $\mathbf{E}^3$ from Eq. (2.5) into Eq. (2.8) to get

$$V_\mathrm{r} = \mathbf{E}^1 \cdot \left(\frac{\mathbf{E}_3 \times \mathbf{E}_1}{V_\mathrm{d}} \times \frac{\mathbf{E}_1 \times \mathbf{E}_2}{V_\mathrm{d}} \right) \tag{2.10}$$

Using a vector identity, Eq. (2.10) may be written as

$$\begin{aligned} V_\mathrm{r} &= \frac{\mathbf{E}^1}{V_\mathrm{d}^2} \cdot (\mathbf{E}_1(\mathbf{E}_3 \cdot (\mathbf{E}_1 \times \mathbf{E}_2)) - \mathbf{E}_3(\mathbf{E}_1 \cdot (\mathbf{E}_1 \times \mathbf{E}_2))) \\ V_\mathrm{r} &= \frac{\mathbf{E}^1 \cdot \mathbf{E}_1}{V_\mathrm{d}} = \frac{1}{V_\mathrm{d}} \\ V_\mathrm{d} V_\mathrm{r} &= 1 \end{aligned} \tag{2.11}$$

2.2.3 Crystal Planes and Directions

A plane in a three-dimensional Cartesian coordinate system is defined by the equation,

$$\frac{x}{A} + \frac{y}{B} + \frac{z}{C} = 1 \tag{2.12}$$

where the numbers A, B, C are the intercepts of the plane on the x, y, z axes, respectively. These numbers may be positive or negative. If the numbers A, B, C are given then a plane is completely defined. In a similar fashion, a plane in a crystal in an oblique coordinate system is defined by its intercepts on the $\mathbf{E}_1, \mathbf{E}_2, \mathbf{E}_3$ axes.

$$\frac{l_1}{A} + \frac{l_2}{B} + \frac{l_3}{C} = 1 \tag{2.13}$$

Now we know that l_1, l_2, l_3 are integers and therefore A, B, C are also integers. Hence if the numbers (A, B, C) are given then the crystal plane is completely specified. We note that these numbers are positive or negative integers. All planes parallel to each other will be denoted by the same set of numbers (A, B, C) except for a multiplication factor. Hence it is customary to denote all parallel planes with that set of numbers (A, B, C), which have no common factor among them. The numbers (A, B, C) are then called the *miller indices* of a plane. A set of parallel planes will be denoted as $[A, B, C]$.

In a three-dimensional Cartesian coordinate system a plane may also be defined in terms of a vector normal to it. From elementary vector calculus we know that a vector normal to the plane given by Eq. (2.12) is

$$\mathbf{N} = \frac{\mathbf{i}}{A} + \frac{\mathbf{j}}{B} + \frac{\mathbf{k}}{C} \tag{2.14}$$

The equation of a plane is then written as

$$\mathbf{N} \cdot \mathbf{r} = d \tag{2.15}$$

where $\mathbf{r} = x\mathbf{i} + y\mathbf{j} + z\mathbf{k}$ and $d/|\mathbf{N}|$ is the shortest distance of the plane from the origin. Thus if the normal vector is known or in other words if the numbers (A, B, C) are known then the plane is completely defined. In a crystal with an oblique coordinate system a plane may be defined by a similar method (Eq. 2.15):

$$\mathbf{N} \cdot \mathbf{r} = d$$

Now we have $\mathbf{r} = l_1\mathbf{E}_1 + l_2\mathbf{E}_2 + l_3\mathbf{E}_3$. The normal vector $\mathbf{N}$ may be written in terms of the covariant basis vectors or equivalently in terms of the contravariant basis vectors. It will be convenient to use the contravariant basis for the normal vector. Let

$$\mathbf{N} = h_1\mathbf{E}^1 + h_2\mathbf{E}^2 + h_3\mathbf{E}^3 \tag{2.16}$$

Substituting the expressions of **N** and **r** into Eq. (2.15) we get

$$\begin{aligned}&(h_1\mathbf{E}^1 + h_2\mathbf{E}^2 + h_3\mathbf{E}^3)\cdot(l_1\mathbf{E}_1 + l_2\mathbf{E}_2 + l_3\mathbf{E}_3) = d\\&h_1l_1 + h_2l_2 + h_3l_3 = d\end{aligned} \tag{2.17}$$

Eq. (2.17) may be rewritten as

$$\frac{l_1}{d/h_1} + \frac{l_2}{d/h_2} + \frac{l_3}{d/h_3} = 1 \tag{2.18}$$

Comparing with Eq. (2.13) we find that

$$A = \frac{d}{h_1} \qquad B = \frac{d}{h_2} \qquad C = \frac{d}{h_3} \tag{2.19}$$

Hence, the numbers (A, B, C) and $(1/h_1, 1/h_2, 1/h_3)$ are in the same ratio and any set of these three numbers can be used to define a plane in a three-dimensional crystal. The number d is such that the numbers h_1, h_2, h_3 are all integers, positive or negative. From this discussion we can conclude that any vector $\mathbf{a} = h_1\mathbf{E}^1 + h_2\mathbf{E}^2 + h_3\mathbf{E}^3$ defined in the reciprocal lattice actually defines a plane in the direct lattice. All parallel planes will be defined by the same vector. In addition, the minimum (perpendicular) distance between any two parallel planes defined by the vector $\mathbf{a} = h_1\mathbf{E}^1 + h_2\mathbf{E}^2 + h_3\mathbf{E}^3$ is given as

$$\delta = \frac{1}{|\mathbf{a}|} \tag{2.20}$$

For a proof of Eq. (2.20) the reader may refer to literature [2].

A direction in the crystal lattice is defined by the vector $\mathbf{r} = l_1\mathbf{E}_1 + l_2\mathbf{E}_2 + l_3\mathbf{E}_3$. All directions parallel to the original one will be defined by the same vector $\mathbf{r} = l_1\mathbf{E}_1 + l_2\mathbf{E}_2 + l_3\mathbf{E}_3$ except for a multiplication factor. Hence, a particular direction will be given by the three numbers (l_1, l_2, l_3) such that there is no common factor among them. A set of parallel directions will be denoted as $\langle l_1, l_2, l_3 \rangle$.

2.3 LATTICE VIBRATION

2.3.1 Chain of Identical Atoms

Atoms/molecules in a three-dimensional crystal are in a constant vibratory motion about their mean positions. This vibratory motion is not in isolation. Each atom/molecule imparts a force on its neighbors and hence all the atoms/molecules in a crystal lattice are vibrating in unison. From the theory of vibrations we know that for such a system we can define normal modes, wavelengths, and so on. To gain an understanding of the behavior of this vibratory motion it is helpful to first consider a one-dimensional chain of atoms and study the characteristics of its vibratory motion.

Consider a chain of atoms each of mass m and separated by a mean distance d. The distance along the chain will be demoted by the coordinate x. Since the atoms are vibrating there will be a displacement of the atom from its mean position. Let that displacement be denoted by y. Then the following relation exists between these variables:

$$x_n = nd + y_n \tag{2.21}$$

The number n denotes a particular atom in the chain starting from 0 for the atom at the origin. Eq. (2.21) is self-evident. The location of the mean position of atom number n is simply nd. Due to vibrations this location may change slightly by an amount y_n, which may be positive or negative (Fig. 2.5).

The vibratory motion is related to the relative positions of the atoms with respect to each other. In this discussion we will assume only nearest neighbor interactions; i.e., an atom will have negligible, direct influence on atoms that are not its immediate neighbors. The relative distance between the nth and the $(n - 1)$th atom is given as

$$r_{n-1,n} = x_n - x_{n-1} = d + y_n - y_{n-1} \tag{2.22}$$

and that between the nth and the $(n + 1)$th atom is given as

$$r_{n,n+1} = x_{n+1} - x_n = d + y_{n+1} - y_n \tag{2.23}$$

The relative distances $r_{n,n-1}$ and $r_{n,n+1}$ are always positive. The potential energy between any two atoms is a function of their relative distance. And the force between two atoms is equal to the negative of the gradient of the potential energy between them. When the atoms occupy their mean positions, the resultant force on any atom is zero, which means that

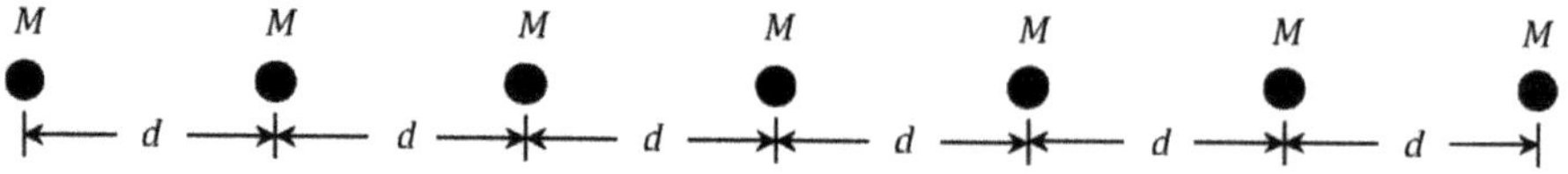

■ FIGURE 2.5 A one-dimensional chain of identical atoms.

the potential energy is a minimum. Let the potential energy be denoted as $U(r)$; then the force on the nth atom may be written as

$$F_n = -\frac{\partial U(r_{n,n-1})}{\partial y_n} - \frac{\partial U(r_{n+1,n})}{\partial y_n} \tag{2.24}$$

If the displacements y_n are small than the potential energy function $U(r)$ can be expanded as a Taylor series as described next.

$$U(r_{n-1,n}) = U(d) + (y_n - y_{n-1})\frac{\partial U}{\partial r_{n-1,n}}\bigg| + \frac{(y_n - y_{n-1})^2}{2}\frac{\partial^2 U}{\partial r^2_{n-1,n}}\bigg|_{r=d} \tag{2.25}$$

$$U(r_{n,n+1}) = U(d) + (y_{n+1} - y_n)\frac{\partial U}{\partial r_{n,n+1}}\bigg| + \frac{(y_{n+1} - y_n)^2}{2}\frac{\partial^2 U}{\partial r^2_{n,n+1}}\bigg|_{r=d} \tag{2.26}$$

In these expansions, only terms of the second order are considered. Since we are interested in the force on the atoms, $U(d)$ can be taken to be zero, because it is not going to contribute to the force. The second term in Eqs. (2.25) and (2.26) vanish; i.e., $\partial U/\partial r_{n-1,n}\big|_{r=d} = \partial U/\partial r_{n,n+1}\big|_{r=d} = 0$. This is because the potential energy is a minimum at the mean position. Denoting $\partial^2 U/\partial r^2_{n-1,n}\big|_{r=d} = \partial^2 U/\partial r^2_{n,n+1}\big|_{r=d}$ as U'', the force on the nth atom may finally be written as

$$F_n = -\frac{\partial}{\partial y_n}\left[\frac{(y_n - y_{n-1})^2}{2}U''\right] - \frac{\partial}{\partial y_n}\left[\frac{(y_{n+1} - y_n)^2}{2}U''\right] \tag{2.27}$$

$$F_n = (y_{n+1} - 2y_n + y_{n-1})U'' \tag{2.28}$$

The equation of motion for the nth atom is now written as

$$m\frac{d^2 y_n}{dt^2} = F_n = (y_{n+1} - 2y_n + y_{n-1})U'' \tag{2.29}$$

Following the development in literature [2], we now assume a wave solution for the displacement y_n:

$$y_n = Ae^{-i2\pi(\nu t - ax_n)} = Ae^{-i2\pi(\nu t - and)} \tag{2.30}$$

In Eq. (2.30), ν is the frequency and a is the wavenumber of the wave solution. The wavelength of the wave is given as $\lambda = 1/a$. Substituting the solution in Eq. (2.30) into Eq. (2.29) and simplifying we get

$$-4A\pi^2\nu^2 me^{-i2\pi(\nu t - and)} = U''(e^{i2\pi ad} - 2 + e^{-i2\pi ad})Ae^{-i2\pi(\nu t - and)} \tag{2.31}$$

$$\pi^2\nu^2 m = U'' \sin^2(\pi ad) \tag{2.32}$$

From Eq. (2.32) we may find the dispersion relation for the chain of atoms. We note that the frequency is a function of the wavenumber and hence of the wavelength. This is given as

$$\nu = \sqrt{\frac{U''}{m}}\frac{\sin(\pi a d)}{\pi} \tag{2.33}$$

The phase velocity V_p is defined and given as

$$V_p = \nu\lambda = \frac{\nu}{a} \tag{2.34}$$

$$V_p = \sqrt{\frac{U''}{m}}\frac{\sin(\pi a d)}{\pi a} \tag{2.35}$$

The group velocity is defined and given as

$$V_p = \frac{\partial\nu}{\partial a} = d\sqrt{\frac{U''}{m}}\cos(\pi a d) \tag{2.36}$$

A very important aspect concerning the vibratory motion of the chain of atoms is the concept of the *cut-off wavelength.* Related to this is the concept of the *cut-off frequency*. The waves in the chain of atoms consist of several atoms vibrating together. These waves are of different wavelengths. Hence in some waves several atoms are there in a single wavelength, whereas in others, only a few atoms may be there. However, we may note that the minimum number of atoms in a wavelength must be two; otherwise the concept of the wavelength will become meaningless. As a result, we conclude that the wave of minimum wavelength that the chain of atoms is capable of sustaining has a wavelength of $2d$. This is called the cut-off wavelength, $\lambda_{\text{cut-off}}$. Any wave excited in the crystal that has a wavelength less than this number will die out quickly. The reciprocal of the cut-off wavelength is the cut-off wavenumber $a_{\text{cut-off}} = 1/\lambda_{\text{cut-off}}$. This concept of the cut-off wavelength is a manifestation of the discrete nature of the chain of atoms. In a continuous medium wavelength of any magnitude is allowed.

$$\lambda_{\text{cut-off}} = 2d, \quad a_{\text{cut-off}} = \frac{1}{2d} \tag{2.37}$$

From Eq. (2.33) we deduce that corresponding to a cut-off wavelength there is also a cut-off frequency, $\nu_{\text{cut-off}}$. This is defined as

$$\nu_{\text{cut-off}} = \sqrt{\frac{U''}{m}}\frac{\sin(\pi a_{\text{cut-off}}\, d)}{\pi} = \frac{1}{\pi}\sqrt{\frac{U''}{m}} \tag{2.38}$$

Waves of frequency greater than the cut-off frequency will not be sustained by the crystal lattice.

2.3.2 Phonons

Real crystals are of finite size. This aspect forces the quantization of the waves, which means that not all wavelengths are allowed in the crystal. Or that the phonon frequency spectrum is discrete instead of continuous. To understand this concept we may again consider the chain of atoms as initially described in Section 2.3.1. Let the total number of atoms in the chain be N. Then the length of the chain is $L = Nd$. Now, only those waves are allowed, the integral multiple of whose wavelength is equal to the length of the chain.

$$n\lambda = L \Rightarrow \lambda_n = \frac{L}{n} \quad n = 1, 2, \ldots n_{\text{cut-off}} \tag{2.39}$$

The electrons emit electromagnetic waves when they undergo a transition from one orbit to another. These waves are of specific wavelengths corresponding to the energy difference in the orbits. Hence, these waves are also treated as a particle emitted by the electron and are called *photons*. Analogously, the waves of allowed wavelengths sustained by the crystal lattice can be treated as a particle and these are called *phonons*. Hence, we talk about the wavelength, wavenumber, and frequency of the phonons.

Eq. (2.39) can be derived in another way. Eq. (2.30) describes a wave propagating through the crystal. If we now demand that the amplitude (considering Eq. (2.30), which is given in Section 2.3.1)

$$y_n = Ae^{-2\pi i(\nu t - and)}$$

be the same at $x = 0$ and at $x = L$ then this constitutes a *periodic boundary condition*. This condition when applied to the amplitude function results in the following outcome:

$$\begin{aligned} y_0 &= y_N \\ Ae^{-2\pi i(\nu t)} &= Ae^{-2\pi i(\nu t - aNd)} \\ 1 &= e^{i2\pi aL} \\ n = a_n L &\Rightarrow a_n = \frac{n}{L} \Rightarrow \lambda_n = \frac{L}{n} \end{aligned} \tag{2.40}$$

In two and three dimensions also, the periodic boundary condition is used to determine the allowed wavelengths and the corresponding allowed frequencies.

2.3.3 Chain of Two Types of Atoms

Real crystals may consist of molecules. In that case, not only is there an interaction between neighboring molecules but there is also an interaction

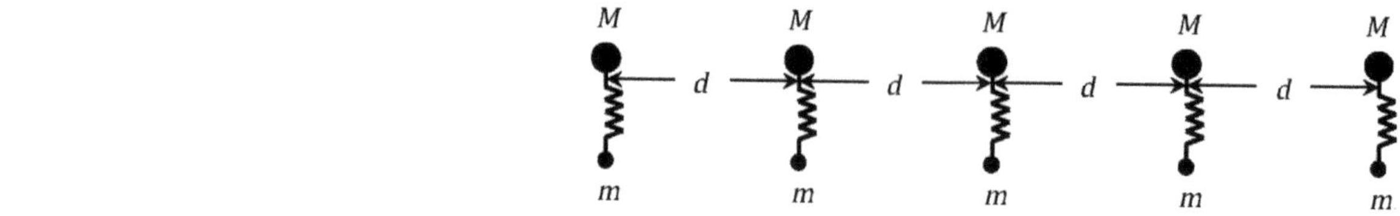

FIGURE 2.6 A one-dimensional chain of identical molecules.

between the atoms constituting the molecule. In other words there is an internal degree of freedom within the molecule. The effect of such interaction may be understood by means of a very simple model. Consider now a one-dimensional chain of molecules as shown in Fig. 2.6.

Following a procedure similar to the one described in Section 2.3.1, the force on the nth larger mass can be written as

$$F_n = -\frac{\partial U(r_{n,n-1})}{\partial y_n} - \frac{\partial U(r_{n+1,n})}{\partial y_n} - \frac{\partial U(r_{n,\bar{n}})}{\partial y_n} \tag{2.41}$$

In Eq. (2.41), the $\bar{n}$ denotes the nth smaller atom, whereas $\bar{y}_{\bar{n}}$ denotes its displacement from the mean position. $r_{n,\bar{n}}$ is the relative distance between the nth larger and the nth smaller atom. The relative distance between these two atoms is written as

$$r_{n,\bar{n}} = x_n - \bar{x}_n = nd + y_n - nd - \bar{y}_{\bar{n}} = y_n - \bar{y}_{\bar{n}} \tag{2.42}$$

Assuming that the displacements y_n and $\bar{y}_{\bar{n}}$ are small then the potential energy function $U(r)$ can be expanded as a Taylor series up to the second order as described here:

$$U(r_{n,\bar{n}}) = U(0) + (y_n - \bar{y}_{\bar{n}})\frac{\partial U}{\partial r_{n,\bar{n}}}\bigg|_{r=0} + \frac{(y_n - \bar{y}_{\bar{n}})^2}{2}\frac{\partial^2 U}{\partial r^2_{n-1,n}}\bigg|_{r=0} \tag{2.43}$$

As before, $U(0)$ can be taken to be zero and the second term in Eq. (2.43) vanishes; i.e., $\partial U/\partial r_{n,\bar{n}}\big|_{r=0} = 0$. Denoting $\partial^2 U/\partial r^2_{n,\bar{n}}\big|_{r=0} = U''$ as U'', the force on the nth atom may finally be written as

$$F_n = (y_{n+1} - 2y_n + y_{n-1})U'' - (y_n - \bar{y}_{\bar{n}})U'' \tag{2.44}$$

The force on the nth smaller atom is given as

$$\overline{F}_{\bar{n}} = (y_n - \bar{y}_{\bar{n}})U'' \tag{2.45}$$

The equations of motion for the nth atoms are now written as

$$M\frac{d^2 y_n}{dt^2} = (y_{n+1} - 2y_n + y_{n-1})U'' - (y_n - \bar{y}_{\bar{n}})U'' \tag{2.46}$$

$$m\frac{d^2\overline{y}_{\overline{n}}}{dt^2} = (y_n - \overline{y}_{\overline{n}})U'' \tag{2.47}$$

Consider now the following wave solutions for the displacements y_n and $\overline{y}_{\overline{n}}$,

$$y_n = Ae^{-2\pi i(\nu t - ax_n)} = Ae^{-2\pi i(\nu t - and)} \tag{2.48}$$

$$\overline{y}_{\overline{n}} = \overline{A}e^{-2\pi i(\nu t - ax_n)} = Ae^{-2\pi i(\nu t - and)} \tag{2.49}$$

After substituting Eqs. (2.48) and (2.49) into Eqs. (2.46) and (2.47) and simplifying, the following algebraic equations are obtained:

$$-M4\pi^2\nu^2 A = -4A\sin^2(\pi ad)U'' - (A - \overline{A})U'' \tag{2.50}$$

$$-m4\pi^2\nu^2\overline{A}e^{-2\pi i(\nu t - and)} = e^{-2\pi i(\nu t - and)}(A - \overline{A})U'' \tag{2.51}$$

These equations are two simultaneous equations in two unknowns A and $\overline{A}$. Since both equations are homogenous, a solution is obtained if the following determinant vanishes:

$$\begin{vmatrix} M4\pi^2\nu^2 - \{1 + 4\sin^2(\pi ad)\}U'' & U'' \\ U'' & m4\pi^2\nu^2 - U'' \end{vmatrix} \tag{2.52}$$

Expanding the determinant in this equation and letting $\omega = 2\pi\nu$ results in the following quadratic equation in the frequency ν:

$$Mm\omega^4 - [M + m\{1 + 4\sin^2(\pi ad)\}]U''\omega^2 + 4\sin^2(\pi ad)U''^2 = 0 \tag{2.53}$$

The solution of this equation is written as

$$\omega^2 = \frac{U''[M+m\{1+4\sin^2(\pi ad)\}] \pm U''\sqrt{[M+m\{1+4\sin^2(\pi ad)\}]^2 - 16mM\sin^2(\pi ad)}}{2mM} \tag{2.54}$$

Eq. (2.54) gives two different frequencies ω as functions of the wave-number a. These functions are plotted in Fig. 2.7.

The lower curve is denoted as the *acoustic branch*, whereas the upper curve is denoted as the *optical branch*. These figures are called *dispersion curves*.

In his book, *Wave Propagation in Periodic Structures*, [2] Brillouin discusses a chain of two atoms with different masses that are arranged in a single chain as shown in Fig. 2.8.

Assuming only nearest-neighbor interactions, the dispersion relation for such a chain of atoms is given as

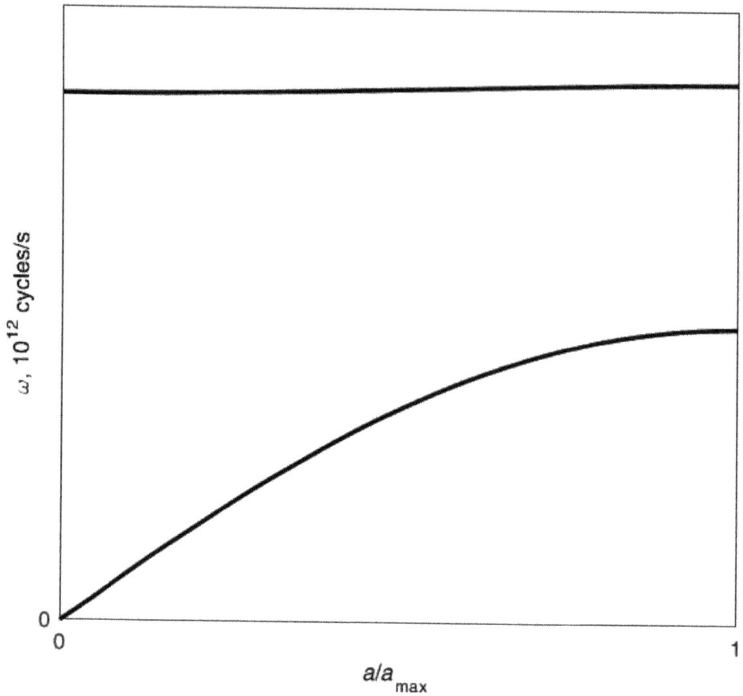

■ **FIGURE 2.7** The dispersion relation for the chain of two types of atoms.

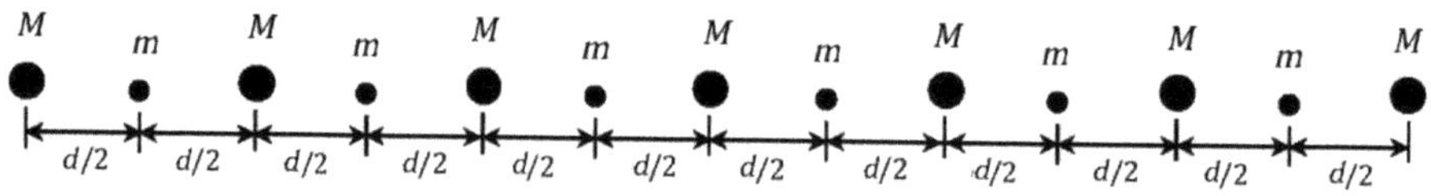

■ **FIGURE 2.8** A one-dimensional chain of identical molecules as discussed by Brillouin [2].

$$\omega^2 = U''\left[\left(\frac{1}{M} + \frac{1}{m}\right) \pm \sqrt{\left(\frac{1}{M} + \frac{1}{m}\right)^2 - \frac{4\sin^2(\pi a d/2)}{Mm}}\right] \tag{2.55}$$

The corresponding dispersion relation curve is presented in Fig. 2.9.

In three dimensions, there are in general three acoustic branches and three optical branches corresponding to the single longitudinal polarization and to the two transverse polarizations. Furthermore, distinct dispersion curves are obtained for distinct wave-propagation directions. It is possible that a particular direction may happen to be a symmetry direction of a crystal. In such cases the two separate transverse branches

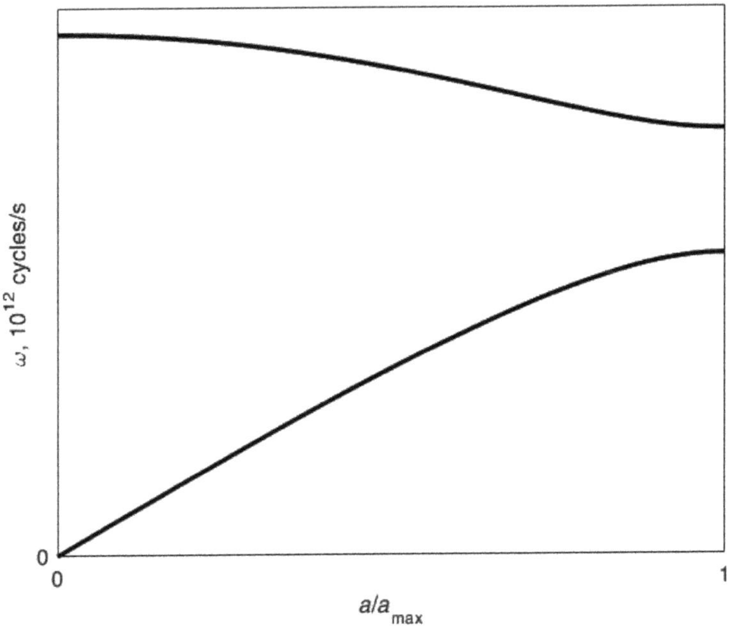

■ **FIGURE 2.9** The dispersion relation for the chain of two types of atoms as discussed by Brillouin [2].

degenerate into a single dispersion curve, both for the acoustic branch as well as for the optical branch. Finally, if there are N atoms in a molecule, then there are N sets of dispersion curves. The first set is called the acoustic branch, and all others are called the optical branch. As an example, the dispersion relations for a silicon crystal at 300K and in the [001] symmetry direction is presented in Fig. 2.10.

2.4 PHONON SCATTERING

The phonons moving in a crystal transfer energy from one point to another. While in motion these phonons also undergo scattering due to various reasons. Due to these scattering mechanisms the energy transport in a crystal has a finite velocity. In a crystal without any scattering mechanism the energy would be transported without hindrance and the thermal conductivity would be infinite. As discussed in later sections, the thermal conductivity in a crystal can be approximated as

$$k = \frac{C\Lambda V}{3} \tag{2.56}$$

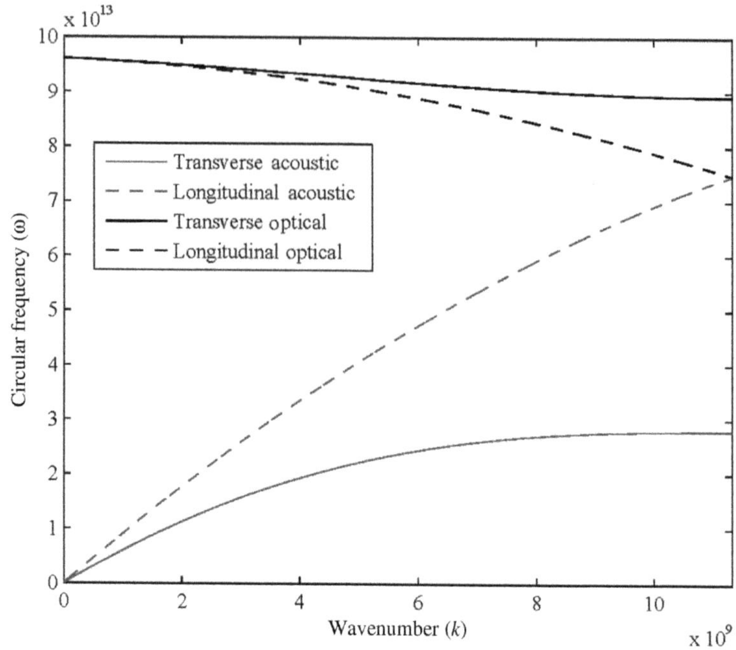

■ **FIGURE 2.10** Experimental dispersion curves for silicon at 300K in the [001] direction [3].

In Eq. (2.56), the volumetric heat capacity C and the phonon group velocity V are not affected by the phonon scattering. However, the phonon mean free path Λ or equivalently the phonon relaxation time τ are highly dependent on phonon scattering and this in turn limits the thermal conductivity k. In this section, we will briefly discuss four phonon scattering mechanisms. Each mechanism is important in a particular temperature range and this is one of the reasons for the temperature dependence of the thermal conductivity.

2.4.1 Phonon Scattering With Impurities and Defects

Crystal imperfections such as impurities and/or defects are a cause of phonon scattering. Remember that phonons are the quantized waves of a chain of crystal atoms. These waves are scattered when an impurity atom and/or a crystal defect blocks their path. This mechanism of phonon scattering is ineffective at low temperatures. This is because at low temperatures, the number of phonons is less and in addition their wavelengths are very large, much larger than the size of the impurity or crystal defect, due to which the probability of such scattering is low.

2.4.2 Phonon Scattering With the Crystal Boundaries

At low temperatures, the primary scattering mechanism involves the crystal boundaries. This is because at low temperatures, low frequency, long wavelength phonons dominate, with a wavelength comparable to the size of the crystal. This gives rise to *size effects*. The mean free path of the phonons becomes comparable to the crystal grain size and thus becomes independent of the temperature. The temperature dependence of the thermal conductivity in the low temperature regime comes entirely through the temperature dependence of the heat capacity.

2.4.3 Phonon–Phonon Scattering

A very important scattering mechanism is the phonon–phonon scattering. Two phonons may interact with each other to produce a third phonon. Or, one high-energy phonon may annihilate into two phonons. These two processes are called three-phonon scattering since three phonons are involved in the process. Scattering processes that involve more than three phonons are also possible but their probability of occurrence is much less than that of the three-phonon scattering processes and therefore it is customary to focus on three-phonon processes only. In these phonon–phonon scattering processes, the energy is conserved; however, the momentum may or may not be conserved. If the momentum is also conserved then the process is called the normal process, otherwise it is called the Umklapp process. The origin of this type of scattering is the anharmonic interaction between the atoms of the crystal. If the interaction between the atoms is limited to the harmonic terms then it is not possible for two phonons to interact. However, at high temperatures when the atoms are vibrating with higher amplitudes, the anharmonic term is not negligible. A phonon passing through a region will produce a strain field. Another phonon passing through the same region will experience this strain field and its motion may get disturbed accordingly and we may say that the two phonons have undergone a scattering process.

2.4.3.1 *Normal Processes*

In these processes, two phonons combine together to produce a phonon whose wavenumber is below the cut-off/maximum wavenumber. Not only does the sum of the initial energies equal the final energy, but the sum of the initial momenta equals the final momentum. Normal processes do not contribute to the thermal conductivity. This type of scattering process is depicted in Fig. 2.11.

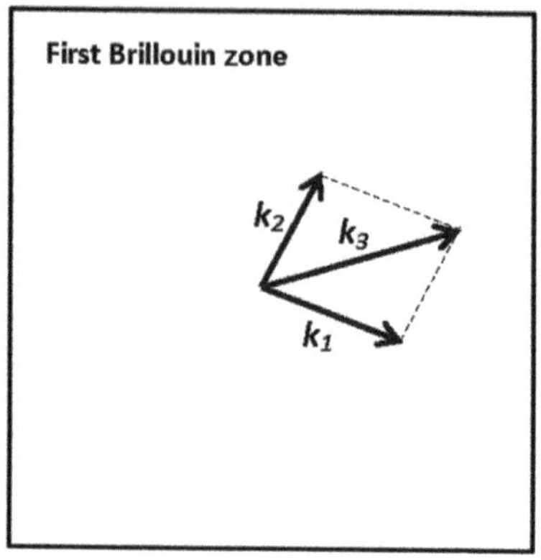

■ FIGURE 2.11 The normal process.

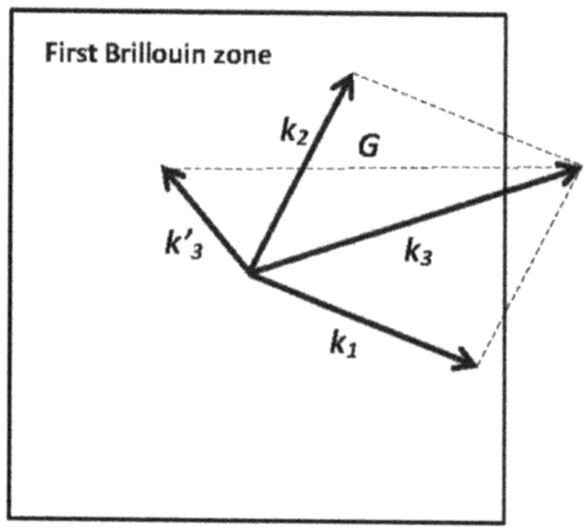

■ FIGURE 2.12 The Umklapp process.

2.4.3.2 *Umklapp Processes*

In these processes, two phonons combine together to produce a phonon whose wavenumber is above the cut-off/maximum wavenumber. This means that the new wavevector will be out of the first Brillouin zone. Since it is not possible to have a wave with a wavenumber more than the cut-off value, the new wavevector is thrown back into the first Brillouin zone by subtracting the reciprocal position vector from the new wavevector. In such a phonon–phonon scattering process the sum of the initial energies is equal to the final energy but the sum of the initial momenta does not equal the final momentum. These processes are a major contributor to the thermal conductivity at high temperatures. This type of scattering process is depicted in Fig. 2.12.

2.5 THERMAL PROPERTIES

2.5.1 The Density of States

In three dimensions the waves in a crystal may propagate in all directions. In this case instead of the wavenumber we may define a wavevector whose magnitude is equal to the reciprocal of the wavelength, and it points in the direction of the propagation of the wave.

$$\mathbf{a} = h_1\mathbf{E}^1 + h_2\mathbf{E}^2 + h_3\mathbf{E}^3, \quad |\mathbf{a}| = a = \frac{1}{\lambda} \tag{2.57}$$

From Eq. (2.39) (which is given in Section 2.3.2) we know that in one dimension only waves of certain wavelengths are allowed:

$$\lambda_n = \frac{L}{n}$$

In terms of the wavevector **a** we may write

$$a_n = \frac{n}{L} \tag{2.58}$$

In three dimensions we can write an expression for the wavenumber that is analogous to Eq. (2.58). That expression is described here:

$$a = \frac{(n_1^2 + n_2^2 + n_3^2)^{1/2}}{V^{1/3}} \tag{2.59}$$

In this equation, the length L in one dimension is replaced by the cube-root of the volume in three dimensions. This is because the shape of the real crystal in three dimensions can be arbitrary; however, the number of allowed wavevectors is so large that the consideration of the exact shape of the crystal is not required.

Consider now that the vectors $(\mathbf{E}^1, \mathbf{E}^2, \mathbf{E}^3)$ are the Cartesian unit vectors $(\mathbf{i}, \mathbf{j}, \mathbf{k})$. The wavevector may then be written as

$$\mathbf{a} = a_x\hat{\mathbf{i}} + a_y\hat{\mathbf{j}} + a_z\hat{\mathbf{k}} \tag{2.60}$$

or

$$|\mathbf{a}|^2 = a^2 = a_x^2 + a_y^2 + a_z^2 = \frac{(n_1^2 + n_2^2 + n_3^2)}{V^{2/3}} \tag{2.61}$$

For a given wavenumber a, a large combination of (n_1, n_2, n_3) can satisfy Eq. (2.61). The larger the magnitude of the wavenumber, the larger the number of valid combinations. The question now arises, how many wavenumbers are there in the interval $(a, a + da)$? The answer to this question is uncomplicated if we assume that the reciprocal space is isotropic;

i.e., all directions of propagation of the waves are equally likely. In this event the total wavenumbers in the interval $(a, a + da)$ are equal to the volume of a spherical shell whose radius is a and thickness is da; i.e., $4\pi a^2 da$.

The density-of-states function is now defined as the number of waves whose frequency lies in the interval $(\nu, \nu + d\nu)$ per unit interval per unit crystal volume. This function comes out to be

$$D(\nu) = 2a^2 \frac{da}{d\nu} \tag{2.62}$$

In calculations, it is more common to use the circular frequency $\omega = 2\pi\nu$ and a different wavenumber $k = 2\pi a$. In terms of these new variables we can write [4]

$$D(\omega) = \frac{k^2}{2\pi^2}\frac{dk}{d\omega} \tag{2.63}$$

2.5.2 **Heat Capacity**

A dielectric crystal stores thermal energy primarily in the wave motion of its atoms. In other words, the thermal energy stored in a crystal is actually the sum of the energy carried by all the waves that are allowed to exist at a particular temperature T. As discussed previously, the quantized waves are treated as quasiparticles and are named *phonons*. The phonons in reality are bosons and have the property that their number is not conserved.

A phonon of frequency ω carries an energy $\hbar\omega$. Also, at any given temperature T, only phonons having a frequency less than a certain maximum frequency are allowed in the crystal. This aspect is described by the fact that the phonons are governed by the Bose–Einstein distribution function. Furthermore, the number of phonons of a particular frequency is not constant but is dependent on the magnitude of the frequency itself and is given by the density-of-states function. Hence the number of equivalent phonons with a larger frequency is more than the number of phonons of a lower frequency. Taking all these factors into account, the internal energy per unit volume of a crystal can be written as

$$u(T) = \int_0^{\omega_{\text{cut-off}}} \hbar\omega f(\omega, T) D(\omega) d\omega \tag{2.64}$$

$$u(T) = \int_0^{\omega_{\text{cut-off}}} \frac{\hbar\omega}{\exp(\hbar\omega/k_B T)} \frac{k^2}{2\pi^2}\frac{dk}{d\omega} d\omega \tag{2.65}$$

Additionally we have to take into account the following two features of wave propagation through the crystals. The first one is that there are three polarizations of the waves so that the internal energy expression has to be written for each polarization. Second, there may be one or more groups of these waves. The waves in the first group are called *acoustic waves* and the waves in all other groups are called *optical waves*. It is possible to have one or more groups of optical waves. Again, the internal energy expression is to be written for each group of waves separately. These deliberations force us to write the internal energy expression as

$$u(T) = \sum_q \sum_p \int_0^{\omega_{\text{cut-off}}} \frac{\hbar\omega}{\exp(\hbar\omega/k_B T)} \frac{k^2}{2\pi^2} \frac{dk}{d\omega} d\omega \tag{2.66}$$

In Eq. (2.66) the summation variable p denotes a particular polarization and takes on the values $1, 2, 3$. The summation variable q denotes a particular phonon group or branch. $q = 1$ for the acoustic branch and is equal to $2, 3, \ldots$ for the optical branches. The wavenumber k is a function of the frequency ω and this dependence comes from the dispersion relation of the crystal. In general, there is a unique dispersion relation for each polarization and for each phonon branch with distinct cut-off frequencies.

If we observe the dispersion relations of crystals we notice that it is more convenient to take the wavenumber as the independent variable and the frequency as the dependent variable. This becomes more evident from the fact that the various phonon polarizations and branches have different cut-off frequencies but the same cut-off wavenumber. These reflections compel us to write the expression for the internal energy per unit volume as

$$u(T) = \sum_q \sum_p \int_0^{k_{\text{cut-off}}} \frac{\hbar\omega}{\exp(\hbar\omega/k_B T)} \frac{k^2}{2\pi^2} dk \tag{2.67}$$

The volumetric heat capacity of a dielectric crystal is defined as the amount of internal energy that can be stored in the crystal per unit volume when its temperature is increased by 1 K Mathematically, the volumetric heat capacity is defined as

$$C = \frac{\partial u}{\partial T} \tag{2.68}$$

$$C(T) = \sum_q \sum_p \frac{\partial}{\partial T} \int_0^{k_{\text{cut-off}}} \frac{\hbar\omega}{\exp(\hbar\omega/k_B T)} \frac{k^2}{2\pi^2} dk \tag{2.69}$$

$$C(T) = \int_0^{k_{\text{cut-off}}} \sum_q \sum_p C_k dk \tag{2.70}$$

where $C_k = \dfrac{k^2}{2\pi^2} \dfrac{\partial}{\partial T} \left(\dfrac{\hbar\omega}{\exp(\hbar\omega/k_B T)} \right)$

2.5.3 Thermal Conductivity

In the theory of conduction heat transfer, the thermal conductivity tensor is used to relate the heat flux vector and the temperature gradient. These vectors, in general, can point in different directions.

$$\mathbf{q}'' = \mathbf{k} \cdot \nabla T \tag{2.71}$$

If the heat flux vector and the temperature gradient point in the same direction, which is the case for isotropic mediums, then the thermal conductivity tensor degenerates into a single number k, the thermal conductivity of the medium. The kinetic theory of dilute gases defines the thermal conductivity (as given in Section 2.4, Eq. 2.56) as

$$k = \frac{C \Lambda V}{3}$$

where C is the volumetric heat capacity, Λ is the mean free path of the gas molecules, and V is the speed of the molecules. In dielectric crystals, the large number of phonons may be viewed as a phonon gas and the kinetic theory of gases can be applied to it after some modifications. The thermal conductivity for a phonon gas can be defined as

$$k = \sum_q \sum_p \int_0^{k_{\max}} \frac{C_k v_k^2 \tau_k}{3} dk \tag{2.72}$$

In Eq. (2.72), the mean free path is replaced by the expression $\Lambda_k = v_k \tau_k$, where v_k is the phonon group velocity and τ_k is the phonon relaxation time. C_k is the volumetric heat capacity. Each of the variables C_k, v_k, and τ_k are functions of the wavenumber k. This functional dependence on k is unique for each phonon polarization p and each phonon branch q. The volumetric heat capacity C_k has already been defined in Eq. (2.69). The phonon group velocity is calculated from the dispersion relation and is defined as

$$v_k = \frac{d\omega}{dk} \tag{2.73}$$

The wavenumber-dependent phonon relaxation times are determined by the particular scattering mechanism. Different expressions exist for impurity scattering, boundary scattering, N-process, and U-process scattering. Here we give an example of the relaxation times of phonons in silicon for the N- and U-process scatterings. Ward et al. [5] determined the phonon relaxation times for silicon and germanium crystals. They gave expressions for the transverse acoustic and longitudinal acoustic phonon branches, both for normal and the Umklapp scattering processes. These two relaxation times are then combined together by means of Mattheissen's rule. These expressions are:

$$\tau_{TA,k}^{(N)} = \frac{2.3 \times 10^{24}}{253,322 \times \omega^2 T[1 - \exp(-3T/\theta_{D,TA})]} \tag{2.74}$$

$$\tau_{TA,k}^{(U)} = \frac{5.3 \times 10^{48}}{2012 \times \omega^4 T[1 - \exp(-3T/\theta_{D,TA})]} \tag{2.75}$$

$$\frac{1}{\tau_{TA,k}} = \frac{1}{\tau_{TA,k}^{(N)}} + \frac{1}{\tau_{TA,k}^{(U)}} \tag{2.76}$$

$$\tau_{LA}^{(N)} = \frac{2.3 \times 10^{24}}{163,921 \times \omega^2 T[1 - \exp(-3T/\theta_{D,LA})]} \tag{2.77}$$

$$\tau_{LA}^{(U)} = \frac{5.3 \times 10^{48}}{507 \times \omega^4 T[1 - \exp(-3T/\theta_{D,LA})]} \tag{2.78}$$

$$\frac{1}{\tau_{LA,k}} = \frac{1}{\tau_{LA,k}^{(N)}} + \frac{1}{\tau_{LA,k}^{(U)}} \tag{2.79}$$

In these expressions the superscripts (N) and (U) represent the normal and Umklapp phonon scattering processes, respectively. θ_D is the Debye temperature and is defined as $\theta_{D,TA} = \hbar\omega_{\text{cut-off},TA}/k_B$ and $\theta_{D,LA} = \hbar\omega_{\text{cut-off},LA}/k_B$. The circular frequency ω is dependent on the wavenumber k and is obtained from the appropriate dispersion relation. $\omega_{\text{cut-off}}$ is also obtained from the dispersion relation and is the frequency corresponding to the maximum/cut-off wavenumber.

2.6 CLOSING REMARKS

The thermophysical properties of a crystal such as its heat capacity, thermal conductivity, and others are directly related to the structure of the crystal and associated with the dynamics of the atoms inside the crystal. To be able to calculate these properties, it is necessary to first describe the crystalline structure mathematically. The description of the dynamics of a crystal is based on this fundamental knowledge of the crystalline structure. Force interactions between neighboring atoms give rise to long-range waves in a crystal. These waves are only allowed to have specific frequencies and wavelengths due to the finite size of the crystal. In addition, there is a maximum allowed frequency that a crystal can support, called the cut-off frequency. This limit on the frequency magnitude is because the crystal structure is discrete. The discrete nature of the crystal structure also gives rise to the dispersion relation. The cut-off frequency is related to the concept of the inelastic scattering that in turn is connected to the thermal conductivity. The knowledge of the crystal structure and its dynamics particularly becomes important when dealing with thin films or similar structures, in which one or more physical dimensions of the crystal are of the same order of magnitude as the lattice parameter of the crystal.

REFERENCES

[1] S. Wang, Fundamentals of Semiconductor Theory and Device Physics, Prentice-Hall Inc., Englewood Cliffs, 1989.

[2] L. Brillouin, Wave Propagation in Periodic Structures, Dover Publications Inc., Mineola, 1953.

[3] B.N. Brockhouse, Lattice vibrations in silicon and germanium, Phys. Rev. Lett. 2 (6) (1959) 256–258.

[4] N.W. Arhcroft, N.D. Mermin, Solid State Physics, Harcourt College Publishers, 1976.

[5] A. Ward, D. Broido, D. Stewart, G. Deinzer, Ab initio theory of the lattice thermal conductivity in diamond, Phys. Rev. B 80 (12) (2009) 125203.

Chapter 3

Some Aspects of Statistical Thermodynamics

3.1 INTRODUCTION

Thermodynamics deals with the macroscopic parameters of matter. The thermodynamic system can be characterized by macroscopic properties like temperature, volume, pressure, and energy. In the earlier stages, the theory of thermodynamics was developed without detailed understanding the microscopic picture of the matter. With the development of the atomic theory, Boltzmann and Gibbs took the first step toward the linkage of microscopic and macroscopic pictures of matter in relation to thermodynamic principles.

The scope of this chapter is to familiarize the reader with some basic concepts and understanding of statistical mechanics that can be used to bridge the gap between the microscopic and macroscopic pictures of the matter. This chapter starts with the introduction of statistical mechanics, which provides some elementary information like the concept of microstates, macrostates, and phase space. After that, an introduction to probability theory is provided. In addition, the concepts of ensemble and statistical distribution function are provided.

3.2 STATISTICAL MECHANICS

Statistical mechanics is the branch of physics in which the physical properties of matter are obtained from the dynamical behavior of its microscopic constituents. The physical properties/behavior of matter can be considered as the macroscopic picture of matter. Therefore the formalism of statistical mechanics is the theoretical bridge between the macroscopic picture and the microscopic picture of the system. Electrons, phonons, protons, neutrons, molecules, crystals, and other structures on an atomic length scale are included under the microscopic picture of any system, whereas the macroscopic pictures consist of things that can be viewed/observed by the naked eye. The structure of microscopic constituents is usually completely hidden from the view perspective. In order

Heat Transport in Micro- and Nanoscale Thin Films.
DOI: http://dx.doi.org/10.1016/B978-0-323-42979-5.00003-7

to determine the physical properties of a system, we need a chain of reasoning that connects the fundamental laws of physics that govern the structure and the interactions of the unseen microscopic constituents.

One mole of any substance conations the 6.02×10^{23} particles. It is very difficult for anyone to record the complete information about each particle, such as its position, temperature, velocity, and so forth. Therefore, in order to determine the macroscopic properties, we may use selected information about the microscopic constitutes. For the predication of system properties using the selected information of microscopic constitutes, we need to use the probability theory. Therefore, in statistical mechanics, probability theory is used for the predication of physical properties.

3.2.1 Microstates and Macrostates

In statistical mechanics, the concept of micro- and macrostates is very important, so we should have a very clear picture about them. Descriptions of micro- and macrostates are given in this section.

3.2.1.1 *Macrostates*

The macrostate can be considered an alternative word for the thermodynamic state of the system [1]. The macrostate provides adequate information about the system so that we can define its thermodynamic state. For many of the thermodynamic systems, by defining pressure (P), volume (V), and temperature (T), we can define the thermodynamic state, which can provide the complete picture of the system. For example, if 1 kg of water in a container has a temperature of 325 K and a pressure of 250 kPa, we can find its thermodynamic state. Using the thermodynamic relations, other thermodynamic properties like internal energy (U), entropy (S), chemical potential (μ), and such can also be calculated. Macrostate gives the overall behavior of the system; the detailed information about all individual particles is not required.

Example 3.1

Consider a system with three distinguishable particles p1, p2, and p3, with two possible energy levels E_1 and E_2. Find the total number of possible macrostates.

Solution

Fig. 3.1 shows all possible macrostates with the arrangement of particles into energy level E_1 and E_2. There are four possible macrostates: (1) all three particles are in energy level E_1, (2) two particles are in E_1 and one is in E_2, (3) one particle is in E_1 and two are in E_2, (4) all three particles are in energy level E_2.

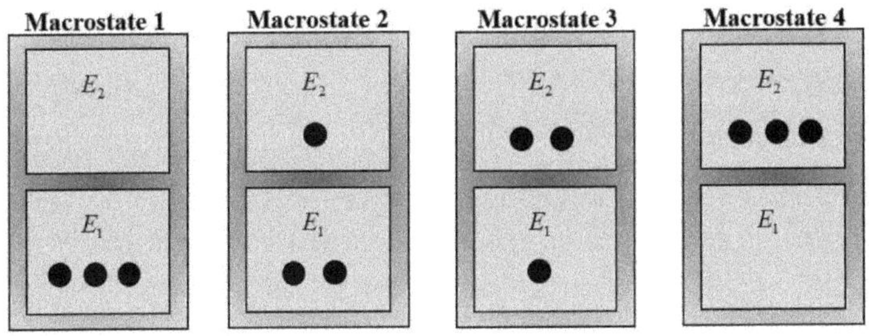

■ **FIGURE 3.1** Four possible macrostates for Example 3.1.

3.2.1.2 *Microstates*

Microstate represents the most detailed specification of the system that can be imagined. For a microstate to be defined, detailed information about each particle is required. One macrostate can have many possible microstates.

Example 3.2

Find the possible number of microstates for macrostate 2 of Example 3.1.

Solution

In the case of the microstate, the information about each particle is required. There are three possible microstates for macrostate 2 (Example 3.1); (1) particle p1 is in energy level E_2 and particles p2 and p3 are in energy level E_1, (2) particle p2 is in energy level E_2 and particles p1 and p3 are in energy level E_1, (3) particle p3 is in energy level E_2 and particles p1 and p2 are in energy level E_2, as shown in Fig. 3.2.

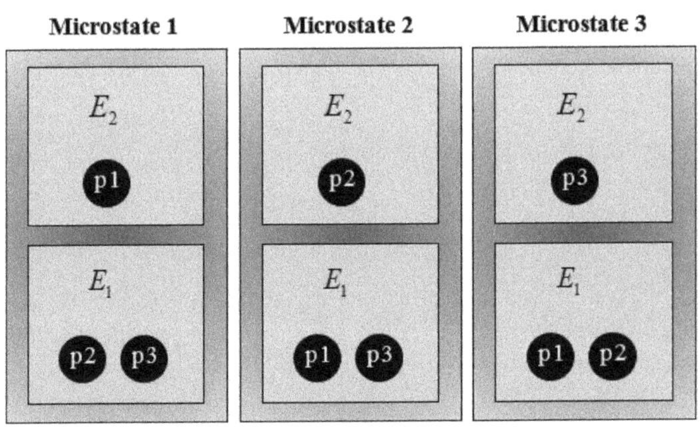

■ **FIGURE 3.2** Possible microstates for macrostate 2 of Example 3.1.

Consider a system that contains N identical microscopic particles. In this system, there are an enormous number of possible microstates. If we employ the kinetic theory of gases on the given system, then three position vectors and three momentums are required to define the dynamical state of each particle; therefore we need $6N$ variables in order to describe the system completely. For 1 mol of any substance the N is equal to Avogadro's number (N_A), which is 6.0×10^{23}. Therefore, for complete description of 1 mol of substance, we need $6 \times 6.02 \times 10^{23}$ data points. The situation for recording the information about all the microstates becomes difficult because the microstates change very rapidly in the system. For example, in 1 second of time duration, 1 mol of gas changes its microstates 10^{32} times [1]. It is very difficult to deal with such a huge information, therefore, the theory of probability is required to handle the information.

Example 3.3

Consider 40 g of helium in a container as shown in Fig. 3.3. Find the number of data points required for a given time for a complete description of all molecules present in the container.

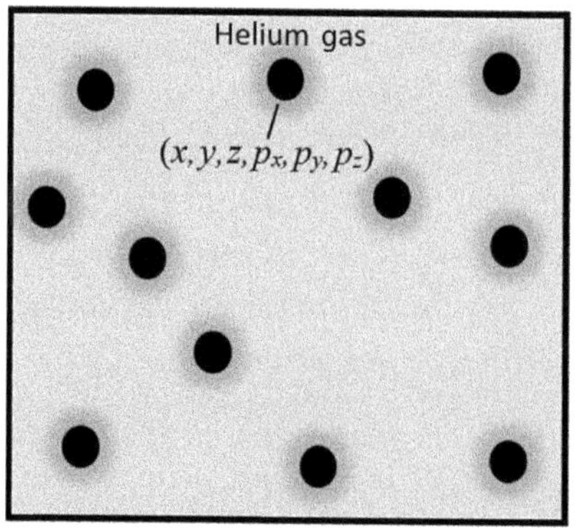

■ FIGURE 3.3 Helium gas molecules in the container.

Solution

Mass of helium gas = 40 g

Molar mass of helium gas = 4 g

Number of moles = Mass of helium gas/Molecular mass of helium gas

Therefore,

Number of moles = 40/4 = 10

Number of molecules in 1 mol = $N_A = 6.02 \times 10^{23}$

Total number of molecules in the box = $10 \times 6.02 \times 10^{23}$

Since the complete description of the dynamical state of each molecule requires three position variables x, y, z and three momentum p_x, p_y, and p_z, to describe the dynamical state of all molecules in the box we need $6 \times 6.02 \times 10^{24}$ data points for a given time.

3.2.2 Probability Theory

Probability is a measure of the likelihood of a random phenomenon or chance behavior. There are two distinct definitions of probability [2]; classical probability and statistical probability.

3.2.2.1 *Classical Probability*

Consider an experiment or measurement with X numbers of different possible outcomes. Then the probability of an event E_i is defined as

$$P(E_i) = \frac{1}{X} \tag{3.1}$$

Example 3.4

Consider an unbiased dice as shown in Fig. 3.4. Find the probability of getting 4 as an outcome.

■ **FIGURE 3.4** Unbiased dice.

Solution

Step 1: Find X.

There are six possible outcomes of the experiment, 1, 2, 3, 4, 5, and 6.

Therefore,

$X = 6$.

Step 2: Find the probability for an event E_4.

The probability of getting 4 as an outcome is

$$P(E_4) = \frac{1}{6} = 0.167$$

There is a 16.7% chance of getting 4 as an outcome when an unbiased dice is rolled.

3.2.2.2 *Statistical Probability*

Consider an experiment or measurement that is conducted N times and an event E_i occurs n_i times. We can write the statistical probability of event E_i as

$$P(E_i) = \lim_{N \to \infty} \frac{n_i}{N} \tag{3.2}$$

For infinite number of trials the statistical probability formula reduces to the classical probability formula. The probability of two mutually exclusive events can be written as

$$P(E_{1,2}) = P(E_1) + P(E_2) \tag{3.3}$$

where $P(E_{1,2})$ represents the combined probability of two events, $P(E_1)$ is the probability of the first event, and $P(E_2)$ is the probability of the second event.

There are two important properties of probability:

1. The value of all probabilities always lies in between 0 and 1.

$$0 \leq P(E_i) \leq 1$$

2. The summation of all probabilities is equal to 1. That is,

$$\sum P(E_i) = 1$$

Example 3.5

Consider an unbiased dice as shown in Fig. 3.4. Find the possibility of getting 4 or 6 as an outcome.

Solution

Step 1: Find *X*.

There are six possible outcomes of the experiment, 1, 2, 3, 4, 5, and 6.

Therefore,

$X = 6.$

Step 2: Find the probability of individual events.

The probability of getting the 4 as an outcome is $P(E_4) = 0.167$ and the probability for getting 6 as an outcome is $P(E_6) = 0.167$.

Step 3: Find the combined probability.

The probability of getting the 4 or 6 as an outcome is

$$P(E_{4,6}) = P(E_4) + P(E_6) = 0.167 + 0.167$$

$$P(E_{4,6}) = 0.334$$

3.2.3 Probability Distributions

There are two different types of probability distribution [2]: (1) discrete distributions and (2) continuous distributions.

3.2.3.1 *Discrete Distributions*

The probability distribution for a random variable describes the distribution of probabilities $P(X_i)$ over the values of the discrete random variable X.

Consider a random variable X, which can have discrete values $X_1, X_2, X_3, \ldots\ldots, X_n$ as an output of an experiment. Let $P(X_1), P(X_2), P(X_3), \ldots\ldots, P(X_n)$ represent the probabilities of discrete variables $X_1, X_2, X_3, \ldots\ldots, X_n$, respectively. The discrete probability distribution $P(X)$ can be expressed as [2]

$$\sum_{i=1}^{n} P(X_i) = 1 \tag{3.4}$$

Eq. (3.4) shows the sum of discrete probability distribution over all the possible values of X.

Example 3.6

Find the probability distribution for an unbiased dice, when the dice is thrown six times.

Solution

Step 1: The outcome and its probability.

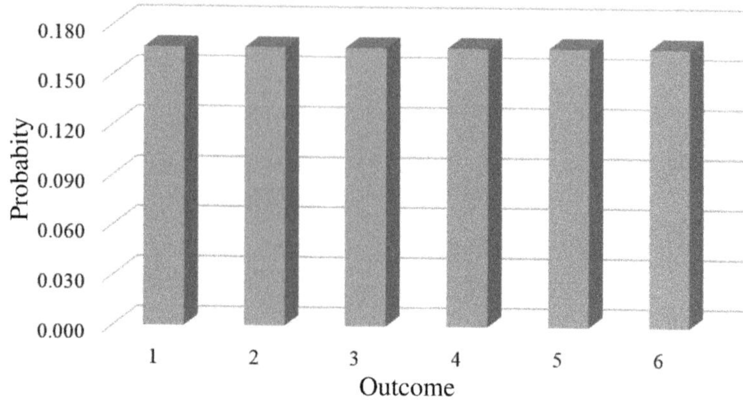

■ **FIGURE 3.5** Discrete probability distribution for the outcome of a single dice is rolled six times.

Since the dice is unbiased, all six faces have equal probability, which is 1/6.

Outcome	1	2	3	4	5	6
Probability	1/6	1/6	1/6	1/6	1/6	1/6

■

Fig. 3.5 shows the probability distribution for the outcome of a single dice that is thrown six times. It represents the uniform probability distribution.

Example 3.7

Consider two unbiased dice as shown in Fig. 3.6. In each dice, the probability of getting any face is 1/6. Find the probability distribution for the sum of outcome when a pair of dice is thrown 36 times.

■ **FIGURE 3.6** Two unbiased dice.

Solution

In the first step all possible outcomes are listed, and then the probability for each outcome is listed, as shown in Table 3.1; Fig. 3.7.

Table 3.1 The Sum of Outcomes and the Probability for the Pair of Dice

Sum of Outcomes	Possible Combinations	Number of Possibilities	Probability P_i
2	1 + 1	1	1/36
3	1 + 2, 2 + 1	2	2/36
4	1 + 3, 3 + 1, 2 + 2	3	3/36
5	1 + 4, 4 + 1, 2 + 3, 3 + 2	4	4/36
6	1 + 5, 5 + 1, 2 + 4, 4 + 2, 3 + 3	5	5/36
7	1 + 6, 6 + 1, 2 + 5, 5 + 2, 3 + 4, 4 + 3	6	6/36
8	2 + 6, 6 + 2, 3 + 5, 5 + 3, 4 + 4	5	5/36
9	3 + 6, 6 + 3, 4 + 5, 5 + 4	4	4/36
10	4 + 6, 6 + 4, 5 + 5	3	3/36
11	5 + 6, 6 + 5	2	2/36
12	6 + 6	1	1/36

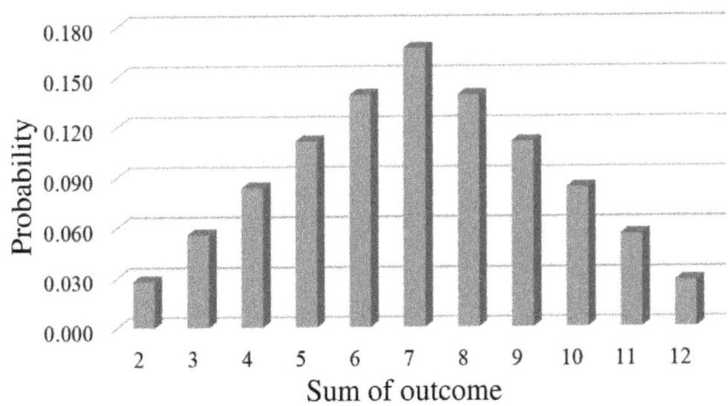

■ **FIGURE 3.7** Probability distribution for the pair of dice for different sum of outcomes.

■

3.2.3.2 *Continuous Distributions*

Consider the random variable x, which is continuous in-between the interval $[a, b]$ with the frequencies $f(x)$.

The continuous distribution can be defined by the probability density function as

$$\rho(x) = \frac{f(x)}{\int_a^b f(x)dx} \tag{3.5}$$

where $\int_a^b \rho(x)dx = 1$.

Example 3.8

The nanosize particles are distributed in the one-dimensional thin film of 4 μm as shown in Fig. 3.8. The frequency of nanosize particles is given by $f(x) = 30x^2$, where x is in μm.

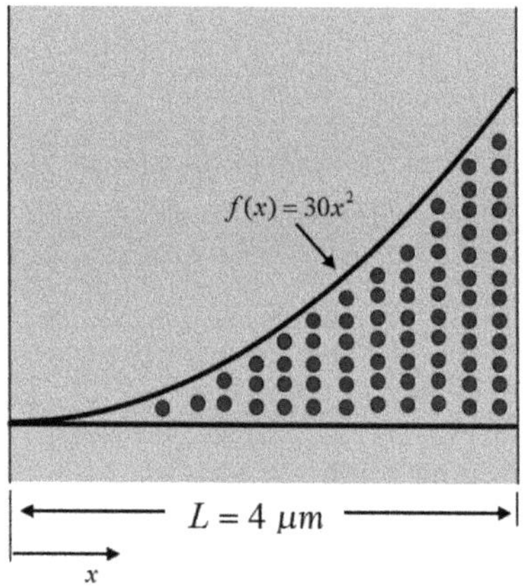

■ **FIGURE 3.8** Thin film with nanosize particle distribution.

Find the following:

1. The total number of nanosize particles in thin film.
2. Probability density function of the distribution.
3. The number of nanosize particles between 1 and 3 μm.

Solution

1. Total number of particles in thin film can be found by performing the integration over the film thickness; i.e.,

$$\text{Total number of particles} = \int_a^b f(x)dx$$

$$= \int_0^{4\ \mu\text{m}} 30x^2 dx$$

$$= \frac{30}{3}\left[x^3\right]_0^4$$

$$= \frac{30}{3}\left[x^3\right]_0^4 = 10(4^3 - 0)$$

$$\text{Total number of particles} = 640$$

2. Using Eq. (3.5) the probability density function $\rho(x)$ can be written as

$$\rho(x) = \frac{f(x)}{\int_a^b f(x)dx} = \frac{30x^2}{640}$$

$$\rho(x) = \frac{3}{64}x^2$$

3. The number of nanosize particles between 1 and 3 μm can be calculated as

$$\int_{1\ \mu\text{m}}^{3\ \mu\text{m}} f(x)dx = \int_{1\ \mu\text{m}}^{3\ \mu\text{m}} 30x^2 dx$$

$$\int_{1\ \mu\text{m}}^{3\ \mu\text{m}} f(x)dx = 280$$

3.2.4 **Phase Space**

A single particle moving in three dimensions (x, y, z) and thus having three degrees of freedom has associated with it three momentums (p_x, p_y, p_z). If the three coordinates of the particle along with the three momentums are prescribed at any given time then the dynamics of the particle is completely prescribed. If these six quantities are given as functions of time then the complete dynamical history of the particle is known. Consider a Cartesian coordinate system with each single axis assigned to each of the six quantities; there are three axes for the three position variables and three axes for the three momentum variables. This is called the phase space of the particle. The space consisting of only three position axes is called the configuration space, and the space

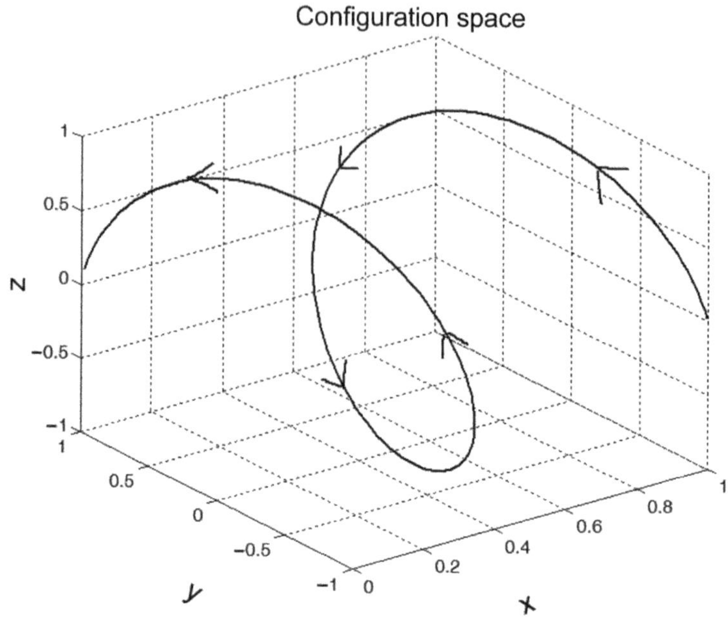

■ **FIGURE 3.9** Configuration space.

consisting of only three momentum axes is called the velocity/momentum space.

A single point in a phase space completely determines the dynamics of the particle at any given time. A trajectory in such a phase space completely determines the dynamical time history of the particle. This trajectory is obviously parameterized by the time variable. Figs. 3.9 and 3.10 show the trajectory of a particle in configuration and momentum space, respectively.

3.3 ENSEMBLES

In order to describe the dynamical behavior of a single particle statistically, we may assume that the phase point corresponding to the dynamics of a particle can be anywhere in the phase space, but with a certain probability. That is, there is a certain probability associated with each location in the phase space. The magnitude of this probability indicates the likelihood of the particle's phase point to be in that location. The probability can be defined through the probability distribution function; i.e., $f = f(x, y, z, p_x, p_y, p_z, t)$.

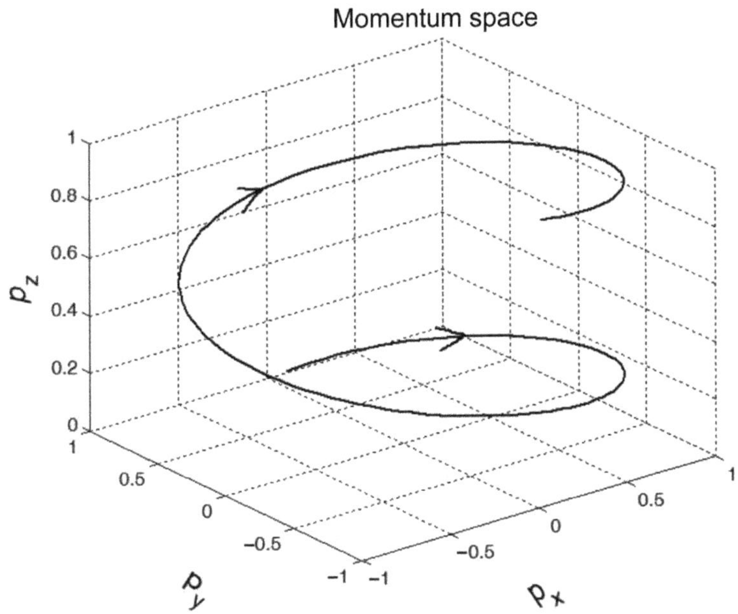

■ FIGURE 3.10 Momentum space.

Another way to look at this feature of statistical mechanics is by the concept of an ensemble. An ensemble is the name given to the collection of a large number of identical mental copies of a single particle. Each mental copy of a particle differs from the other mental copies in position, momentum, or both.

There are three different types of ensembles:

1. Microcanonical ensemble
2. Canonical ensemble
3. Grandcanonical ensemble

3.3.1 The Microcanonical Ensemble

Consider an isolated system that is controlled or defined by using extensive variables and properties like the number of particles in the system N, the volume of the system V, and the energy of the system E as shown in Fig. 3.11. Because the system is isolated, it neither gives its energy to the surroundings nor receives any energy from the surroundings. In this case the energy of the system E remains constant.

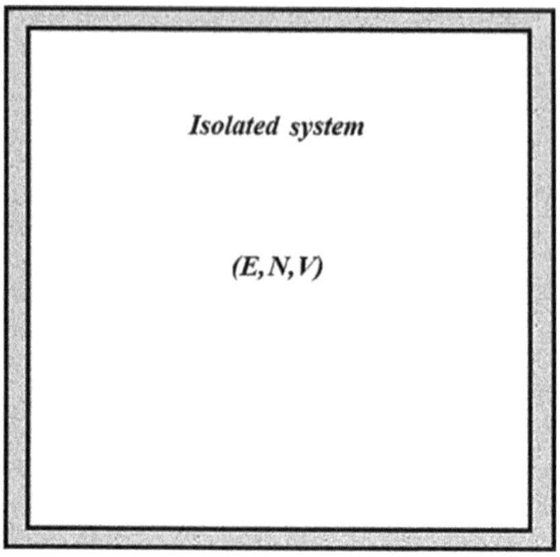

■ **FIGURE 3.11** Microcanonical ensemble.

For any given time, we need a set of variables in order to define the state of the particular particle completely. A particular isolated particle requires its momentum vector in order to define its state.

We cannot use the energy to define the state of the particle because one energy value can represent many different particles. The two different particles can have the same value of absolute momentum but can have a different momentum vector. The microscopic particles change their states rapidly with time.

To define the macrostate of the system, we need information about all the particles within the system. Although the particles change their state with time, here we are assuming that the overall macrostate of the system remains the same; i.e., the values of E, V, and N remain constant for the macrostate. Here we can consider the large number of microparticles that are having the same state. The group of such identical particles is known as the microcanonical ensemble.

Example 3.9

Consider a system with three particles p1, p2, and p3 with the total amount of system energy $E = 3\ hv$ (where h is Planck's constant and v is the frequency). In addition, consider that particles can have four energy levels 0, 1 hv, 2 hv, and 3 hv. Find the possible combination of microstates.

Solution

There are three sets, as shown in Fig. 3.12.

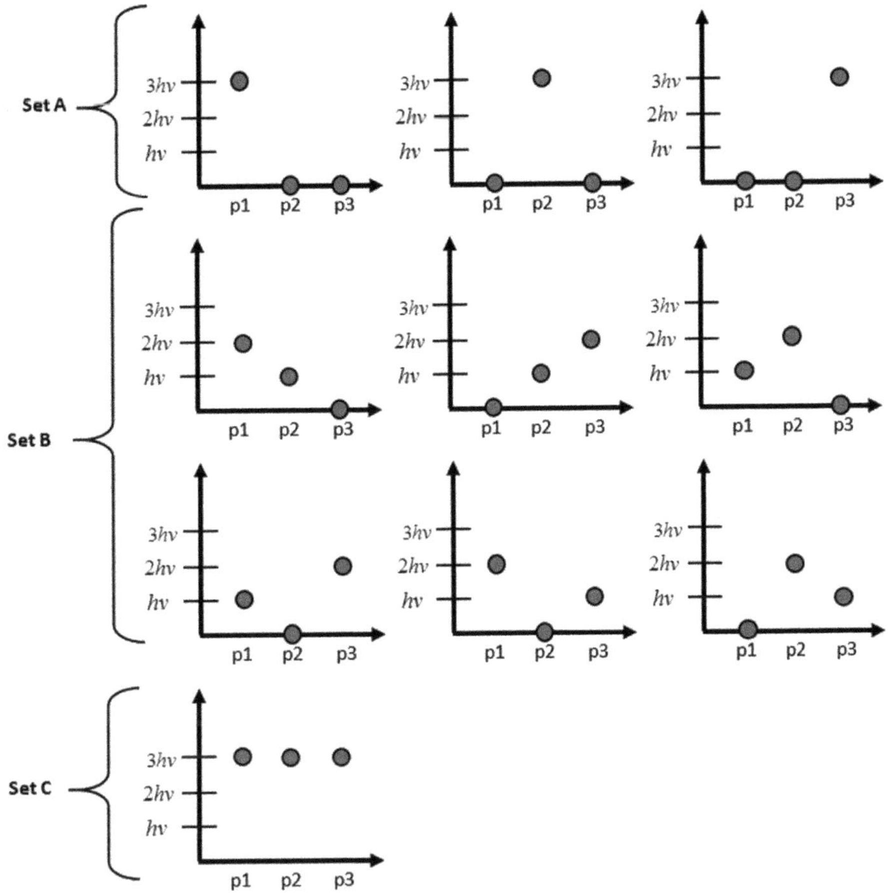

■ **FIGURE 3.12** Three particles and four different energy levels.

Set A: One particle with the energy 3 hv and other two with 0 energy.

Here we can have three different possibilities, as shown in Fig. 3.12.

This can be written as

$$W_A = 3$$

where W represent the number of microstates.

Set B: One particle with the energy hv, one with 2 hv, and the third with 0 energy.

For this set, there are six different ways in which the particles can be arranged (as shown in Fig. 3.12) in the system:

$$W_B = 6$$

Set C: All particles have energy hv.

There is only one possibility for this set; therefore,

$$W_C = 1$$

We can generalize the formula for the number of microstates in each set as follows:

$$W = \frac{N!}{n!} \tag{3.6}$$

where N is the total number of particles and n represents the number of particles in the same energy state.

The combination of these microcanonical ensembles corresponds to the same macrostate of the system.

A number of functions exist that relate the different thermodynamics properties, known as the thermodynamic potential. Whenever perturbation occurs in the thermodynamic system while the chosen/given properties are kept constant, the equilibrium state is reached for the maximum value of the potential.

The entropy S of the system can be considered a thermodynamic potential, which can be determined by using the volume, energy, and number of particles in the system; i.e., $S = fn(N, V, E)$.

The differential form of energy E can be written as [3]

$$dE = TdS - PdV + \mu dN \tag{3.7}$$

Rearranging this equation for dS,

$$dS = dE/T + (P/T)dV - (\mu/T)dN \tag{3.8}$$

where T represents the temperature, P represents the pressure, and μ represents the chemical potential.

Now for constant value of V and N we get the following [4]:

$$(\partial S/\partial E)_{V,N} = 1/T \tag{3.9}$$

There are three postulates related to the ensemble.

3.3.1.1 ***Postulate 1: The Probability for All Microstates Are Equal***

The probability P_i, to find a particular system of the ensemble in a given microstate (labeled i) is the same for all the microstates (W). As is discussed earlier, $0 \leq P_i \leq 1$ and $\sum P_i = 1$. When W represents the total number of microstates, then the probability of a particular system can be written as

$$P_i = \frac{1}{W} \tag{3.10}$$

Since the probability of all microstates are equal for the given macrostate, no preferential microstate exists for the given macrostate.

3.3.1.2 ***Postulate 2: Boltzmann Entropy Formula***

The entropy S of the system is given by the Boltzmann Entropy formula [3]:

$$S = k_B \ln W \tag{3.11}$$

where k_B is the Boltzmann constant and $k_B = 1.38 \times 10^{-23}\ \mathrm{J\ K^{-1}}$.

3.3.1.3 ***Postulate 3: Largest Value of the Entropy Represents the Equilibrium State***

For the case of a closed system, the equilibrium state occurs when the entropy of the system is at its maximum. Example 3.10 can explain this postulate.

Example 3.10

Consider the four coins in a row as shown in Fig. 3.13. Find the combination of heads and tails that represents the equilibrium state.

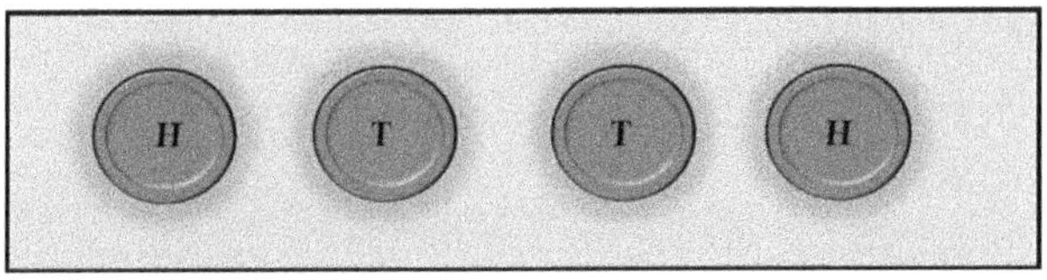

FIGURE 3.13 Four coins in a row.

Solution

The number of microstates can be given by

$$W = \frac{4!}{n_H! n_T!} \tag{3.12}$$

where n_H and n_T represent the number of heads and tails in the row, respectively.

In this combination, the largest number of microstates represents the equilibrium state, and the larger the number of microstates, the larger the entropy (Eq. 3.11). Table 3.2 shows the number of microstates for different numbers of heads.

Table 3.2 Number of Microstates for Different Numbers of Heads

n_T	0	1	2	3	4
W	1	4	6	4	1

The values of the number of states W indicate that the equilibrium state is achieved for the combination in which two heads and two tails are in the row.

In this example, although we use a coin to explain the postulate, a similar idea can be applied to an actual system to get the information about the equilibrium state.

3.3.2 Canonical Ensembles

Consider a system with N numbers of particles and having volume V. The system is in contact with a thermal reservoir in such a way that it can exchange heat with the thermal reservoir. This is shown in Fig. 3.14. The size of a thermal reservoir is much larger than the size of the system and the energy of the thermal reservoir E_R is much larger than the energy of the system E_S.

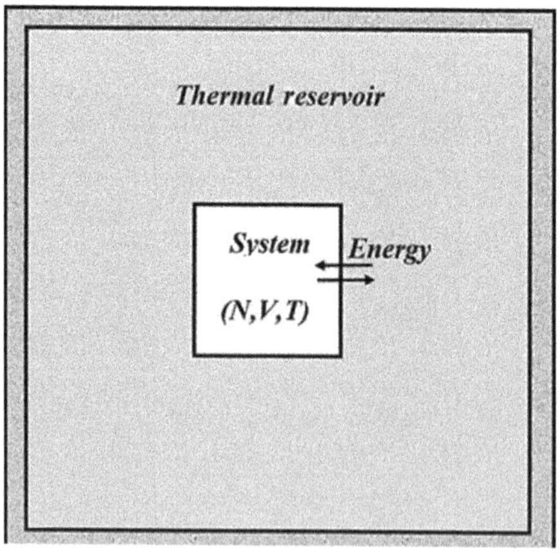

■ **FIGURE 3.14** Canonical ensembles.

In addition, we also consider that the combination of the system and the thermal reservoir results in a larger system that is completely isolated from its surrounding. The overall energy and number of particles in the combined larger system (system and thermal reservoir) is kept constant.

Consider E_S as the energy of the system and E_0 as the energy of the combined system, which consists of the original system and reservoir. Therefore, the energy of the reservoir is $E_0 - E_S$. Now the probability of a microstate having the energy E_S is given by

$$P_S = \frac{W_R(E_0 - E_S)}{W_{CS}(E_0)} \quad (3.13)$$

where W_R is the number of microstates in the reservoir and W_{CS} is the number of microstates in the combined system.

The entropies of combined system S_{CS} and reservoir S_R can be written as

$$S_{CS} = k_B \ln W_{CS}(E_0) \quad (3.14)$$

$$S_R = k_B \ln W_R(E_0 - E_S) \quad (3.15)$$

Now, the number of microstates for the combined system and reservoir from Eqs. (3.14) and (3.15) can be expressed as

$$W_{CS}(E_0) = \exp[S_{CS}(E_0)/k_B] \quad (3.16)$$

$$W_R(E_0 - E_S) = \exp[S_R(E_0 - E_S)/k_B] \quad (3.17)$$

Although the thermal reservoir is in thermal contact with the system, it is much larger than the system, and therefore, the reservoir practically remains unperturbed by the system.

Now, inserting Eqs. (3.16) and (3.17) into Eq. (3.13) results in

$$P_S = \frac{\exp[S_R(E_0 - E_S)/k_B]}{\exp[S_{CS}(E_0)/k_B]} \quad (3.18)$$

The term, $S_R(E_0 - E_S)$ can be expanded by using the Taylor series,

$$S_R(E_0 - E_S) = S_R(E_0) - E_S(\partial S_R/\partial E)_0 + (E_S^2/2)\left(\partial^2 S_R/\partial E^2\right)_0 + \cdots\cdots \quad (3.19)$$

As $E_0 \gg E_S$ therefore, neglecting the second and other higher order term we can get

$$S_R(E_0 - E_S) = S_R(E_0) - E_S(\partial S_R/\partial E)_0 \quad (3.20)$$

Since $(\partial S_R/\partial E) = 1/T$,

$$S_R(E_0 - E_S) = S_R(E_0) - E_S/T \quad (3.21)$$

The probability P_S can be written as

$$P_S = \frac{\exp[S_R(E_0)/k_B - E_S/k_BT]}{\exp[S_{CS}(E_0)/k_B]} \quad (3.22)$$

or

$$P_S = \frac{\exp[S_R(E_0)/k_B]\exp[-E_S/k_BT]}{\exp[S_{CS}(E_0)/k_B]} \quad (3.23)$$

As $\sum_S P_S = 1$, therefore,

$$\sum_S P_S = \sum_S \frac{\exp[S_R(E_0)/k_B]\exp[-E_S/k_BT]}{\exp[S_{CS}(E_0)/k_B]} = 1 \quad (3.24)$$

Eq. (3.24) results in the following:

$$\exp[S_{CS}(E_0)/k_B] = \exp[S_R(E_0)/k_B]\sum_S \exp[-E_S/k_BT] \quad (3.25)$$

Now inserting Eq. (3.25) into Eq. (3.23) gives

$$P_S = \exp[-E_S/k_BT]/\sum_S \exp[-E_S/k_BT] \quad (3.26)$$

The term $\sum_S \exp[-E_S/k_BT]$ is known as the partition function, and can be expressed as

$$Z = \sum_S \exp[-E_S/k_BT] \quad (3.27)$$

In canonical ensemble, the partition function plays an important role for the linkage between the microscopic and the macroscopic viewpoints. The partition function Z is the function of T, V, and N, which is used to define the system.

The inverse of k_BT is defined as β [5]:

$$\beta = \frac{1}{k_BT} \quad (3.28)$$

Now partition function Z can be written as

$$Z = \sum_S \exp[-\beta E_S] \quad (3.29)$$

More than one microstate can have the same energy, which is known as degeneracy of state, defined by the function $g(E_S)$. The partition function for the sum over the energies can be written as

$$Z = \sum_E g(E_S)\exp[-\beta E_S] \quad (3.30)$$

3.3.3 **Grand Canonical Ensemble**

Consider the system in contact with the constant temperature reservoir, which is much larger than the system. The system and the reservoir can exchange the energy and particles as shown in Fig. 3.15. The reservoir is much larger than the system. The temperature T, volume V, and chemical potential μ are used to define the system.

Consider that a large number of identical microscopic systems constitute the ensemble. We have to find the probability of a system having the energy E_S and n_S number of particles. Similar to Eq. (3.13), we can write the equation for probability of the system, which can be given by

$$P_S = \frac{W_R(E_0 - E_S, N_0 - n_S)}{W_{CS}(E_0, N_0)} \tag{3.31}$$

Here the number of microstates is the function of energy and the number of particles. N_0 represents the number of particles in the combined system, which consists of the system and the thermal reservoir and n_S is the number of particles in the system. Therefore, $N_0 - n_S$ is the number of particles in the thermal reservoir.

The entropies of the combined system S_{CS} and the reservoir S_R can be written as

$$S_{CS} = k_B \ln W_{CS}(E_0, N_0) \tag{3.32}$$

$$S_R = k_B \ln W_R(E_0 - E_S, N_0 - n_S) \tag{3.33}$$

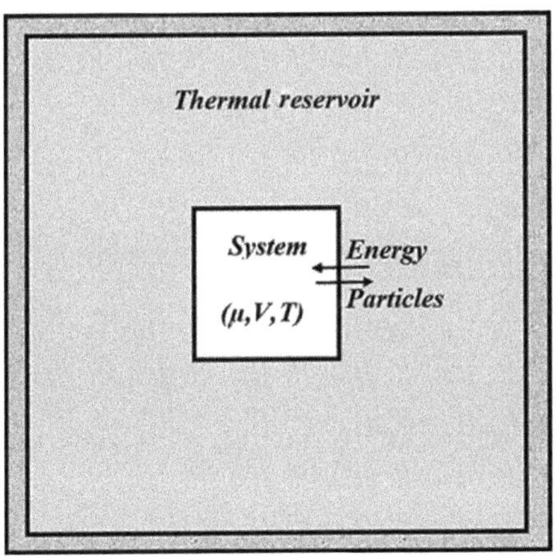

■ **FIGURE 3.15** Grand canonical ensemble.

Now, the number of microstates for combined system and reservoir from Eqs. (3.32) and (3.33) can be expressed as

$$W_{CS}(E_0, N_0) = \exp\left[\frac{S_{CS}(E_0, N_0)}{k_B}\right] \tag{3.34}$$

$$W_R(E_0 - E_S, N_0 - n_S) = \exp\left[\frac{S_R(E_0 - E_S, N_0 - n_S)}{k_B}\right] \tag{3.35}$$

Using the thermodynamics relation

$$E = TS - PV + \mu N \tag{3.36}$$

The entropy S can be written as

$$S = \frac{PV}{T} + \frac{E - \mu N}{T} \tag{3.37}$$

Using Eq. (3.37) into Eq. (3.34) results in

$$\begin{aligned} W_{CS}(E_0, N_0) &= \exp\left[\frac{P_{CS}V_{CS}}{k_B T} + \frac{E_0 - \mu N_0}{k_B T}\right] \\ &= \exp\left[\frac{P_{CS}V_{CS}}{k_B T}\right]\exp\left[\frac{E_0 - \mu N_0}{k_B T}\right] \\ &= A_{CS}\exp\left[\frac{E_0 - \mu N_0}{k_B T}\right] \end{aligned} \tag{3.38}$$

where $A_{CS} = \exp\left[\frac{P_{CS}V_{CS}}{k_B T}\right]$.

Similarly for $W_R(E_0 - E_S, N_0 - n_S)$, we can get the following,

$$W_R(E_0 - E_S, N_0 - n_S) = A_R \exp\left[\frac{E_0 - \mu N_0}{k_B T}\right]\exp\left[-\frac{E_S - \mu n_S}{k_B T}\right] \tag{3.39}$$

where $A_R = \exp\left[\frac{P_R V_R}{k_B T}\right]$.

Using Eqs. (3.38) and (3.39) in Eq. (3.31) results in

$$P_S = \frac{A_R}{A_{CS}}\exp\left[-\frac{E_0 - \mu n_S}{k_B T}\right] \tag{3.40}$$

The probabilities sum over the microstates is unity, that is $\sum_S P_S = 1$, therefore,

$$\sum_S P_S = \frac{A_R}{A_{CS}}\sum_S \exp\left[-\frac{E_0 - \mu n_S}{k_B T}\right] = 1 \tag{3.41}$$

This gives

$$\frac{A_R}{A_{CS}} = \frac{1}{\sum_S \exp\left[-\frac{E_0 - \mu n_S}{k_B T}\right]} \tag{3.42}$$

Therefore, now substituting Eq. (3.42) in Eq. (3.40) gives

$$P_S = \frac{\exp\left[-\frac{E_0 - \mu n_S}{k_B T}\right]}{\sum_S \exp\left[-\frac{E_0 - \mu n_S}{k_B T}\right]} \tag{3.43}$$

The mean value of energy can be determined by taking the sum over all the possible microstates of the system, which can be expressed as

$$E = \sum_S P_S E_S \tag{3.44}$$

Similarly, the number of particles can be expressed as

$$N = \sum_S P_S n_S \tag{3.45}$$

Similarly, in the case of the canonical ensemble, we can define the grand partition function Z_G as

$$Z_G = \sum_S \exp[-\beta(E_S - \mu n_S)] \tag{3.46}$$

where $\beta = 1/k_B T$. In reality, Eq. (3.46) represents the sum over the microstates and the number of particles. Also, $Z_G = fn(T, V, \mu)$.

3.4 STATISTICAL DISTRIBUTIONS

In this section we will describe the statistical distributions: Maxwell–Boltzmann, Fermi–Dirac, and Bose–Einstein. These will be used to describe the ideal gas, the electrons, and the phonons, respectively.

3.4.1 Maxwell–Boltzmann Distribution

The Maxwell–Boltzmann distribution is used to describe the behavior of identical and indistinguishable particles, like an ideal gas system.

Consider the thermodynamic system shown in Fig. 3.16, which consists of large number of identical and distinguishable particles. The system is in equilibrium; it contains N number of particles, the volume of the system is V, and the total energy of the system is E. The N, V, and E remain constant.

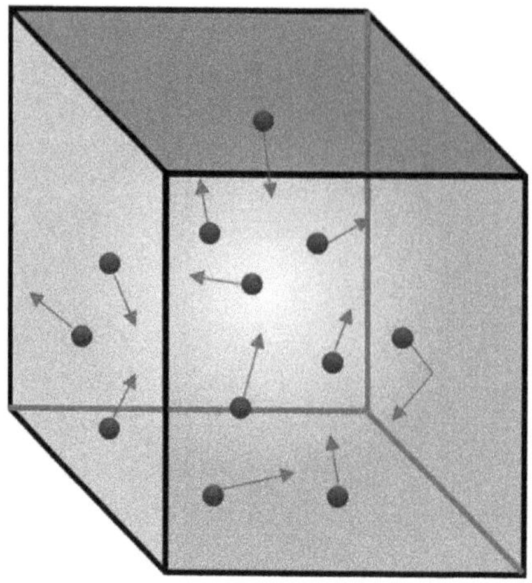

■ **FIGURE 3.16** A thermodynamic system.

The system consists of r number of energy levels; that is, $\varepsilon_1, \varepsilon_2, \varepsilon_3 \ldots \varepsilon_r$. Let $n_1, n_2, n_3 \ldots n_r$ represent the number of particles in energy levels $\varepsilon_1, \varepsilon_2, \varepsilon_3 \ldots \varepsilon_r$, respectively. In the system, energy levels are fixed but the number of particles in these energy levels are variable.

The summation of particles $n_1, n_2, n_3 \ldots n_r$ in all energy levels are equal to the total number of particles in the system.

$$N = n_1 + n_2 + n_3 \cdots + n_r \tag{3.47}$$

$$\sum_{i=0}^{r} n_i = N \tag{3.48}$$

The total energy of the system is constant and equal to E, therefore,

$$E = n_1\varepsilon_1 + n_2\varepsilon_2 + n_3\varepsilon_3 \cdots + n_r\varepsilon_r \tag{3.49}$$

$$\sum_{i=0}^{r} n_i\varepsilon_i = E \tag{3.50}$$

The total number of number of microstates W in the system can be calculated as

$$W = \frac{N!}{n_1!n_2!n_3! \ldots n_r!} \tag{3.51}$$

$$W = \frac{N!}{\prod_{i=0}^{r} n_i!} \tag{3.52}$$

where $\prod$ represents the product $n_1!n_2! \ldots n_r!$.

To find the equilibrium state of the system, which represents the most probable state, we need to find the combination of $n_1, n_2, n_3 \ldots n_r$, which can maximize the entropy of the system.

Since $S = k_B \ln W$ (Eq. 3.11), the maximum value of W will give the maximum entropy of the system.

Taking the natural logarithm of both sides of Eq. (3.52),

$$\ln W = \ln N! - \sum_{i=0}^{r} \ln n_i! \tag{3.53}$$

Stirling's approximation is given as

$$\ln A! = A \ln A - A \tag{3.54}$$

Applying Stirling's approximation in Eq. (3.53) gives:

$$\ln W = N \ln N - N - \sum_{i=0}^{r} [n_i \ln n_i - n_i] \tag{3.55}$$

Taking the differential of Eq. (3.55),

$$\partial \ln W = \partial N \ln N + N \frac{1}{N} \partial N - \partial N - \sum_{i=0}^{r} \left[\partial n_i \ln n_i + n_i \frac{1}{n_i} \partial n_i - \partial n_i \right] \tag{3.56}$$

$$d \ln W = \ln N dN + N \frac{dN}{N} - dN - \sum_{i=0}^{r} \left[\ln n_i dn_i + n_i \frac{dn_i}{n_i} - dn_i \right] \tag{3.57}$$

The total number of particles N in the system is constant, therefore,

$$dN = 0 \tag{3.58}$$

Eq. (3.56) becomes

$$d \ln W = - \sum_{i=0}^{r} [\ln n_i \, dn_i] \tag{3.59}$$

To find the maximum value of W, the left-hand side is equated to zero; i.e.,

$$d \ln \mathrm{W} = 0 \tag{3.60}$$

$$\sum_{i=0}^{r} [\ln n_i dn_i] = 0 \tag{3.61}$$

The number of particles in the system is constant, therefore, from Eq. (3.48),

$$\sum_{i=0}^{r} dn_i = 0 \tag{3.62}$$

As the total energy of the system is constant, Eq. (3.50) gives

$$\sum_{i=0}^{r} \varepsilon_i dn_i = 0 \tag{3.63}$$

The most probable combination is given by Eq. (3.61), with the constraints that the number of particles and the total energy of the system remain constant, as given in Eqs. (3.62) and (3.63). The Lagrange method of undetermined multiplier is now used. Multiply Eq. (3.62) by α, multiply Eq. (3.63) by β, and add to Eq. (3.61):

$$\sum_{i=0}^{r} [\ln n_i + \alpha + \beta\varepsilon_i] dn_i = 0 \tag{3.64}$$

Since dn_i represents the change in the number of particles in a particular energy level (which cannot be equal to zero), to make sure that the left-hand side of Eq. (3.64) is equal to zero, the factor $(\ln n_i + \alpha + \beta\varepsilon_i)$ should be equal to zero; i.e.,

$$[\ln n_i + \alpha + \beta\varepsilon_i] = 0 \tag{3.65}$$

$$n_i = e^{-\alpha} e^{-\beta\varepsilon_i} \tag{3.66}$$

Eq. (3.66) represents the number of particles n_i in energy level ε_i, and α and β are the constants.

$$\sum_{i=0}^{r} n_i = e^{-\alpha} \sum_{i=0}^{r} e^{-\beta\varepsilon_i} = N \tag{3.67}$$

$$e^{-\alpha} = \frac{N}{\sum_{i=0}^{r} e^{-\beta\varepsilon_i}} \tag{3.68}$$

Therefore,

$$n_i = \frac{N}{\sum_{i=0}^{r} e^{-\beta\varepsilon_i}} e^{-\beta\varepsilon_i} \tag{3.69}$$

The constant β is given as [5]:

$$\beta = \frac{1}{k_B T} \tag{3.70}$$

The reader can find the details in the literature [5]. Now, Eq. (3.69) becomes

$$n_i = \frac{N}{\sum_{i=0}^{r} e^{-\varepsilon_i/k_B T}} e^{-\varepsilon_i/k_B T} \quad (3.71)$$

Let

$$A = \frac{N}{\sum_{i=0}^{r} e^{-\varepsilon_i/k_B T}} \quad (3.72)$$

Here, A is a constant. Now, Eq. (3.71) can be written as

$$n(\varepsilon_i) = Ae^{-\varepsilon_i/k_B T} \quad (3.73)$$

Eq. (3.73) is known as the Maxwell–Boltzmann distribution.

Fig. 3.17 shows the distribution of ideal gas molecules in the system at different temperatures. The plotting is done using Eq. (3.73) while incorporating $A = 1$.

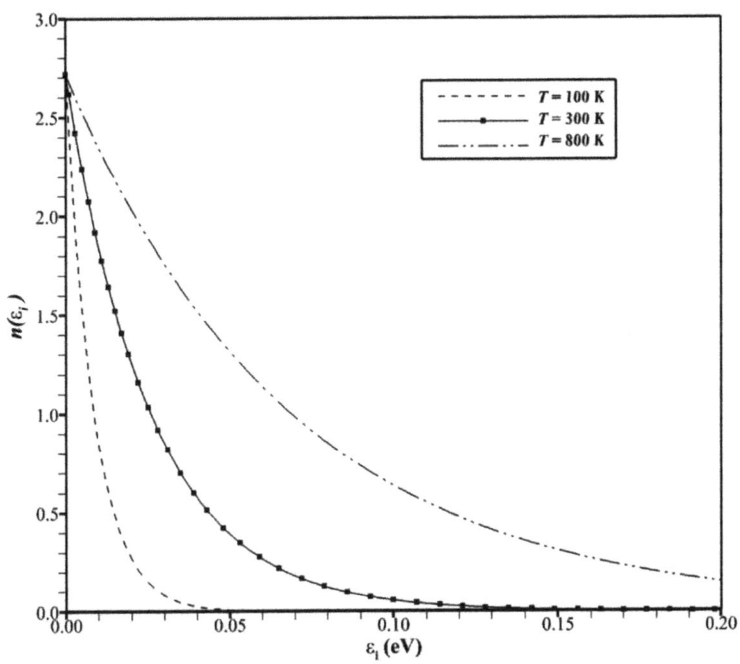

■ **FIGURE 3.17** Maxwell–Boltzmann distribution for ideal gas molecules for different temperatures.

3.4.2 Fermi–Dirac Distribution

The Fermi–Dirac distribution can be used to describe the behavior of indistinguishable particles, which are the fermions. Fermions obey the Pauli Exclusion principle, which states that two fermions cannot have the same quantum state. As the electron obeys the Pauli Exclusion principle, therefore, electron characteristics can also be defined by using the Fermi–Dirac distribution.

When the fermions are added to the energy bands, they fill the energy levels in ascending order. This means that the lowest energy state is filled first, then the next energy level is filled, and then the next. The energy level filled by fermions at absolute zero Kelvin, $T = 0$ K, is known as the Fermi energy level, E_f.

Consider the thermodynamic system that consists of a larger number of fermions. The number of fermions N and the total energy of the system E remain constant. The system has energy levels $\varepsilon_1, \varepsilon_2, \varepsilon_3 \ldots \varepsilon_r$. Let $g_1, g_2, g_3 \ldots g_r$ represent the number of states in the energy levels $\varepsilon_1, \varepsilon_2, \varepsilon_3 \ldots \varepsilon_r$, respectively. The energy levels and the number of states in each energy level are fixed for the system. Let $n_1, n_2, n_3 \ldots n_r$ be the number of fermions in energy levels $\varepsilon_1, \varepsilon_2, \varepsilon_3 \ldots \varepsilon_r$, respectively.

The total number of particles in the system is N, which remains constant,

$$\sum_{i=0}^{r} n_i = N \tag{3.74}$$

The total energy of the system is constant E, therefore,

$$\sum_{i=0}^{r} n_i \varepsilon_i = E \tag{3.75}$$

The number of possible microstates W_i, also known as configurations, can be given as

$$W_i = \frac{g_i!}{n_i!(g_i - n_i)!} \tag{3.76}$$

Numbering all states, and exchanging the states instead of fermions, will provide $g_i!$ configurations. All the empty states $(g_i - n_i)$ are identical, therefore, the factor $(g_i - n_i)!$ comes in the denominator. Also, the filled states are indistinguishable; therefore, we divide by $n_i!$.

The total number of possible configurations can be obtained as:

$$W = \frac{g_1!}{n_1!(g_1 - n_1)!} \cdot \frac{g_2!}{n_2!(g_2 - n_2)!} \cdot \frac{g_3!}{n_3!(g_3 - n_3)!} \cdots \frac{g_r!}{n_r!(g_r - n_r)!} \tag{3.77}$$

$$W = \prod_{i=1}^{r} \frac{g_i!}{n_i!(g_i - n_i)!} \tag{3.78}$$

where $\prod$ represents the product in Eq. (3.75).

Taking the natural logarithm of both sides of Eq. (3.78),

$$\ln W = \sum_{i=1}^{r} [\ln g_i! - \ln n_i! - ln(g_i - n_i)!] \tag{3.79}$$

Applying Stirling's approximation in Eq. (3.77) results in

$$\ln W = \sum_{i=1}^{r} [g_i \ln g_i - g_i - \{n_i \ln n_i - n_i\} - \{(g_i - n_i)\ln(g_i - n_i) - (g_i - n_i)\}] \tag{3.80}$$

Upon simplification,

$$\ln W = \sum_{i=1}^{r} [g_i \ln g_i - n_i \ln n_i - (g_i - n_i)\ln(g_i - n_i)] \tag{3.81}$$

The most probable configuration, which corresponds to the equilibrium state, can be found by maximizing Eq. (3.81). Taking the differential of Eq. (3.81) and using the relation $dg_i = 0$ as the states are fixed, gives

$$d \ln W = \sum_{i} [-dn_i \ln n_i + dn_i \ln(g_i - n_i)] = 0 \tag{3.82}$$

$$\sum_{i=1}^{r} dn_i \left[\ln \frac{(g_i - n_i)}{n_i}\right] = 0 \tag{3.83}$$

Eq. (3.83) is subject to two constraints: fixed number of particles and constant total energy. Taking the differentials of Eqs. (3.74) and (3.75) gives

$$\sum_{i=0}^{r} \partial n_i = 0 \tag{3.84}$$

and

$$\sum_{i=0}^{r} \varepsilon_i \partial n_i = 0 \tag{3.85}$$

Using the Lagrange method of undetermined multiplier, multiply Eq. (3.84) by α and Eq. (3.85) by β and then subtract them from Eq. (3.83). This gives

$$\sum_{i} \left[-\ln \frac{(g_i - n_i)}{n_i} + \alpha + \beta\varepsilon_i\right] dn_i = 0 \tag{3.86}$$

In order to ensure that the left-hand side of Eq. (3.86) is zero, the term $-\ln((g_i - n_i)/n_i) + \alpha + \beta\varepsilon_i$ should be zero; i.e.,

$$\ln\frac{(g_i - n_i)}{n_i} = \alpha + \beta\varepsilon_i \tag{3.87}$$

Upon simplification,

$$\frac{n_i}{g_i} = \frac{1}{1 + e^{\alpha+\beta\varepsilon_i}}$$

Here [5],

$$\beta = \frac{1}{k_B T} \quad \text{and} \quad \alpha = -\frac{E_f}{k_B T} \tag{3.88}$$

Therefore,

$$\frac{n_i}{g_i} = \frac{1}{1 + e^{(\varepsilon_i - E_f)/k_B T}} \tag{3.89}$$

Since $n_i = P(\varepsilon_i)g_i$, where $P(\varepsilon_i)$ is the probability of having fermions in energy level ε_i,

$$P(\varepsilon_i) = \frac{1}{1 + e^{(\varepsilon_i - E_f)/k_B T}} \tag{3.90}$$

This expression in known as the Fermi–Dirac distribution.

Fig. 3.18 shows the Fermi–Dirac distribution for three different temperatures. When the temperature is 0 K, the Fermi–Dirac distribution behaves like a step function. The probability of having fermions below E_f is 1, which means that no fermions can have energy more than E_f. As the temperature increases some fermions become thermally excited and move above the fermi energy level.

3.4.3 **Bose–Einstein Distribution**

Bose–Einstein statistics describe the behavior of indistinguishable particles with no limit on the number of particles per energy level. These particles are known as bosons. Bosons don't obey the Pauli Exclusion principle since a particular energy level can have a number of bosons; therefore, the Bose–Einstein distribution is different from the Fermi–Dirac distribution. Due to this behavior of bosons, at very low temperature an unlimited number of them accumulates into the same energy state; this phenomenon is known as condensation. Phonons and photons are considered bosons.

Consider the thermodynamic system, which consists of N number of bosons. The total energy of the system is E. The system is having

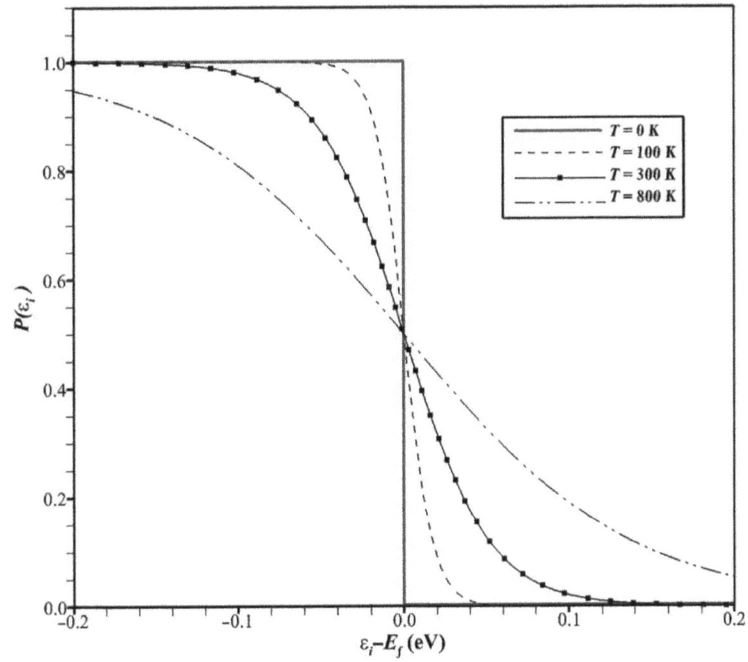

FIGURE 3.18 Fermi—Dirac distribution for different temperatures.

$\varepsilon_1, \varepsilon_2, \varepsilon_3 \ldots \varepsilon_r$ energy levels. Let $n_1, n_2, n_3 \ldots n_r$ represent the number of bosons and $g_1, g_2, g_3 \ldots g_r$ represent the degeneracy in $\varepsilon_1, \varepsilon_2, \varepsilon_3 \ldots \varepsilon_r$ energy levels, respectively.

The system consists of N number of particles,

$$\sum_{i=0}^{r} n_i = N \tag{3.91}$$

E, which is the total energy of the system, can be given as

$$\sum_{i=0}^{r} n_i \varepsilon_i = E \tag{3.92}$$

The number of possible microstates W_i can be given as

$$W_i = \frac{(n_i + g_i - 1)!}{n_i!(g_i - 1)!} \tag{3.93}$$

The total number of possible microstates W is given as

$$W = \prod_{i=1}^{r} \frac{(n_i + g_i - 1)!}{n_i!(g_i - 1)!} \tag{3.94}$$

where $\prod$ represents the product $W_1 W_2 \ldots W_r$.

Now, taking the natural logarithm of both sides,

$$\ln W = \sum_{i=0}^{r} [\ln (n_i + g_i - 1)! - \ln n_i! - \ln (g_i - 1)!] \tag{3.95}$$

Considering $n_i + g_i - 1 \approx n_i + g_i$,

$$\ln W = \sum_{i=0}^{r} [\ln (n_i + g_i)! - \ln n_i! - \ln (g_i - 1)!] \tag{3.96}$$

Using the Stirling's approximation in Eq. (3.96),

$$\ln W = \sum_{i} \begin{bmatrix} (n_i + g_i)\ln (n_i + g_i) - (n_i + g_i) - \{n_i \ln n_i - n_i\} \\ - \{(g_i - 1)\ln (g_i - 1) - (g_i - 1)\} \end{bmatrix} \tag{3.97}$$

On simplification,

$$\ln W = \sum_{i=1}^{r} [(n_i + g_i)\ln (n_i + g_i) - n_i \ln n_i - (g_i - 1)\ln (g_i - 1) - 1] \tag{3.98}$$

The system is in equilibrium state when the entropy of the system is maximum. To determine the maxima, take the differential of Eq. (3.98) and equate it to zero. In addition, taking $dg_i = 0$ results in

$$d \ln W = \sum_{i=1}^{r} [dn_i\{\ln (n_i + g_i) - \ln g_i\}] = 0 \tag{3.99}$$

$$\sum_{i=1}^{r} dn_i \left[\ln \left(\frac{n_i + g_i}{g_i} \right) \right] = 0 \tag{3.100}$$

Eq. (3.100) contains two constraints: the energy of the system and the number of particles in the system are constant, which can be given as

$$\sum_{i=0}^{r} dn_i = 0 \tag{3.101}$$

and

$$\sum_{i=0}^{r} \varepsilon_i dn_i = 0 \tag{3.102}$$

We now use the Lagrange method of undetermined multipliers to solve Eq. (3.100) under the constraints of Eqs. (3.101) and (3.102). Multiply Eq. (3.101) by α and Eq. (3.102) with β and subtract them from Eq. (3.100).

$$\sum_i \left[-\ln\left(\frac{n_i + g_i}{g_i}\right) + \alpha + \beta\varepsilon_i \right] \partial n_i = 0 \quad (3.103)$$

When the term $-\ln((g_i + n_i)/n_i) + \alpha + \beta\varepsilon_i$ is zero, it will make the left-hand side of Eq. (3.103) equal to zero:

$$-\ln\left(\frac{n_i + g_i}{g_i}\right) + \alpha + \beta\varepsilon_i = 0 \quad (3.104)$$

This results in the following:

$$\frac{n_i}{g_i} = \frac{1}{e^{\alpha+\beta\varepsilon_i} - 1} \quad (3.105)$$

Here the α and β are given as [5]

$$\beta = \frac{1}{k_B T} \quad \text{and} \quad \alpha = -\frac{E_f}{k_B T} \quad (3.106)$$

Eq. (3.105) becomes

$$\frac{n_i}{g_i} = \frac{1}{e^{(\varepsilon_i - E_f)/k_B T} - 1} \quad (3.107)$$

$n_i = P(\varepsilon_i) g_i$, where $P(\varepsilon_i)$ is the probability of having bosons in energy level ε_i; therefore,

$$P(\varepsilon_i) = \frac{1}{e^{(\varepsilon_i - E_f)/k_B T} - 1} \quad (3.108)$$

Eq. (3.108) is known as the Bose–Einstein distribution.

Fig. 3.19 shows the distribution of bosons at different temperatures. This figure is generated using Eq. (3.108).

3.5 CLOSING REMARKS

To bridge the gap between the macroscopic and the microscopic pictures related to a system, some basics of statistical mechanics are introduced. It is a well-established fact that oscillators such as electrons, phonons, protons, neutrons, molecules, crystals, and other atomic length scale structures are covered within a frame of microscopic picture of any given system. The structure of microscopic constituents is usually completely hidden from the view perspective. On the other hand, the macroscopic picture involved with physical processes and phenomenon could be observable at much larger scales, such as by the naked eye. To assess and measure or predict the physical aspects and properties of a given system, a chain of reasoning needs to considered, which connects the

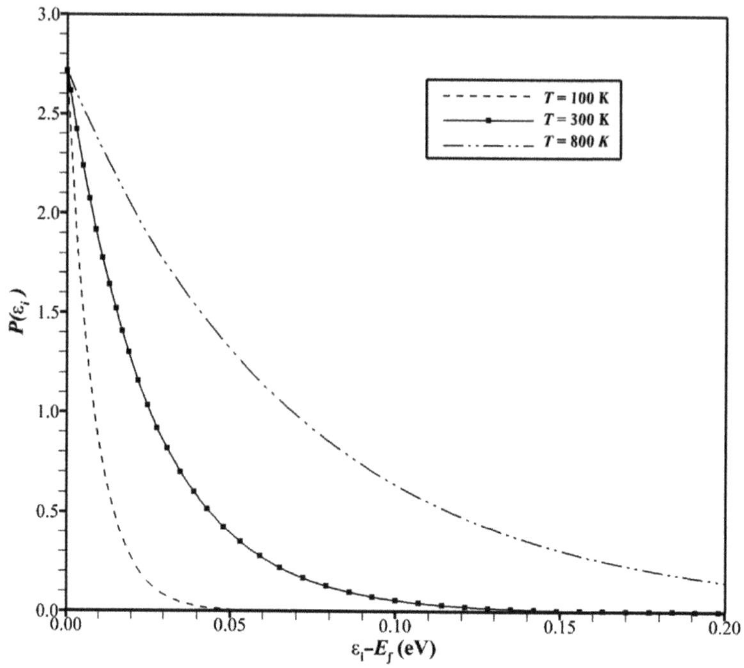

■ **FIGURE 3.19** Bose–Einstein distribution for different temperatures.

fundamental laws of physics associated with the structure and the interactions of the unseen microscopic constituents. Consequently, in this chapter, a brief introduction to the statistical analysis was introduced while incorporating the micro/macrostates and probability functions. Later, phase space and ensemble were explained and some examples were provided. Finally, statistical distributions of energy states of the oscillators and waves while incorporating Maxwell–Boltzmann, Fermi–Dirac, and Bose–Einstein distributions was presented.

REFERENCES

[1] T. Guenault, Statistical Physics, second ed., Springer, Netherlands, 2007.
[2] Y. Kaznessis, Statistical Thermodynamics and Stochastic Kinetics an Introduction for Engineers, first ed., Cambridge University Press, New York, 2011.
[3] L.G. Benguigui, Statistical Mechanics for Beginners, World Scientific Publishing Company, Singapore, 2010.
[4] D.V. Schroeder, An Introduction to Thermal Physics, Addison Wesley, San Francisco, 2000.
[5] R. Sonntag, G. Wylen, Introduction to Thermodynamics Classical & Statistical, second ed., John Wiley & Sons Ltd., New York, 1982.

Chapter 4

Analysis of Energy Transport Equations at Micro/Nanoscales

4.1 INTRODUCTION

In this chapter the formulation of a heat equation in hyperbolic form and derived from electron kinetic theory is introduced; practical applications in terms of short-pulse heating situations are presented first. Thermal stress field development is also included in the analysis. The fundamentals of wave nature of thermal energy transport pertinent to thin films is considered and the relevant mathematical formulations for energy transport are described in detail. To model energy transport in the thin films, the equation of phonon radiative transfer (EPRT) is introduced. As in the dielectric thin film, mainly the phonons are considered to be responsible for energy transport; therefore, EPRT has been incorporated to account for the phonon transport. In metals, the thermal transport involves the electron and lattice subsystems; therefore, a modified version of EPRT is incorporated in both subsystems through introducing the coupling term in the governing equations. In addition, the modified two-equation system is presented for metallic thin films. The models used and the associated initial, boundary, and interface conditions are presented. Casimir limit for radiative transport and the fundamentals of thermal boundary resistance (TBR) for interface conditions of thin films pairs are described.

4.2 HYPERBOLIC HEAT EQUATION AND APPLICATIONS

Due to the nonequilibrium nature of the heat transfer in micro/nanoscale devices during a short interaction period, heat transport within the medium is governed by the hyperbolic heat equation. In this case, anomalies associated with the Fourier model are avoided, by considering the heat wave propagation toward the substrate material. On the other hand, the thermal expansion of the solid substrate under high temperature gradients gives rise to formation of thermal stresses in the heated region.

Heat Transport in Micro- and Nanoscale Thin Films.
DOI: http://dx.doi.org/10.1016/B978-0-323-42979-5.00004-9

However, to determine the thermal stress field, temperature variation needs to be known in the heated region. The analytical solution for the heating problem can be presented via the solution of the Cattaneo equation with the appropriate heat-transferring and stress-boundary conditions. In this section, analytical solution of the Cattaneo equation for a short-pulse heating situation is presented in line with the previous study [1,2].

4.2.1 Analysis and Solution of Hyperbolic Heat Equation

The Cattaneo equation can be presented in terms of the heat flux:

$$\varepsilon \frac{\partial q}{\partial t} = -q - K\nabla T \tag{4.1}$$

where ε is the relaxation time, K is the thermal conductivity, q is the heat flux vector, and T is temperature. The Cattaneo equation yields to the hyperbolic heat conduction equation for the temperature field, which is propagative with a speed $c = \sqrt{\frac{K}{\varepsilon \rho C_p}}$, where ρ is the density and C_p is the specific heat. Although the thermal properties of the substrate material are temperature dependent, the constant properties are assumed in the equation due to the simplicity. It should be noted that the use of the variable properties results in a nonlinear equation of energy transport, which may not be solved analytically to obtain the closed form solution for temperature and stress distributions. Combining Eq. (4.1) with the energy equation $\left(\rho C_p \frac{\partial T}{\partial t} + \frac{\partial q}{\partial x} = 0\right)$, the resulting hyperbolic equation (telegraph) can be written in terms of temperature:

$$\alpha \frac{\partial^2 T}{\partial x^2} = \frac{\partial T}{\partial t} + \varepsilon \frac{\partial^2 T}{\partial t^2} \tag{4.2}$$

with initial and boundary conditions:

$$T = 0 \quad at \quad t = 0 \tag{4.3}$$

$$q = f(t) \quad at \quad x = 0 \tag{4.4}$$

Taking the Laplace transformation of the governing equation (Eq. 4.2), initial and boundary conditions (Eqs. 4.3 and 4.4) result in:

$$\alpha \frac{\partial^2 \hat{T}}{\partial x^2} - s^2 \varepsilon \hat{T} - s\hat{T} = 0 \tag{4.5}$$

$$\hat{q} = -K \frac{\partial \hat{T}}{\partial x} = F(s) \quad at \quad x = 0 \tag{4.6}$$

The heating pulse can be presented in terms of exponentially decaying intensity [1]:

$$f(t) = e^{-\beta_1 t} - e^{-\beta_2 t} \tag{4.7}$$

Since the model is linear, we can use the superposition principle to compute the temperature distribution due to the two terms of the pulse function separately and add them to get the final solution. Fig. 4.1 shows the pulse used in the analysis. In the sequel, one term, namely, $e^{-\beta t}$, is considered as a pulse source and then the solution for temperature yields

$$T = T\big|_{e^{-\beta_1 t}} - T\big|_{e^{-\beta_2 t}} \tag{4.8}$$

Temperature distribution can be obtained using the Laplace transformation method [1]. Therefore, the solution of the governing equation in the s-domain (Laplace domain) can be represented as

$$\hat{T}(x,s) = C_1 e^{-\frac{x}{\sqrt{\alpha}}\sqrt{\varepsilon s^2 + s}} + C_2 e^{\frac{x}{\sqrt{\alpha}}\sqrt{\varepsilon s^2 + s}} \tag{4.9}$$

For the solution to be bounded C_2 must vanish.

By applying the transformed boundary condition, the constant C_1 can be determined and the solution in the transformed domain can be finally written as

$$\hat{T}(x,s) = \frac{\sqrt{\varepsilon\alpha}}{K(s+\beta)\sqrt{s(s+1/\varepsilon)}} e^{-\frac{x}{\sqrt{\alpha/\varepsilon}}\sqrt{s(s+1/\varepsilon)}} \tag{4.10}$$

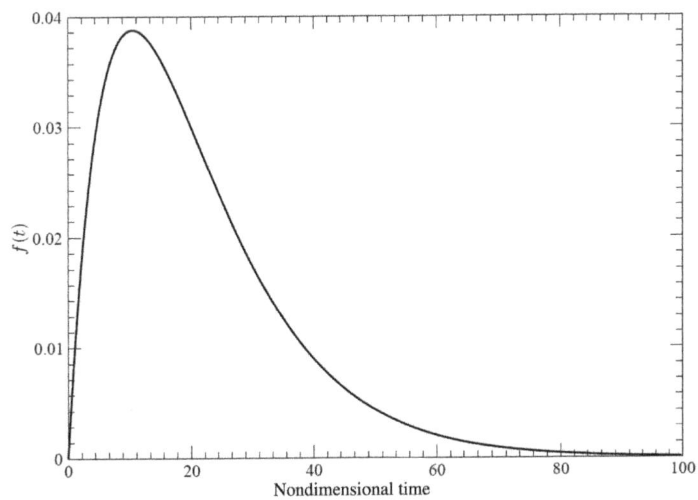

FIGURE 4.1 Temporal variation of exponential pulse distribution.

The temperature distribution can be written as a multiplication of two functions as

$$\hat{T}(x,s) = \hat{T}_1(x,s)\hat{T}_2(x,s) \tag{4.11}$$

The inverse of the two terms can individually be written as

$$\begin{aligned} £^{-1}\hat{T}(x,s) &= £^{-1}\left(\frac{1}{\sqrt{s(s+1/\varepsilon)}}e^{-\frac{x}{\sqrt{\alpha/\varepsilon}}\sqrt{s(s+1/\varepsilon)}}\right) \\ &= e^{-\frac{t}{2\varepsilon}}I_o\left(\frac{1}{2\varepsilon}\sqrt{t^2-\frac{x^2}{\alpha/\varepsilon}}\right)U\left(t-\frac{x}{\sqrt{\alpha/\varepsilon}}\right) \end{aligned} \tag{4.12}$$

and

$$£^{-1}(\hat{T}_2(x,s)) = £^{-1}\left(\frac{\sqrt{\alpha\varepsilon}}{K(s+\beta)}\right) = \frac{\sqrt{\alpha\varepsilon}}{K}e^{-\beta t} \tag{4.13}$$

where I_o is the modified Bessel function of the first kind and U is the unit step function.

The overall temperature distribution is obtained using the convolution

$$T(x,t) = \int_0^t \hat{T}_1(r)\hat{T}_2(t-\tau)d\tau \tag{4.14}$$

The thermoelasticity is a temperature rate dependent by including temperature rate among the constitutive variables [3]. However, for most of the materials, the strain rate is of the same order as the temperature rate; in this case, the thermomechanical coupling becomes small for the small displacements and it can be neglected [4–6]. This simplifies the physical problem to yield the closed form solution for the thermal stress field in the solid substrate. After considering the homogenous isotropic structure, the temperature distribution of the solid media in the Laplace domain (s-domain) can be represented in a nondimensional form:

$$\hat{T}(x,s) = \frac{(s+2)e^{-\sqrt{s^2+2s}x}}{(s+\beta)\sqrt{s^2+2s}} \tag{4.15}$$

The corresponding nondimensional stress field equation in the s-domain is given by

$$\hat{\sigma}(x) - \frac{s^2\hat{\sigma}(x)}{c_1^2} = \frac{c_2 s^2(s+2)}{(s+b)\sqrt{s^2+2s}}e^{-\sqrt{s(s+2)x}} \tag{4.16}$$

and c_1 and c_2 are the nondimensional wave speed and the nondimensional thermal modulus, respectively.

$$\hat{\sigma} = B_1 e^{\frac{sx}{c_1}} + B_2 e^{-\frac{sx}{c_1}} + \frac{c_2 s^2(s+2)}{(s+b)\sqrt{s^2+2s}} \frac{c_1^2 e^{-\sqrt{s(s+2)}x}}{s(c_1^2(s+2)-s)} \tag{4.17}$$

For boundness of the solution $B_1 = 0$, and the condition of free stress (i.e., $\sigma = 0$) on the surface is imposed at this stage to calculate the other coefficient B_2 as

$$B_2 = -\frac{c_1^2}{s(c_1^2 s + 2c_1^2 - s)} \frac{c_2 s^2(s+2)}{(s+b)\sqrt{s^2+2s}} \tag{4.18}$$

Therefore,

$$\hat{\sigma}(x,s) = -\frac{c_1^2 c_2 s(s+2) e^{-\frac{sx}{c_1}}}{\sqrt{s^2+2s}(c_1^2 s + 2c_1^2 - s)(s+\beta)} + \frac{c_1^2 c_2 s(s+2) e^{-\sqrt{s(s+2)s}}}{\sqrt{s^2+2s}(c_1^2(s+2)-s)(s+\beta)} \tag{4.19}$$

where $\hat{\sigma}(x,s) = \hat{\sigma}_h(s,x) + \hat{\sigma}_p(x,s)$. The stress component $\hat{\sigma}_h$ can be written as a multiplication of two subcomponents as $\hat{\sigma}_h(x,s) = \hat{\sigma}_h^{(1)}(s,x)\hat{\sigma}_h^{(2)}(x,s)$. The inverse Laplace transform of the two subcomponents is carried out as the following:

$$\sigma_h^{(1)}(x,t) = L^{-1}\left(\hat{\sigma}_h^{(1)}\right) = \frac{c_1^2 c_2}{c_1^2(\beta-2)-\beta}\left(\beta(\beta-2)e^{\frac{x\beta}{c_1}-t\beta} - \frac{4c_1^2}{(c_1^2-1)^2} e^{\frac{2c_1(x-c_1 t)}{c_1^2-1}}\right)$$
$$\times U\left[t - \frac{x}{c_1}\right] - \frac{c_1^2 c_2}{c_1^2-1}\delta\left[t-\frac{x}{c_1}\right] \tag{4.20}$$

$$\sigma_h^{(2)}(x,t) = L^{-1}\left(\hat{\sigma}_h^{(2)}\right) = e^{-t} J_0[it] \tag{4.21}$$

We can get $\hat{\sigma}_h(x,t)$ by employing the convolution theorem:

$$\sigma_h(x,t) = \int_0^t \hat{\sigma}_h^{(1)}(\tau)\hat{\sigma}_h^{(2)}(t-\tau)d\tau \tag{4.22}$$

Similarly, the second component of the stresses $\hat{\sigma}_p$ can be written as a multiplication of two subcomponents:

$$\hat{\sigma}_p(x,s) = \hat{\sigma}_p^{(1)}(x,s)\hat{\sigma}_p^{(2)}(x,s) \tag{4.23}$$

The inverse Laplace transform is carried out for both subcomponents:

$$\begin{aligned}\sigma_p^{(1)}(x,t) &= \pounds^{-1}\left(\hat{\sigma}_p^{(1)}\right)\\ &= c_1^2 c_2\left(\frac{1}{c_1^2-1}\delta[t]+\frac{(2\beta-\beta^2)}{c_1^2\beta-2c_1^2-\beta}e^{-\beta t}+\frac{4c_1^2}{\left(c_1^2-1\right)^2\left(c_1^2\beta-2c_1^2-\beta\right)}e^{-\frac{2c_1^2}{c_1^2-1}t}\right)\end{aligned} \tag{4.24}$$

$$\sigma_p^{(2)}(x,t) = \pounds^{-1}\left(\hat{\sigma}_p^{(2)}\right) = e^{-t} I_0\left(\sqrt{t^2-x^2}\right) U[t-x] \tag{4.25}$$

We can get $\hat{\sigma}_p(x,t)$ by employing the convolution theorem:

$$\sigma_p(x,t) = \int_0^t \hat{\sigma}_p^{(1)}(\tau)\hat{\sigma}_p^{(2)}(t-\tau)d\tau \tag{4.26}$$

All components of the stresses have been transformed to (x,t) domain, and we are ready to get the stresses as

$$\sigma(x,t) = \sigma_h(x,t) + \sigma_p(x,t) \tag{4.27}$$

A computer code is developed to compute temperature and stress distributions from Eqs. (4.14) to (4.27), respectively.

4.2.2 Perturbation Solution for Hyperbolic Heat Equation

High intensity irradiation of thin metallic films results in thermal separation of electron and lattice subsystems. The collisional process taking place between electron and lattice subsystems defines the energy transport between the subsystems. Electrons receive energy from the irradiation source, such as laser beam, resulting in increased electron excess energy in the electron subsystem. During the heating period, electrons transfer some fraction of their excess energy to lattice subsystem, by thermal communication of electron-lattice subsystems, via electron–phonon coupling. This results in lattice site energy increase; consequently, lattice site temperature increases gradually as the collisional process for energy transport from electron to lattice subsystems continues. Since the electron and lattice subsystem separates thermally, modeling of the energy transport on the microscopic scale becomes necessary. In this section, temperature field in the electron and lattice subsystem is presented using the perturbation method in line with the previous study [2].

The energy transport equation for electron and lattice subsystems in small size substrates, such as thin films, can be written as [7]

$$C_{\mathrm{E}}\frac{\partial T_{\mathrm{E}}}{\partial \bar{t}} = k\frac{\partial^2 T_{\mathrm{E}}}{\partial \bar{x}^2} - G(T_{\mathrm{E}} - T_{\mathrm{L}}) \tag{4.28}$$

$$C_{\mathrm{L}}\frac{\partial T_{\mathrm{L}}}{\partial \bar{t}} = G(T_{\mathrm{E}} - T_{\mathrm{L}}) \tag{4.29}$$

where T_{E} and T_{L} are the electron and lattice site temperatures, respectively. C_{E} is the electron heat capacity, C_{L} is the lattice heat capacity, G is the electron–phonon coupling factor, and k is the thermal conductivity. $\bar{x}$ is the dimensional lattice depth and $\bar{t}$ is the dimensional time. If the dimensionless quantities are defined as

$$\theta_{\mathrm{E,L}} = \frac{T_{\mathrm{E,L}}}{T_{\mathrm{o}}}, \quad x = \bar{x}d, \quad t = \frac{\bar{t}}{C_{\mathrm{E}}/G} \tag{4.30}$$

Eqs. (4.28) and (4.29) reduce to the following form:

$$\frac{\partial \theta_{\mathrm{E}}}{\partial t} = \alpha\frac{\partial^2 \theta_{\mathrm{E}}}{\partial x^2} - (\theta_{\mathrm{E}} - \theta_{\mathrm{L}}) \tag{4.31}$$

$$\frac{\partial \theta_{\mathrm{L}}}{\partial t} = \varepsilon(\theta_{\mathrm{E}} - \theta_{\mathrm{L}}) \tag{4.32}$$

where

$$\alpha = \frac{kd^2}{G}, \quad \varepsilon = \frac{C_{\mathrm{E}}}{C_{\mathrm{L}}} \tag{4.33}$$

In the previous formulation, $d = 1/\lambda$ and λ is the mean free path of electrons.

Fig. 4.2 shows the schematic view of the thin film geometry. The initial and boundary conditions for the problem are:

$$\theta_{\mathrm{L}}(x,0) = 0, \quad \theta_{\mathrm{E}}(x,\infty) = 0, \quad \frac{\partial \theta_{\mathrm{E}}}{\partial x}(0,t) = S_0 = I_0 e^{-k_1 t}, \quad \frac{\partial \theta_{\mathrm{E}}}{\partial x}(Ld,t) = 0 \tag{4.34}$$

where the dimensionless source amplitude is related to one-dimensional through the relation $I_0 = \bar{I}_0/(kdT_0)$. Note that at one of the surfaces of the layer, there exists an exponentially decaying heat source, whereas at the other surface the insulated boundary is considered.

In metallic materials, the dimensionless thermal conductivity $\alpha\ \left(=\frac{kd^2}{G}\right)$ is of order 1 and the ratio of heat capacities ε is small compared to order 1 terms. Therefore, a perturbation type of solution in terms of the small

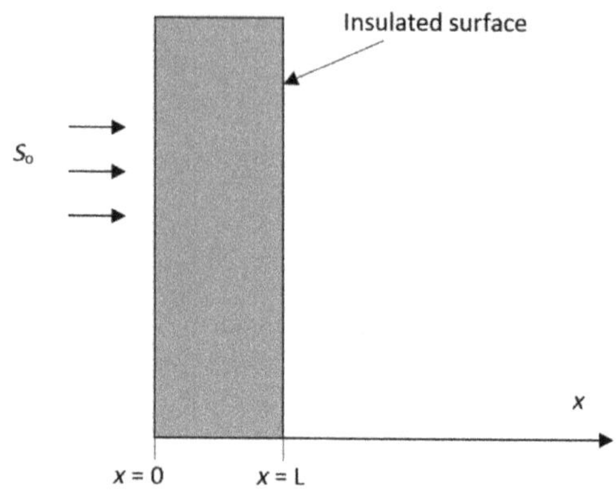

■ **FIGURE 4.2** Schematic view of a solid thin film [2].

parameter ε is possible. The method of strained parameters [8] is incorporated, which is involved with finding a parameter in Eqs. (4.31) and (4.32) that is altered by the perturbations. It then expands both the dependent variables as well as this parameter using an appropriate sequence of gage functions derived from the powers of the strength of the perturbation. The perturbations in the ε parameter are then chosen to render the expansion uniformly valid. This technique is called the method of strained parameter, which can be effectively used in the solution of differential equations, which may be difficult to solve analytically by other methods such as Fourier and Laplace transform methods.

Consequently, the expansions can be written as

$$\begin{aligned} \theta_E &= \theta_{E0} + \varepsilon\theta_{E1} + \cdots \\ \theta_L &= \theta_{L0} + \varepsilon\theta_{L1} + \cdots \\ \alpha &= \alpha_0 + \varepsilon\alpha_1 + \cdots \end{aligned} \tag{4.35}$$

Substituting the expansions into Eqs. (4.31), (4.32), and (4.34), separating at each order of approximation, we can have

Order 1:

$$\frac{\partial\theta_{E0}}{\partial t} = \alpha_0\frac{\partial^2\theta_{E0}}{\partial x^2} - (\theta_{E0} - \theta_{L0}) \tag{4.36}$$

$$\frac{\partial\theta_{L0}}{\partial t} = 0 \tag{4.37}$$

$$\theta_{L0}(x,0) = 0, \quad \theta_{E0}(x,\infty) = 0, \quad \frac{\partial\theta_{E0}}{\partial x}(0,t) = I_0e^{-k_1 t}, \quad \frac{\partial\theta_{E0}}{\partial x}(Ld,t) = 0 \tag{4.38}$$

Order ε:

$$\frac{\partial \theta_{E1}}{\partial t} = \alpha_0 \frac{\partial^2 \theta_{E1}}{\partial x^2} - (\theta_{E1} - \theta_{L1}) + \alpha_1 \frac{\partial^2 \theta_{E0}}{\partial x^2} \tag{4.39}$$

$$\frac{\partial \theta_{L1}}{\partial t} = \theta_{E0} - \theta_{L0} \tag{4.40}$$

$$\theta_{L1}(x,0) = 0, \quad \theta_{E1}(x,\infty) = 0, \quad \frac{\partial \theta_{E1}}{\partial x}(0,t) = 0, \quad \frac{\partial \theta_{E1}}{\partial x}(Ld,t) = 0 \tag{4.41}$$

Introducing the method of separation of variables to order 1 equations, for the given boundary conditions at this order, the solutions are as follows:

$$\theta_{L0} = 0 \tag{4.42}$$

$$\theta_{E0} = \sqrt{\frac{\alpha_0}{k_1 - 1}} I_0 e^{-k_1 t} \left\{ \cot\left(\sqrt{\frac{k_1 - 1}{\alpha_0}} Ld \right) \cos\left(\sqrt{\frac{k_1 - 1}{\alpha_0}} x \right) + \sin\left(\sqrt{\frac{k_1 - 1}{\alpha_0}} x \right) \right\} \tag{4.43}$$

At the second level of approximation, we can insert Eqs. (4.42) and (4.43) into Eq. (4.40). A solution satisfying the initial condition at this level of approximation becomes

$$\theta_{L1} = \frac{1}{k_1} \sqrt{\frac{\alpha_0}{k_1 - 1}} I_0 (1 - e^{-k_1 t}) \times \left\{ \cot\left(\sqrt{\frac{k_1 - 1}{\alpha_0}} Ld \right) \cos\left(\sqrt{\frac{k_1 - 1}{\alpha_0}} x \right) + \sin\left(\sqrt{\frac{k_1 - 1}{\alpha_0}} x \right) \right\} \tag{4.44}$$

Solutions of Eqs. (4.43) and (4.44) are finally substituted into Eq. (4.39), and, at this level, only the secularities are eliminated, which become

$$\alpha_1 = -\frac{\alpha_0}{k_1 (k_1 - 1)} \tag{4.45}$$

Hence, the approximate solutions for the problem are as follows:

$$\theta_L = \frac{\varepsilon}{k_1} \sqrt{\frac{\alpha_0}{k_1 - 1}} I_0 (1 - e^{-k_1 t}) \times \left\{ \cot\left(\sqrt{\frac{k_1 - 1}{\alpha_0}} Ld \right) \cos\left(\sqrt{\frac{k_1 - 1}{\alpha_0}} x \right) + \sin\left(\sqrt{\frac{k_1 - 1}{\alpha_0}} x \right) \right\} + O(\varepsilon^2) \tag{4.46}$$

$$\theta_E = \sqrt{\frac{\alpha_0}{k_1 - 1}} I_0 e^{-k_1 t} \times \left\{ \cot\left(\sqrt{\frac{k_1 - 1}{\alpha_0}} Ld\right) \cos\left(\sqrt{\frac{k_1 - 1}{\alpha_0}} x\right) + \sin\left(\sqrt{\frac{k_1 - 1}{\alpha_0}} x\right) \right\} + O(\varepsilon) \tag{4.47}$$

The expansion for thermal conductivity to ensure regular solutions is

$$\alpha = \alpha_0 - \varepsilon \frac{\alpha_0}{k_1(k_1 - 1)} + O(\varepsilon^2) \tag{4.48}$$

Since ε is the ratio of electron-to-lattice heat capacities, the dimensionless thermal conductivity has some dependence on this parameter.

4.2.3 Findings and Discussions

The findings of the solution of the hyperbolic heat equation in relation to a nonequilibrium heating situation are presented together with thermal stress analysis in line with the previous studies [1,2].

4.2.3.1 *Temperature and Stress Fields*

Fig.,4.3 shows dimensionless temperature distribution inside the substrate material for different heating periods. Temperature decays sharply in the substrate material. As the distance below the surface of the substrate material increases beyond the dimensionless distance $x = 4$, temperature decay becomes gradual. The sharp decay of temperature in the surface region is associated with the energy transfer from the surface vicinity to the solid bulk. In this case, the high temperature gradient generated in the surface vicinity is responsible for the heat diffusion from the surface vicinity to the solid bulk. Since the heating duration is short, the amount of heat transferred through conduction from the surface region to the solid bulk is not considerable. This, in turn, suppresses the energy transfer from the surface region to the solid bulk. Consequently, temperature gradient remains high in the region next to the surface vicinity. This situation is true for all the heating periods shown in the figure. However, the gradual decay in temperature at some depth below the surface is associated with the diffusional energy transfer in this region. The amount of heat transfer in this region is small during the short heating duration. This, in turn, lowers the temperature gradient and the diffusional energy transfer in this region.

Fig. 4.4 shows dimensionless stress distribution inside the substrate material for different heating periods. It is evident that in the early

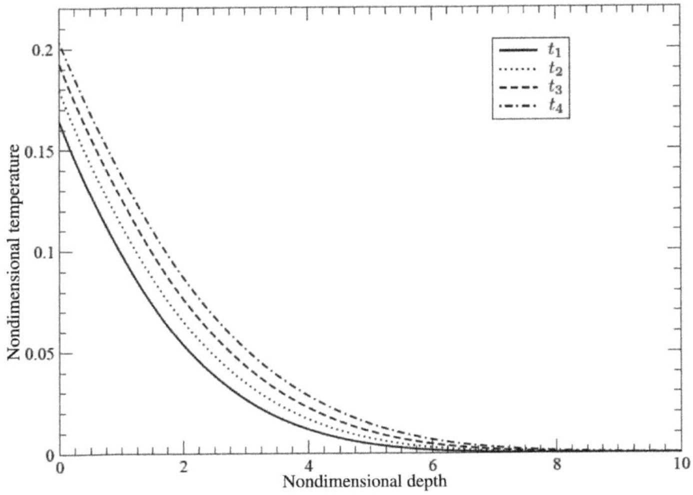

FIGURE 4.3 Temperature distribution versus depth at different times of dimensionless heating durations [1].

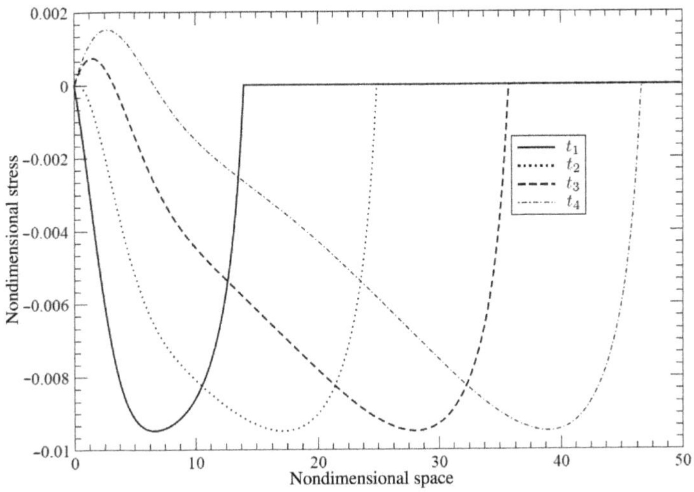

FIGURE 4.4 Nondimensional stresses with distance for different dimensionless heating durations [1].

heating periods, a compressive stress is generated in the surface region and the maximum compressive stress amplitude occurs at some depth below the surface. The location of the maximum stress amplitude moves away from the surface with progressing time. Moreover, the behavior of the compressive stress in the early heating period is slightly different;

the compressive stress decays sharply, resulting in high stress gradient in the surface region. As the heating period progresses, the stress field below the surface is modified due to the change of the temperature gradient below the surface. Consequently, gradually decaying of the thermal stress takes place in the region next to the surface vicinity. Moreover, as the heating period progresses further, thermal stress generated in the surface region becomes tensile. This is associated with the thermal expansion of the surface after the initial compression. However, the amplitude of the tensile wave is considerably smaller than that of the compressive wave. In addition, the tensile wave is only limited to the surface vicinity and as the depth below the surface increases, the thermal stress becomes compressive. This indicates that small expansion of the surface takes place during the heating period while generating the tensile stress in the surface region. However, the material below the surface remains in compression, causing a compressive thermal stress development in this region.

4.2.3.2 ***Perturbation Solution of Temperature Field***

The solutions of temperature field in the electron and lattice subsystems inside the solid substrate are presented for temporarily decaying power intensity (Fig. 4.5) of the substrate free surface (Eqs. 4.46 and 4.47). A computer code is developed to demonstrate the temperature filed inside the gold film. The thermophysical properties used in the analysis are given in Table 4.1.

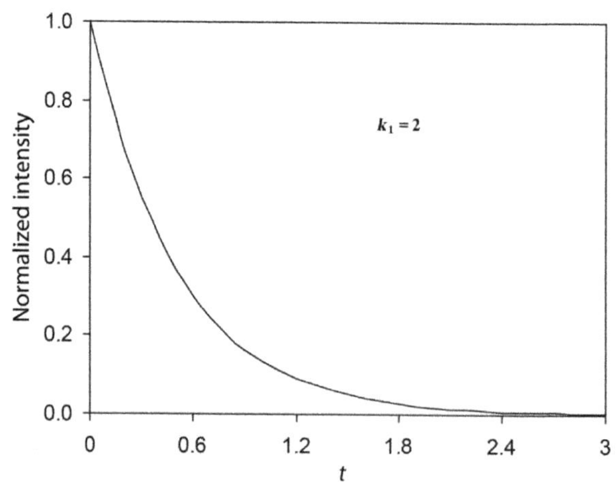

■ **FIGURE 4.5** Temporal variation of pulse intensity along the dimensionless time [2].

Table 4.1 Thermophysical Properties of Gold at 300K

Property	Numerical Value
C_L	$2.5\ 10^6\ J\ m^{-3}K$
C_E	$2.1\ 10^4\ J\ m^{-3}K$
G	$2.6\ 10^{16}\ W\ m^{-3}K$
k	$315\ W\ m^{-1}K$
d	$10^9\ m^{-1}$
I_0	31.75
T_o	300K

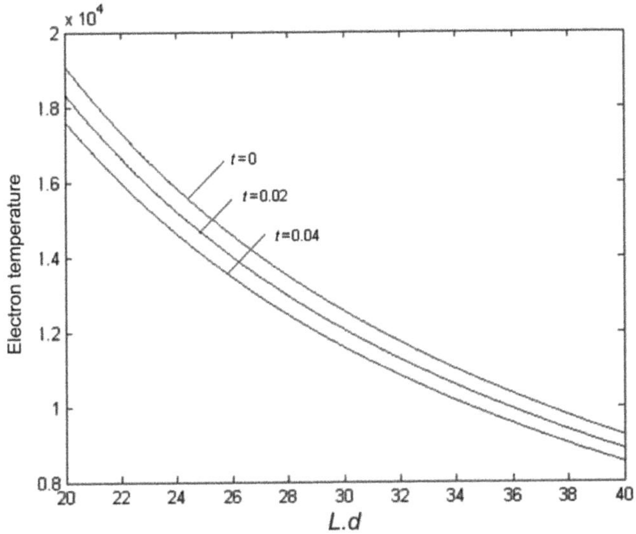

■ **FIGURE 4.6** Dimensionless electron temperature inside the thin film for various dimensionless heating durations [2].

Fig. 4.6 shows electron temperature at the surface with layer thickness for different heating periods. Increasing layer thickness lowers electron temperature at the surface. This is true for all heating periods. In this case, energy intensity inputted at the surface is absorbed by electrons and distributed in the layer. Consequently, increasing the layer thickness lowers electron energy in the surface region, resulting in low electron temperature at the surface. Moreover, increasing the heating period lowers the magnitude of the power intensity at the layer surface; hence, the magnitude of energy inputted at the surface reduces with progressing time. Therefore, electron temperature becomes low with increasing layer thickness and heating period.

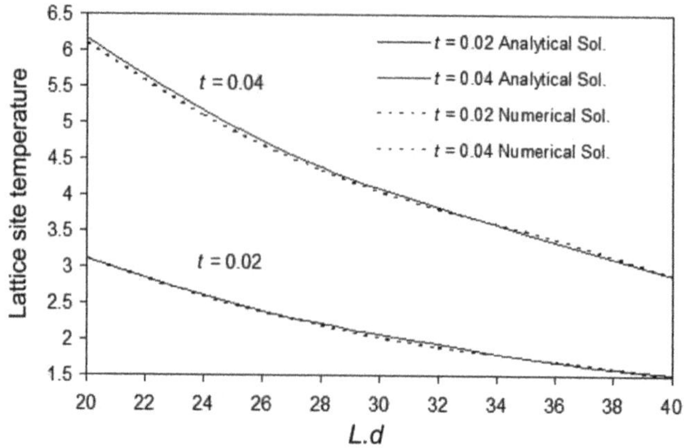

■ **FIGURE 4.7** Dimensionless lattice temperature inside the thin film for two dimensionless heating durations. Analytical and numerical solutions are given for comparison [2].

Fig. 4.7 shows lattice site temperature at the surface with different layer thickness for different heating periods due to analytical and numerical solutions. It should be noted that the results obtained from the analytical solution and numerical predictions are in good agreement. Therefore, the analytical solution obtained in the present study gives accurate results for the temperature distributions within the domain of the levels of the parameters selected in the present study. Moreover, increasing layer thickness lowers lattice site temperature at the surface. This is because of the energy transfer from electrons to lattice site. Since electrons in the surface vicinity have less energy for thick layers, the amount of energy transfer from electrons to lattice site is low, resulting in low lattice site temperatures with increasing layer thickness. The decay rate of lattice site temperatures with increasing layer thickness is almost the same for different heating periods.

4.3 ELECTRON KINETIC THEORY APPROACH FOR ENERGY TRANSFER IN METALLIC FILMS

When the metallic film is subjected to heat transfer, the heating duration and the size of the film become critical for the thermal energy transport across the film. In this case, nonequilibrium thermal energy transport takes place for the heating duration less than the thermalization time and for the film size comparable to the mean free path of the film material. The thermal energy transport equation becomes the hyperbolic nature and the parabolic heating model fails to describe the heating process accurately inside the film material. In order to account for the nonequilibrium heating situation, an electron kinetic theory approach is introduced. The electron kinetic theory approach considers that nonequilibrium transport causes thermal separation of electron and lattice subsystems inside the film. The thermal communication of both subsystems through electron–phonon coupling gives rise to the thermal energy transport. In order to assess the energy transport via thermal communication in between the electron and lattice subsystems, a probabilistic approach is incorporated, formulating the rate of energy transport through electron–phonon collisions. To generalize the formulation of energy transfer inside the film, a volumetric heat source is incorporated, which resembles the high intensity beam, such as laser beam, interaction with the substrate material. The formulation of the electron kinetic theory is given next, in line with the previous study [9].

4.3.1 Formulation of Microscopic Energy Transport in Metallic Substrates

Because of the presence of a volumetric irradiated source, such as laser irradiation, electrons in the surface vicinity gain excess energy from the irradiated field. They undergo several collisions with lattice phonons and transfer the fraction of their excess energy. This in turn gives rise to lattice site energy increase with progressing collisions. However, electron excess energy gain takes place at a faster rate than that of the lattice site energy gain. However, energy transfer via collision process can be formulated by the consideration of the electron–phonon coupling. Since the specific heat capacity of an electron is much smaller than its counterpart corresponding to the lattice site, electron temperature increases rapidly while lattice site temperature increase is gradual during the short heating duration. The temperature differential between electron and lattice subsystems results in nonequilibrium energy transport in the substrate material. Moreover, the energy distribution of the excited electrons may not be uniform in the surface region and also varies with time. This causes temperature differential occurring in the electron subsystem.

In order to formulate the electron kinetic theory approach for the volumetric source of irradiated energy, such as the laser short-pulse heating process, the behavior of electron motion in the surface region of the metallic substrates becomes critical. The electron motion in the surface region of the metals due to the irradiated field can be shown schematically in Fig. 4.8.

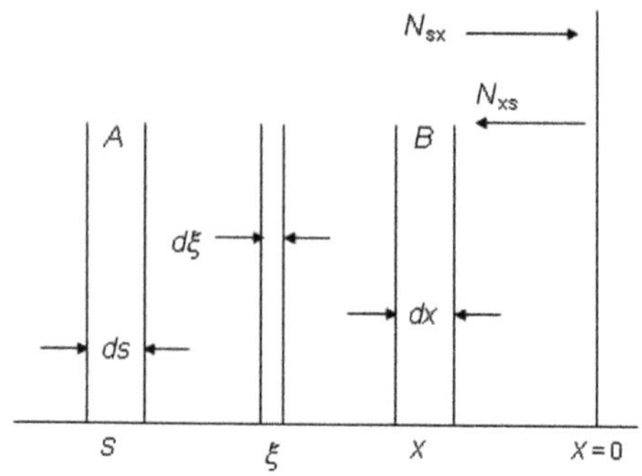

FIGURE 4.8 Electron movement in the surface region ($x = 0$ is the free surface) [9].

Electrons travel from surface to solid bulk as well as bulk to solid. Electrons reaching the surface can escape once their energy exceeds the surface potential barrier. Consequently, the number of electrons with ψ fraction reaching the surface can manage to escape. In order to account for the reflected electrons from the surface, a mirror image at the surface is considered [10]. Consider the location A in Fig. 4.8; the net flow of N_{SX} ($\sim\frac{N}{6}$) number of electrons toward the surface occurs and ψ fraction of these electrons may escape from the surface. The situation, which occurs at location A in Fig. 4.8, is an exact mirror image of that corresponding to point B $(1-\psi)$ N_{SX} number electrons still flow toward the surface from location B $(1-\psi)$; N_{SX} electrons flow to the right and $\frac{N}{6}$ to the left at location B (Fig. 4.8). In the case of location A approaches to location B, the number of electrons N_{SX}, which flow from s to x changes discontinuously at $s = 0$ and at $s = x$. These changes can be summarized as follows:

$$\begin{aligned}&\text{In the negative direction of } x\\&-\infty < s < x : N_{XS} = \frac{N}{6}\\&x < s < 0 : N_{XS} = (1-\psi)\frac{N}{6}\\&0 < s < \infty : N_{XS} = \frac{N}{6}\end{aligned} \tag{4.49}$$

and

$$\begin{aligned}&\text{In the positive direction of } x\\&-\infty < s < 0 : N_{SX} = \frac{N}{6}\\&0 < s < x : N_{SX} = (1-\psi)\frac{N}{6}\\&x < s < \infty : N_{SX} = \frac{N}{6}\end{aligned} \tag{4.50}$$

Moreover, at all locations:

$$N_{sx} + N_{xs} = (2 - \psi)\frac{N}{6} \tag{4.51}$$

where N_{sx} is the number of electrons that flow from x to s.

It should be noted that changes in N_{sx} follow a distribution, which can be described by a rectangle function of unit height and width x centered on the positive $s = \frac{x}{2}$. Consequently, the rectangle function can be written as:

$$\prod \frac{\left(s - \frac{x}{2}\right)}{x} = \prod \frac{(2s - x)}{2x} \tag{4.52}$$

where

$$\prod \frac{(2s - x)}{2x} = 0 \quad \text{for} \quad |2s - x| > \frac{1}{2} \tag{4.53}$$

and

$$\prod \frac{(2s - x)}{2x} = 1 \quad \text{for} \quad |2s - x| < [x] \tag{4.54}$$

Therefore, the electron distribution can be described as

$$N_{sx} = \frac{N}{6}\left(1 - \psi \prod \frac{(2s - x)}{2x}\right) \quad \text{with} \quad N_{sx} + N_{xs} = (1 - \psi)\frac{N}{6} \tag{4.55}$$

It should be noted that electron energy, which is characterized by temperature $T_e(s, t)$, is augmented from the initial $T_e(s, t)$ by an amount equal to that absorbed in traveling from s to x. The total amount of energy, which is absorbed in an element $d\xi$, area A, in time δt is

$$I_o A f'(\xi) dt \, d\xi \tag{4.56}$$

where I_o and $f'(\xi)$ are the laser peak power intensity and the intensity distribution function in the absorption depth of the solid material, since all the beam energy is absorbed in the x axis. The electron density can vary along the x axis; in particular, the number of electrons traveling from ds to dx may not be the same as that from dx to ds. Therefore, the portion of energy absorbed by electrons that travel from ds to dx in dt is

$$I_o A \, dt \, d\xi f'(\xi) \frac{N_{sx}}{N_{sx} + N_{xs}} \tag{4.57}$$

where N_{sx} and N_{xs} are the number of electrons that travel from s to x and from x to s, respectively. The total number of electrons that travel from ds to dx in this time is

$$N_{sx} A \overline{V} dt \tag{4.58}$$

where $\overline{V}$ is electron mean velocity. Hence, the average energy absorbed by one electron in $d\xi$ in time dt is

$$I_o \frac{f'(\xi)}{(N_{sx} + N_{xs})\overline{V}} d\xi \tag{4.59}$$

and the total amount of energy absorbed by this electron from dx to ds is

$$\int_x^s I_o \frac{f'(\xi)}{(N_{sx} + N_{xs})\overline{V}} d\xi \tag{4.60}$$

This expression gives the extra energy gain by the electrons traveling from ds to dx.

Electrons receive energy from the irradiated field and make collisions among themselves as well as lattice site ions, and they transfer some fraction of their excess energy through the collisional process; i.e., after the first collision electrons scatter and make further collisions with less energetic electrons and lattice site ions. The energy exchange between energetic electrons, due to absorption of the irradiated field, and other species can be formulated after considering the collision probability of energetic electrons.

Consider the probability of electrons traveling a distance x without making a collision [11]:

$$\exp\left(-\frac{x}{\lambda}\right) \tag{4.61}$$

where $x < 2\lambda$ and λ is the mean free path of the electrons. Consider Fig. 4.8; the probability of electrons that collide in B can be written as

$$1 - \exp\left(-\frac{x}{\lambda}\right) \tag{4.62}$$

or

$$1 - \left(1 - \frac{dx}{\lambda} + \cdots\right) \tag{4.63}$$

or

$$\sim \frac{dx}{\lambda} \tag{4.64}$$

provided that $x < 2\lambda$. The probability of electrons that last collided in B now colliding in A is

$$\frac{ds}{\lambda} \exp\left(-\frac{|x-s|}{\lambda}\right) \frac{dx}{\lambda} \tag{4.65}$$

However, the number of electrons (N_{sx}) leaving location A in Fig. 4.8, area A, in time dt is $N_{sx}A\overline{V}dt$, where N_{sx} is the number density of electrons that transfer energy from dx to ds, and $\overline{V}$ is the mean electron velocity. The number of these electrons that have just collided in location A is

$$N_{sx}A\overline{V}dt\frac{ds}{\lambda}\frac{dx}{\lambda}\exp\left(-\frac{|x-s|}{\lambda}\right) \tag{4.66}$$

where $\frac{ds}{\lambda}\frac{dx}{\lambda}\exp\left(-\frac{|x-s|}{\lambda}\right)$ is the probability of electrons just collided in location A.

If the temperature of the lattice site in dx is $T_l(x,t)$ and the temperature of the electrons when they arrive at dx is $T_e(s,t)$ (allowing for absorption on the way), then the energy transfer to the lattice site in dx from collisions with electrons in which the electrons give up a fraction, f, of their excess energy is

$$N_{sx}A\overline{V}dt\frac{ds}{\lambda}\frac{dx}{\lambda}\exp\left(-\frac{|x-s|}{\lambda}\right)f(E_e-E_l) \tag{4.67}$$

where E_e and E_l are the energy of electron and lattice ion, respectively. The analysis related to f is given as follows.

The fraction of electron excess energy transfer during the time comparable or slightly greater than the electron–phonon collision time (τ_p) can be written in terms of the energy balance across the section dx in the substrate material; i.e.,

$$f=\frac{(\text{Electron energy})_{in}-(\text{Electron energy})_{out}}{(\text{Electron energy})_{in}-(\text{Phonon energy})}\quad\text{or}\quad f=\frac{(T_e)_{in}-(T_e)_{out}}{(T_e)_{in}-T_l}, \tag{4.68}$$

where $(T_e)_{in}$ is the temperature of an electron entering the section, $(T_e)_{out}$ is the temperature of an electron leaving the section, and T_l is the phonon temperature. f takes the values $0\le f\le 1$.

Integrating the contributions from all such infinitely small strips as to the energy in location B (Fig. 4.8) gives

$$\int_{-\infty}^{\infty}N_{sx}A\overline{V}dt\frac{ds}{\lambda}\frac{dx}{\lambda}\exp\left(-\frac{|x-s|}{\lambda}\right)f(E_e-E_l)ds \tag{4.69}$$

In this case, energy transfer during Δt ($\Delta t\ge\tau_p$), where τ_p is the electron–phonon collision time) due to absorption of irradiated field and the collisional process can be written after incorporating electron distribution function [12]:

$$\frac{\Delta E_{\text{trans}}}{Adx\Delta t} = \int_{-\infty}^{\infty} \frac{\overline{V}fk_B}{\lambda^2} N_{\text{sx}} \left(1 - \psi\prod\frac{(2s-x)}{2x}\right) \exp\left(-\frac{|x-s|}{\lambda}\right) T_e(s,t)ds$$
$$- \int_{-\infty}^{\infty} \frac{\overline{V}fk_B}{\lambda^2} N_{\text{sx}} \left(1 - \psi\prod\frac{(2s-x)}{2x}\right) \exp\left(-\frac{|x-s|}{\lambda}\right) T_l(x,t)ds$$
$$+ \int_{-\infty}^{\infty} \frac{I_o f}{\lambda^2} \frac{N_{\text{sx}}}{N_{\text{sx}} + N_{\text{xs}}} \exp\left(-\frac{|x-s|}{\lambda}\right) \int_x^s f'(\xi)d\xi ds \tag{4.70}$$

where

$$\frac{N_{\text{sx}}}{N_{\text{sx}} + N_{\text{xs}}} = \frac{\left(1 - \psi\prod\frac{(2s-x)}{x}\right)}{2-\psi} \tag{4.71}$$

where f is the fraction of electron excess energy, which transfers to the lattice site during a single electron lattice site collision. The first term on the left-hand side of Eq. (4.70) is energy gain by the substrate material through the collisional process; the first and second terms on the right-hand side represent electron and lattice energies; the third term on the right-hand side is the energy gain of the electrons due to the irradiated field.

The final temperature of the electrons in dx after the collisional process can be readily found from the conservation of energy; i.e.,

Total electron energy after collision = Total electron energy during dt−Change of lattice energy

Total electron energy after a collision is

$$\int_{-\infty}^{\infty} \frac{\overline{V}k_B}{\lambda^2} N_{\text{sx}} \left(1 - \psi\prod\frac{(2s-x)}{2x}\right) \exp\left(-\frac{|x-s|}{\lambda}\right) (T_e(s,t) - fT_l(x,t))ds \tag{4.72}$$

Total electron energy carried into dx during dt is

$$\int_{-\infty}^{\infty} \frac{\overline{V}k_B(1-f)}{\lambda^2} N_{\text{sx}} \left(1 - \psi\prod\frac{(2s-x)}{2x}\right) \exp\left(-\frac{|x-s|}{\lambda}\right) T_e(s,t)ds$$
$$+ \int_{-\infty}^{\infty} \frac{I_o(1-f)}{\lambda^2} \frac{\left(1 - \psi\prod\frac{(2s-x)}{2x}\right)}{2-\psi} \exp\left(-\frac{|x-s|}{\lambda}\right) \int_x^s f'(\xi)d\xi ds \tag{4.73}$$

Therefore, the conservation of energy yields

$$
\begin{aligned}
&\int_{-\infty}^{\infty} \frac{\overline{V} k_B}{\lambda^2} N_{sx} \left(1 - \psi \prod \frac{(2s-x)}{2x}\right) \exp\left(-\frac{|x-s|}{\lambda}\right) (T_e(s,t) - f T_l(x,t)) ds \\
&= \int_{-\infty}^{\infty} \frac{\overline{V} k_B (1-f)}{\lambda^2} N_{sx} \left(1 - \psi \prod \frac{(2s-x)}{2x}\right) \exp\left(-\frac{|x-s|}{\lambda}\right) T_e(s,t) ds \\
&+ \int_{-\infty}^{\infty} \frac{I_o(1-f)}{\lambda^2} \frac{\left(1 - \psi \prod \frac{(2s-x)}{2x}\right)}{2-\psi} \exp\left(-\frac{|x-s|}{\lambda}\right) \int_x^s f'(\xi) d\xi ds
\end{aligned}
\tag{4.74}
$$

Eqs. (4.70) and (4.74) can be rewritten after considering the electron distributions for the lattice element dx apart and for electrons passing an area A:

$$
\begin{aligned}
\frac{\Delta E_{trans}}{A dx \Delta t} &= \int_{-\infty}^{\infty} \frac{fk}{\lambda^3} \exp\left(-\frac{|x-s|}{\lambda}\right) T_e(s,t) ds - \int_{-\infty}^{\infty} \frac{fk}{\lambda^3} \exp\left(-\frac{|x-s|}{\lambda}\right) T_l(x,t) ds \\
&+ \int_0^{|x|} \psi \frac{fk}{\lambda^3} \exp\left(-\frac{|x-s|}{\lambda}\right) T_l(x,t) ds - \int_0^{|x|} \psi \frac{fk}{\lambda^3} \exp\left(-\frac{|x-s|}{\lambda}\right) T_e(s,t) ds \\
&+ \int_{-\infty}^{\infty} \frac{I_o f}{\lambda^2} \frac{1}{(2-\psi)} \exp\left(-\frac{|x-s|}{\lambda}\right) \int_x^s f'(\xi) d\xi ds \\
&- \int_{-\infty}^{\infty} \frac{I_o f}{\lambda^2} \frac{\psi}{(2-\psi)} \exp\left(-\frac{|x-s|}{\lambda}\right) \int_x^s f'(\xi) d\xi ds
\end{aligned}
\tag{4.75}
$$

and

$$
\begin{aligned}
&\int_{-\infty}^{\infty} \frac{k}{\lambda^2} \exp\left(-\frac{|x-s|}{\lambda}\right) (T_e(s,t) - f T_l(x,t)) ds \\
&- \int_0^{|x|} \frac{\psi k}{\lambda^2} \exp\left(-\frac{|x-s|}{\lambda}\right) (T_e(s,t) - f T_l(x,t)) ds \\
&= \int_{-\infty}^{\infty} \frac{k(1-f)}{\lambda^2} \exp\left(-\frac{|x-s|}{\lambda}\right) T_e(s,t) ds \\
&- \int_0^{|x|} \frac{k(1-f)}{\lambda^2} \exp\left(-\frac{|x-s|}{\lambda}\right) T_e(s,t) ds \\
&+ \int_{-\infty}^{\infty} \frac{I_o(1-f)}{\lambda^2 (2-\psi)} \exp\left(-\frac{|x-s|}{\lambda}\right) \int_x^s f'(\xi) d\xi ds \\
&- \int_0^{|x|} \frac{I_o(1-f)}{\lambda^2 (2-\psi)} \exp\left(-\frac{|x-s|}{\lambda}\right) \int_x^s f'(\xi) d\xi ds
\end{aligned}
\tag{4.76}
$$

where k is the thermal conductivity, which makes use of the simple kinetic theory result for the electron thermal conductivity [13]:

$$k = \frac{N\overline{V}k_B\lambda}{3} \tag{4.77}$$

The energy content of the small lattice site element dx apart can be written as

$$\Delta E = AdxC_l T_l(x,t) \tag{4.78}$$

where $C_l = \rho Cp$. The energy gain of the small lattice element during the small time interval δt is $\Delta E/\delta t$. The expansion of $\Delta E/\delta t$ yields

$$\frac{\Delta E}{\delta t} = \frac{1}{\delta t}\left[E(t) + \delta t E'(t) + \frac{(\delta t)^2}{2!}E''(t) + \cdots\cdots - E(t)\right] \tag{4.79}$$

or

$$\frac{\Delta E}{\delta t} = E'(t) + \frac{(\delta t)}{2!}E''(t) + \cdots. \tag{4.80}$$

Combining Eqs. (4.78) and (4.80) yields

$$\frac{\Delta E}{\delta t} = A\ dx\left(C_l\frac{\partial}{\partial t}[T_l(x,t)] + C_l\frac{(\delta t)}{2!}\frac{\partial^2}{\partial t^2}[T_l(x,t)] + \cdots\right) \tag{4.81}$$

When the time increment approaches thermal relaxation time ($\delta t \rightarrow \tau_s$), Eq. (4.81) reduces to

$$\frac{\partial e}{\partial t} = \frac{\Delta E}{\delta t A dx} \simeq C_l\frac{\partial}{\partial t}\left[T_l(x,t) + \tau_s\frac{\partial}{\partial t}(T_l(x,t))\right] \tag{4.82}$$

where e is the volumetric energy content of the lattice site. The energy gain of the small lattice element through collisional energy transport can also be written as

$$\frac{\Delta E}{Adx\Delta t} = \frac{1}{Adx}\left[\frac{\Delta E_{\text{trans}}}{\Delta t} + \tau_s\frac{\partial}{\partial t}\left(\frac{\Delta E_{\text{trans}}}{\Delta t}\right)\right] \tag{4.83}$$

Substituting Eq. (4.75) into Eq. (4.83) yields the change of lattice site energy, which is

$$
\begin{aligned}
\frac{\Delta E}{A\delta x \Delta t} = {} & \left(\begin{aligned} & \int_{-\infty}^{\infty} \frac{fk}{\lambda^3} \exp\left(-\frac{|x-s|}{\lambda}\right) T_e(s,t)ds \\ & - \int_{-\infty}^{\infty} \frac{fk}{\lambda^3} \exp\left(-\frac{|x-s|}{\lambda}\right) T_l(x,t)ds \\ & + \int_0^{|x|} \psi \frac{fk}{\lambda^3} \exp\left(-\frac{|x-s|}{\lambda}\right) T_l(x,t)ds \\ & - \int_0^{|x|} \psi \frac{fk}{\lambda^3} \exp\left(-\frac{|x-s|}{\lambda}\right) T_e(s,t)ds \\ & + \int_{-\infty}^{\infty} \frac{I_o f}{\lambda^2} \frac{1}{(2-\psi)} \exp\left(-\frac{|x-s|}{\lambda}\right) \int_x^s f'(\xi)d\xi ds \\ & - \int_{-\infty}^{\infty} \frac{I_o f}{\lambda^2} \frac{\psi}{(2-\psi)} \exp\left(-\frac{|x-s|}{\lambda}\right) \int_x^s f'(\xi)d\xi ds \end{aligned} \right) \\
& + \tau_s \frac{\partial}{\partial t} \left(\begin{aligned} & \int_{-\infty}^{\infty} \frac{fk}{\lambda^3} \exp\left(-\frac{|x-s|}{\lambda}\right) T_e(s,t)ds \\ & - \int_{-\infty}^{\infty} \frac{fk}{\lambda^3} \exp\left(-\frac{|x-s|}{\lambda}\right) T_l(x,t)ds \\ & + \int_0^{|x|} \psi \frac{fk}{\lambda^3} \exp\left(-\frac{|x-s|}{\lambda}\right) T_l(x,t)ds \\ & - \int_0^{|x|} \psi \frac{fk}{\lambda^3} \exp\left(-\frac{|x-s|}{\lambda}\right) T_e(s,t)ds \\ & + \int_{-\infty}^{\infty} \frac{I_o f}{\lambda^2} \frac{1}{(2-\psi)} \exp\left(-\frac{|x-s|}{\lambda}\right) \int_x^s f'(\xi)d\xi ds \\ & - \int_{-\infty}^{\infty} \frac{I_o f}{\lambda^2} \frac{\psi}{(2-\psi)} \exp\left(-\frac{|x-s|}{\lambda}\right) \int_x^s f'(\xi)d\xi ds \end{aligned} \right)
\end{aligned}
\tag{4.84}
$$

Eqs. (4.76) and (4.84) are the energy transport equations of interest for the laser short-pulse heating process. However, a small rise of electron temperature during the low intensity laser heating pulse prevents electron escape from the surface. Consequently, the term ψ in Eqs. (4.76) and (4.84) becomes zero.

Eqs. (4.76) and (4.84) can be transformed into differential equations. The method of solution to be used in the following analysis is the transformation of the simultaneous differential–integral Eqs. (4.76) and (4.84) using the Fourier integral transformation, with respect to x [11]. This is due to the fact that the resultant ordinary differential equations may then be handled much more conveniently. Consider first reduction of the set of equations to the differential equation of heat conduction.

The Fourier transformation of a function $f(x)$ is defined by

$$F[f(x)] = \int_{-\infty}^{\infty} \exp(-i\omega x)f(x)dx = F(\omega) \tag{4.85}$$

and the Fourier inversion by

$$f(x) = \frac{1}{2\pi}\int_{-\infty}^{\infty} F(\omega)(\exp(-i\omega x)d\omega \tag{4.86}$$

The Fourier transformation of the convolution integral

$$\int_{-\infty}^{\infty} f(\xi)g(x-s)ds \tag{4.87}$$

is the product of the transforms

$$\bar{f}(\omega)\cdot\bar{g}(\omega) \tag{4.88}$$

and the transform of function $\exp\left(-\frac{|x|}{\lambda}\right)$ is

$$\frac{2\lambda}{1+\omega^2\lambda^2} \tag{4.89}$$

Therefore, the Fourier transform of the function

$$IX = \int_{-\infty}^{\infty} \frac{k}{\lambda^3}\exp\left(-\frac{|x-s|}{\lambda}\right)T_1(x,t)ds \tag{4.90}$$

will be a constant factor (the value of integral) multiplying the transform of the function $T_1(x, t)$; i.e.,

$$F[IX] = \frac{kf}{\lambda^3}\overline{T_1}F\left\{\int_{-\infty}^{\infty} \exp\left(-\frac{|x-s|}{\lambda}\right)ds\right\} \tag{4.91}$$

or

$$F[IX] = \frac{kf}{\lambda^3}\overline{T_1}F\left\{\int_{-\infty}^{\infty} \exp\left(-\frac{|x-s|}{\lambda}\right)H(|s|)ds\right\} \tag{4.92}$$

where $H(|s|) = 1$ for $-\infty < s < \infty$. Therefore,

$$F[IX] = \frac{kf}{\lambda^3}\overline{T_1}F\left\{\int_{-\infty}^{\infty} \exp\left(-\frac{|x-s|}{\lambda}\right)\right\}F\{H(|s|)ds\} \le \frac{kf}{\lambda^3}\overline{T_1}\frac{2\lambda}{\omega^2\lambda^2+1}\delta(\omega) \tag{4.93}$$

where $\delta(\omega)$ is the Dirac delta function. Since this function only has a value of 1 at $\omega = 0$, then the transform is

$$\frac{kf}{\lambda^2}\overline{T_1} \tag{4.94}$$

Using these results, Eqs. (4.76) and (4.84) can be Fourier transformed, the result of which is

$$\frac{\partial e}{\partial t} = \frac{kf}{\lambda^3}\left[\frac{2\lambda}{\omega^2\lambda^2+1}\overline{T_e}\right] - \frac{kf}{\lambda^2}\overline{T_1} + \left[\frac{I_o\delta f}{2\lambda}\right]\left[\frac{2\lambda}{\omega^2\lambda^2+1}\right]\left[\frac{2\delta}{\delta^2+\omega^2}\right] \times \tau_p\frac{\partial}{\partial t}\left[\frac{kf}{\lambda^3}\left[\frac{2\lambda}{\omega^2\lambda^2+1}\overline{T_e}\right] - \frac{kf}{\lambda^2}\overline{T_1} + \left[\frac{I_o\delta f}{2\lambda}\right]\left[\frac{2\lambda}{\omega^2\lambda^2+1}\right]\left[\frac{2\delta}{\delta^2+\omega^2}\right]\right] \tag{4.95}$$

and

$$\frac{k}{\lambda^2}\left[\overline{T_e} - f\overline{T_1}\right] = \left[\frac{k(1-f)}{\lambda^3}\right]\left[\frac{2\lambda}{\omega^2\lambda^2+1}\right]\overline{T_e} + \frac{I_o\delta(1-f)}{2\lambda}\left[\frac{2\lambda}{\omega^2\lambda^2+1}\right]\left[\frac{2\delta}{\delta^2+\omega^2}\right] \tag{4.96}$$

If the transform function $\overline{T_e}$ is obtained from Eq. (4.95) using Eq. (4.96), the result is

$$\left[f+\omega^2\lambda^2\right]\frac{\partial e}{\partial t} = -\omega^2 kf\overline{T_1} - \tau_p\frac{\partial}{\partial t}\left(\omega^2 kf\overline{T_1}\right) + I_o\delta f\left[\frac{2\delta}{\delta^2+\omega^2}\right] + \tau_p\frac{\partial}{\partial t}\left[I_o\delta f\left[\frac{2\delta}{\delta^2+\omega^2}\right]\right] \tag{4.97}$$

Insertion of $\partial e/\partial t$ in terms of T_1 and multiplication of Eq. (4.97), which is in the transform domain, by $(i\omega)^2$ corresponds to second order differential in the real plane. Hence the inversion of the previous equation gives

$$\left[\left(1+\tau_s\frac{\partial}{\partial t}\right) - \frac{\lambda^2}{f}\frac{\partial^2}{\partial x^2}\right]C_1\frac{\partial T_1}{\partial t} = k\frac{\partial^2 T_1}{dx^2} + \tau_p\frac{\partial}{\partial t}\left(\frac{\partial^2 T_1}{dx^2}\right) + I_o\delta f\exp(-\delta|x|) + \tau_p\frac{\partial}{\partial t}\left[I_o\delta f\exp(-\delta|x|)\right] \tag{4.98}$$

Eq. (4.98) is similar to that obtained from the quasiballistic approach [14]. It should be noted that the time derivative of diffusion $\left(\tau_p \frac{\partial}{\partial t}\left(\frac{\partial^2 T_l}{dx^2}\right)\right)$ and source $\left(\frac{\partial}{\partial t}\left[I_o \delta f \exp(-\delta|x|)\right]\right)$ terms are included in Eq. 4.98) as similar to that obtained from the quasiballistic approach [14].

4.3.2 Parabolic Heating Model

First, let us consider the two-equation model presented previously [7]. The two-equation model representing the energy exchange mechanism during phonon absorption and electron–phonon coupling after one-dimensional consideration can be written as [7]

$$\begin{aligned} C_e \frac{\partial T_e(s,t)}{\partial t} &= \nabla \cdot (k \nabla T_e(s,t)) - G[T_e(s,t) - T_l(x,t)] + S \\ C_l \frac{\partial T_l(x,t)}{\partial t} &= G[T_e(s,t) - T_l(x,t)] \end{aligned} \tag{4.99}$$

$T_e(s,t)$ and $T_l(x,t)$ are the electron and lattice site temperatures, s is the laser source term $(I_o \delta \exp(-\delta x))$, and C_e and C_l are the electron and lattice heat capacities, respectively. G is the electron–phonon coupling factor, given by [15]

$$G = \frac{\pi^2 m_e N \overline{V}^2}{6 \tau_p T_e(s,t)} \tag{4.100}$$

where m_e, N, $\overline{V}$, and τ_p are electron mass, electron number density, electron drift velocity, and the electron–phonon collision time, respectively.

Now, let us consider the equation derived from the electron kinetic theory approach (Eq. 4.98). If the terms $\tau_p \frac{\partial}{\partial t}\left(\frac{\partial^2 T_l}{dx^2}\right)$ and $\tau_p \frac{\partial}{\partial t}\left[I_o \delta f \exp(-\delta|x|)\right]$ are neglected in Eq. (4.98) for all t values, Eq. (4.98) becomes:

$$\left[\left(1 + \tau_s \frac{\partial}{\partial t}\right) - \frac{\lambda^2}{f} \frac{\partial^2}{\partial x^2}\right] C_l \frac{\partial T_l}{\partial t} = k \frac{\partial^2 T_l}{dx^2} + I_o \delta f \exp(-\delta|x|) \tag{4.101}$$

which is the same as the previously formulated kinetic theory model [16]; i.e., the time derivative of diffusion and source terms are omitted in the previously formulated kinetic theory approach. Eq. (4.101) can be rewritten as

$$C_l \frac{\partial T_l}{\partial t} = k \frac{\partial^2 T_l}{dx^2} + f \frac{\lambda^2 \partial^2}{\partial x^2}\left(\rho Cp \frac{\partial T_l}{\partial t}\right) - C_l \tau_s \frac{\partial^2 T_l}{\partial t^2} + I_o \delta \exp(-\delta|x|) \tag{4.102}$$

Eq. (4.102) is a third-order partial differential equation, which can be decomposed into second- and first-order two-differential equations;

i.e., when Eq. (4.102) is decomposed into two equations, the resulting probable differential equations are

$$
\begin{aligned}
&A\frac{\partial T_e}{\partial t} = B\frac{\partial^2 T_e}{dx^2} - C[T_e - T_l] + I_o\delta \exp(-\delta|x|) \\
&D\frac{\partial T_l}{\partial t} = C[T_e - T_l]
\end{aligned}
\tag{4.103}
$$

where A, B, C, and D are the coefficients. To find the values of A, B, C, and D, the following procedure is adopted:

$$D\frac{\partial^2 T_l}{\partial t^2} = C\left[\frac{\partial T_e}{\partial t} - \frac{\partial T_l}{\partial t}\right] \tag{4.104}$$

or

$$\frac{\partial T_e}{\partial t} = \frac{D}{C}\frac{\partial^2 T_l}{\partial t^2} + \frac{\partial T_l}{\partial t} \tag{4.105}$$

Similarly,

$$D\frac{\partial^2}{\partial x^2}\left(\frac{\partial T_l}{\partial t}\right) = C\left[\frac{\partial^2 T_e}{\partial x^2} - \frac{\partial^2 T_l}{\partial x^2}\right] \tag{4.106}$$

or

$$\frac{\partial^2 T_e}{\partial x^2} = \frac{D}{C}\frac{\partial^2}{\partial x^2}\left(\frac{\partial T_l}{\partial t}\right) + \frac{\partial^2 T_l}{\partial x^2} \tag{4.107}$$

Substitution of Eqs. (4.107) and (4.105) into Eq. (4.103), and inserting $C[T_e - T_l] = D\frac{\partial T_l}{\partial t}$ into Eq. (4.103), yields

$$(D + A)\frac{\partial T_l}{\partial t} = \frac{BD}{C}\frac{\partial^2}{\partial x^2}\left(\frac{\partial T_l}{\partial t}\right) + B\frac{\partial^2 T_l}{\partial x^2} - \frac{AD}{C}\frac{\partial^2 T_l}{\partial t^2} + I_o\delta \exp(-\delta|x|) \tag{4.108}$$

After equating Eqs. (4.102) and (4.108), the coefficients A, B, C, and D can be calculated:

$$
\begin{aligned}
&A = \frac{fk\tau_s}{\lambda^2} \\
&B = k \\
&C = \frac{fk}{\lambda^2}\left(1 - \frac{fk\tau_s}{\rho Cp\lambda^2}\right) \\
&D = \rho Cp - \frac{fk\tau_s}{\lambda^2}
\end{aligned}
\tag{4.109}
$$

Eq. (4.103) is identical to Eq. (4.99) given in the two-equation model. Consequently, setting the coefficients in Eqs. (4.103) and (4.99) yields

$$\frac{fk\tau_s}{\lambda^2} = C_e$$
$$\frac{fk}{\lambda^2}\left(1 - \frac{fk\tau_s}{\rho Cp\lambda^2}\right) = G \tag{4.110}$$
$$\rho Cp - \frac{fk\tau_s}{\lambda^2} = C_l$$

where $t_s = C_e/G$ [7] and $C_e = \gamma T_e$ (where γ is constant; e.g., $\gamma = 96.6\ \mathrm{Jm^{-3}K^{-2}}$ for copper). Moreover, the electron–phonon coupling factor is temperature dependent, which can be written as [15]

$$G = \frac{\pi^2 mN\overline{V}^2}{6\tau_p(T_e)T_e} \tag{4.111}$$

Eq. (4.111) can be used for lattice site temperature less than the Debye temperature. Since the electron–phonon collision time is $\tau_p \sim \frac{1}{T_e}$, the electron–phonon coupling factor becomes constant as temperature increases [17,18]; i.e., it becomes independent of temperature.

The electron mean free path (λ) can be written as [19]

$$\lambda = \overline{V}\tau \tag{4.112}$$

where τ is the electron relaxation time, which is [20]

$$\tau = \frac{3mk}{\pi^2 Nk_B^2 T_e} \tag{4.113}$$

where N and m are electron number density and effective mass of free electrons, respectively. k_B is Boltzmann's constant.

4.3.3 Application of Laser Short-Pulse Heating

One of the applications of the electron kinetic theory is laser short-pulse heating. The heating analysis via application of the electron kinetic theory is given, in line with the previous study [21].

Since the laser short-pulse source term can be expressed in the exponential function, the laser short-pulse formulation, which is used in the previous study [17], is introduced to formulate the volumetric source term in Eq. (4.101). The laser source term is [17]

$$S_o = \frac{0.94(1-r_f)}{t_p} J\delta \exp\left[-x\delta - 2.77\left(\frac{t}{t_p}\right)^2\right] \tag{4.114}$$

where t_p is full-width-at-half-maximum duration of the laser pulse, and $t=0$ is defined at the moment when the peak of a laser pulse arrives at the metal surface. J is the total energy carried by a laser pulse divided by the laser spot cross-section. Table 4.2 gives the laser pulse parameters used in the simulations. It should be noted that the laser source term used in the simulations is similar to that described in the previous study [17], provided that t_p is selected as 0.96 picosecond.

Since Eq. (4.101) is involved in nonlinear treatment, the analytical solution becomes difficult to achieve. Therefore, the numerical scheme is introduced, which uses a finite difference scheme, and is well established in the literature [17]. In order to obtain accurate results, the convergence criteria should be met. The stability criteria for the heating model are as follows:

$$\begin{aligned} 1 \geq ABS&\left[\frac{f\rho Cp}{\Delta t} + 2\lambda^2\rho Cp\left[\frac{1}{\Delta t(\Delta x)^2}\right] - 2kf\left[\frac{1}{(\Delta x)^2}\right]\right] \\ &+ ABS\left[\frac{kf}{(\Delta x)^2} - \frac{2\lambda^2\rho Cp}{\Delta t(\Delta x)^2}\right] + ABS\left[\frac{2\lambda^2\rho Cp}{\Delta t(\Delta x)^2}\right] \\ &- ABS\left[\frac{f\rho Cp}{\Delta t} + 2\lambda^2\rho Cp\left[\frac{1}{\Delta t(\Delta x)^2}\right]\right] \end{aligned} \tag{4.115}$$

where Δx is spatial increments in the x axis while Δt is the time increment.

In the simulations gold and copper are used as the metallic substrates and the thermosphysical properties of gold and copper are given in Table 4.3. In the numerical solution of Eq. (4.101), the coefficients are considered constants.

Table 4.2 Laser Pulse Parameters Used in the Simulations

t_p (ps)	J (m J m^{-2})	$(1-r_f)$
0.96	100	0.06

Table 4.3 Thermophysical Properties of Layer Materials in the Assembly

	δ (1 m^{-1}) $\times 10^7$	k (Wm^{-1}K)	C_e (J m^{-3}K) $\times 10^4$	C_l (Jm^{-3}K) $\times 10^6$	G (Wm^{-3}K) $\times 10^{16}$	τ_s (ps)	E_f (eV)	χ
Gold	6.16	315	2.1	2.8	2.8	0.7438	5.53	−1.48
Copper	6.16	386	2.1	4.8	4.8	0.4348	7.01	−1.79

4.3.4 Findings of Numerical Simulations

Since the simulations are carried out for electron and lattice subsystems, the findings are presented according to the subsubsystems in light of the previous study [21].

Fig. 4.9 shows electron temperature distribution inside the substrate material while Fig. 4.10 shows the electron temperature gradient for different heating periods. Electron temperature attains high values in the surface region during the early heating period ($t = 0.5$ ps) and it decays sharply as the distance from the surface increases toward the solid bulk. This is because of the laser pulse intensity distribution, which is high in

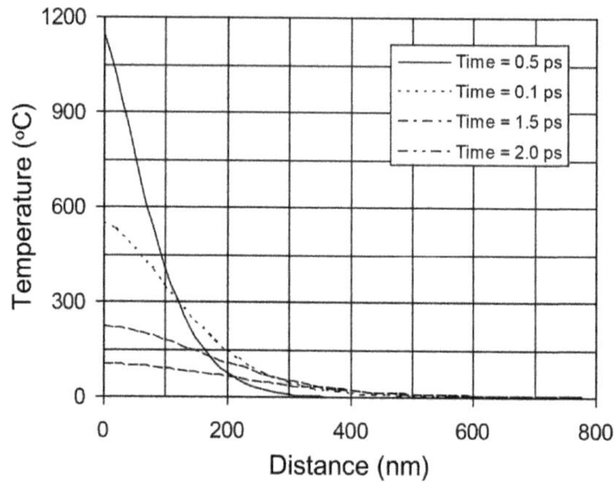

■ **FIGURE 4.9** Electron temperature at different heating times [21].

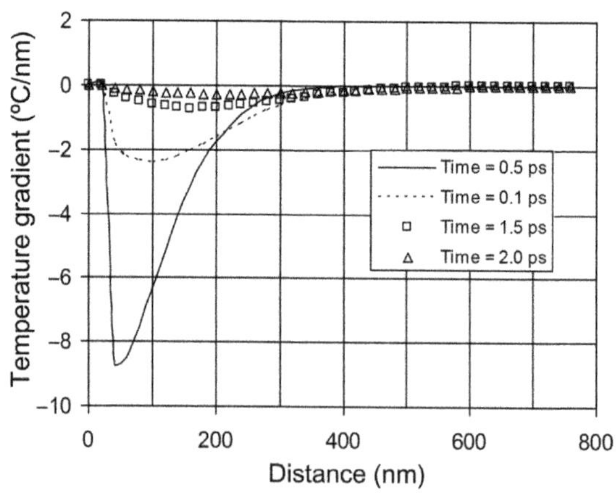

■ **FIGURE 4.10** Electron temperature gradient at different heating times [21].

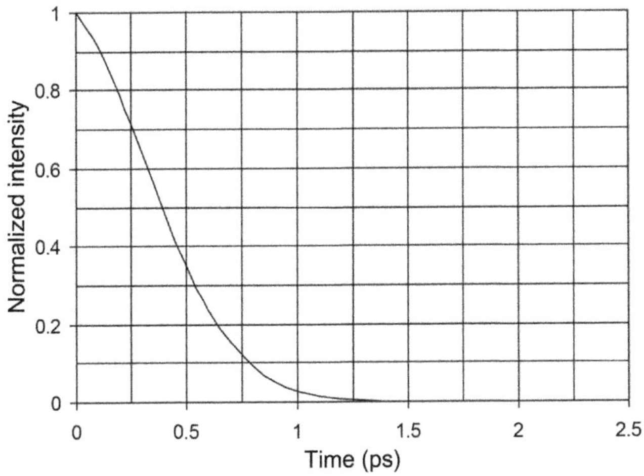

■ **FIGURE 4.11** Normalized laser pulse intensity [21].

the early heating period (Fig. 4.11). Moreover, electrons gain excess energy from the irradiated field in the surface region and as the distance increases further away from the surface, the amount of electron excess energy gain reduces (Lambert's law). Consequently, high magnitude of electron temperature gradient occurs in the surface region, particularly at the early heating period. The maximum location of electron temperature gradient moves away from the surface with progressing time. This indicates that electron excess energy loss through electron-electron and electron-lattice collisions become significant with progressing time. It should be noted that electron excess energy gain from the irradiated field reduces significantly at late heating periods, since laser power intensity decays sharply with time. In addition, electron excess energy gain from an irradiated field during the early heating period is dissipated through the collisional process as time progresses. Moreover, electron temperature variation across the layers is not noticeable; i.e., no deflection or jump occurs across the layers.

Fig. 4.12 shows lattice site temperature distribution inside the substrate material, and Fig. 4.13 shows the temperature gradient at different heating periods. Lattice site temperature attains high values in the surface region, which is more pronounced as the heating period progresses. The number of collisions, taking place between electrons and lattice site, increases with progressing time. This, in turn, enhances the electron excess energy transfer to the lattice site. Consequently, the lattice site temperature increases while electron temperature reduces with

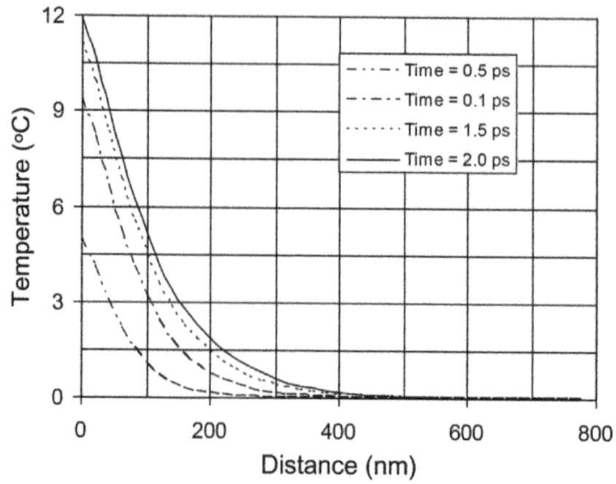

■ **FIGURE 4.12** Lattice site temperature for different heating times [21].

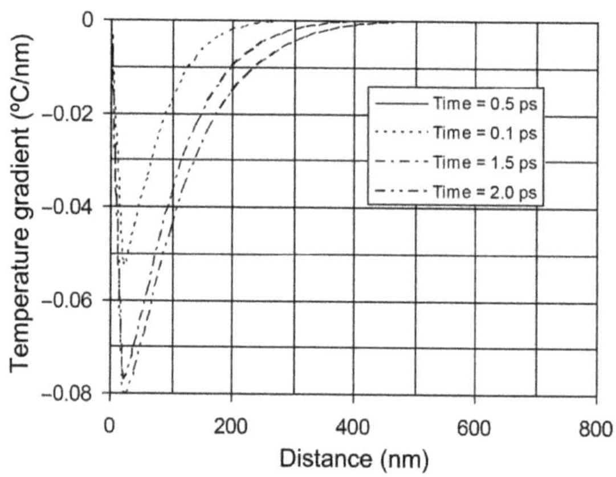

■ **FIGURE 4.13** Lattice site temperature gradient at different times [21].

progressing time. The lattice temperature gradient is large in the surface region at early heating period despite the fact that lattice site temperature is low in this period; i.e., collisional energy transport to lattice site is high in the surface region and energy transfer to the solid bulk due to diffusion is low, resulting in high lattice site temperature gradient in the surface region. Lattice site temperature distribution across the layers is smooth and no jump or abrupt change in the temperature gradient is observed in this region. This is because the electron–phonon coupling constant for gold and copper has close values for both metals (Table 4.3).

4.4 EQUATION OF PHONON RADIATIVE TRANSFER

EPRT was first derived and presented by Majumdar in his 1993 paper [22]. The basic idea behind the derivation is that since the phonons are similar to the photons, the transfer of phonons (i.e., the transfer of heat through conduction) between two parallel faces of a thin film may be modeled as a type of radiation exchange between the two surfaces, exactly analogous to the radiation heat transfer between two heated surfaces separated either by a vacuum or by a participating medium, such as a semitransparent gas.

When the thin film thickness approaches zero, the steady-state radiation exchange of phonons between the two faces of the thin film approaches the limit of radiation exchange between two surfaces separated by a vacuum. This is the Casimir limit [23], and in this case the phonon radiative heat transfer for low temperatures is given by a law similar to the Stefan–Boltzmann law. Specifically,

$$q'' = \sigma(T_1^4 - T_2^4) \tag{4.116}$$

In this equation, σ is the Stefan–Boltzmann constant for phonons and T_1 and T_2 are the surface temperatures in Kelvin. The equation implies that during the transfer of phonons from one surface to the other, there is no scattering of the phonons. The phonons will only be scattered at the surfaces, where they may be scattered completely diffusely, completely specularly, or some combination of the two.

In the case when the film thickness is less than or comparable to the mean free path of the phonons, the space between the two faces of the thin film cannot be considered a vacuum. The phonon radiation exchange now has to be modeled analogous to the photon radiation exchange between two surfaces with an absorbing, transmitting, and scattering medium in between them, such as a semitransparent gas. In this case, the intensity distribution inside the gas is calculated from the equation of radiative transfer [24]. For the case of the phonon radiative transfer a similar equation can be derived from the Boltzmann transport equation (BTE), resulting in the EPRT.

4.4.1 Transport Properties of a Dielectric Material

To calculate the transport properties of a dielectric material in local thermodynamic equilibrium, the following picture of the lattice vibrations is assumed. The atoms or molecules are arranged in a regular lattice structure, and there is a very long-range order in the material. These atoms or molecules vibrate about their mean positions in the lattice at different frequencies, so that we may think about long-range longitudinal and transverse waves propagating inside the lattice. These waves are actually responsible for the storage and transport of thermal energy. According to quantum mechanics, waves can also be thought of as particles with momentum according to the relation $p = h/\lambda$. The particles associated with the lattice vibrations are phonons and they obey the Bose–Einstein statistics. In a thick dielectric material, these phonons experience considerable scattering during motion and hence produce a local thermodynamic equilibrium inside the material. Hence, a temperature gradient may be defined at each location. On the basis of these assumptions the transport properties of a dielectric material in local thermodynamic equilibrium may be calculated through the methods of statistical mechanics. Specifically, we may use the relaxation time approximation of the BTE, together with the Bose–Einstein distribution, as the equilibrium phonon distribution in phase space, to calculate the nonequilibrium probability distribution function f. The appropriate moments of this distribution function will result in the relevant transport properties.

4.4.2 Heat Transfer Mechanism in a Thin Dielectric Film

In thin dielectric films, in which the film thickness is comparable to or smaller than the mean free path of the phonons, the phonons do not experience enough scattering events and hence local thermodynamic equilibrium does not prevail. It may happen that some phonons, which originate at a boundary, may reach the other boundary without being scattered. This is called ballistic transport in contrast to diffusive transport, which involves scattering. When the film thickness is comparable to the mean free path of the phonons, diffusive as well as ballistic transport of phonons occurs in the dielectric thin film. As the film thickness approaches zero, ballistic transport of phonons prevail and we reach the Casimir limit. In these situations, temperature cannot be defined in the usual sense. It therefore becomes necessary to solve the BTE to calculate the correct transport characteristics as well as the correct transport properties. The BTE may be used to analyze equilibrium as well as nonequilibrium transport.

4.4.3 Boltzmann Transport Equation

The BTE was originally developed to model a dilute gas in which the gas particles interact only through two-particle collisions and do not interact through the attractive/repulsive forces [25]. This representation implies a straight-line trajectory of the gas particles through relatively empty space interrupted occasionally by abrupt collisions. The BTE is derived from Liouville's equation, which governs the time evolution of the N particle distribution function of a collection of particles in $6N$ dimensional phase space [26]. The distribution function $f(x_1 \ldots x_N, y_1 \ldots y_N, Z_1 \ldots Z_N, p_{x1} \ldots p_{xN}, p_{y1} \ldots p_{yN}, p_{z1} \ldots p_{zN}, t)$ (where x, y, and z are the space coordinates and p_x, p_y, and p_z are the momentum coordinates) is related to the density of the phases in an ensemble. This is the normalized density function in phase space and the total number of particles in the collection is N. Moreover, Liouville's equation is applicable to a solid, liquid, or gas. For a gas, if it is dilute, only two-particle collisions are the only modes of energy and momentum transfer, it is possible to simplify Liouville's equation. Suppose that the gas consists of a single particle. In this case the one-particle distribution function becomes $f^{(1)}(x, y, Z, p_x, p_y, p_z, t)$. It is now assumed that the N particles are completely identical and they all have the same single-particle distribution function. It is then possible to simply find $f^{(1)}(x, y, Z, p_x, p_y, p_z, t)$ and derive the relevant transport properties from this function. Liouville's equation can be modified to yield an equation for $f^{(1)}(x, y, Z, p_x, p_y, p_z, t)$ and this is actually the BTE [27]. Due to the assumption that a single distribution function represents the N particles, it becomes necessary to include the scattering or collision term in the modified Liouville's equation. The three-dimensional BTE is described as [28]

$$\frac{\partial f}{\partial t} + \frac{1}{m}\left(p_x \frac{\partial f}{\partial x} + p_y \frac{\partial f}{\partial y} + p_z \frac{\partial f}{\partial z}\right) + \left(\dot{p}_x \frac{\partial f}{\partial p_x} + \dot{p}_y \frac{\partial f}{\partial p_y} + \dot{p}_z \frac{\partial f}{\partial p_z}\right) = \left(\frac{\partial f}{\partial t}\right)_{\text{collision}} \tag{4.117}$$

$$\frac{\partial f}{\partial t} + \frac{p}{m} \cdot \nabla_r f + \dot{p} \cdot \nabla_p f = \left(\frac{\partial f}{\partial t}\right)_{\text{collision}} \tag{4.118}$$

where f is the distribution function; x, y, and z are the space coordinates; p_x, p_y, and p_z are the momentum coordinates; t is time; m is the mass; $\dot{p}_x$, $\dot{p}_y$, and $\dot{p}_z$ are rate of momentum.

The collision term is a complicated expression accounting for the effects of collisions between the particles. It may be replaced by a simpler expression by assuming the relaxation time approximation, which was explained in Chapter 1, Introduction. The preceding equation may then be written as

$$\frac{\partial f}{\partial t} + \frac{p}{m} \cdot \nabla_r f + \dot{p} \cdot \nabla_p f = -\frac{(f - f^0)}{\tau} \tag{4.119}$$

where f^0 is the equilibrium probability distribution defined by different distribution functions, which are given in Table 4.4.

Table 4.4 Equilibrium Probability Distribution

f^0	Application
Maxwell–Boltzmann distribution	Ordinary or ideal gas
Fermi–Dirac distribution	Fermion particles (such as electrons)
Bose–Einstein distribution	Boson particles (such as phonons)

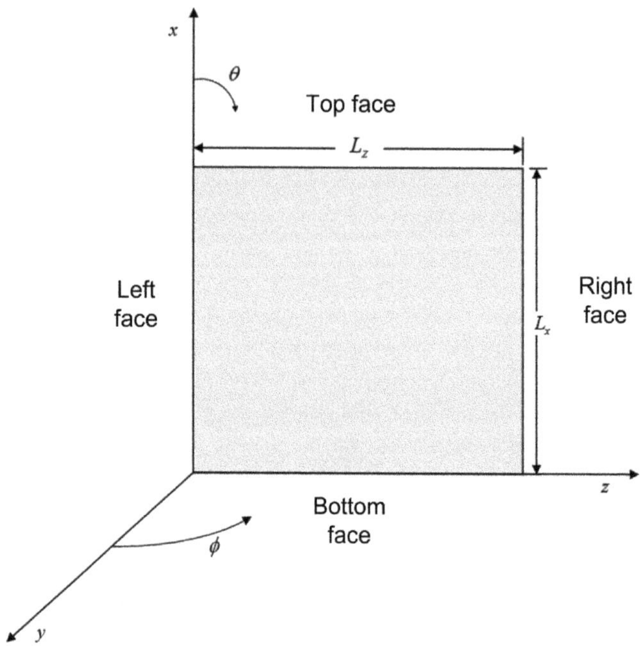

■ **FIGURE 4.14** A schematic showing the coordinate systems for two-dimensional thin films.

4.4.4 Equation of Phonon Radiative Transfer for Two-Dimensional Dielectric Thin Films

A thin two-dimensional dielectric film has a thickness and its width is very small as compared to its other third dimension. These lengths are comparable to or less than the phonon mean free path in the film. For the two-dimensional thin dielectric films as shown in Fig. 4.14, the frequency-dependent BTE with relaxation time approximation simplifies to

$$\frac{\partial f_\omega}{\partial t} + v_{x,\omega}\frac{\partial f_\omega}{\partial x} + v_{z,\omega}\frac{\partial f_\omega}{\partial z} + \dot{p}_{x,\omega}\frac{\partial f_\omega}{\partial p_{x,\omega}} + \dot{p}_{y,\omega}\frac{\partial f_\omega}{\partial p_{y,\omega}} + \dot{p}_{z,\omega}\frac{\partial f_\omega}{\partial p_{z,\omega}} = -\frac{(f_\omega - f_\omega^0)}{\tau_\omega} \tag{4.120}$$

In this equation, f is the single-particle distribution function of the phonons in the phase space and ω in subscript showing the frequency dependency. f_ω^0 is the frequency-dependent equilibrium intensity distribution and τ_ω is the frequency-dependent relaxation time. Here f is considered as function of all three momentum variables because phonons travel in all directions [25].

The momentums (p) for a single phonon using the de Broglie formula can written as [29]

$$p_\omega = \frac{h}{\lambda_\omega} = \hbar k(\omega) \tag{4.121}$$

In this equation; λ_ω is the frequency dependent wave length; $k(\omega)$ is called the wave number ($k(\omega) = 2\pi/\lambda_\omega$), which is the function of frequency; and $\hbar$ is the reduced Planck's constant ($\hbar = h/2\pi$). In three dimensions the three components of the momentum are related to the wave vector k. These are given as

$$p_{x,\omega} = \hbar k_x(\omega), \;\; p_{y,\omega} = \hbar k_y(\omega), \;\; p_{z,\omega} = \hbar k_z(\omega), \quad \text{or} \quad p_\omega = \hbar k(\omega) \tag{4.122}$$

also, $k(\omega) = (2\pi/\lambda_\omega)\hat{n}$, where $\hat{n}$ is the unit vector in the direction of propagation of the wave.

The momentum and the wave vector may be used interchangeably. It is assumed that the phonons are not under the influence of any attractive/repulsive force, therefore [25],

$$\dot{p}_{x,\omega} = \dot{p}_{y,\omega} = \dot{p}_{z,\omega} = 0 \tag{4.123}$$

Substituting the preceding equations in Eq. (4.120) yields

$$\frac{\partial f_\omega}{\partial t} + v_{x,\omega}\frac{\partial f_\omega}{\partial x} + v_{z,\omega}\frac{\partial f_\omega}{\partial z} = -\frac{(f_\omega - f_\omega^0)}{\tau_\omega} \tag{4.124}$$

where $f_\omega(x, y, Z, p_{x,\omega}, p_{y,\omega}, p_{z,\omega}, t)$ may also write equivalently, $f_\omega(x, y, Z, k_x, k_y, k_z, t)$. The phonons are pictured as moving among the faces of the thin film. These phonons travel in all directions with a speed and momentum that is dependent on the corresponding frequency of the lattice vibration. This picture is exactly that of photons, when they are transferring thermal energy among different faces in a vacuum. This suggests that for radiative energy transport through phonons, an equation completely analogous to the equation of radiative transfer, which is used for radiative energy transport through photons, may be derived. This equation is called the EPRT [22].

In Eq. (4.124), $v_{x,\omega}$ and $v_{z,\omega}$ are the frequency dependent velocity components in x and z coordinate, respectively. This can be written as

$$v_{x,\omega} = v_\omega \cos\theta, \quad v_{z,\omega} = v_\omega \sin\theta \sin\phi \tag{4.125}$$

where θ is the polar angle and ϕ is azimuthal angle, as shown in Fig. 4.14. Eq. (4.124) becomes

$$\frac{\partial f_\omega}{\partial t} + v_\omega \cos\theta \frac{\partial f_\omega}{\partial x} + v_\omega \sin\theta \sin\phi \frac{\partial f_\omega}{\partial z} = \frac{f_\omega^0 - f_\omega}{\tau_\omega} \tag{4.126}$$

Dividing this equation by v_ω on both sides results in the following:

$$\frac{1}{v_\omega}\frac{\partial f_\omega}{\partial t} + \cos\theta \frac{\partial f_\omega}{\partial x} + \sin\theta \sin\phi \frac{\partial f_\omega}{\partial z} = \frac{f_\omega^0 - f_\omega}{v_\omega \tau_\omega} \tag{4.127}$$

Here the product of frequency dependent group velocity and the relaxation time is equal to the frequency dependent mean free path that is $\Lambda_\omega = v_\omega \tau_\omega$. Now Eq. (4.127) becomes

$$\frac{1}{v_\omega}\frac{\partial f_\omega}{\partial t} + \cos\theta \frac{\partial f_\omega}{\partial x} + \sin\theta \sin\phi \frac{\partial f_\omega}{\partial z} = \frac{f_\omega^0 - f_\omega}{\Lambda_\omega} \tag{4.128}$$

Multiplying Eq. (4.128) with $v_\omega \hbar\omega D(\omega)/4\pi$ on both sides and rearranging,

$$\begin{aligned} &\frac{1}{v_\omega}\frac{\partial v_\omega f_\omega \hbar\omega D(\omega)/4\pi}{\partial t} + \cos\theta \frac{\partial v_\omega f_\omega \hbar\omega D(\omega)/4\pi}{\partial x} \\ &\quad + \sin\theta \sin\phi \frac{\partial v_\omega f_\omega \hbar\omega D(\omega)/4\pi}{\partial z} = \frac{v_\omega f_\omega^0 \hbar\omega D(\omega)/4\pi - v_\omega f_\omega \hbar\omega D(\omega)/4\pi}{\Lambda_\omega} \end{aligned} \tag{4.129}$$

where $D(\omega)$ is the density of state, which is the number of phonons/electrons per unit volume in a unit energy interval [26]. It is given by [30]

$$D(\omega) = \frac{k^2}{2\pi^2}\frac{dk}{d\omega} \tag{4.130}$$

In Eq. (4.130), frequency is the function of wave number, $\omega = \omega(k)$, which describes the effect of dispersion in a medium on the properties of a traveling wave inside the medium. Dispersion relations are obtained from the experimental studies. Dispersion relations for silicon and diamond are shown in Figs. 4.15 and 4.16, respectively.

In the theory of thermal radiation transfer, the spectral intensity is defined as the amount of energy transferred per unit time, per unit perpendicular area, per unit solid angle, per unit wavelength interval; that is,

$$I_\lambda = I_\lambda(x, y, Z, \theta, \phi, t) \tag{4.131}$$

A similar spectral intensity may be defined for the phonons. Note that the dependence on (θ, ϕ) is actually dependence on the solid angle (Ω), which

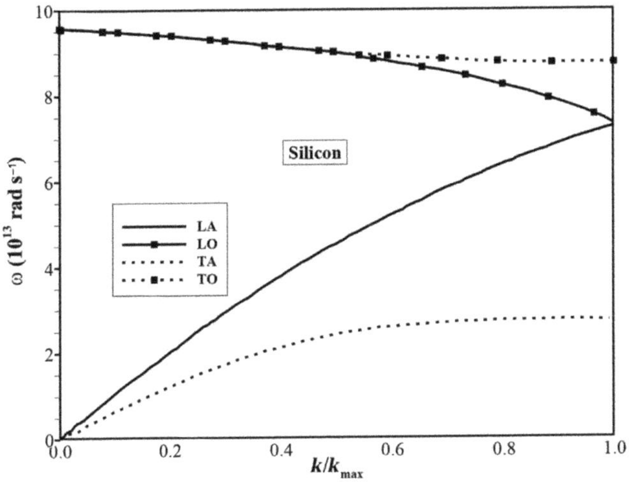

■ **FIGURE 4.15** Dispersion relation for silicon [32].

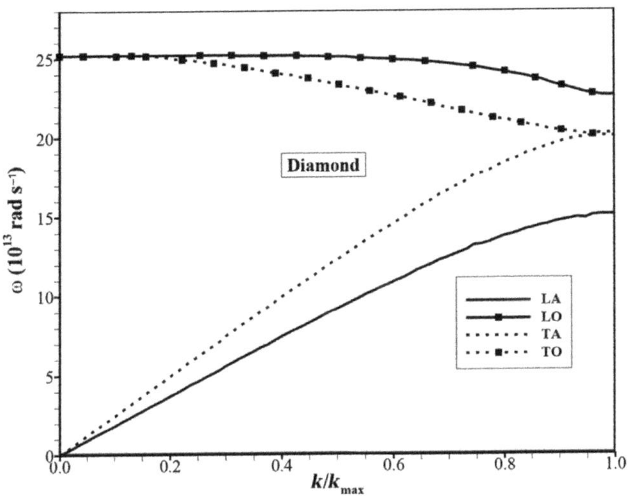

■ **FIGURE 4.16** Dispersion relation for diamond [33].

in turn implies dependence on the direction. Similarly dependence on wavelength (λ) implies a dependence on the wave number or the frequency, which in turn implies dependence on the momentum and the energy of the phonons. The phonon spectral intensity may be related to the phonon distribution function, which is given as [31]

$$I_{\omega}(x, z, t, \theta, \phi) = \frac{1}{4\pi} f_{\omega} v_{\omega} \hbar \omega D(\omega) \tag{4.132}$$

Eq. (4.129) can now be written in terms of phonon intensities, given by

$$\frac{1}{v_\omega}\frac{\partial I_\omega}{\partial t} + \cos\theta\frac{\partial I_\omega}{\partial x} + \sin\theta\sin\phi\frac{\partial I_\omega}{\partial z} = \frac{I_\omega^0 - I_\omega}{\Lambda_\omega}$$
$$t>0,\ 0<x<L_x,\ 0<z<L_z,\ 0<\theta<\pi,\ 0<\phi<2\pi \tag{4.133}$$

where ω is frequency (in subscript it shows the frequency dependency), I_ω is frequency dependent intensity in phase space, I_ω^0 is frequency dependent equilibrium probability distribution function in phase space, t is the time, x is the distance in the x direction, z is the distance in the z direction, θ is the polar angle, ϕ is the azimuthal angle, τ_ω is the frequency-dependent relaxation time, v_ω is the frequency-dependent phonon group velocity, and $\Lambda_\omega = v_\omega\tau_\omega$ is the frequency-dependent mean free path. The phonon intensity is a function of five independent variables, $I_\omega = I_\omega(x, z, \theta, \phi, t)$.

Eq. (4.133) is known as the EPRT.

For the frequency-independent case, the EPRT (Eq. 4.133) simplifies into the following:

$$\frac{1}{v}\frac{\partial I}{\partial t} + \cos\theta\frac{\partial I}{\partial x} + \sin\theta\sin\phi\frac{\partial I}{\partial z} = \frac{I^0 - I}{\Lambda} \tag{4.134}$$

4.4.4.1 *Heat Fluxes*

The heat flux vector of phonon for any frequency can be given as [22]

$$q''_\omega(x, y, z, t) = N\iiint v_\omega f_\omega \hbar\omega dp_x dp_y dp_z \tag{4.135}$$

In order to find the total heat flux by all phonons, the previous expression needs to be integrated over the whole range of frequencies. Therefore, expression for the total heat flux can be written as:

$$q''(x, y, z, t) = N\int_0^{\omega_{\max}}\left(\iiint v_\omega f_\omega \hbar\omega dp_x dp_y dp_z\right)d\omega$$

Instead of the momentum, the frequency and the solid angle can be used as the independent variables. We may then write

Infinitesimal solid angle $d^2\Omega = \sin\theta d\phi d\theta$

Heat flux vector

$$q''(x,y,z,t) = \int_0^{\omega_{\max}}\left(\iiint v_\omega f_\omega \hbar\omega D(\omega) d\omega d^2\Omega\right)d\omega$$
$$q''(x,y,z,t) = \int_0^{\omega_{\max}}\left(\iiint v_\omega f_\omega \hbar\omega D(\omega)\sin\theta d\omega d\phi d\theta\right)d\omega \tag{4.136}$$

In the two-dimensional domain heat flux in the x direction and the z direction is defined as follows.

Heat flux in the x direction (q''_x) is written as

$$q''_x(x,z,t) = \int_0^{\omega_{max}} \left(\iiint v_{x,\omega} f_\omega \hbar\omega D(\omega) \sin\theta d\omega d\phi d\theta \right) d\omega \tag{4.137}$$

Using $v_{x,\omega} = v_\omega \cos\theta$ and the definition of the frequency-dependent phonon intensity (Eq. 4.132), the previous equations can be written as

$$q''_x(x,z,t) = \int_0^{\omega_{max}} \int_0^{2\pi} \int_0^{\pi} I_\omega \cos\theta \sin\theta d\theta d\phi d\omega \tag{4.138}$$

For the frequency-independent case, Eq. (4.138) can be written as

$$q''_x(x,z,t) = \int_0^{2\pi} \int_0^{\pi} I \cos\theta \sin\theta d\theta d\phi \tag{4.139}$$

Heat flux in the z direction (q''_z) is written as

$$q''_z(x,z,t) = \int_0^{\omega_{max}} \left(\iiint v_{z,\omega} f_\omega \hbar\omega D(\omega) \sin\theta d\omega d\phi d\theta \right) d\omega \tag{4.140}$$

Using $v_{z,\omega} = v_\omega \sin\theta \sin\phi$ and the definition of the frequency-dependent phonon intensity (Eq. 4.132), the previous equations can be written as

$$q''_z(x,z,t) = \int_0^{\omega_{max}} \int_0^{2\pi} \int_0^{\pi} I_\omega \sin\phi \sin^2\theta d\theta d\phi d\omega \tag{4.141}$$

For the frequency-independent case, Eq. (4.141) can be written as

$$q''_z(x,z,t) = \int_0^{2\pi} \int_0^{\pi} I \sin\phi \sin^2\theta d\theta d\phi \tag{4.142}$$

In the vector form the heat flux is given as

$$\mathbf{q}'' = q''_x \hat{\mathbf{i}} + q''_z \hat{\mathbf{k}} \tag{4.143}$$

4.4.4.2 ***Equilibrium Intensity Calculation***

I^0_ω is the equilibrium intensity, corresponding to blackbody intensity at a temperature below the Debye temperature [25]. Debye temperature is the highest temperature that can be achieved in a normal mode of vibration. In order to determine equilibrium intensity I^0_ω, first we multiply

Eq. (4.133) by $\sin\theta\, d\theta\, d\phi$ and integrate over a solid angle and over the complete range of frequencies.

$$\frac{\partial}{\partial t}\int_0^{\omega_{\max}}\int_0^{2\pi}\int_0^{\pi}\frac{1}{v_\omega}I_\omega \sin\theta d\theta d\phi d\omega + \frac{\partial}{\partial x}\int_0^{\omega_{\max}}\int_0^{2\pi}\int_0^{\pi} I_\omega \cos\theta \sin\theta d\theta d\phi d\omega$$
$$+\frac{\partial}{\partial z}\int_0^{\omega_{\max}}\int_0^{2\pi}\int_0^{\pi} I_\omega \sin\phi \sin^2\theta d\theta d\phi d\omega = \int_0^{\omega_{\max}}\frac{1}{\Lambda_\omega}\left[\begin{array}{l}\int_0^{2\pi}\int_0^{\pi} I_\omega^o \sin\theta d\theta d\phi \\ -\int_0^{2\pi}\int_0^{\pi} I_\omega \sin\theta d\theta d\phi\end{array}\right]d\omega \tag{4.144}$$

Using the definition of heat flux from Eqs. (4.138) and (4.141) into Eq. (4.144) results in

$$\frac{\partial}{\partial t}\int_0^{\omega_{\max}}\int_0^{2\pi}\int_0^{\pi}\frac{1}{v_\omega}I_\omega \sin\theta d\theta d\phi d\omega + \frac{\partial q''_x}{\partial x} + \frac{\partial q''_z}{\partial z} = \int_0^{\omega_{\max}}\frac{1}{\Lambda_\omega}\left[\begin{array}{l}\int_0^{2\pi}\int_0^{\pi} I_\omega^o \sin\theta d\theta d\phi \\ -\int_0^{2\pi}\int_0^{\pi} I_\omega \sin\theta d\theta d\phi\end{array}\right]d\omega \tag{4.145}$$

To simplify the first term of left-hand side using the definition of phonon intensity from Eq. (4.132),

$$\int_0^{\omega_{\max}}\int_0^{2\pi}\int_0^{\pi}\frac{1}{v_\omega}I_\omega \sin\theta d\theta d\phi\, d\omega = \int_0^{\omega_{\max}}\int_0^{2\pi}\int_0^{\pi}\frac{f_\omega \hbar\omega D(\omega)}{4\pi}\sin\theta d\theta d\phi\, d\omega \tag{4.146}$$

The energy of each phonon is $\hbar\omega$. Therefore, on the right-hand side of the previous equation, the product of phonon energy, density of state, and nonequilibrium probability distribution, integrated over a complete solid angle and frequency spectrum gives the internal energy density of phonons; i.e.,

$$u(x,z,t) = \int_0^{\omega_{\max}}\int_0^{2\pi}\int_0^{\pi}\frac{f_\omega \hbar\omega D(\omega)}{4\pi}\sin\theta d\theta d\phi d\omega \tag{4.147}$$

Therefore, Eq. (4.146) represents the internal energy density of phonons. Utilizing the definition of divergence and the previous equation into Eq. (4.145) results in

$$\frac{\partial u}{\partial t} + \nabla\cdot q'' = \int_0^{\omega_{\max}}\frac{1}{\Lambda_\omega}\left[\begin{array}{l}\int_0^{2\pi}\int_0^{\pi} I_\omega^0 \sin\theta d\theta d\phi \\ -\int_0^{2\pi}\int_0^{\pi} I_\omega \sin\theta d\theta d\phi\end{array}\right]d\omega \tag{4.148}$$

The law of energy conservation with no internal heat generation is $\partial u/\partial t + \nabla \cdot q'' = 0$. Now, as in the present dissertation, heat generation is not included, therefore Eq. (4.148) should satisfy the conservation law of energy without internal heat generation, which leads to the following:

$$\int_0^{\omega_{\max}} \frac{1}{\Lambda_\omega} \left[\int_0^{2\pi} \int_0^{\pi} I_\omega^o \sin\theta d\theta d\phi - \int_0^{2\pi} \int_0^{\pi} I_\omega \sin\theta d\theta d\phi \right] d\omega = 0 \tag{4.149}$$

The equilibrium intensity is only a function of spatial coordinate and time; i.e., $I_\omega^0 = I_\omega^0(x, z, t)$, therefore,

$$\int_0^{\omega_{\max}} \frac{1}{\Lambda_\omega} \left[I_\omega^o - \frac{1}{4\pi} \int_0^{2\pi} \int_0^{\pi} I_\omega \sin\theta d\theta d\phi \right] d\omega = 0 \tag{4.150}$$

Here, an assumption is made that radiative equilibrium is satisfied at all frequencies, therefore at each frequency the integrand in Eq. (4.150) vanished. This yields

$$I_\omega^0 = \frac{1}{4\pi} \int_0^{2\pi} \int_0^{\pi} I_\omega \sin\theta d\theta d\phi \tag{4.151}$$

Eq. (4.133) can now be written as

$$\frac{1}{v_\omega}\frac{\partial I_\omega}{\partial t} + \cos\theta \frac{\partial I_\omega}{\partial x} + \sin\theta \sin\phi \frac{\partial I_\omega}{\partial z} = \frac{\frac{1}{4\pi} \int_0^{2\pi} \int_0^{\pi} I_\omega \sin\theta d\theta d\phi - I_\omega}{\Lambda_\omega} \tag{4.152}$$

4.4.4.3 *Physical Significance of Each Term in EPRT*

Each term of the EPRT has a physical significance, which is given here:

$$\frac{1}{v_\omega}\frac{\partial I_\omega}{\partial t} + \cos\theta \frac{\partial I_\omega}{\partial x} + \sin\theta \sin\phi \frac{\partial I_\omega}{\partial z} = \frac{\frac{1}{4\pi} \int_0^{2\pi} \int_0^{\pi} I_\omega \sin\theta d\theta d\phi - I_\omega}{\Lambda_\omega} \tag{4.153}$$

The first term on the left-hand side $\frac{1}{v_\omega}\frac{\partial I_\omega}{\partial t}$ accounts for the fact that the thermal energy transport through lattice vibration takes place at a finite speed. Therefore at any point in the film the intensity changes until a steady state is reached. The second and third term on the left-hand side represents the intensity gradient, which is necessary to balance the effect of phonon scattering, emission, and transmittance. The right-hand side is the scattering term associated with the phonon mean free path and film thickness. For the larger values of mean free path as compared to film thickness, scattering becomes smaller.

4.4.4.4 ***Polarization and Modes of Phonons***

There are three different polarizations and two different modes of vibrations of the phonons. This information is collected through the dispersion relations of the dielectric materials. Polarization means that waves can oscillate in more than one orientation. Waves can propagate in a dielectric crystal in three different polarizations, one longitudinal and two transverse polarizations. The two transverse waves coincide and therefore only one dispersion curve is required to represent the two transverse waves for each mode of vibration. As the vibration of two atoms at each lattice point can be in-phase or out-of-phase; therefore, there are two modes of lattice in dielectric crystal [27]. The first mode is called acoustic waves, and represents the in-phase lattice vibration; the second mode is called optical waves, and represents the out-of-phase vibration. Hence there are six different types of waves but since the transverse waves are identical, actually different types of waves are present, therefore four different curves in the dispersion relation are shown in Figs. 4.15 and 4.16 for silicon and diamond.

The EPRT given in Eq. (4.152) has to be applied to each of these waves individually. These EPRTs are given as follows.

4.4.4.4.1 **Longitudinal Acoustic**

$$\frac{1}{v_{\mathrm{LA},\omega}}\frac{\partial I_{\mathrm{LA},\omega}}{\partial t}+\cos\theta\frac{\partial I_{\mathrm{LA},\omega}}{\partial x}+\sin\theta\cos\phi\frac{\partial I_{\mathrm{LA},\omega}}{\partial z}=\frac{\frac{1}{4\pi}\int_0^{2\pi}\int_0^{\pi}I_{\mathrm{LA},\omega}\sin\theta d\theta d\phi-I_{\mathrm{LA},\omega}}{\Lambda_{\mathrm{LA},\omega}} \tag{4.154}$$

4.4.4.4.2 **Transverse Acoustic**

$$\frac{1}{v_{\mathrm{TA},\omega}}\frac{\partial I_{\mathrm{TA},\omega}}{\partial t}+\cos\theta\frac{\partial I_{\mathrm{TA},\omega}}{\partial x}+\sin\theta\cos\phi\frac{\partial I_{\mathrm{TA},\omega}}{\partial z}=\frac{\frac{1}{4\pi}\int_0^{2\pi}\int_0^{\pi}I_{\mathrm{TA},\omega}\sin\theta d\theta d\phi-I_{\mathrm{TA},\omega}}{\Lambda_{\mathrm{TA},\omega}} \tag{4.155}$$

4.4.4.4.3 **Longitudinal Optical**

$$\frac{1}{v_{\mathrm{LO},\omega}}\frac{\partial I_{\mathrm{LO},\omega}}{\partial t}+\cos\theta\frac{\partial I_{\mathrm{LO},\omega}}{\partial x}+\sin\theta\cos\phi\frac{\partial I_{\mathrm{LO},\omega}}{\partial z}=\frac{\frac{1}{4\pi}\int_0^{2\pi}\int_0^{\pi}I_{\mathrm{LO},\omega}\sin\theta d\theta d\phi-I_{\mathrm{LO},\omega}}{\Lambda_{\mathrm{LO},\omega}} \tag{4.156}$$

4.4.4.4.4 **Transverse Optical**

$$\frac{1}{v_{TO,\omega}}\frac{\partial I_{TO,\omega}}{\partial t}+\cos\theta\frac{\partial I_{TO,\omega}}{\partial x}+\sin\theta\cos\phi\frac{\partial I_{TO,\omega}}{\partial z}=\frac{\frac{1}{4\pi}\int_0^{2\pi}\int_0^{\pi} I_{TO,\omega}\sin\theta d\theta d\phi - I_{TO,\omega}}{\Lambda_{TO,\omega}} \tag{4.157}$$

where LA represents the longitudinal acoustic, TA corresponds to transverse acoustic, LO is the longitudinal optical, and TO is transverse optical.

4.4.4.5 ***Equilibrium Equivalent Temperature***

This temperature defined anywhere in the dielectric thin layer is analogous to the usual thermodynamic temperature defined in the diffusive limit in any medium. It represents the average energy of all phonons around a local point and is equivalent to the equilibrium temperature of phonons when they redistribute adiabatically to an equilibrium state. It is defined as [34]

$$T(x,z,t)=\frac{\int_0^{\omega_{max}}\int_0^{2\pi}\int_0^{\pi} I_\omega \sin\theta d\theta d\phi d\omega}{\int_0^{\omega_{max}}\left(C_{LA,\omega}v_{LA,\omega}+2C_{TA,\omega}v_{TA,\omega}+C_{LO,\omega}v_{LO,\omega}+2C_{TO,\omega}v_{TO,\omega}\right)d\omega} \tag{4.158}$$

In this equation, $C_{LA,\omega}$ is the frequency-dependent volumetric specific heat capacity, and $v_{LA,\omega}$ is the frequency-dependent phonon group velocity for the longitudinal acoustic branch for the dielectric material. Similar definitions apply for the transverse acoustic, transverse optical, and longitudinal optical branches of phonons.

The equation just derived is equally valid when written as a function of wavenumber (k) instead of frequency (ω).

For the frequency-independent case, the equilibrium equivalent temperature in Eq. (4.158) becomes

$$T(x,z,t)=\frac{\int_0^{2\pi}\int_0^{\pi} I\sin\theta d\theta d\phi}{Cv} \tag{4.159}$$

4.4.5 Equation of Phonon Radiative Transfer for One-Dimensional Dielectric Thin Films

For one-dimensional film, the EPRT derived in Eq. (4.133) for two dimensional film reduces to

$$\frac{1}{v_\omega}\frac{\partial I_\omega}{\partial t} + \cos\theta\frac{\partial I_\omega}{\partial x} = \frac{I_\omega^0 - I_\omega}{\Lambda_\omega} \tag{4.160}$$

Now, introducing $\mu = \cos\theta$ in Eq. (4.160) we get

$$\frac{1}{v_\omega}\frac{\partial I_\omega}{\partial t} + \mu\frac{\partial I_\omega}{\partial x} = \frac{I_\omega^0 - I_\omega}{\Lambda_\omega} \quad t>0,\ 0<x<L,\ \ -1<\mu<1 \tag{4.161}$$

The equilibrium intensity in one-dimensional film can be calculated from the following relation [22]:

$$I_\omega^0 = \frac{1}{2}\int_{-1}^{1} I_\omega d\mu \tag{4.162}$$

For the frequency-independent case, which is the gray medium, EPRT simplifies to

$$\frac{1}{v}\frac{\partial I}{\partial t} + \mu\frac{\partial I}{\partial x} = \frac{I^0 - I}{\Lambda} \quad 0<x<L,\ \ -1<\mu<1 \tag{4.163}$$

The thermal properties of dielectric materials, which is required for the solution of Eq. (4.163), are given in Table 4.5.

4.4.5.1 *Equilibrium Equivalent Temperature*

Equilibrium equivalent temperature for one-dimensional dielectric thin film can be written as

$$T(x,t) = \frac{2\pi\int_0^{\omega_{\max}}\int_{-1}^{1}(I_{\mathrm{LA},\omega} + 2I_{\mathrm{TA},\omega} + I_{\mathrm{LO},\omega} + 2I_{\mathrm{TO},\omega})d\mu d\omega}{\int_0^{\omega_{\max}}(C_{\mathrm{LA},\omega}v_{\mathrm{LA},\omega} + 2C_{\mathrm{TA},\omega}v_{\mathrm{TA},\omega} + C_{\mathrm{LO},\omega}v_{\mathrm{LO},\omega} + 2C_{\mathrm{TO},\omega}v_{\mathrm{TO},\omega})d\omega} \tag{4.164}$$

Table 4.5 Properties of Dielectric Materials Used in the Numerical Simulations of the Frequency-Independent Solution of the Phonon Radiative Transport Equation

	Thermal Conductivity ($Wm^{-1}K^{-1}$)	Volumetric Specific Heat ($Jm^{-3}K^{-1}$)	Phonon Speed (ms^{-1})	Mean Free Path (nm)
Silicon	157	1.653×10^6	8430	33.8
Diamond	3323	1.815×10^6	12,288	447

Thermal conductivity can be determined from the kinetic theory formula $k = Cv\Lambda/3$.

Equilibrium equivalent temperature for the one-dimensional frequency-independent case simplifies as

$$T(x,t) = \frac{2\pi \int_{-1}^{1} I\, d\mu}{Cv} \tag{4.165}$$

4.4.5.2 ***Heat Flux***

The total heat flux due to the contribution of all branches of the acoustic and optical phonons for one-dimensional dielectric thin film is given as

$$q''(x,t) = 2\pi \int_0^{\omega_{\max}} \int_{-1}^{1} I_\omega \mu d\mu d\omega \tag{4.166}$$

The total heat flux for the one-dimensional frequency independent case can be given as

$$q''(x,t) = 2\pi \int_{-1}^{1} I\mu d\mu \tag{4.167}$$

4.4.5.3 ***Entropy Generation***

The entropy generation rate per unit volume is given as [35]

$$S_{\text{gen}} = \frac{q}{T^2}(\nabla T) \tag{4.168}$$

Using Eqs. (4.166) and (4.168) yields

$$S_{\text{gen}} = 2\pi \frac{1}{T^2(x)} \frac{\partial T(x)}{\partial x} \int_0^{\omega_{\max}} \int_{-1}^{1} I_\omega \mu d\mu d\omega \tag{4.169}$$

Using Eqs. (4.166) and (4.168) yields

$$S_{\text{gen}} = 2\pi \frac{1}{T^2(x)} \frac{\partial T(x)}{\partial x} \int_0^{\omega_{\max}} \int_{-1}^{1} I_\omega \mu d\mu d\omega \tag{4.170}$$

Eq. (4.170) can be used to calculate the entropy generation due to the frequency-dependent phonon transport in the thin film after integrating over the volume; i.e.,

$$S_{\text{gen},tot} = \iiint S_{\text{gen}}(x) dV = 2\pi A \int_0^L \left(\frac{1}{T^2(x)} \frac{\partial T(x)}{\partial x} \int_0^{\omega_{\max}} \int_{-1}^{1} I_\omega \mu d\mu \, d\omega \right) dx \tag{4.171}$$

where $S_{\text{gen},tot}$ is the total entropy generation rate, V is the volume, and A is the area normal to the x direction.

The expression for total entropy generation rate derived in Eq. (4.171) can be simplified for the frequency-independent case, which is given as

$$S_{gen,tot} = \iiint S_{gen}(x)dV = 2\pi A \int_0^L \left(\frac{1}{T^2(x)} \frac{\partial T(x)}{\partial x} \int_{-1}^{1} I\mu d\,\mu \right) dx \qquad (4.172)$$

4.4.6 One-Dimensional Metallic Thin Films

4.4.6.1 *Modified Two-Equation Model*

In metal films, heat is predominantly transported by means of the electron motion. Hence, electrons are the principal heat carriers. However, a small fraction of the total thermal energy (approximately 5%) is also transported by the phonons because of the mean free path of the electrons, which is extremely smaller than that of the phonons. Thin metallic films used in practice always have a thickness that is much larger than the electron mean free path. Therefore, in a metal, the electrons are always under a state of local thermodynamic equilibrium and a temperature gradient is always established. This means that the Fourier law of heat conduction is applicable for the description of heat transport in thin metallic films. Usually, the electrons and phonons in a metal are always under thermodynamic equilibrium. This means that they both have a single temperature. However, there are circumstances when this is no longer true and the electrons and phonons are no longer in thermodynamic equilibrium. In these situations, we have to define two temperatures, one for the electron and one for the phonons. When a metal substrate is heated very rapidly, such as with ultrashort laser pulses whose pulse length is shorter than the phonon relaxation time, the electrons gets heated and their temperature rises sharply. However, the heating time is so short that the electrons can only transfer a slight amount of thermal energy to the phonons and hence we define an electron temperature and a phonon temperature. When a very thin metal film is in contact with a very thin dielectric film it is possible that the thermodynamic equilibrium between the electrons and phonons is destroyed in a very small region near the metal-dielectric interface, and again we have to define separate temperatures. In these scenarios it is necessary to use the two-equation model, or its modified form. The original two-equation model does not account for heat transport through the phonons but since we are studying coupled heat transport in a metal and dielectric film, it is necessary to incorporate the diffusion terms in the two-equation model. This is presented as

$$C_e \frac{\partial T_e}{\partial t} = k_e \frac{\partial^2 T_e}{\partial x^2} - G(T_e - T_{ph}) + S \quad 0 < x < L, \ t > 0 \qquad (4.173)$$

$$C_{ph} \frac{\partial T_{ph}}{\partial t} = k_{ph} \frac{\partial^2 T_{ph}}{\partial x^2} - G(T_{ph} - T_e) \quad 0 < x < L, \ t > 0 \qquad (4.174)$$

In these equations, the subscript e stands for electron and the subscript ph stands for phonon. C is the volumetric specific heat capacity, k is the thermal conductivity, G is the electron–phonon coupling constant, and S is the volumetric heat generation rate. Since laser heating is being studied, the heat source term appears only in the electron heat conduction equation, because the laser radiation is absorbed only by the electrons. The volumetric heat generation can be an incident radiation, such as laser irradiation, and it can be expressed as

$$S = I_0\alpha \exp[-\alpha x - 2.77(t/t_p)^2] \tag{4.175}$$

In this expression, I_0 is the intensity of the laser radiation in units of W m^{-2}, α is the absorption coefficient in the metal film in units of 1 m^{-1}, and t_p is a characteristic pulse interval in units of S.

4.4.6.2 *Electron–Phonon Coupling Resistance*

In a metal, near the interface of a metal-dielectric film, the electrons and phonons are not in thermal equilibrium. Their temperatures, therefore, diverge. This nonequilibrium condition constitutes a resistance to heat flow, which is in addition to the TBR discussed earlier. The reason for this extra resistance is not difficult to understand. Consider a metal layer in contact with a dielectric layer, both having infinite thickness. Assume that the heat is flowing from the metal to the dielectric layer. Far away from the interface, the electrons and phonons have the same temperature. This means that the heat is flowing inside the metal toward the dielectric layer, through the electrons as well as through the phonons. Now, at the interface, the phonons are coupled to the phonons of the dielectric and hence transfer their thermal energy to them. However, the electrons are not coupled to the heat carriers of the dielectric layer and hence their only mechanism of heat transfer is to the phonons. This can only occur if the electron and phonon temperatures are different. This nonequilibrium condition is essentially different from that which occurs due to the ultra-fast heating of the metal layer and constitutes an additional resistance to the heat flow. A mathematical expression for this resistance was derived by Majumdar and Reddy [36]; the derivation is given as follows.

Consider the schematic drawing in Fig. 4.17. The steady-state, two-equation model in the metal is given as

$$k_e \frac{d^2T_e}{dy^2} - G(T_e - T_{ph}) = 0 \tag{4.176}$$

$$k_{ph} \frac{d^2T_{ph}}{dy^2} - G(T_{ph} - T_e) = 0 \tag{4.177}$$

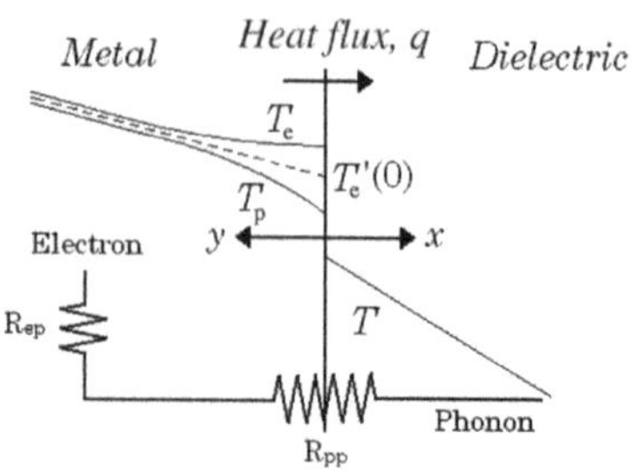

■ **FIGURE 4.17** Schematic of a metal-dielectric interface [36].

Let $\theta = T_e - T_p$; subtracting the phonon equation from the electron equation yields

$$\frac{d^2\theta}{dy^2} - \left[\frac{1}{k_e/G} + \frac{1}{k_{ph}/G}\right]\theta = 0 \tag{4.178}$$

If $l_e^2 = \frac{k_e}{G}$ and $l_{ph}^2 = \frac{k_{ph}}{G}$, then we may write $\frac{1}{l^2} = \frac{1}{l_e^2} + \frac{1}{l_{ph}^2}$. The previous equation may then be written as

$$\frac{d^2\theta}{dy^2} - \frac{\theta}{l^2} = 0 \tag{4.179}$$

The solution of this equation is

$$\theta = Ae^{+y/l} + Be^{-y/l} \tag{4.180}$$

For the solution to remain bounded as $y \to \infty$, $A = 0$. We finally get

$$\theta = Be^{-y/l} \tag{4.181}$$

Substituting this in the equation of the heat conduction for electron and solving yields

$$T_e = B\frac{l^2}{l_e^2}e^{-y/l} + Cy + D \tag{4.182}$$

$$T_{ph} = B\left(\frac{l^2}{l_e^2} - 1\right)e^{-y/l} + Cy + D \tag{4.183}$$

At $y = 0$, $dT_e/dy = 0$. This yields

$$C = B\frac{l}{l_e^2} \tag{4.184}$$

The steady-state temperature distribution in the dielectric layer will follow a linear profile. This is written as

$$T = Ex + F \tag{4.185}$$

At the interface, the heat flux continuity condition is satisfied. This is written as follows.

$$q'' = k_{ph}\frac{dT_{ph}}{dy}\bigg|_{y=0} = \frac{T_{ph}(y=0) - T(x=0)}{R} = -k\frac{dT}{dx}\bigg|_{x=0} \tag{4.186}$$

Physically, this equation means that the heat flux at the interface is proportional to the first derivative of the phonon temperature at the interface, to the difference in the phonon and dielectric temperatures at the interface, and to the first derivative of the dielectric temperature at the interface. This interface condition is simplified as

$$q'' = k_{ph}\frac{B}{l} = \frac{B\left(\frac{l^2}{l_e^2} - 1\right) + D - F}{R} = -kE \tag{4.187}$$

which yields

$$E = -\frac{k_{ph}}{k}\frac{B}{l} \quad F = B\left(\frac{l^2}{l_e^2} - 1 - \frac{k_{ph}R}{l}\right) + D \tag{4.188}$$

Classically we assume that there is a single temperature for the electron and phonons throughout the metal layer. The single interface temperature is then found from extrapolating the linear electron–phonon temperature profile, which exits far away from the interface. This is seen in Fig. 4.17. From the electron and phonon distributions found earlier, this extrapolated temperature is found to be

$$T'_e(y=0) = T'_{ph}(y=0) = D \tag{4.189}$$

The total resistance, heat flow experiences is then defined as

$$q'' = \frac{T'_e(y=0) - T(y=0)}{R_{total}} \tag{4.190}$$

$$R_{total} = -\left(\frac{l^3}{l_e^2 k_{ph}} - \frac{l}{k_{ph}} - R\right) \tag{4.191}$$

Now, $k_{ph} \ll k_e$. Therefore,

$$\frac{1}{l^2} = G\left(\frac{1}{k_e} + \frac{1}{k_{ph}}\right) \approx \frac{G}{k_{ph}} \tag{4.192}$$

The total resistance may then be approximated as

$$R_{total} = \sqrt{\frac{1}{Gk_{ph}}} + R \tag{4.193}$$

This expression clearly demonstrates that in addition to the TBR R, there is another resistance, $1/\sqrt{Gk_{ph}}$, which is there because of the nonequilibrium between the electron and phonon subsystems in a metal, near the interface. The two resistances are seen to be in series. Hence, it may be concluded that the heat flow through a metal–dielectric interface experiences resistance due to the electron–phonon nonequilibrium as well as the phonon reflection as the interface.

4.4.6.3 *Equation of Phonon Radiative Transfer for Metallic Films*

In the case of metal thin films, thermal separation of electron and phonon subsystems occur, which requires incorporating the restructuring of the energy equations in line with the Boltzmann equation. This can be achieved through formulating energy transport in each subsystem using the BTE. In this case, separate EPRTs for each subsystem need to be developed. Due to the presence of thermal communication of both subsystems during the energy transport, the electron–phonon coupling term should be included in the EPRTs of each subsystem. The functional form of the electron–phonon coupling term should be in such a way that in the diffusive limit, the EPRT should reduce to the standard two-equation model. The resulting EPRTs using the Boltzmann equation for each subsystem are given next.

4.4.6.3.1 **EPRT for Lattice Subsystem**

In order to account for the frequency-dependent formulation of the lattice subsystem, the experimental dispersion relations for aluminum is introduced [37] as shown in Fig. 4.18, which provides three different polarizations for acoustic mode of vibrations of phonons. However, two transverse waves are degenerate and, therefore, we can have only one dispersion curve for two transverse waves; the other is for the longitudinal wave for aluminum. The constant relaxation time is assumed for all modes in aluminum; the value $\tau = 2.73$ ps is chosen to yield the phonon thermal conductivity around 11.3 ($\mathrm{Wm^{-1}K^{-1}}$) [38].

Due to the high speed of electrons, the electron subsystem is assumed to be a gray body and the radiative transport is considered to be frequency independent [9]. Therefore, the term contains electron–phonon coupling constant calculated separately and, then, it is redistributed over all polarizations and frequencies. Hence, the electron–phonon coupling term yields

$$G_{\mathrm{k}} = \frac{G}{4\pi}\frac{(T_{\mathrm{p}} - T_{\mathrm{e}})}{3k_{\max}} \tag{4.194}$$

where G is the electron–phonon coupling constant; T_{p} is the phonon subsystem temperature $\left(T_{\mathrm{p}} = \frac{2\pi}{C_{\mathrm{p}}v_{\mathrm{p}}}\int_{-1}^{1} I_{\mathrm{p}}d\mu\right)$; T_{e} is the electron subsystem

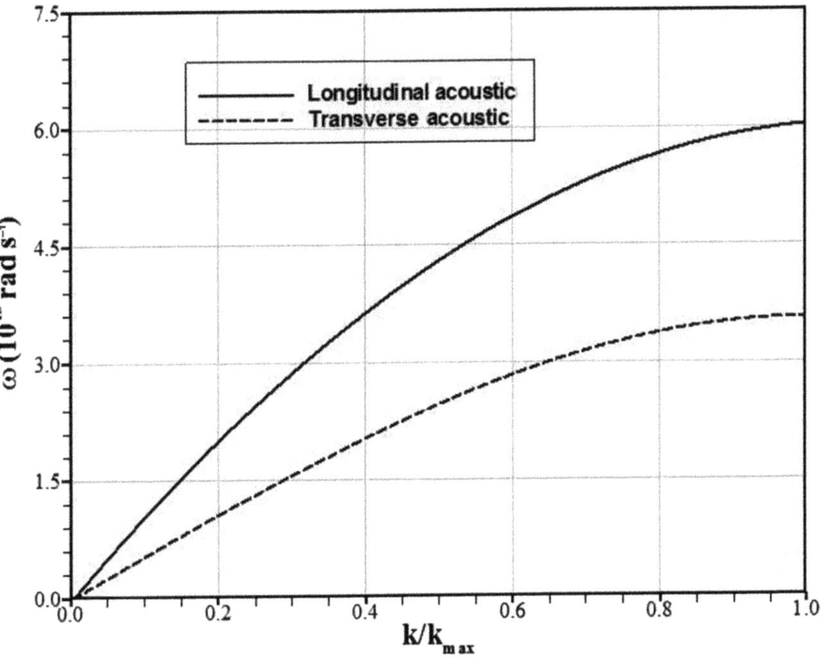

■ **FIGURE 4.18** Dispersion relations of aluminum used in the simulations [37].

temperature $\left(T_e = \frac{2\pi}{C_e v_e}\int_{-1}^{1} I_e d\mu\right)$; k_{max} is the maximum wavenumber; and I_p, v_p, Λ_p, and C_p are the phonon intensity, speed, mean free path, and volumetric specific heat, respectively. These variables are defined for the electron with subscript *e*.

The frequency-dependent, transient phonon radiative transport for lattice subsystem is given by following equations:

4.4.6.3.2 **Longitudinal Acoustic**

$$\frac{1}{v_{p,LA,k}}\frac{\partial I_{p,LA,k}}{\partial t} + \mu\frac{\partial I_{p,LA,k}}{\partial x} = \frac{\frac{1}{2}\int_{-1}^{1} I_{p,LA,k}d\mu - I_{p,LA,k}}{\Lambda_{p,LA,k}} - G_k \tag{4.195}$$

4.4.6.3.3 **Transverse Acoustic**

$$\frac{1}{v_{p,TA,k}}\frac{\partial I_{p,TA,k}}{\partial t} + \mu\frac{\partial I_{p,TA,k}}{\partial x} = \frac{\frac{1}{2}\int_{-1}^{1} I_{p,TA,k}d\mu - I_{p,TA,k}}{\Lambda_{p,TA,k}} - G_k \tag{4.196}$$

For the frequency-independent case the equation for lattice subsystem becomes

$$\frac{1}{v_p}\frac{\partial I_p}{\partial t} + \mu\frac{\partial I_p}{\partial x} = \frac{\frac{1}{2}\int_{-1}^{1} I_p d\mu - I_p}{\Lambda_p} - \frac{G}{4\pi}(T_p - T_e) \tag{4.197}$$

4.4.6.3.4 **EPRT for Electron Subsystem**

The proposed equation for the energy transport in the electron subsystem should satisfy the conservation of energy, and it can be written as

$$\frac{1}{v_e}\frac{\partial I_e}{\partial t} + \mu\frac{\partial I_e}{\partial x} = \frac{\frac{1}{2}\int_{-1}^{1} I_e d\mu - I_e}{\Lambda_e} - \frac{G}{4\pi}(T_e - T_p) \tag{4.198}$$

4.4.6.3.5 **Definition of Temperature**

Phonon temperature for the frequency-dependent case is defined as

$$T_p(x,t) = \frac{2\pi\int_0^{k_{max}}\int_{-1}^{1}(2I_{p,TA,k} + I_{p,LA,k})d\mu dk}{\int_0^{k_{max}}(2C_{p,TA,k}v_{p,TA,k} + C_{p,LA,k}v_{p,LA,k})dk} \tag{4.199}$$

For the frequency-independent case, Eq. (4.199) can be given as

$$T_p(x,t) = \frac{2\pi\int_{-1}^{1} I\, d\mu}{C_p v_p} \tag{4.200}$$

Electron temperature can be defined as

$$T_e(x,t) = \frac{2\pi}{C_e v_e}\int_{-1}^{1} I_e\, d\mu \tag{4.201}$$

4.4.6.3.6 **Heat Flux**

Heat flux for the phonon frequency-dependent case is defined as

$$q''_p(x,t) = 2\pi\int_0^{k_{max}}\int_{-1}^{1}(2I_{p,TA,k} + I_{p,LA,k})\mu d\mu dk \tag{4.202}$$

For the frequency-independent case, Eq. (4.202) simplifies to the following:

$$q''_p(x,t) = 2\pi\int_{-1}^{1} I_p \mu d\mu \tag{4.203}$$

Table 4.6 Properties of Aluminum Used in Numerical Simulations

	Aluminum, Phonons	Aluminum, Electrons
Thermal conductivity ($Wm^{-1}K^{-1}$)	$11.3 = 0.05 \times k_e$	225.7
Volumetric specific heat ($Jm^{-3}K^{-1}$)	2,294,396.6	23,659
Speed (ms^{-1})	6420	2.03×10^6
Mean free path $\Lambda = 3k/Cv$, (m)	2.3×10^{-9}	14.1×10^{-9}
Relaxation time $\tau = \Lambda/v$, (s)	3.58×10^{-13}	6.945×10^{-15}
Electron–Phonon coupling constant in aluminum ($Wm^{-3}K^{-1}$)		2.4642×10^{17}

Electron heat flux can be defined as

$$q''_e(x,t) = 2\pi \int_{-1}^{1} I_e \mu d\mu \tag{4.204}$$

The properties of electron and lattice subsystems are given in Table 4.6 for demonstration purposes.

4.4.6.4 *Further Analysis of Phonon Radiative Transport in Metallic Thin Films*

EPRT for the lattice and electron subsystems can be reduce to the standard two-equation model. Consider EPRTs for the frequency-independent lattice subsystem and electron subsystem; i.e., Eqs. (4.197) and (4.198).

$$\frac{1}{v_p}\frac{\partial I_p}{\partial t} + \mu \frac{\partial I_p}{\partial x} = \frac{\frac{1}{2}\int_{-1}^{1} I_p d\mu - I_p}{\Lambda_p} - \frac{G}{4\pi}(T_p - T_e) \tag{4.205}$$

$$\frac{1}{v_e}\frac{\partial I_e}{\partial t} + \mu \frac{\partial I_e}{\partial x} = \frac{\frac{1}{2}\int_{-1}^{1} I_e d\mu - I_e}{\Lambda_e} - \frac{G}{4\pi}(T_e - T_p) \tag{4.206}$$

Eqs. (4.205) and (4.206) satisfy the law of conservation of energy, which can be verified as follows. Multiplying Eqs. (4.205) and (4.206) throughout by $2\pi\, d\mu$ and then integrating from -1 to 1, yields

$$\frac{2\pi}{v_p}\int_{-1}^{1}\frac{\partial I_p}{\partial t}d\mu + 2\pi\int_{-1}^{1}\mu\frac{\partial I_p}{\partial x}d\mu = \frac{2\pi\left(\int_{-1}^{1}d\mu\right)\left(\frac{1}{2}\int_{-1}^{1}I_p d\mu\right) - 2\pi\int_{-1}^{1}I_p d\mu}{\Lambda_p} - 2\pi\frac{G}{4\pi}(T_p - T_e)\left(\int_{-1}^{1}d\mu\right) \tag{4.207}$$

$$\frac{2\pi}{v_e}\int_{-1}^{1}\frac{\partial I_e}{\partial t}d\mu + 2\pi\int_{-1}^{1}\mu\frac{\partial I_e}{\partial x}d\mu = \frac{2\pi\left(\int_{-1}^{1}d\mu\right)\left(\frac{1}{2}\int_{-1}^{1}I_e d\mu\right) - 2\pi\int_{-1}^{1}I_e d\mu}{\Lambda_p} - 2\pi\frac{G}{4\pi}(T_e - T_p)\left(\int_{-1}^{1}d\mu\right) \tag{4.208}$$

The phonon/electron heat flux is defined as $q''_{p/e} = 2\pi\int_{-1}^{1}\mu I_{p/e}d\mu$ (these definitions are to be used in the second term on the left-hand side of Eqs. (4.207) and (4.208)). Now, Eqs. (4.207) and (4.208) are simplified into the following:

$$C_p\frac{\partial T_p}{\partial t} + \frac{\partial q''_p}{\partial x} = \frac{2\pi\int_{-1}^{1}I_p d\mu - 2\pi\int_{-1}^{1}I_p d\mu}{\Lambda_p} - G(T_p - T_e) \tag{4.209}$$

and

$$C_e\frac{\partial T_e}{\partial t} + \frac{\partial q''_e}{\partial x} = \frac{2\pi\int_{-1}^{1}I_e d\mu - 2\pi\int_{-1}^{1}I_e d\mu}{\Lambda_e} - G(T_e - T_p) \tag{4.210}$$

or finally,

$$C_p\frac{\partial T_p}{\partial t} + \frac{\partial q''_p}{\partial x} = -G(T_p - T_e) \tag{4.211}$$

and

$$C_e\frac{\partial T_e}{\partial t} + \frac{\partial q''_e}{\partial x} = -G(T_e - T_p) \tag{4.212}$$

However, Eqs. (4.211) and (4.212) represent the modified two-equation model, which satisfies the conservation of energy equations [7]. Therefore, Eqs. (4.207) and (4.208) intrinsically satisfy the principle of conservation of energy.

In the limit when the Knudsen number $\mathrm{Kn} = \Lambda/L \rightarrow 0$ (where Λ is the mean free path and L is the film thickness) [39], Eqs. (4.207) and (4.208) should reduce to the two-equation model. This can be verified as follows.

After multiplying Eqs. (4.207) and (4.208) throughout by $2\pi\,\mu\,d\mu$ and then integrating from -1 to 1, it yields

$$\frac{2\pi}{v_p}\int_{-1}^{1}\mu\frac{\partial I_p}{\partial t}d\mu + 2\pi\int_{-1}^{1}\mu^2\frac{\partial I_p}{\partial x}d\mu = \frac{2\pi\left(\int_{-1}^{1}\mu d\mu\right)\left(\frac{1}{2}\int_{-1}^{1}I_p d\mu\right) - 2\pi\int_{-1}^{1}\mu I_p d\mu}{\Lambda_p} - 2\pi\frac{G}{2}\left(\frac{1}{C_p v_p}\int_{-1}^{1}I_p d\mu - \frac{1}{C_e v_e}\int_{-1}^{1}I_e d\mu\right)\left(\int_{-1}^{1}\mu d\mu\right) \tag{4.213}$$

$$\frac{2\pi}{v_e}\int_{-1}^{1}\mu\frac{\partial I_e}{\partial t}d\mu + 2\pi\int_{-1}^{1}\mu^2\frac{\partial I_e}{\partial x}d\mu = \frac{2\pi\left(\int_{-1}^{1}\mu d\mu\right)\left(\frac{1}{2}\int_{-1}^{1}I_e d\mu\right) - 2\pi\int_{-1}^{1}\mu I_e d\mu}{\Lambda_e} - 2\pi\frac{G}{2}\left(\frac{1}{C_e v_e}\int_{-1}^{1}I_e d\mu - \frac{1}{C_p v_p}\int_{-1}^{1}I_p d\mu\right)\left(\int_{-1}^{1}\mu d\mu\right) \tag{4.214}$$

To simplify Eqs. (4.213) and (4.214), we should use the definitions introduced earlier. The second term on the left-hand side needs some elaboration. In the diffusive limit, the behavior of $I_{e,p}(x,\mu,t)$ is mainly linear in μ. This means that we may neglect the second and higher order terms in the Taylor series expansion of $I_{e,p}(x,\mu,t)$ in μ.

$$\begin{aligned} I_{e,p}(x,\mu,t) &= I_{e,p}(x,0,t) + \mu\left.\frac{\partial I_{e,p}}{\partial \mu}\right|_{\mu=0} + O(\mu) \\ &\approx I^o_{e,p}(x,t) + \mu\left.\frac{\partial I_{e,p}}{\partial \mu}\right|_{\mu=0} \end{aligned} \tag{4.215}$$

This is because,

$$I^o_{e,p}(x,t) = \frac{1}{2}\int_{-1}^{1}I_{e,p}(x,\mu,t)d\mu \approx I_{e,p}(x,0,t) + \frac{1}{2}\left(\int_{-1}^{1}\mu d\mu\right)\left.\frac{\partial I_{e,p}}{\partial \mu}\right|_{\mu=0} = I_{e,p}(x,0,t)$$

Then,

$$\begin{aligned} 2\pi\int_{-1}^{1}\mu^2\frac{\partial I_{e,p}}{\partial x}d\mu &= 2\pi\int_{-1}^{1}\mu^2\frac{\partial}{\partial x}\left(I^o_{e,p}(x,t) + \mu\left.\frac{\partial I_{e,p}}{\partial \mu}\right|_{\mu=0}\right)d\mu \\ &= 2\pi\int_{-1}^{1}\mu^2\frac{\partial I^o_{e,p}(x,t)}{\partial x}d\mu + 2\pi\frac{\partial}{\partial x}\left(\left.\frac{\partial I_{e,p}}{\partial \mu}\right|_{\mu=0}\right)\int_{-1}^{1}\mu^3 d\mu \\ &= 2\pi\frac{2}{3}\frac{\partial I^o_{e,p}(x,t)}{\partial x} \end{aligned} \tag{4.216}$$

Since, $\frac{C_{e,p}v_{e,p}T_{e,p}}{4\pi} = I^o_{e,p}(x,t) = \frac{1}{2}\int_{-1}^{1} I_{e,p}(x,\mu,t)d\mu$, so that finally,

$$2\pi\int_{-1}^{1}\mu^2\frac{\partial I_{e,p}}{\partial x}d\mu = 2\pi\frac{2}{3}\frac{dI^o_{e,p}}{dT_{e,p}}\frac{\partial T_{e,p}}{\partial x} \tag{4.217}$$

Eqs. (4.213) and (4.214) are now simplified to

$$\frac{1}{v_p}\frac{\partial q''_p}{\partial t} + \frac{4\pi}{3}\frac{dI^o_p}{dT_p}\frac{\partial T_p}{\partial x} = \frac{0 - q''_p}{\Lambda_p} - 0 \tag{4.218}$$

and

$$\frac{1}{v_e}\frac{\partial q''_e}{\partial t} + \frac{4\pi}{3}\frac{dI^o_e}{dT_e}\frac{\partial T_e}{\partial x} = \frac{0 - q''_e}{\Lambda_e} - 0 \tag{4.219}$$

or,

$$\frac{\Lambda_p}{v_p}\frac{\partial q''_p}{\partial t} + q''_p = -\frac{C_p v_p \Lambda_p}{3}\frac{\partial T_p}{\partial x} \tag{4.220}$$

and

$$\frac{\Lambda_e}{v_e}\frac{\partial q''_e}{\partial t} + q''_e = -\frac{C_e v_e \Lambda_e}{3}\frac{\partial T_e}{\partial x} \tag{4.221}$$

The kinetic theory formula for the thermal conductivity is given as

$$k_p = \frac{C_p v_p \Lambda_p}{3} \quad k_e = \frac{C_e v_e \Lambda_e}{3} \tag{4.222}$$

Hence, Eqs. (4.220) and (4.221) can be transformed into

$$\frac{\Lambda_p}{v_p}\frac{\partial q''_p}{\partial t} + q''_p = -k_p\frac{\partial T_p}{\partial x} \tag{4.223}$$

and

$$\frac{\Lambda_e}{v_e}\frac{\partial q''_e}{\partial t} + q''_e = -k_e\frac{\partial T_e}{\partial x} \tag{4.224}$$

The law of energy conservation is given by Eqs. (4.211) and (4.212). To eliminate q''_p and q''_p between Eqs. (4.211) and (4.223) as well as (4.212) and (4.224), we can proceed as follows. Differentiating Eqs. (4.223) and (4.224) once with respect to x gives

$$\frac{\Lambda_p}{v_p}\frac{\partial}{\partial t}\left(\frac{\partial q''_p}{\partial x}\right) + \frac{\partial q''_p}{\partial x} = -k_p\frac{\partial^2 T_p}{\partial x^2} \tag{4.225}$$

and

$$\frac{\Lambda_e}{v_e}\frac{\partial}{\partial t}\left(\frac{\partial q''_e}{\partial x}\right) + \frac{\partial q''_e}{\partial x} = -k_e\frac{\partial^2 T_e}{\partial x^2} \tag{4.226}$$

Substitute for $\partial q''_{\mathrm{p}}/\partial x$ & $\partial q''_{\mathrm{e}}/\partial x$ from Eqs. (4.211) and (4.212) into Eqs. (4.225) and (4.226), respectively, and simplify to obtain

$$\frac{\Lambda_{\mathrm{p}}}{v_{\mathrm{p}}}\frac{\partial}{\partial t}\left[C_{\mathrm{p}}\frac{\partial T_{\mathrm{p}}}{\partial t}+G(T_{\mathrm{p}}-T_{\mathrm{e}})\right]+C_{\mathrm{p}}\frac{\partial T_{\mathrm{p}}}{\partial t}=k_{\mathrm{p}}\frac{\partial^2 T_{\mathrm{p}}}{\partial x^2}-G(T_{\mathrm{p}}-T_{\mathrm{e}}) \tag{4.227}$$

and

$$\frac{\Lambda_{\mathrm{e}}}{v_{\mathrm{e}}}\frac{\partial}{\partial t}\left[C_{\mathrm{e}}\frac{\partial T_{\mathrm{e}}}{\partial t}+G(T_{\mathrm{e}}-T_{\mathrm{p}})\right]+C_{\mathrm{e}}\frac{\partial T_{\mathrm{e}}}{\partial t}=k_{\mathrm{e}}\frac{\partial^2 T_{\mathrm{e}}}{\partial x^2}-G(T_{\mathrm{e}}-T_{\mathrm{p}}) \tag{4.228}$$

Noting that $\tau_{\mathrm{p}}=\Lambda_{\mathrm{p}}/v_{\mathrm{p}}$ & $\tau_{\mathrm{e}}=\Lambda_{\mathrm{e}}/v_{\mathrm{e}}$ are the relaxation times of the phonons and electrons, respectively, these equations may be written as

$$\tau_{\mathrm{p}}\frac{\partial}{\partial t}\left[C_{\mathrm{p}}\frac{\partial T_{\mathrm{p}}}{\partial t}+G(T_{\mathrm{p}}-T_{\mathrm{e}})\right]+C_{\mathrm{p}}\frac{\partial T_{\mathrm{p}}}{\partial t}=k_{\mathrm{p}}\frac{\partial^2 T_{\mathrm{p}}}{\partial x^2}-G(T_{\mathrm{p}}-T_{\mathrm{e}}) \tag{4.229}$$

and

$$\tau_{\mathrm{e}}\frac{\partial}{\partial t}\left[C_{\mathrm{e}}\frac{\partial T_{\mathrm{e}}}{\partial t}+G(T_{\mathrm{e}}-T_{\mathrm{p}})\right]+C_{\mathrm{e}}\frac{\partial T_{\mathrm{e}}}{\partial t}=k_{\mathrm{e}}\frac{\partial^2 T_{\mathrm{e}}}{\partial x^2}-G(T_{\mathrm{e}}-T_{\mathrm{p}}) \tag{4.230}$$

or,

$$\tau_{\mathrm{p}}C_{\mathrm{p}}\frac{\partial^2 T_{\mathrm{p}}}{\partial t^2}+(C_{\mathrm{p}}+\tau_{\mathrm{p}}G)\frac{\partial T_{\mathrm{p}}}{\partial t}-\tau_{\mathrm{p}}G\frac{\partial T_{\mathrm{e}}}{\partial t}=k_{\mathrm{p}}\frac{\partial^2 T_{\mathrm{p}}}{\partial x^2}-G(T_{\mathrm{p}}-T_{\mathrm{e}}) \tag{4.231}$$

and

$$\tau_{\mathrm{e}}C_{\mathrm{e}}\frac{\partial^2 T_{\mathrm{e}}}{\partial t^2}+(C_{\mathrm{e}}+\tau_{\mathrm{e}}G)\frac{\partial T_{\mathrm{e}}}{\partial t}-\tau_{\mathrm{e}}G\frac{\partial T_{\mathrm{p}}}{\partial t}=k_{\mathrm{e}}\frac{\partial^2 T_{\mathrm{e}}}{\partial x^2}-G(T_{\mathrm{e}}-T_{\mathrm{p}}) \tag{4.232}$$

It should be noted that when the overall time scale of the heat transfer process is much larger than the relaxation time, Eqs. (4.229) and (4.230) or (4.231) and (4.232) can be further simplified to

$$C_{\mathrm{p}}\frac{\partial T_{\mathrm{p}}}{\partial t}=k_{\mathrm{p}}\frac{\partial^2 T_{\mathrm{p}}}{\partial x^2}-G(T_{\mathrm{p}}-T_{\mathrm{e}}) \tag{4.233}$$

$$C_{\mathrm{e}}\frac{\partial T_{\mathrm{e}}}{\partial t}=k_{\mathrm{e}}\frac{\partial^2 T_{\mathrm{e}}}{\partial x^2}-G(T_{\mathrm{e}}-T_{\mathrm{p}}) \tag{4.234}$$

Eqs. (4.233) and (4.234) constitute the familiar two-equation model [9].

The further verifications of reduced equation of phonon transport to two-equation model derivation is possible. In this case, the simplification of Eqs. (4.213) and (4.214) yields the following EPRT in the phonon and electron subsystems. These can be rewritten as

$$\frac{1}{v_{\mathrm{p}}}\frac{\partial I_{\mathrm{p}}}{\partial t}+\mu\frac{\partial I_{\mathrm{p}}}{\partial x}=\frac{\frac{1}{2}\int_{-1}^{1}I_{\mathrm{p}}d\mu-I_{\mathrm{p}}}{\Lambda_{\mathrm{p}}}+\frac{3G}{2\Lambda_{\mathrm{p}}}\mu\int_{o}^{x}\int_{-1}^{1}\left(\frac{I_{\mathrm{p}}}{C_{\mathrm{p}}v_{\mathrm{p}}}-\frac{I_{\mathrm{e}}}{C_{\mathrm{e}}v_{\mathrm{e}}}\right)d\mu dy \tag{4.235}$$

EPRT for the electron subsystem can be written as

$$\frac{1}{v_e}\frac{\partial I_e}{\partial t}+\mu\frac{\partial I_e}{\partial x}=\frac{\frac{1}{2}\int_{-1}^{1}I_e d\mu-I_e}{\Lambda_e}+\frac{3G}{2\Lambda_e}\mu\int_o^x\int_{-1}^{1}\left(\frac{I_e}{C_e v_e}-\frac{I_p}{C_p v_p}\right)d\mu dy \tag{4.236}$$

It is important to demonstrate that the EPRTs (Eqs. 4.235 and 4.236) can be reduced to the two-equation formulation under the specified conditions. In this case, the first verification can be introduced under the followings steps.

Multiply Eq. (4.235) throughout by $2\pi\,\mu\,d\mu$ and then integrate from -1 to 1. This yields

$$\begin{aligned}\frac{2\pi}{v_p}\int_{-1}^{1}\mu\frac{\partial I_p}{\partial t}d\mu+2\pi\int_{-1}^{1}\mu^2\frac{\partial I_p}{\partial x}d\mu\\=\frac{2\pi\left(\int_{-1}^{1}\mu d\mu\right)\left(\frac{1}{2}\int_{-1}^{1}I_p d\mu\right)-2\pi\int_{-1}^{1}\mu I_p d\mu}{\Lambda_p}\\+\frac{3\pi G}{\Lambda_p}\left(\int_{-1}^{1}\mu^2 d\mu\right)\left[\int_o^x\int_{-1}^{1}\left(\frac{I_p}{C_p v_p}-\frac{I_e}{C_e v_e}\right)d\mu dy\right]\end{aligned} \tag{4.237}$$

To further simplify Eq. (4.237), note the following points:

- The phonon heat flux is defined as $q_p''=2\pi\int_{-1}^{1}\mu I_p d\mu$. (This definition is to be used in the first term on the right-hand side of Eq. 4.237.)
- The phonon and electron temperatures are defined as $C_p v_p T_p/4\pi=\frac{1}{2}\int_{-1}^{1}I_p d\mu$ and $C_e v_e T_e/4\pi=\frac{1}{2}\int_{-1}^{1}I_e d\mu$. (These definitions are to be used in the second term on the right-hand side of Eq. 4.237.)
- In the diffusive limit, $\partial I_p/\partial x\approx\partial I_p^o/\partial x$, where I_p^o is the equilibrium intensity at temperature T_p and is defined as $I_p^o=C_p v_p T_p/4\pi$. (This approximation is to be used in the first term on the left-hand side of Eq. 4.237.)

Using these definitions and approximation, the EPRT for the phonon subsystem (Eq. 4.237) is simplified as

$$\frac{1}{v_p}\frac{\partial q_p''}{\partial t}+2\pi\frac{\partial I_p^o}{\partial T_p}\frac{\partial T_p}{\partial x}\int_{-1}^{1}\mu^2 d\mu=\frac{0-q_p''}{\Lambda_p}+\frac{G}{\Lambda_p}\int_o^x(T_p-T_e)dy \tag{4.238}$$

or

$$\frac{1}{v_p}\frac{\partial q_p''}{\partial t}+\frac{C_p v_p}{3}\frac{\partial T_p}{\partial x}=\frac{0-q_p''}{\Lambda_p}+\frac{G}{\Lambda_p}\int_o^x(T_p-T_e)dy \tag{4.239}$$

or

$$\frac{\Lambda_p}{v_p}\frac{\partial q_p''}{\partial t} + q_p'' = -\frac{C_p v_p \Lambda_p}{3}\frac{\partial T_p}{\partial x} + G\int_o^x (T_p - T_e)dy \tag{4.240}$$

Using the thermal conductivity formula given in Eq. (4.222) transforms Eq. (4.240) into

$$\frac{\Lambda_p}{v_p}\frac{\partial q_p''}{\partial t} + q_p'' = -k_p\frac{\partial T_p}{\partial x} + G\int_o^x (T_p - T_e)dy \tag{4.241}$$

The energy conservation equation is written as

$$C_p\frac{\partial T_p}{\partial t} + \frac{\partial q_p''}{\partial x} = 0 \tag{4.242}$$

To eliminate q_p'' between Eqs. (4.241) and (4.242), we can proceed as follows. Differentiate Eq. (4.242) once with respect to time (t). This gives

$$C_p\frac{\partial^2 T_p}{\partial t^2} + \frac{\partial}{\partial x}\left(\frac{\partial q_p''}{\partial t}\right) = 0 \tag{4.243}$$

Substitute for $\partial q_p''/\partial t$ from Eq. (4.241) into Eq. (4.243) to yield

$$\frac{C_p\Lambda_p}{v_p}\frac{\partial^2 T_p}{\partial t^2} - \frac{\partial q_p''}{\partial x} - k_p\frac{\partial^2 T_p}{\partial x^2} + G(T_p - T_e) = 0 \tag{4.244}$$

Substituting for $\partial q_p''/\partial x$ from Eq. (4.242) into Eq. (4.244) yields:

$$\frac{C_p\Lambda_p}{v_p}\frac{\partial^2 T_p}{\partial t^2} + C_p\frac{\partial T_p}{\partial t} = k_p\frac{\partial^2 T_p}{\partial x^2} - G(T_p - T_e) \tag{4.245}$$

Dividing this equation throughout by C_p, noting that $\alpha_p = k_p/C_p$ is the thermal diffusivity and $\tau_p = \Lambda_p/v_p$ is the relaxation time of the phonons, yields

$$\tau_p\frac{\partial^2 T_p}{\partial t^2} + \frac{\partial T_p}{\partial t} = \alpha_p\frac{\partial^2 T_p}{\partial x^2} - \frac{G}{C_p}(T_p - T_e) \tag{4.246}$$

A similar procedure with the electron EPRT yields

$$\tau_e\frac{\partial^2 T_e}{\partial t^2} + \frac{\partial T_e}{\partial t} = \alpha_e\frac{\partial^2 T_e}{\partial x^2} - \frac{G}{C_e}(T_e - T_p) \tag{4.247}$$

Eqs. (4.246) and (4.247) are the modified two-equation model, which are reduced from the EPRTs for phonon and electron subsystems (Eqs. 4.235 and 4.236). When the relaxation time is very small or when the overall time scale of the heat transfer is much larger than the relaxation time, Eqs. (4.246) and (4.247) may be further simplified into

$$C_p\frac{\partial T_p}{\partial t} = k_p\frac{\partial^2 T_p}{\partial x^2} - G(T_p - T_e) \tag{4.248}$$

and

$$C_e\frac{\partial T_e}{\partial t}=k_e\frac{\partial^2 T_e}{\partial x^2}-G(T_e-T_p) \tag{4.249}$$

Eqs. (4.248) and (4.249) are the familiar standard two-equation model. Eq. (4.235) may also be written implicitly as

$$\frac{1}{v_p}\frac{\partial I_p}{\partial t}+\mu\frac{\partial I_p}{\partial x}=\frac{\frac{1}{2}\int_{-1}^{1}I_p d\mu-I_p}{\Lambda_p}+\frac{3G}{4\pi\Lambda_p}\mu\int_o^x(T_p-T_e)dy \tag{4.250}$$

where $C_p v_p T_p/4\pi=\frac{1}{2}\int_{-1}^{1}I_p d\mu$ and $T_p/2\pi=\int_{-1}^{1}I_p/C_p v_p d\mu$

Eq. (4.236) is similarly written as

$$\frac{1}{v_e}\frac{\partial I_e}{\partial t}+\mu\frac{\partial I_e}{\partial x}=\frac{\frac{1}{2}\int_{-1}^{1}I_e d\mu-I_e}{\Lambda_e}+\frac{3G}{4\pi\Lambda_e}\mu\int_o^x(T_e-T_p)dy \tag{4.251}$$

where $C_e v_e T_e/4\pi=\frac{1}{2}\int_{-1}^{1}I_e d\mu$ and $T_e/2\pi=\int_{-1}^{1}I_e/C_e v_e d\mu$

The second verification can follow these steps:

Multiply Eq. (4.235) throughout by $2\pi\,d\mu$ and then integrate from -1 to 1. This yields

$$\begin{aligned}&\frac{2\pi}{v_p}\int_{-1}^{1}\frac{\partial I_p}{\partial t}d\mu+2\pi\int_{-1}^{1}\mu\frac{\partial I_p}{\partial x}d\mu\\&=\frac{2\pi\left(\int_{-1}^{1}d\mu\right)\left(\frac{1}{2}\int_{-1}^{1}I_p d\mu\right)-2\pi\int_{-1}^{1}I_p d\mu}{\Lambda_p}+\\&+\frac{6\pi^2 G}{\Lambda_p}\left(\int_{-1}^{1}\mu d\mu\right)\left[\int_o^x\int_{-1}^{1}\left(\frac{I_p}{C_p v_p}-\frac{I_e}{C_e v_e}\right)d\mu dy\right]\end{aligned} \tag{4.252}$$

Using the definitions introduced earlier, Eq. (4.252) is simplified to

$$C_p\frac{\partial T_p}{\partial t}+\frac{\partial q''_p}{\partial x}=\frac{2\pi\int_{-1}^{1}I_p d\mu-2\pi\int_{-1}^{1}I_p d\mu}{\Lambda_p}+0 \tag{4.253}$$

which is the same as Eq. (4.242) ($C_p\frac{\partial T_p}{\partial t}+\frac{\partial q''_p}{\partial x}=0$)

Hence, the EPRT intrinsically satisfies the principle of conservation of energy.

4.4.6.4.1 The Steady-State Case

The explicit and implicit forms of the phonon and electron EPRTs for the steady-state case are given as follows.

For the phonon subsystem:

$$\mu\frac{\partial I_p}{\partial x} = \frac{\frac{1}{2}\int_{-1}^{1} I_p d\mu - I_p}{\Lambda_p} + \frac{3G}{2\Lambda_p}\mu\int_o^x\int_{-1}^{1}\left(\frac{I_p}{C_p v_p} - \frac{I_e}{C_e v_e}\right)d\mu dy \quad (4.254)$$

or

$$\mu\frac{\partial I_p}{\partial x} = \frac{\frac{1}{2}\int_{-1}^{1} I_p d\mu - I_p}{\Lambda_p} + \frac{3G}{4\pi\Lambda_p}\mu\int_o^x (T_p - T_e)dy \quad (4.255)$$

For the electron subsystem:

$$\mu\frac{\partial I_e}{\partial x} = \frac{\frac{1}{2}\int_{-1}^{1} I_e d\mu - I_e}{\Lambda_e} + \frac{3G}{2\Lambda_e}\mu\int_o^x\int_{-1}^{1}\left(\frac{I_e}{C_e v_e} - \frac{I_p}{C_p v_p}\right)d\mu dy \quad (4.256)$$

or

$$\mu\frac{\partial I_e}{\partial x} = \frac{\frac{1}{2}\int_{-1}^{1} I_e d\mu - I_e}{\Lambda_e} + \frac{3G}{4\pi\Lambda_e}\mu\int_o^x (T_e - T_p)dy \quad (4.257)$$

In the implicit forms of the equations, $C_p v_p T_p/4\pi = \frac{1}{2}\int_{-1}^{1} I_p d\mu$ & $C_e v_e T_e/4\pi = \frac{1}{2}\int_{-1}^{1} I_e d\mu$

4.4.7 **Casimir Limit**

When the film thickness reduces to the mean free path of the film material, phonon radiative transport replaces the radiative transport in line with the Casimir limit.

For the numerical solution of the one-dimensional EPRT, the phonon intensities are broken into two parts. For $0<\mu<1$, the intensity denoted as I^+ and for $-1<\mu<0$, it is denoted by I^-. I^+ is corresponds to those phonons that are traveling from the left face toward the right face, and I^- corresponds to those phonons that are traveling from the right face toward the left face of the film, as shown in Fig. 4.19. The separate EPRT may be written for each intensity.

Consider, the one-dimensional, frequency-independent, steady-state EPRT can be written as

$$\mu\frac{\partial I^+}{\partial x} = \frac{\frac{1}{2}\int_0^1 I^+ d\mu - I^+}{\Lambda} \qquad \mu\frac{\partial I^-}{\partial x} = \frac{\frac{1}{2}\int_{-1}^{0} I^- d\mu - I^-}{\Lambda} \quad (4.258)$$

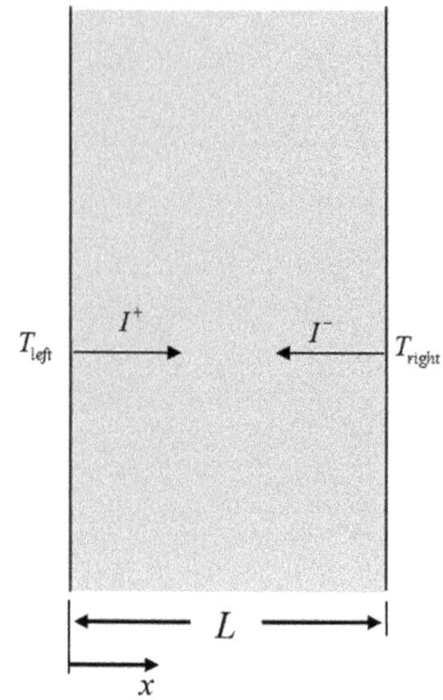

■ **FIGURE 4.19** Schematic of one-dimensional thin films.

The limit when the film thickness L approaches zero may be modeled in this equation by letting the mean free path approach infinity. This yields

$$\mu\frac{\partial I^+}{\partial x}=0\Rightarrow I^+=I^+(\mu)\quad \mu\frac{\partial I^-}{\partial x}=0\Rightarrow I^-=I^-(\mu) \tag{4.259}$$

This means that the intensity is not attenuated between the two surfaces. If it is further assumed that the surfaces are diffuse then we obtain

$$I^+=\text{constant},\quad I^-=\text{constant} \tag{4.260}$$

This relation indicates that whatever the magnitude of the intensity at the boundary, this magnitude prevails throughout the thin film. In this case, the heat flux expression is simplified as

$$\begin{aligned} q'' &= 2\pi\int_{-1}^{1}\mu I d\mu \\ &= 2\pi\left[\int_0^1 \mu I^+ d\mu + \int_{-1}^{0}\mu I^- d\mu\right] \\ &= \pi(I^+ - I^-) \end{aligned} \tag{4.261}$$

For high temperatures, we can obtain

$$q''=\frac{Cv}{4}(T_{\text{left}}-T_{\text{right}}) \tag{4.262}$$

For low temperatures, we can get

$$q'' = \sigma(T_1^4 - T_2^4) \tag{4.263}$$

Eq. (4.263) becomes exactly the Stefan–Boltzmann Law of radiation exchange. This proves the Casimir limit reported in the early study [23].

4.4.8 **Large Film Thickness**

On the other extreme, when the film thickness becomes much more than the phonon mean free path, the EPRT approaches the hyperbolic heat conduction equation. To confirm this assertion we multiply the one-dimensional frequency-independent EPRT (Eq. 4.163) by $2\pi\mu$ and integrate from $-1 < \mu < 1$ to yield

$$\frac{1}{v}\frac{\partial}{\partial t}2\pi\int_{-1}^{1}\mu I d\mu + 2\pi\int_{-1}^{1}\mu^2\frac{\partial I}{\partial x}d\mu = \frac{\left(\frac{1}{2}\int_{-1}^{1} I d\mu\right)2\pi\int_{-1}^{1}\mu d\mu - 2\pi\int_{-1}^{1}\mu I d\mu}{\Lambda} \tag{4.264}$$

The heat flux is defined as $q'' = 2\pi\int_{-1}^{1}\mu I d\mu$. Also for large film thickness, there is considerable phonon scattering and therefore the thermodynamic temperature can be defined anywhere in the film under local thermodynamic equilibrium conditions [40]. In this case the actual intensity $I = I(x, \mu, t)$ at any point in the film is very nearly equal to the equilibrium intensity $I^o = I^o(T)$ at that point. The temperature is a function of the position and time; i.e., $T = T(x, t)$. Therefore, we may write $I \approx I^o$ and this implies that $\frac{\partial I}{\partial x} \approx \frac{\partial I^o}{\partial x} = \frac{dI^o}{dT}\frac{\partial T}{\partial x}$. Also from [41], we may write $\frac{dI^o}{dT} = \frac{Cv}{4\pi}$. These considerations simplify this equation as follows:

$$\frac{1}{v}\frac{\partial q''}{\partial t} + \frac{4\pi}{3}\frac{dI^o}{dT}\frac{\partial T}{\partial x} = \frac{0 - q''}{\Lambda} \tag{4.265}$$

$$\tau\frac{\partial q''}{\partial t} + q'' = -\frac{Cv\Lambda}{3}\frac{\partial T}{\partial x} \tag{4.266}$$

$$\tau\frac{\partial q''}{\partial t} + q'' = -k\frac{\partial T}{\partial x} \tag{4.267}$$

If this equation is now substituted in the energy conservation equation,

$$C\frac{\partial T}{\partial t} + \frac{\partial q''}{\partial x} = 0 \tag{4.268}$$

we obtain the hyperbolic heat conduction equation:

$$\tau \frac{\partial^2 T}{\partial t^2} + \frac{\partial T}{\partial t} = \alpha \frac{\partial^2 T}{\partial x^2} \tag{4.269}$$

4.4.9 Large Film Thickness; Large Characteristic Time

With the exception of ultrafast heat transfer processes such as ultrashort pulse laser heating of metal substrates, the characteristic time in any heat transfer process is usually large compared to the relaxation time τ. In this situation the hyperbolic heat conduction equation may further simplify to the parabolic heat conduction equation:

$$\frac{\partial T}{\partial t} = \alpha \frac{\partial^2 T}{\partial x^2} \tag{4.270}$$

From the preceding derivations it is clear that the EPRT gives the correct description of the phenomenon of thermal transport at extreme length scales as well as at extreme time scales. For film thicknesses much smaller than the phonon mean free path, the EPRT reproduces the Stefan–Boltzmann law of phonon transport. For film thicknesses larger than the phonon free path, the EPRT reproduces the hyperbolic heat conduction equation. For large film thicknesses and large process characteristic times, the EPRT reproduces the parabolic heat conduction equation. Consequently, it is conceivable that the EPRT gives the correct description of thermal transport at all lengths and time scales, especially those comparable to the mean free path of the phonons.

4.4.10 Initial Conditions and Boundary Conditions

4.4.10.1 *Initial Conditions*

In order to perform the simulation, we need to define the initial conditions. The intensities in the whole domain can be initialized corresponding to some initial temperature (T_{initial}). For example, if we need to simulate the two-dimensional dielectric films as shown in Fig. 4.20, the initial condition can be given as

$$\begin{aligned}
I_{\text{LA,k}}(x,z,\theta,\phi,t=0) &= \frac{C_{\text{LA,k}} v_{\text{LA,k}} T_{\text{initial}}}{4\pi} \\
I_{\text{TA,k}}(x,z,\theta,\phi,t=0) &= \frac{C_{\text{TA,k}} v_{\text{TA,k}} T_{\text{initial}}}{4\pi} \qquad 0<x<L_x,\ 0<z<L_z \\
I_{\text{LO,k}}(x,z,\theta,\phi,t=0) &= \frac{C_{\text{LO,k}} v_{\text{LO,k}} T_{\text{initial}}}{4\pi} \qquad 0<\theta<\pi,\ 0<\phi<2\pi \\
I_{\text{TO,k}}(x,z,\theta,\phi,t=0) &= \frac{C_{\text{TO,k}} v_{\text{TO,k}} T_{\text{initial}}}{4\pi}
\end{aligned} \tag{4.271}$$

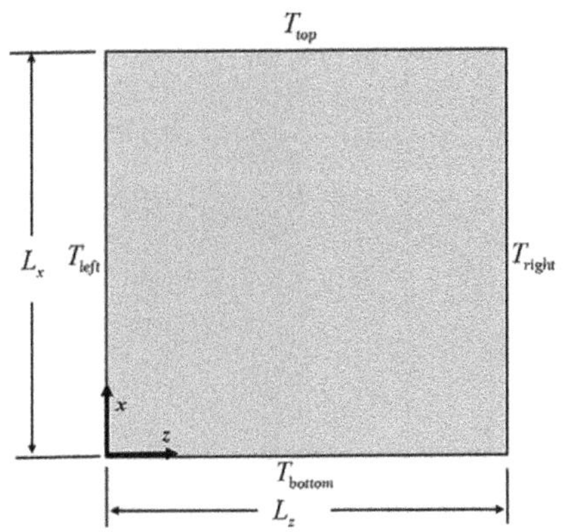

■ **FIGURE 4.20** Schematic of two-dimensional thin films.

4.4.10.2 *Boundary Conditions*

The boundary conditions need to be introduced to simulate the energy transport characteristics. The phonon intensity specification at the boundaries should be prescribed to complete the problem specification. Consider the two-dimensional dielectric films as shown in Fig. 4.20. The boundary can be given as follows.

4.4.10.2.1 **Bottom Boundary**

$$\begin{aligned}
I_{\mathrm{LA,k}}(x=0,z,\theta,\phi,t) &= \frac{C_{\mathrm{LA,k}}v_{\mathrm{LA,k}}T_{\mathrm{bottom}}}{4\pi} \\
I_{\mathrm{TA,k}}(x=0,z,\theta,\phi,t) &= \frac{C_{\mathrm{TA,k}}v_{\mathrm{TA,k}}T_{\mathrm{bottom}}}{4\pi} \qquad t>0,\ 0<z<L_z, \\
I_{\mathrm{LO,k}}(x=0,z,\theta,\phi,t) &= \frac{C_{\mathrm{LO,k}}v_{\mathrm{LO,k}}T_{\mathrm{bottom}}}{4\pi} \qquad 0<\theta<\pi/2,\ 0<\phi<2\pi \\
I_{\mathrm{TO,k}}(x=0,z,\theta,\phi,t) &= \frac{C_{\mathrm{TO,k}}v_{\mathrm{TO,k}}T_{\mathrm{bottom}}}{4\pi}
\end{aligned} \tag{4.272}$$

4.4.10.2.2 **Top Boundary**

$$\begin{aligned}
I_{\mathrm{LA,k}}(x=0,z,\theta,\phi,t) &= \frac{C_{\mathrm{LA,k}}v_{\mathrm{LA,k}}T_{\mathrm{top}}}{4\pi} \\
I_{\mathrm{TA,k}}(x=0,z,\theta,\phi,t) &= \frac{C_{\mathrm{TA,k}}v_{\mathrm{TA,k}}T_{\mathrm{top}}}{4\pi} \qquad t>0,\ 0<z<L_z, \\
I_{\mathrm{LO,k}}(x=0,z,\theta,\phi,t) &= \frac{C_{\mathrm{LO,k}}v_{\mathrm{LO,k}}T_{\mathrm{top}}}{4\pi} \qquad 0<\theta<\pi,\ 0<\phi<\pi \\
I_{\mathrm{TO,k}}(x=0,z,\theta,\phi,t) &= \frac{C_{\mathrm{TO,k}}v_{\mathrm{TO,k}}T_{\mathrm{top}}}{4\pi}
\end{aligned} \tag{4.273}$$

4.4.10.2.3 **Left Boundary**

$$
\left.
\begin{aligned}
I_{\mathrm{LA,k}}(x, z=0, \theta, \phi, t) &= \frac{C_{\mathrm{LA,k}} v_{\mathrm{LA,k}} T_{\mathrm{left}}}{4\pi} \\
I_{\mathrm{TA,k}}(x, z=0, \theta, \phi, t) &= \frac{C_{\mathrm{TA,k}} v_{\mathrm{TA,k}} T_{\mathrm{left}}}{4\pi} \\
I_{\mathrm{LO,k}}(x, z=0, \theta, \phi, t) &= \frac{C_{\mathrm{LO,k}} v_{\mathrm{LO,k}} T_{\mathrm{left}}}{4\pi} \\
I_{\mathrm{TO,k}}(x, z=0, \theta, \phi, t) &= \frac{C_{\mathrm{TO,k}} v_{\mathrm{TO,k}} T_{\mathrm{left}}}{4\pi}
\end{aligned}
\right.
\quad
\begin{aligned}
&0<x<L_x,\ t>0 \\
&\pi/2<\theta<\pi,\ 0<\phi<2\pi
\end{aligned}
\tag{4.274}
$$

4.4.10.2.4 **Right Boundary**

$$
\left.
\begin{aligned}
I_{\mathrm{LA,k}}(x, z=L_z, \theta, \phi, t) &= \frac{C_{\mathrm{LA,k}} v_{\mathrm{LA,k}} T_{\mathrm{right}}}{4\pi} \\
I_{\mathrm{TA,k}}(x, z=L_z, \theta, \phi, t) &= \frac{C_{\mathrm{TA,k}} v_{\mathrm{TA,k}} T_{\mathrm{right}}}{4\pi} \\
I_{\mathrm{LO,k}}(x, z=L_z, \theta, \phi, t) &= \frac{C_{\mathrm{LO,k}} v_{\mathrm{LO,k}} T_{\mathrm{right}}}{4\pi} \\
I_{\mathrm{TO,k}}(x, z=L_z, \theta, \phi, t) &= \frac{C_{\mathrm{TO,k}} v_{\mathrm{TO,k}} T_{\mathrm{right}}}{4\pi}
\end{aligned}
\right.
\quad
\begin{aligned}
&0<x<L_x,\ t>0 \\
&0<\theta<\pi,\ \pi<\phi<2\pi
\end{aligned}
\tag{4.275}
$$

For the case of the electrons, the insulated boundary condition is considered. This is due to the fact that the electrons themselves do not take part in the transfer of thermal energy to, or from, the phonons of a different material. This is true for the case when the bounding surface is exposed to the atmospheric air. When the bounding surface is actually an interface between the metal and the dielectric thin film, then again the electrons do not transfer to, or receive thermal energy from, the phonons of the dielectric material. Hence, we may conclude that the electron subsystem is actually insulated at the boundaries. Let us consider Fig. 4.19, showing the metallic film. In this case, the boundary condition for the electron subsystem can be given as follows.

4.4.10.2.5 **Left Boundary**

$$
q''_{\mathrm{e}}(x=0, t) = 2\pi \int_{-1}^{1} \mu I_{\mathrm{e}}(x=0, \mu, t) d\mu = 0 \tag{4.276}
$$

4.4.10.2.6 **Right Boundary**

$$
q''_{\mathrm{e}}(x=L, t) = 2\pi \int_{-1}^{1} \mu I_{\mathrm{e}}(x=L, \mu, t) d\mu = 0 \tag{4.277}
$$

4.4.11 Interface Conditions

On the other hand, in the case of cross-plane energy transport across the thin films pair, interface conditions need to be defined. In this case, for a dielectric–dielectric interface or a metal–dielectric interface, in general, the temperatures of the two films and the common heat flux in the two films are unknown. If there is no TBR at the interface, then the temperature of the first film is equal to that of the second film. If TBR exits at the interface then there will be a small temperature difference between the two films. Furthermore, whether the TBR exits at the boundary or not, the heat flux should be the same in the two films.

At a metal–dielectric or dielectric–dielectric interface a temperature discontinuity is present, even if there is a heat flux across the interface. This temperature difference is proportional to the heat flux across the interface and is given as

$$q'' = \frac{\Delta T}{R} \tag{4.278}$$

In this expression, R stands for TBR and is given in units of $\mathrm{Km^2W^{-1}}$. It is a function of the interface temperature and varies with the third power of temperature; i.e., $R \propto T^3$. The TBR has nothing to do with imperfect contact of the two surfaces. Even if the surfaces are highly smooth and have perfect contact, TBR will exist. The reason behind such a resistance is the scattering of phonons at an interface. At an interface, not all phonons traveling in a metal or a dielectric will pass through an interface. This results in a resistance to heat flow and hence there is a temperature drop at the interface in the direction of the heat flow. The interface conditions can be calculated by means of three models:

1. Acoustic mismatch model
2. Diffuse mismatch model
3. Cut-off mismatch model

4.4.11.1 *Acoustic Mismatch Model*

In this model, the phonons are governed by continuum acoustics (phonons can treated as plane waves) and the interface can be treated as a plane perfect interface. This continuum approximation can be considered as accurate for the phonons with the wavelength more than the interatomic spacing. When the phonon waves are incident of the interface it may specularly reflect, reflect with mode conversion, refract, or refract with mode conversion as shown in Fig. 4.21A. In this model, any possible elastic or

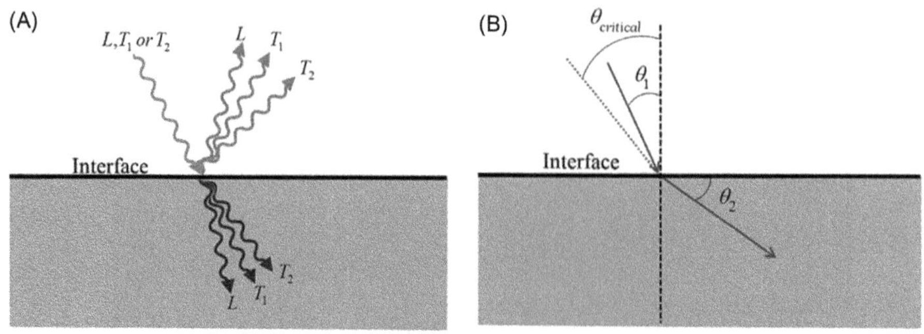

■ **FIGURE 4.21** (A) Schematic of all possibilities for incident phonon in the acoustic mismatch model; (B) Incident and transmitted phonon angles.

inelastic scatterings are ignored. The angles and probabilities of reflection or refraction, with or without mode conversion, can be calculated using the acoustic analog of Snell's law for electromagnetic waves. Let's consider the longitudinal phonon is incident on the interface at angle θ_1 with the velocity v_1^L (Fig. 4.21(B)). If the inside the medium incident phonon is transmitted as longitudinal phonon with the velocity v_2^L, then the angle of transmission θ_2 can be calculated using Snell's law [42]:

$$\sin\theta_2 = \frac{v_2^L}{v_1^L}\sin\theta_1 \tag{4.279}$$

The similar relation can also be written for the angle of transmitted phonon when mode conversion takes place during the transmission. Since the maximum transmitted angle (θ_2) is 90 degree, when the sine of the incident angle ($\sin\theta_1$) is equal to v_1^L/v_2^L or v_1^L/v_2^T, the probability of transmission to a longitudinal or transverse phonon vanishes, respectively. These angles are known as critical angles. Incident phonons can be transmitted only when the incident angle is less than the critical angle.

In order to define the probability for the energy transmission $\alpha_{1\to 2}$ in the simplest way, we can use the acoustic impedance (Z_i), which is the product of mass density (ρ_i) and the phonon velocity (v_i) (i.e., $Z_i = \rho_i v_i$).

The $\alpha_{1\to 2}$ can be given as

$$\alpha_{1\to 2} = \frac{4Z_2Z_1}{(Z_1+Z_2)^2} \tag{4.280}$$

4.4.11.2 ***Diffusive Mismatch Model***

In the diffusive mismatch model, it is assumed that all the incident phonon on the interface are diffusely scattered. Due to this scattering, the outgoing phonon forgets where it came from. The schematic diagram for diffusive mismatch is shown in Fig. 4.22. In this model, the intensities are integrated over all frequencies and then the energy balance at the interface [42] is applied. In this case, energy transfer is considered independent of frequencies [42]. Therefore, all phonons at the first film interface can possibly transfer their energy to the second film interface irrespective of the phonon frequency mismatch. The energy balance should be satisfied at the interface of two mediums.

Let us consider the two thin films pair (silicon and diamond) as shown in Fig. 4.23. The details to calculate the interface conditions are as follows.

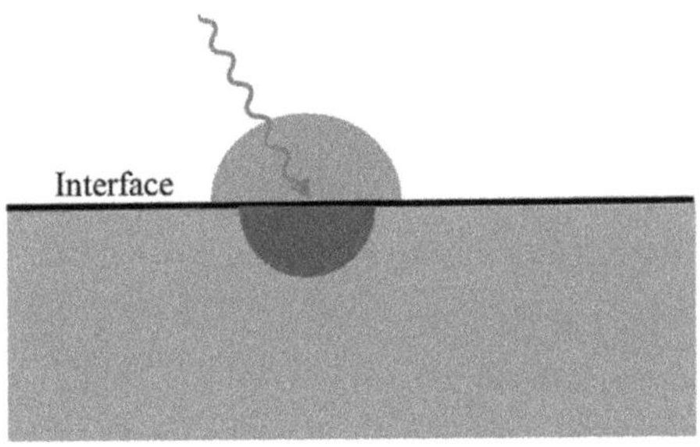

■ **FIGURE 4.22** Schematic of diffusive mismatch model.

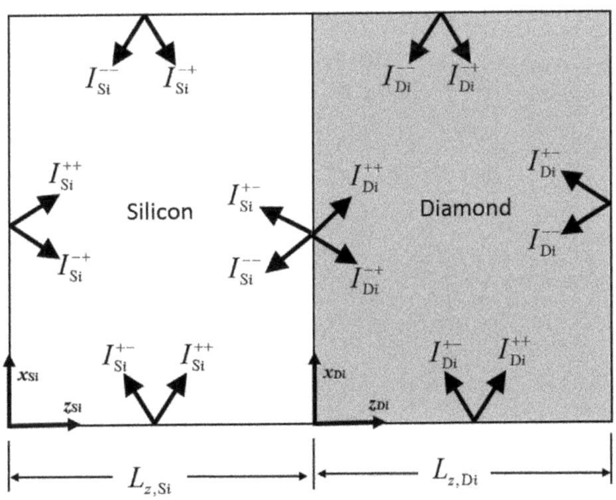

■ **FIGURE 4.23** Schematic of silicon and diamond thin film pairs with the direction of interface intensities.

Heat fluxes at the interface can be written as

$$q''^{+}_{z,\mathrm{Si}}(x_{\mathrm{Si}}, z_{\mathrm{Si}} = L_{z,\mathrm{Si}}, t) = \int_0^{\omega_{\max,Si}} \left(\begin{array}{l} \int_0^{\pi}\int_0^{\pi/2} I^{++}_{\mathrm{Si},\omega} \sin\phi \sin^2\theta d\theta d\phi \\ + \int_0^{\pi}\int_{\pi/2}^{\pi} I^{-+}_{\mathrm{Si},\omega} \sin\phi \sin^2\theta d\theta d\phi \end{array} \right) d\omega \quad (4.281)$$

and

$$q''^{-}_{z,\mathrm{Si}}(x_{\mathrm{Si}}, z_{\mathrm{Si}} = L_{z,\mathrm{Si}}, t) = \int_0^{\omega_{\max,Si}} \left(\begin{array}{l} \int_{\pi}^{2\pi}\int_0^{\pi/2} I^{+-}_{\mathrm{Si},\omega} \sin\phi \sin^2\theta d\theta d\phi \\ + \int_{\pi}^{2\pi}\int_{\pi/2}^{\pi} I^{--}_{\mathrm{Si},\omega} \sin\phi \sin^2\theta d\theta d\phi \end{array} \right) d\omega \quad (4.282)$$

and

$$q''^{+}_{z,\mathrm{Di}}(x_{\mathrm{Di}}, z_{\mathrm{Di}} = 0, t) = \int_0^{\omega_{\max,Di}} \left(\begin{array}{l} \int_0^{\pi}\int_0^{\pi/2} I^{++}_{\mathrm{Di},\omega} \sin\phi \sin^2\theta d\theta d\phi \\ + \int_0^{\pi}\int_{\pi/2}^{\pi} I^{-+}_{\mathrm{Di},\omega} \sin\phi \sin^2\theta d\theta d\phi \end{array} \right) d\omega \quad (4.283)$$

and

$$q''^{-}_{z,\mathrm{Di}}(x_{\mathrm{Di}}, z_{\mathrm{Di}} = 0, t) = \int_0^{\omega_{\max,Di}} \left(\begin{array}{l} \int_{\pi}^{2\pi}\int_0^{\pi/2} I^{+-}_{\mathrm{Di},\omega} \sin\phi \sin^2\theta d\theta d\phi \\ + \int_{\pi}^{2\pi}\int_{\pi/2}^{\pi} I^{--}_{\mathrm{Di},\omega} \sin\phi \sin^2\theta d\theta d\phi \end{array} \right) d\omega \quad (4.284)$$

where $I^{++/--/+-/-+}_{\mathrm{Si/Di},\omega}$ is sum of all frequency branches and modes. For example $I^{++}_{\mathrm{Si},\omega}$ is

$$I^{++}_{\mathrm{Si},\omega} = I^{++}_{\mathrm{LA,Si},\omega} + 2I^{++}_{\mathrm{TA,Si},\omega} + I^{++}_{\mathrm{LO,Si},\omega} + 2I^{++}_{\mathrm{TO,Si},\omega} \quad (4.285)$$

Using the energy balance at the interface, we can write the outgoing heat fluxes at the interface in terms of incoming heat fluxes:

$$q''^{-}_{z,\mathrm{Si}} = T_{\mathrm{Di-Si}}\, q''^{-}_{z,\mathrm{Di}} - R_{\mathrm{Si-Di}}\, q''^{+}_{z,\mathrm{Si}} \quad (4.286)$$

$$q''^{+}_{z,\mathrm{Di}} = T_{\mathrm{Si-Di}}\, q''^{+}_{z,\mathrm{Si}} - R_{\mathrm{Di-Si}}\, q''^{-}_{z,\mathrm{Di}} \quad (4.287)$$

where T is the transmittance and R is used for the reflectance at the interface. Here, the subscript Si–Di indicates from silicon to diamond and Di–Si indicates from diamond to silicon.

Considering the diffuse scattering and utilizing Eqs. (4.286) and (4.287) with Eqs. (4.282) and (4.283) we can get the following relations:

$$I_{\mathrm{Si},\omega}^{--} = I_{\mathrm{Si},\omega}^{+-} = \frac{1}{\pi\omega_{\mathrm{max,Si}}}\left[-T_{\mathrm{Di-Si}}\, q''^{-}_{z,\mathrm{Di}} + R_{\mathrm{Si-Di}}\, q''^{+}_{z,\mathrm{Si}}\right] \tag{4.288}$$

and

$$I_{\mathrm{Di},\omega}^{++} = I_{\mathrm{Di},\omega}^{-+} = \frac{1}{\pi\omega_{\mathrm{max,Di}}}\left[T_{\mathrm{Si-Di}} q''^{+}_{z,\mathrm{Si}} - R_{\mathrm{Di-Si}} q''^{-}_{z,\mathrm{Di}}\right] \tag{4.289}$$

The outgoing intensities $I_{\mathrm{Si},\omega}^{--/+-}$ and $I_{\mathrm{Di},\omega}^{++/-+}$, which can be calculated from Eqs. (4.288) and (4.289), are distributed among different frequency modes and branches using

$$I_{\mathrm{Si},ij,\omega}^{--/+-} = \frac{C_{\mathrm{Si},ij,\omega} v_{\mathrm{Si},ij,\omega} \omega_{\mathrm{max,Si}}}{\langle C_{\mathrm{Si}} v_{\mathrm{Si}} \rangle} I_{\mathrm{Si},\omega}^{--/+-} \tag{4.290}$$

where

$$\langle C_{\mathrm{Si}} v_{\mathrm{Si}} \rangle = \int_0^{\omega_{\max}} \begin{pmatrix} C_{\mathrm{Si,LA}}(\omega) v_{\mathrm{Si,LA}}(\omega) + 2C_{\mathrm{Si,TA}}(\omega) v_{\mathrm{Si,TA}}(\omega) \\ + C_{\mathrm{Si,LO}}(\omega) v_{\mathrm{Si,LO}}(\omega) + 2C_{Si,\mathrm{LO}}(\omega) v_{\mathrm{Si,LO}}(\omega) \end{pmatrix} d\omega \tag{4.291}$$

and

$$I_{\mathrm{Di},ij,\omega}^{++/-+} = \frac{C_{\mathrm{Di},ij,\omega} v_{\mathrm{Di},ij,\omega} \omega_{\mathrm{max,Di}}}{\langle C_{\mathrm{Di}} v_{\mathrm{Di}} \rangle} I_{\mathrm{Di},\omega}^{++/-+} \tag{4.292}$$

where

$$\langle C_{\mathrm{Di}} v_{\mathrm{Di}} \rangle = \int_0^{\omega_{\max}} \begin{pmatrix} C_{\mathrm{Di,LA}}(\omega) v_{\mathrm{Di,LA}}(\omega) + 2C_{\mathrm{Di,TA}}(\omega) v_{\mathrm{Di,TA}}(\omega) \\ + C_{\mathrm{Di,LO}}(\omega) v_{\mathrm{Di,LO}}(\omega) + 2C_{\mathrm{Di,LO}}(\omega) v_{\mathrm{Di,LO}}(\omega) \end{pmatrix} d\omega \tag{4.293}$$

where, in Eqs. (4.290) and (4.292), $i = T$ represents the transverse mode and $i = L$ stands for longitudinal mode. In addition, $j = A$ corresponds to acoustic and $j = O$ stands for optical phonons.

The transmittance and reflectance are calculated based on the energy conservation at the interface. In this case, energy conservation gives

$$T_{\mathrm{Si-Di}} \langle C_{\mathrm{Si}} v_{\mathrm{Si}} \rangle = T_{\mathrm{Di-Si}} \langle C_{\mathrm{Di}} v_{\mathrm{Di}} \rangle \tag{4.294}$$

The diffuse mismatch model gives

$$T_{\mathrm{Si-Di}} = 1 - T_{\mathrm{Di-Si}} \tag{4.295}$$

Simultaneous solution of the previous two equations yields the transmissions

$$T_{\text{Si-Di}} = \frac{\langle C_{\text{Di}} v_{\text{Di}} \rangle}{\langle C_{\text{Di}} v_{\text{Di}} \rangle + \langle C_{\text{Si}} v_{\text{Si}} \rangle} \tag{4.296}$$

and

$$T_{\text{Si-Di}} = \frac{\langle C_{\text{Si}} v_{\text{Si}} \rangle}{\langle C_{\text{Di}} v_{\text{Di}} \rangle + \langle C_{\text{Si}} v_{\text{Si}} \rangle} \tag{4.297}$$

The reflectance can be written as

$$R_{\text{Di-Si}} = 1 - T_{\text{Di-Si}} \tag{4.298}$$

and

$$R_{\text{Si-Di}} = 1 - T_{\text{Si-Di}} \tag{4.299}$$

4.4.11.3 *Cut-Off Mismatch Model*

In the cut-off mismatch model, the mismatch due to the frequency and branch is considered [43]. Fig. 4.24 is showing the optical branches frequencies of silicon not matching with the optical branches' frequencies of diamond. Therefore longitudinal and transverse optical branches phonons of both layers are reflected from the interface; i.e., transmittance becomes zero. Similarly, above the cut-off frequency, the transmittance remains zero for longitudinal and transverse acoustic branches of phonons in the diamond thin film, as shown in Fig. 4.24. We can, therefore, apply the energy balance at each frequency.

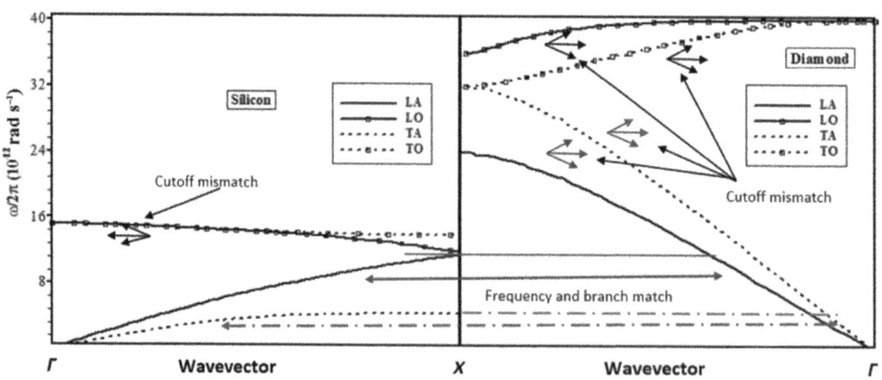

■ **FIGURE 4.24** Dispersion relations for silicon and diamond films and wavevectors for cut-off mismatch [32] and [33].

Heat fluxes at the interface can be written as

$$q''^{+}_{z,\mathrm{Si},ij,\omega}(x_{\mathrm{Si}}, z_{\mathrm{Si}} = L_{z,\mathrm{Si}}, t) = \begin{cases} \int_0^{\pi}\int_0^{\pi/2} I^{++}_{\mathrm{Si},ij,\omega} \sin\phi \sin^2\theta d\theta d\phi \\ + \int_0^{\pi}\int_{\pi/2}^{\pi} I^{-+}_{\mathrm{Si},ij,\omega} \sin\phi \sin^2\theta d\theta d\phi \end{cases} \tag{4.300}$$

and

$$q''^{-}_{z,\mathrm{Si},ij,\omega}(x_{\mathrm{Si}}, z_{\mathrm{Si}} = L_{z,\mathrm{Si}}, t) = \begin{cases} \int_{\pi}^{2\pi}\int_0^{\pi/2} I^{+-}_{\mathrm{Si},ij,\omega} \sin\phi \sin^2\theta d\theta d\phi \\ + \int_{\pi}^{2\pi}\int_{\pi/2}^{\pi} I^{--}_{\mathrm{Si},ij,\omega} \sin\phi \sin^2\theta d\theta d\phi \end{cases} \tag{4.301}$$

and

$$q''^{+}_{z,\mathrm{Di},ij,\omega}(x_{\mathrm{Di}}, z_{\mathrm{Di}} = 0, t) = \begin{cases} \int_0^{\pi}\int_0^{\pi/2} I^{++}_{\mathrm{Di},ij,\omega} \sin\phi \sin^2\theta d\theta d\phi \\ + \int_0^{\pi}\int_{\pi/2}^{\pi} I^{-+}_{\mathrm{Di},ij,\omega} \sin\phi \sin^2\theta d\theta d\phi \end{cases} \tag{4.302}$$

and

$$q''^{-}_{z,\mathrm{Di},ij,\omega}(x_{\mathrm{Di}}, z_{\mathrm{Di}} = 0, t) = \begin{cases} \int_{\pi}^{2\pi}\int_0^{\pi/2} I^{+-}_{\mathrm{Di},ij,\omega} \sin\phi \sin^2\theta d\theta d\phi \\ + \int_{\pi}^{2\pi}\int_{\pi/2}^{\pi} I^{--}_{\mathrm{Di},ij,\omega} \sin\phi \sin^2\theta d\theta d\phi \end{cases} \tag{4.303}$$

where subscript ω is used to indicate that these equations written for each frequency of the phonons, $i = T$ represents the transverse mode, and $i = L$ stands for longitudinal mode. In addition, $j = A$ corresponds to acoustic and $j = O$ stands for optical phonons.

Using the energy balance at the interface, we can write the outgoing heat fluxes at the interface for each frequency in terms of incoming heat fluxes; i.e.,

$$q''^{-}_{z,\mathrm{Si},ij,\omega} = T_{\mathrm{Di-Si},ij,\omega}\, q''^{-}_{z,\mathrm{Di},ij,\omega} - R_{\mathrm{Si-Di},ij,\omega}\, q''^{+}_{z,\mathrm{Si},ij,\omega} \tag{4.304}$$

and

$$q''^{+}_{z,\mathrm{Di},ij,\omega} = T_{\mathrm{Si-Di},ij,\omega}\, q''^{+}_{z,\mathrm{Si},ij,\omega} - R_{\mathrm{Di-Si},ij,\omega}\, q''^{-}_{z,\mathrm{Di},ij,\omega} \tag{4.305}$$

where T is the transmittance and R is used for the reflectance at the interface. Here, the subscript Si–Di indicates from silicon to diamond and Di–Si indicates from diamond to silicon.

Considering the diffuse scattering and utilizing the Eqs. (4.304) and (4.305) along with Eqs. (4.301) and (4.302) one can get the following relations:

$$I^{--}_{\text{Si},ij,\omega} = I^{+-}_{\text{Si},ij,\omega} = \frac{1}{\pi}\left[-T_{\text{Di-Si},ij,\omega}\, q''^{-}_{z,ij,\text{Di},\omega} + R_{\text{Si-Di},ij,\omega}\, q''^{+}_{z,\text{Si},ij,\omega}\right] \tag{4.306}$$

and

$$I^{++}_{\text{Di},ij,\omega} = I^{-+}_{\text{Di},ij,\omega} = \frac{1}{\pi}\left[T_{\text{Si-Di},ij,\omega}\, q''^{+}_{z,\text{Si},ij,\omega} - R_{\text{Di-Si},ij,\omega}\, q''^{-}_{z,\text{Di},ij,\omega}\right] \tag{4.307}$$

The transmittance and reflectance can be calculated from the following relations:

$$T_{\text{Si-Di},ij,\omega} = \frac{C_{\text{Di},ij,\omega} v_{\text{Di},ij,\omega}}{C_{\text{Di},ij,\omega} v_{\text{Di},ij,\omega} + C_{\text{Si},ij,\omega} v_{\text{Si},ij,\omega}} \tag{4.308}$$

and

$$T_{\text{Di-Si},ij,\omega} = \frac{C_{\text{Si},ij,\omega} v_{\text{Si},ij,\omega}}{C_{\text{Di},ij,\omega} v_{\text{Di},ij,\omega} + C_{\text{Si},ij,\omega} v_{\text{Si},ij,\omega}} \tag{4.309}$$

The reflectance can be determined from

$$R_{\text{Di-Si},ij,\omega} = 1 - T_{\text{Di-Si},ij,\omega} \tag{4.310}$$

and

$$R_{\text{Si-Di},ij,\omega} = 1 - T_{\text{Si-Di},ij,\omega} \tag{4.311}$$

4.5 THERMAL STRESSES IN SOLIDS DURING SHORT HEATING DURATIONS AND APPLICATIONS

Rapid rise of temperature takes place inside the substrate material during the high intensity heating periods. In the case of irradiative heating, in which energy is supplied by high intensity beam, the depth of absorption of the irradiated energy at wavelengths within the near infrared region is considerably small, i.e., it is in the order of sub-microns in metallic materials. This causes the attainment of the high temperatures in the irradiated region. Therefore, high temperature gradients generate large thermal strain and stress levels in the heated region. During the short heating period, which is in the order of few picoseconds, the phase change may occur inside the substrate material. Since the thermal strain is only significant in the solid phase of the substrate material, temperature gradient in the solid phase becomes important in terms of the thermal stress analysis. In this section, thermal stress developed during two consecutive high intensity pulse repetition is presented in the light of the previous study [44].

4.5.1 Analysis of Thermal Stress Under Irradiation Pulse

In order to determine the thermal stress filed, temperature rise inside the irradiated solid substrate needs to be formulated. The consideration of the small absorption depth of the irradiated energy, which is assumed to be much smaller than the film thickness, enables us to simplify the heating and secure the analytical solution for the thermal stress developed inside the solid substrate.

In the present formulation, a single-step input pulse of the irradiated energy is considered. The pulse shape used in the analysis is shown in Fig. 4.25. The step input pulse can be constructed through the consideration of taking the differences of the two-step functions. In this case, the first step function will start at $t = 0$ and the other will start at $t + \Delta t$. The step intensity pulse is therefore

$$SP(t) = 1(t) - 1(t - \Delta t) \tag{4.312}$$

where

$$1(t) = \left\{ \begin{matrix} 1, & t > 0 \\ 0, & t < 0 \end{matrix} \right\} \tag{4.313}$$

and

$$1(t - \Delta t) = \left\{ \begin{matrix} 1, & t > \Delta t \\ 0, & t < \Delta t \end{matrix} \right\} \tag{4.314}$$

Since the heating duration considered is greater than the thermalization time of the substrate material, the Fourier heat transfer equation can be considered for a step input pulse heating of the solid surface. The heat

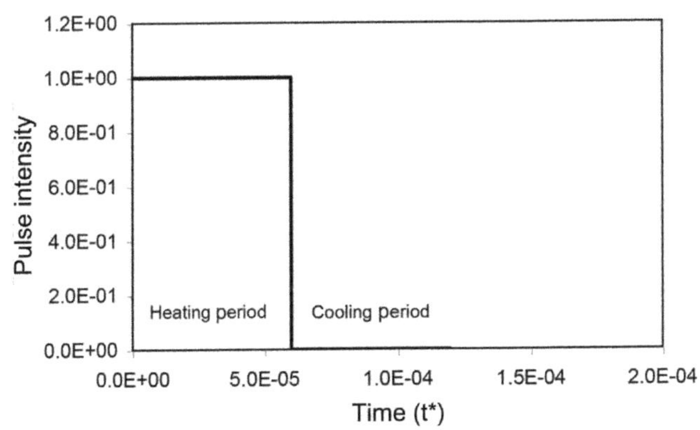

■ **FIGURE 4.25** A dimensionless step input pulse of irradiated energy used in the analysis and in the simulations [44].

equation incorporating the volumetric source term resembling the absorption of the incident radiation energy can be written as

$$\frac{\partial^2 T}{\partial x^2} + \frac{I_1 \delta}{k}\left(C_1 * SP(t)\right)e^{-\delta x} = \frac{1}{\alpha}\frac{\partial T}{\partial t} \tag{4.315}$$

The volumetric heat source term is

$$I_1 = (1 - r_{\mathrm{f}}) \tag{4.316}$$

where I_o is the volumetric heat source term at the surface and I_1 represents the actual energy absorbed inside the substrate material; r_{f} is the surface reflection.

Initially, it is considered that the substrate material is at a constant uniform temperature; therefore, the initial condition yields

$$\text{At } t = 0 \Rightarrow T(x,0) = T_0 \tag{4.317}$$

The radiation and convective losses from the surface are assumed to be negligible; therefore, the corresponding boundary condition becomes

$$\text{At } x = 0 \Rightarrow \left[\frac{\partial T}{\partial x}\right]_{x=0} = 0 \tag{4.318}$$

Since the depth of the irradiated region is limited with the depth of absorption, the size of the absorption depth is considerably smaller than the depth of the solid film; therefore, assumption of a semi-infinite solid enables us to obtain the closed form solution for temperature field. This assumption leads to a boundary condition of constant temperature at the film surface at some depth below; i.e., the influence of the heating pulse is negligible at a depth where film thickness ceases. Therefore, the boundary condition yields

$$\text{At } x = \infty \Rightarrow T(\infty, t) = 0 \tag{4.319}$$

The Laplace transform of Eq. (4.315) with respect to t, results:

$$\frac{\partial^2 \overline{T}}{\partial x^2} + \frac{I_1 \delta}{k}(C_1 \times SP(s))e^{-\delta x} = \frac{1}{\alpha}\left[s\overline{T}(x,s) - T(x,0)\right] \tag{4.320}$$

where

$$SP(s) = \frac{1}{s} - \frac{e^{-(\Delta t)s}}{s} \tag{4.321}$$

Introducing the initial condition and rearranging Eq. (4.320) yields

$$\frac{\partial^2 \overline{T}}{\partial x^2} - \lambda^2 \overline{T} = -\frac{I_1 \delta}{k}(C_1 \times SP(s))e^{-\delta x} - \frac{T_0}{\alpha} \tag{4.322}$$

where $\lambda^2 = s/\alpha$ and s is the transform variable. Eq. (4.322) has the solution [44]:

$$T(x,s) = Ae^{\lambda x} + Be^{-\lambda x} - \frac{I_1\delta}{k}\frac{(C_1 \times SP(s))}{(\delta^2 - s/\alpha)}e^{-\delta x} + \frac{T_0}{s} \tag{4.323}$$

where A and B are constants. Introducing the boundary conditions determines the constants A and B:

$$B = 0, \quad \text{and} \quad A = \frac{I_1\delta^2(C_1 \times SP(s))}{k\lambda(\delta^2 - s/\alpha)} \tag{4.324}$$

Substitution of A and B in Eq. (4.323) yields

$$\overline{T}(x,s) = \frac{I_1\delta^2\sqrt{\alpha}(C_1 \times SP(s))e^{-\lambda x}}{k\sqrt{s}(\delta^2 - \lambda^2)} - \frac{I_1\delta(C_1 \times SP(s))e^{-\delta x}}{k(\delta^2 - \lambda^2)} + \frac{T_0}{s} \tag{4.325}$$

which gives the thermal solution in the Laplace transform domain.

The inverse Laplace transform of Eq. (4.325) gives the temperature distribution inside the substrate material in nondimensional form. This is possible by defining dimensionless quantities, which are

$$x^* = x\delta : t^* = \alpha\delta^2 t : T^\bullet = \frac{Tk\delta}{I_1} \tag{4.326}$$

The dimensionless temperature distribution becomes [45]

$$T^\bullet = T_0^\bullet + C_1\left\{\begin{array}{l} e^{-x^\bullet}\left(e^{t^\bullet} - U[1]\right) - \dfrac{e^{t^\bullet}}{2}\left[\begin{array}{l} e^{-x^\bullet}\operatorname{erfc}\left(-\sqrt{t^\bullet} + \dfrac{x^\bullet}{2\sqrt{t^\bullet}}\right) \\ -e^{-x^\bullet}\operatorname{erfc}\left(\sqrt{t^\bullet} + \dfrac{x^\bullet}{2\sqrt{t^\bullet}}\right) \end{array}\right] \\ + \left[\dfrac{2\sqrt{t^\bullet}}{\sqrt{\pi}}e^{-\frac{x^{\bullet 2}}{4t^\bullet}} - x^\bullet\operatorname{erfc}\left(\dfrac{x^\bullet}{2\sqrt{t^\bullet}}\right)\right] \end{array}\right\}$$

$$- C_1\left\{\begin{array}{l} e^{-x^\bullet}\left(e^{t^\bullet - \Delta t^\bullet} - U[1 - \Delta t^\bullet]\right) \\ -\dfrac{e^{t^\bullet - \Delta t^\bullet}}{2}\left[\begin{array}{l} e^{-x^\bullet}\operatorname{erfc}\left(-\sqrt{t^\bullet - \Delta t^\bullet} + \dfrac{x^\bullet}{2\sqrt{t^\bullet - \Delta t^\bullet}}\right) \\ -e^{-x^\bullet}\operatorname{erfc}\left(\sqrt{t^\bullet - \Delta t^\bullet} + \dfrac{x^\bullet}{2\sqrt{t^\bullet - \Delta t^\bullet}}\right) \end{array}\right] \\ + \left[\dfrac{2\sqrt{t^\bullet - \Delta t^\bullet}}{\sqrt{\pi}}e^{-\frac{x^{\bullet 2}}{4(t^\bullet - \Delta t^\bullet)}} - x^\bullet\operatorname{erfc}\left(\dfrac{x^\bullet}{2\sqrt{t^\bullet - \Delta t^\bullet}}\right)\right] \end{array}\right\} \tag{4.327}$$

where erfc is the complementary error function, which is

$$\operatorname{erfc}(z) = 1 - \operatorname{erf}(z) \quad \text{and} \quad \operatorname{erfc}(z) = \frac{2}{\sqrt{\pi}} \int_0^z e^{-t^2} dt \tag{4.328}$$

Eq. (4.327) is the closed form solution for temperature distribution.

The momentum equation for a linear elastic substrate can be considered to formulate the thermal stress in the heated region [46]. The momentum equation yields [46]

$$\frac{\partial^2 \sigma_x}{\partial x^2} - \frac{1}{c_1{}^2} \frac{\partial^2 \sigma_x}{\partial t^2} = c_2 \frac{\partial^2 T}{\partial t^2} \tag{4.329}$$

where c_1 is the wave speed in the solid

$$c_1 = \sqrt{\frac{E}{\rho}} \quad \text{and} \quad c_2 = \frac{1 + \upsilon}{1 - \upsilon} \rho \alpha_T \tag{4.330}$$

where υ is Poisson's ratio, ρ is the density of the solid, and α_T is the thermal expansion coefficient of the solid.

Initial and boundary conditions for the temperature field in Eq. (4.329) are similar to those given for Eq. (4.315). Initially substrate material is considered as free from the stress. In addition, as time extends to infinity, the stress field vanishes in the substrate material. Therefore, the initial and final conditions for the stress field become:

$$\text{At } t = 0 \Rightarrow \sigma_x = 0 \tag{4.331}$$

and

$$\text{At } t = \infty \Rightarrow \sigma_x = 0 \tag{4.332}$$

The consideration of no mechanical force at the surface prior to laser irradiation pulse leads to stress-free boundary conditions at the surface. In addition, the effective depth of laser irradiation is considerably smaller than the work piece thickness. Therefore, the assumption of a semi-infinite body holds in the stress analysis. In this case, the thermal strain disappears at a depth infinitely long from the surface. Consequently, the corresponding boundary conditions yield

$$\text{At } x = 0 \Rightarrow \sigma_x = 0 \tag{4.333}$$

and

$$\text{At } x = \infty \Rightarrow \sigma_x = 0 \tag{4.334}$$

Taking the Laplace transform of Eq. (4.329) with respect to time yields

$$\frac{\partial^2 \overline{\sigma}_x}{\partial x^2} - \frac{1}{c_1{}^2}\left[s^2\overline{\sigma}_x(x,s) - s\sigma_x(x,0) - \dot{\sigma}_x(x,0)\right] = c_2\left[s^2\overline{T}(x,s) - sT(x,0) - \dot{T}(x,0)\right] \tag{4.335}$$

where $\overline{\sigma}_x(x,s)$ and $\overline{T}(x,s)$ are the Laplace transforms of thermal stress and temperature, respectively, in the x and s domains.

By substituting the initial conditions, Eq. (4.335) reduces to

$$\frac{\partial^2 \overline{\sigma}_x}{\partial x^2} - \frac{s^2}{c_1{}^2}\overline{\sigma}_x(x,s) = c_2 s^2\overline{T}(x,s) - c_2 sT_0 \tag{4.336}$$

Considering the temperature distribution for a pulse with exponential temporal variation, Eq. (4.325), and substituting into Eq. (4.336), and solving for the stress field, yields:

$$\frac{\partial^2 \overline{\sigma}_x}{\partial x^2} - \frac{s^2}{c_1{}^2}\overline{\sigma}_x = c_2 s^2\left[\frac{I_1\delta^2\sqrt{\alpha}(C_1 \times SP(s))e^{-\lambda x}}{k\sqrt{s}(\delta^2-\lambda^2)} - \frac{I_1\delta(C_1 \times SP(s))e^{-\delta x}}{k(\delta^2-\lambda^2)} + \frac{T_0}{s}\right] - c_2 sT_0 \tag{4.337}$$

The complementary and the particular solutions for Eq. (4.337) are:

$$(\overline{\sigma}_x)_c = D_1 e^{\frac{sx}{c_1}} + D_2 e^{\frac{-sx}{c_1}} \tag{4.338}$$

While the particular solution has two parts, the first part is

$$(\overline{\sigma}_x)_{p_1} = D_3 e^{-\sqrt{\frac{s}{\alpha}}x} \tag{4.339}$$

The second part of the particular solution is

$$(\overline{\sigma}_x)_{p_2} = D_4 e^{-\delta x} \tag{4.340}$$

Solving for the particular solutions yields

$$D_3 = \frac{c_2 s^2 I_1\delta^2(C_1 \times SP(s))\sqrt{\alpha}}{k\sqrt{s}(\delta^2-\lambda^2)(\lambda^2 - s^2/c_1^2)} \tag{4.341}$$

and

$$D_4 = \frac{c_2 s^2 I_1\delta(C_1 \times SP(s))}{k(\delta^2-\lambda^2)(s^2/c_1^2 - \delta^2)} \tag{4.342}$$

So, the general solution for the stress field becomes

$$(\overline{\sigma}_x)_g = D_1 e^{\frac{sx}{c_1}} + D_2 e^{\frac{-sx}{c_1}} + D_3 e^{-gx} + D_4 e^{-\delta x} \tag{4.343}$$

From the boundary condition ($x = \infty \Rightarrow \sigma_x = 0$), it yields $D_1 = 0$.

Then,

$$(\overline{\sigma}_x)_g = D_2 e^{\frac{-sx}{c_1}} + D_3 e^{-\sqrt{\frac{s}{\alpha}}x} + D_4 e^{-\delta x} \tag{4.344}$$

Consider the boundary condition at the surface, where at $x = 0 \Rightarrow \overline{\sigma}_x = 0$, the constant in Eq. (4.344) becomes

$$D_2 = -D_3 - D_4 \tag{4.345}$$

Then,

$$\overline{\sigma}_x(x,s) = D_3 e^{-\sqrt{\frac{s}{\alpha}}x} - D_3 e^{\frac{-sx}{c_1}} + D_4 e^{-\delta x} - D_4 e^{\frac{-sx}{c_1}} \tag{4.346}$$

Finding the solution for σ_x in the x and t domain, we should take the inverse Laplace transform for each term in Eq. (4.346). To accomplish this, the following definitions are introduced:

$$\begin{aligned} &\text{Term 1} = D_3 e^{-\sqrt{\frac{s}{\alpha}}x} \quad &&\text{Term 2} = -D_3 e^{\frac{-sx}{c_1}} \\ &\text{Term 3} = D_4 e^{-\delta x} \quad &&\text{Term 4} = -D_4 e^{\frac{-sx}{c_1}} \end{aligned} \tag{4.347}$$

Consequently, the solution for the stress distribution becomes the summation of the inverse Laplace transforms of these terms.

Therefore, the inverse of Laplace transform of Terms (Term1, Term2, Term3, Term4) are

$$\text{Term 1} = \frac{I_1\delta^2 C_1 c_2\sqrt{\alpha}}{k}\left[\frac{e^{-\sqrt{\frac{s}{\alpha}}x}}{\sqrt{s}(\delta^2 - s/\alpha)(1/\alpha - s/c_1^2)} - \frac{e^{-\sqrt{\frac{s}{\alpha}}x}e^{-(\Delta t)s}}{\sqrt{s}(\delta^2 - s/\alpha)(1/\alpha - s/c_1^2)}\right] \tag{4.348}$$

After partial fraction of the preceding equation,

$$\text{Term 1} = \frac{I_1\delta^2 C_1 c_2\sqrt{\alpha}}{k}\left[\begin{aligned} &\frac{\alpha e^{-\sqrt{\frac{s}{\alpha}}x}}{\sqrt{s}\delta^2} - \frac{\sqrt{s}\alpha^4 e^{-\sqrt{\frac{s}{\alpha}}x}}{(\alpha^2\delta^2 - c_1^2)(s\alpha - c_1^2)} \\ &- \frac{c_1^2\alpha\sqrt{s}e^{-\sqrt{\frac{s}{\alpha}}x}}{\delta^2(\alpha^2\delta^2 - c_1^2)(\alpha\delta^2 - s)} - \frac{\alpha e^{-\sqrt{\frac{s}{\alpha}}x}e^{-(\Delta t)s}}{\sqrt{s}\delta^2} \\ &+ \frac{\sqrt{s}\alpha^4 e^{-\sqrt{\frac{s}{\alpha}}x}e^{-(\Delta t)s}}{(\alpha^2\delta^2 - c_1^2)(s\alpha - c_1^2)} + \frac{c_1^2\alpha\sqrt{s}e^{-\sqrt{\frac{s}{\alpha}}x}e^{-(\Delta t)s}}{\delta^2(\alpha^2\delta^2 - c_1^2)(\alpha\delta^2 - s)} \end{aligned}\right] \tag{4.349}$$

let $G_1 = \frac{I_1\delta^2 C_1 c_2\sqrt{\alpha}}{k}$, then,

$$\text{Term 1} = [G_1\ \text{Term 11}] + [G_1\ \text{Term 21}] + [G_1\ \text{Term 31}] \tag{4.350}$$

where

$$\text{Term } 11 = \frac{\alpha e^{-\sqrt{\frac{s}{\alpha}}s}\left[1-e^{-\Delta ts}\right]}{\sqrt{s}\delta^2},\quad \text{Term } 21 = \frac{\sqrt{s}\alpha^4 e^{-\sqrt{\frac{s}{\alpha}}x}\left[1-e^{-\Delta ts}\right]}{(\alpha^2\delta^2-c_1{}^2)(s\alpha-c_1{}^2)}$$
$$\text{and Term } 31 = \frac{c_1{}^2\alpha\sqrt{s}e^{-\sqrt{\frac{s}{\alpha}}x}\left[1-e^{-\Delta ts}\right]}{s^2(\alpha^2\delta^2-c_1{}^2)(\alpha\delta^2-s)} \tag{4.351}$$

The inverse of Laplace transform of Term1 is £$^{-1}$[Term1]; i.e.,

$$£^{-1}[\text{Term } 1] = £^{-1}[\text{Term } 11] + £^{-1}[\text{Term } 21] + £^{-1}[\text{Term } 31] \tag{4.352}$$

The inverse of Laplace transform of terms becomes

$$£^{-1}[G_1\,\text{Term } 11] = G_1\left[\frac{\alpha}{\delta^2}\left(\frac{1}{\sqrt{\pi t}}\exp\left(-\frac{x^2}{4\alpha t}\right)-\frac{1}{\sqrt{\pi(t-\Delta t)}}\exp\left(-\frac{x^2}{4\alpha(t-\Delta t)}\right)\right)\right] \tag{4.353}$$

$$£^{-1}[G_1\,\text{Term } 21] = G_1\frac{\alpha^3}{2(\alpha^2\delta^2-c_1^2)}\left[\begin{array}{l}-\frac{2}{\sqrt{\pi t}}\exp\left(\frac{-x^2}{4\alpha t}\right)-\frac{c_1}{\sqrt{\alpha}}e^{\frac{c_1^2 t}{\alpha}-\frac{c_1 x}{\alpha}}\operatorname{erfc}\left(-c_1\sqrt{\frac{t}{\alpha}}+\frac{x}{2\sqrt{\alpha t}}\right)\\ +\frac{c_1}{\sqrt{\alpha}}e^{\frac{c_1^2 t}{\alpha}+\frac{c_1 x}{\alpha}}\operatorname{erfc}\left(c_1\sqrt{\frac{t}{\alpha}}+\frac{x}{2\sqrt{\alpha t}}\right)\\ +\frac{2}{\sqrt{\pi(t-\Delta t)}}\exp\left(\frac{-x^2}{4\alpha(t-\Delta t)}\right)\\ +\frac{c_1}{\sqrt{\alpha}}e^{\frac{c_1^2(t-\Delta t)}{\alpha}-\frac{c_1 x}{\alpha}}\operatorname{erfc}\left(-c_1\sqrt{\frac{(t-\Delta t)}{\alpha}}+\frac{x}{2\sqrt{\alpha(t-\Delta t)}}\right)\\ -\frac{c_1}{\sqrt{\alpha}}e^{\frac{c_1^2(t-\Delta t)}{\alpha}+\frac{c_1 x}{\alpha}}\operatorname{erfc}\left(c_1\sqrt{\frac{(t-\Delta t)}{\alpha}}+\frac{x}{2\sqrt{\alpha(t-\Delta t)}}\right)\end{array}\right] \tag{4.354}$$

$$£^{-1}[G_1\,\text{Term } 31] = G_1\frac{c_1^2\alpha}{2\delta^2(c_1^2-\alpha^2\delta^2)}\left[\begin{array}{l}-\frac{2}{\sqrt{\pi t}}\exp\left(\frac{-x^2}{4\alpha t}\right)\\ -\sqrt{\alpha}\delta e^{\alpha\delta^2 t-\delta x}\operatorname{erfc}\left(-\delta\sqrt{\alpha t}+\frac{x}{2\sqrt{\alpha t}}\right)\\ +\sqrt{\alpha}\delta e^{\alpha\delta^2 t+\delta x}\operatorname{erfc}\left(\delta\sqrt{\alpha t}+\frac{x}{2\sqrt{\alpha t}}\right)\\ +\frac{2}{\sqrt{\pi(t-\Delta t)}}\exp\left(\frac{-x^2}{4\alpha(t-\Delta t)}\right)\\ +\sqrt{\alpha}\delta e^{\alpha\delta^2(t-\Delta t)-\delta x}\operatorname{erfc}\left(-\delta\sqrt{\alpha(t-\Delta t)}+\frac{x}{2\sqrt{\alpha(t-\Delta t)}}\right)\\ -\sqrt{\alpha}\delta e^{\alpha\delta^2(t-\Delta t)+\delta x}\operatorname{erfc}\left(\delta\sqrt{\alpha(t-\Delta t)}+\frac{x}{2\sqrt{\alpha(t-\Delta t)}}\right)\end{array}\right] \tag{4.355}$$

However, Term 2 is

$$\text{Term 2} = -\frac{c_2\delta^2 I_1 C_1\sqrt{\alpha}}{k}\left[\frac{e^{-\frac{s}{c_1}x}}{\sqrt{s}(\delta^2 - s/\alpha)(1/\alpha - s/c_1^2)} - \frac{e^{-\frac{s}{c_1}x}e^{-(\Delta t)s}}{\sqrt{s}(\delta^2 - s/\alpha)(1/\alpha - s/c_1^2)}\right] \tag{4.356}$$

Let

$$G_2 = -\frac{c_2\delta^2 I_1 C_1\sqrt{\alpha}}{k} \tag{4.357}$$

After the partial fraction, Term2 can be written as

$$\text{Term 2} = G_2[\text{Term 12} + \text{Term 22} + \text{Term 32}] \tag{4.358}$$

where

$$\text{Term 12} = \frac{\alpha}{\delta^2}\frac{e^{-\frac{s}{c_1}x}}{\sqrt{s}}: \text{Term 22} = \frac{{c_1}^2\sqrt{s}\alpha e^{-\frac{s}{c_1}x}}{\delta^2({c_1}^2 - \alpha^2\delta^2)(\alpha\delta^2 - s)}$$
$$\text{and Term 32} = \frac{\sqrt{s}\alpha^4 e^{-\frac{s}{c_1}x}}{\delta^2(\alpha^2\delta^2 - {c_1}^2)(\alpha s - {c_1}^2)} \tag{4.359}$$

The inverse of Laplace transform of Term 2 comes out to be:

$$£^{-1}[\text{Term 2}] = £^{-1}[G_2\,\text{Term 12}] + £^{-1}[G_2\,\text{Term 22}] + £^{-1}[G_2\,\text{Term 32}] \tag{4.360}$$

or

$$£^{-1}[G_2\,\text{Term 12}] = G_2\frac{\alpha}{\delta^2}\left[\frac{1}{\sqrt{\pi(t - x/c_1)}}\cdot 1\left(t - \frac{x}{c_1}\right) - \frac{1}{\sqrt{\pi(t - \Delta t - x/c_1)}}\cdot 1\left(t - \Delta t - \frac{x}{c_1}\right)\right] \tag{4.361}$$

and

$$£^{-1}[G_2\,\text{Term 22}] = G_2\frac{c_1^2\alpha}{\delta^2(\alpha^2\delta^2 - c_1^2)}\left[\begin{array}{l}\left(\frac{1}{\sqrt{\pi(t - x/c_1)}} + \sqrt{\alpha}\delta e^{\alpha\delta^2 t}\operatorname{erf}\left(\sqrt{\alpha}\delta\sqrt{t - \frac{x}{c_1}}\right)\right) \\ \times 1\left(t - \frac{x}{c_1}\right) - \left(\frac{1}{\sqrt{\pi(t - \Delta t - x/c_1)}}\right. \\ \left. + \sqrt{\alpha}\delta e^{\alpha\delta^2(t - \Delta t)}\operatorname{erf}\left(\sqrt{\alpha}\delta\sqrt{t - \Delta t - \frac{x}{c_1}}\right)\right) \\ \times 1\left(t - \Delta t - \frac{x}{c_1}\right)\end{array}\right] \tag{4.362}$$

and

$$\pounds^{-1}[G_2\,\mathrm{Term\,32}] = G_2\frac{\alpha^3}{(\alpha^2\delta^2 - c_1^2)}\left[\begin{array}{l} -\left(\dfrac{1}{\sqrt{\pi(t-x/c_1)}} + \dfrac{c_1}{\sqrt{\alpha}}e^{\frac{c_1^2}{\alpha}t}\mathrm{erf}\left(\dfrac{c_1}{\sqrt{\alpha}}\sqrt{t-\dfrac{x}{c_1}}\right)\right) \\ \times 1\left(t-\dfrac{x}{c_1}\right) + \left(\dfrac{1}{\sqrt{\pi(t-\Delta t-x/c_1)}}\right. \\ \left. + \dfrac{c_1}{\sqrt{\alpha}}e^{\frac{c_1^2}{\alpha}(t-\Delta t)}\mathrm{erf}\left(\dfrac{c_1}{\sqrt{\alpha}}\sqrt{t-\Delta t-\dfrac{x}{c_1}}\right)\right)\cdot 1\left(t-\Delta t-\dfrac{x}{c_1}\right) \end{array}\right] \tag{4.363}$$

where erf (y) is the error function of the variable y.

However, Term 3 is

$$\mathrm{Term\ 3} = \frac{c_2 I_1 \delta C_1}{k}\left[\frac{se^{-\delta x}}{(s^2/c_1^2 - \delta^2)(\delta^2 - s/\alpha)} - \frac{se^{-\delta x}e^{-(\Delta t)s}}{(s^2/c_1^2 - \delta^2)(\delta^2 - s/\alpha)}\right] \tag{4.364}$$

Let

$$G_3 = \frac{c_2 I_1 \delta C_1}{k} \tag{4.365}$$

Then Term 3 can be written as

$$\mathrm{Term\ 3} = [G_3\ \mathrm{Term\ 13}] + [G_3\ \mathrm{Term\ 23}] + [G_3\ \mathrm{Term\ 33}] \tag{4.366}$$

After the partial fraction, Term 3 becomes

$$\begin{aligned} &\mathrm{Term\ 13} = \frac{{c_1}^2\alpha e^{-\delta x}}{2\delta(s - c_1\delta)(c_1 - \alpha\delta)} : \mathrm{Term\ 23} = \frac{{c_1}^2\alpha e^{-\delta x}}{2\delta(s + c_1\delta)(c_1 + \alpha\delta)} \\ &\text{and Term 13} = \frac{{c_1}^2\alpha^2 e^{-\delta x}}{({c_1}^2 - \alpha^2\delta^2)(s - \alpha\delta^2)} \end{aligned} \tag{4.367}$$

The inverse of Laplace transform of Term 3 turns out to be

$$\pounds^{-1}[\mathrm{Term\ 3}] = \pounds^{-1}[G_3\ \mathrm{Term\ 13}] + \pounds^{-1}[G_3\ \mathrm{Term\ 23}] + \pounds^{-1}[G_3\ \mathrm{Term\ 33}] \tag{4.368}$$

where

$$\pounds^{-1}[G_3\ \mathrm{Term\ 13}] = G_3\frac{c_1^2\alpha e^{-\delta x}}{2\delta(c_1 - \alpha\delta)}\left[e^{c_1\delta(t-\Delta t)} - e^{c_1\delta t}\right] \tag{4.369}$$

and

$$\pounds^{-1}[G_3\ \mathrm{Term\ 23}] = G_3\frac{c_1^2\alpha e^{-\delta x}}{2\delta(c_1 + \alpha\delta)}\left[e^{c_1\delta t} - e^{c_1\delta(t-\Delta t)}\right] \tag{4.370}$$

and

$$£^{-1}[G_3 \text{ Term } 33] = G_3 \frac{c_1^2\alpha^2 e^{-\delta x}}{(c_1^2 - \alpha^2\delta^2)}\left[e^{\alpha\delta^2 t} - e^{\alpha\delta^2(t-\Delta t)}\right] \tag{4.371}$$

However, Term 4 is

$$\text{Term } 4 = -\frac{c_2 I_1 \delta C_1}{k}\left[\frac{se^{-\frac{s}{c_1}x}}{(s^2/c_1^2 - \delta^2)(\delta^2 - s/\alpha)} - \frac{se^{-\frac{s}{c_1}x}e^{-(\Delta t)s}}{(s^2/c_1^2 - \delta^2)(\delta^2 - s/\alpha)}\right] \tag{4.372}$$

Let

$$G_4 = -\frac{c_2 I_1 \delta C_1}{k} \tag{4.373}$$

Term 4 can be written as

$$\text{Term } 4 = [G_4 \text{ Term } 14] + [G_4 \text{ Term } 24] + [G_4 \text{ Term } 34] \tag{4.374}$$

After partial fraction of Term 4, it yields

$$\text{Term } 14 = \frac{c_1{}^2\alpha e^{-\frac{s}{c_1}x}}{2\delta(s - c_1\delta)(c_1 - \alpha\delta)} : \text{Term } 24 = \frac{c_1{}^2\alpha e^{-\frac{s}{c_1}x}}{2\delta(s + c_1\delta)(c_1 + \alpha\delta)} :$$
$$\text{and} \quad \text{Term } 34 = \frac{c_1{}^2\alpha^2 e^{-\frac{s}{c_1}x}}{(c_1{}^2 - \alpha^2\delta^2)(s - \alpha\delta^2)} \tag{4.375}$$

The inversion of Term 4 is

$$£^{-1}[\text{Term } 4] = £^{-1}[G_4 \text{ Term } 14] + £^{-1}[G_4 \text{ Term } 24] + £^{-1}[G_4 \text{ Term } 34] \tag{4.376}$$

where

$$£^{-1}[G_4 \text{ Term } 14]$$
$$= G_4 \frac{c_1^2\alpha}{2\delta(c_1 - \alpha\delta)}\left[e^{c_1\delta\left(t-\Delta t-\frac{x}{c_1}\right)} \cdot 1\left(t - \Delta t - \frac{x}{c_1}\right) - e^{c_1\delta\left(t-\frac{x}{c_1}\right)} \cdot 1\left(t - \frac{x}{c_1}\right)\right] \tag{4.377}$$

and

$$£^{-1}[G_4 \text{ Term } 24]$$
$$= G_4 \frac{c_1^2\alpha}{2\delta(c_1 + \alpha\delta)}\left[e^{-c_1\delta\left(t-\frac{x}{c_1}\right)} \cdot 1\left(t - \frac{x}{c_1}\right) - e^{-c_1\delta\left(t-\Delta t-\frac{x}{c_1}\right)} \cdot 1\left(t - \Delta t - \frac{x}{c_1}\right)\right] \tag{4.378}$$

and

$$£^{-1}[G_4 \text{ Term } 34] = G_4 \frac{c_1^2\alpha^2}{(c_1^2 - \alpha^2\delta^2)}\left[e^{\alpha\delta^2\left(t-\frac{x}{c_1}\right)}\cdot 1\left(t-\frac{x}{c_1}\right) - e^{\alpha\delta^2\left(t-\Delta t-\frac{x}{c_1}\right)}\cdot 1\left(t-\Delta t-\frac{x}{c_1}\right)\right] \tag{4.379}$$

where $1\left(t-\frac{x}{c_1}\right)$ is a unit step function.

The closed form solution of stress distribution can be written as

$$\sigma_x(x,t) = £^{-1}[\text{Term 1}] + £^{-1}[\text{Term 2}] + £^{-1}[\text{Term 3}] + £^{-1}[\text{Term 4}] \tag{4.380}$$

where $£^{-1}$ represents the inverse sign of Laplace transform.

Presenting the stress distribution in dimensionless form, the additional dimensionless quantities are defined:

$$c_1^\bullet = \frac{c_1}{\alpha\delta} \quad \text{and} \quad \sigma_x^\bullet = \frac{k\sigma_x}{I_1\delta\alpha^2 c_2 C_1} \tag{4.381}$$

U [1] is the unit step function, which is $U[1] = \left(t^\bullet - \frac{x^\bullet}{c_1^\bullet}\right)$, and $U[1-\Delta t^\bullet]$ is the unit step function for the time shift, $U[1-\Delta t^\bullet] = 1\left(t^\bullet - \Delta t^\bullet - \frac{x^\bullet}{c_1^\bullet}\right)$.

Therefore, for the dimensionless stress distribution, the following results are obtained:

$$(\sigma_x^\bullet)_1 = (\sigma_x^\bullet)_{11} + (\sigma_x^\bullet)_{21} + (\sigma_x^\bullet)_{31} \tag{4.382}$$

where

$$(\sigma_x^\bullet)_{11} = \frac{1}{\sqrt{\pi t^\bullet}}\exp\left(-\frac{x^{\bullet 2}}{4t^\bullet}\right) - \frac{1}{\sqrt{\pi(t^\bullet - \Delta t^\bullet)}}\exp\left(-\frac{x^{\bullet 2}}{4(t^\bullet - \Delta t^\bullet)}\right) \tag{4.383}$$

$$(\sigma_x^\bullet)_{21} = \frac{1}{2(1-c_1^{\bullet 2})}\left[\begin{array}{l} -\frac{2}{\sqrt{\pi t^\bullet}}\exp\left(\frac{-x^{\bullet 2}}{4t^\bullet}\right) - c_1^\bullet e^{c_1^{\bullet 2}t^\bullet - c_1^\bullet x^\bullet}\operatorname{erfc}\left(-c_1^\bullet\sqrt{t^\bullet} + \frac{x^\bullet}{2\sqrt{t^\bullet}}\right) \\ + c_1^\bullet e^{c_1^{\bullet 2}t^\bullet + c_1^\bullet x^\bullet}\operatorname{erfc}\left(c_1^\bullet\sqrt{t^\bullet} + \frac{x^\bullet}{2\sqrt{t^\bullet}}\right) \\ + \frac{2}{\sqrt{\pi(t^\bullet - \Delta t)}}\exp\left(\frac{-x^{\bullet 2}}{4(t^\bullet - \Delta t^\bullet)}\right) \\ + c_1^\bullet e^{c_1^{\bullet 2}(t^\bullet - \Delta t^\bullet) - c_1^\bullet x^\bullet}\operatorname{erfc}\left(-c_1^\bullet\sqrt{(t^\bullet - \Delta t^\bullet)} + \frac{x^\bullet}{2\sqrt{(t^\bullet - \Delta t^\bullet)}}\right) \\ - c_1^\bullet e^{c_1^{\bullet 2}(t^\bullet - \Delta t^\bullet) + c_1^\bullet x^\bullet}\operatorname{erfc}\left(c_1^\bullet\sqrt{(t^\bullet - \Delta t^\bullet)} + \frac{x^\bullet}{2\sqrt{(t^\bullet - \Delta t^\bullet)}}\right) \end{array}\right] \tag{4.384}$$

and

$$(\sigma_x^\bullet)_{31} = -\frac{c_1^{\bullet 2}}{2(1-c_1^{\bullet 2})}\left[\begin{array}{l} -\frac{2}{\sqrt{\pi t^\bullet}}\exp\left(\frac{-x^{\bullet 2}}{4t^\bullet}\right) - e^{t^\bullet - x^\bullet}\operatorname{erfc}\left(-\sqrt{t^\bullet}+\frac{x^\bullet}{2\sqrt{t^\bullet}}\right) \\ +e^{t^\bullet + x^\bullet}\operatorname{erfc}\left(\sqrt{t^\bullet}+\frac{x^\bullet}{2\sqrt{t^\bullet}}\right) \\ +\frac{2}{\sqrt{\pi(t^\bullet-\Delta t^\bullet)}}\exp\left(\frac{-x^{\bullet 2}}{4(t^\bullet-\Delta t^\bullet)}\right) \\ +e^{(t^\bullet-\Delta t^\bullet)-x^\bullet}\operatorname{erfc}\left(-\sqrt{(t^\bullet-\Delta t^\bullet)}+\frac{x^\bullet}{2\sqrt{(t^\bullet-\Delta t^\bullet)}}\right) \\ -e^{(t^\bullet-\Delta t^\bullet)+x^\bullet}\operatorname{erfc}\left(\sqrt{(t^\bullet-\Delta t^\bullet)}+\frac{x^\bullet}{2\sqrt{(t^\bullet-\Delta t^\bullet)}}\right) \end{array}\right] \tag{4.385}$$

However, $(\sigma_x^\bullet)_2$ is

$$(\sigma_x^\bullet)_2 = (\sigma_x^\bullet)_{12} + (\sigma_x^\bullet)_{22} + (\sigma_x^\bullet)_{32} \tag{4.386}$$

where

$$(\sigma_x^\bullet)_{12} = -\frac{1}{\sqrt{\pi(t^\bullet - x^\bullet/c_1^\bullet)}}\cdot U[1] + \frac{1}{\sqrt{\pi(t^\bullet - \Delta t^\bullet - x^\bullet/c_1^\bullet)}}\cdot U[1-\Delta t^\bullet] \tag{4.387}$$

and

$$(\sigma_x^\bullet)_{22} = -\frac{c_1^{\bullet 2}}{(1-c_1^{\bullet 2})}\left[\begin{array}{l} \left(\frac{1}{\sqrt{\pi(t^\bullet - x^\bullet/c_1^\bullet)}} + e^{t^\bullet}\operatorname{erf}\left(\sqrt{t^\bullet - \frac{x^\bullet}{c_1^\bullet}}\right)\right)\cdot U[1] \\ -\left(\frac{1}{\sqrt{\pi(t^\bullet-\Delta t^\bullet - x^\bullet/c_1^\bullet)}}\right. \\ \left. +e^{(t^\bullet-\Delta t^\bullet)}\operatorname{erf}\left(\sqrt{t^\bullet-\Delta t^\bullet-\frac{x^\bullet}{c_1^\bullet}}\right)\right)\cdot U[1-\Delta t^\bullet] \end{array}\right] \tag{4.388}$$

and

$$(\sigma_x^\bullet)_{32} = -\frac{1}{(1-c_1^{\bullet 2})}\left[\begin{array}{l} -\left(\frac{1}{\sqrt{\pi(t^\bullet - x^\bullet/c_1^\bullet)}} + c_1^\bullet e^{c_1^{\bullet 2}t^\bullet}\operatorname{erf}\left(c_1^\bullet\sqrt{t^\bullet - \frac{x^\bullet}{c_1^\bullet}}\right)\right)\cdot U[1] \\ +\left(\frac{1}{\sqrt{\pi(t^\bullet-\Delta t^\bullet - x^\bullet/c_1^\bullet)}}\right. \\ \left. +c_1^\bullet e^{c_1^{\bullet 2}(t^\bullet-\Delta t^\bullet)}\operatorname{erf}\left(c_1^\bullet\sqrt{t^\bullet-\Delta t^\bullet-\frac{x^\bullet}{c_1^\bullet}}\right)\right)\cdot U[1-\Delta t^\bullet] \end{array}\right] \tag{4.389}$$

The term $(\sigma_x^{\bullet})_3$ is

$$(\sigma_x^{\bullet})_3 = (\sigma_x^{\bullet})_{13} + (\sigma_x^{\bullet})_{23} + (\sigma_x^{\bullet})_{33} \tag{4.390}$$

where

$$(\sigma_x^{\bullet})_{13} = \frac{c_1^{\bullet 2} e^{-x^{\bullet}}}{2(c_1^{\bullet} - 1)} \left[e^{c_1^{\bullet}(t^{\bullet} - \Delta t^{\bullet})} - e^{c_1^{\bullet} t^{\bullet}} \right] \tag{4.391}$$

and

$$(\sigma_x^{\bullet})_{23} = \frac{c_1^{\bullet 2} e^{-x^{\bullet}}}{2(c_1^{\bullet} + 1)} \left[e^{-c_1^{\bullet} t^{\bullet}} - e^{-c_1^{\bullet}(t^{\bullet} - \Delta t^{\bullet})} \right] \tag{4.392}$$

and

$$(\sigma_x^{\bullet})_{33} = \frac{c_1^{\bullet 2} e^{-x^{\bullet}}}{(c_1^{\bullet 2} - 1)} \left[e^{t^{\bullet}} - e^{(t^{\bullet} - \Delta t^{\bullet})} \right] \tag{4.393}$$

The term $(\sigma_x^{\bullet})_4$ is

$$(\sigma_x^{\bullet})_4 = (\sigma_x^{\bullet})_{14} + (\sigma_x^{\bullet})_{24} + (\sigma_x^{\bullet})_{34} \tag{4.394}$$

where

$$(\sigma_x^{\bullet})_{14} = \frac{c_1^{\bullet 2}}{2(c_1^{\bullet} - 1)} \left[e^{c_1^{\bullet}\left(t^{\bullet} - \Delta t^{\bullet} - \frac{x^{\bullet}}{c_1^{\bullet}}\right)} \cdot U[1 - \Delta t^{\bullet}] - e^{c_1^{\bullet}\left(t^{\bullet} - \frac{x^{\bullet}}{c_1^{\bullet}}\right)} \cdot U[1] \right] \tag{4.395}$$

and

$$(\sigma_x^{\bullet})_{24} = \frac{c_1^{\bullet 2}}{2(c_1^{\bullet} + 1)} \left[e^{-c_1^{\bullet}\left(t^{\bullet} - \frac{x^{\bullet}}{c_1^{\bullet}}\right)} \cdot U[1] - e^{-c_1^{\bullet}\left(t^{\bullet} - \Delta t^{\bullet} \frac{x^{\bullet}}{c_1^{\bullet}}\right)} \cdot U[1 - \Delta t^{\bullet}] \right] \tag{4.396}$$

and

$$(\sigma_x^{\bullet})_{34} = \frac{c_1^{\bullet 2}}{(c_1^{\bullet 2} - 1)} \left[e^{\left(t^{\bullet} - \frac{x^{\bullet}}{c_1^{\bullet}}\right)} \cdot U[1] - e^{\left(t^{\bullet} - \Delta t^{\bullet} - \frac{x^{\bullet}}{c_1^{\bullet}}\right)} \cdot U[1 - \Delta t] \right] \tag{4.397}$$

Consequently, the dimensionless form of the stress equation becomes

$$\sigma_x^* = (\sigma_x^*)_1 + (\sigma_x^*)_2 + (\sigma_x^*)_3 + (\sigma_x^*)_4 \tag{4.398}$$

Eq. (4.398) gives the thermal stress inside the heat region when the solid substrate is subjected to the irradiated energy, which is absorbed in the skin of the substrate material.

4.5.2 Findings and Discussion

The analytical solution of temperature distribution is used when formulating the stress field [44,45]. The thermal stress equation (Eq. 4.380) is nondimensionalized to obtain Eq. (4.398), which generalizes the solution independent of the material properties. The stress field in the solid phase of the substrate material is modeled, although the vapor and gas phases may be developed under the high-intensity irradiated energy. However, the thermal strains generated in the gas and liquid phases are insignificant due to the free molecular activities in these phases. Therefore, the thermal stress formulation can also be applicable for surface melting and evaporation processes, provided that the step input intensity is incorporated in the analysis. A computer program is used to solve Eqs. (4.327) and (4.398) to obtain numerical values for temperature and stress fields. The findings are presented later in the light of the previous study, thermal stress development due to laser step [44]. Table 4.7 gives the properties of the irradiated pulse used in the simulations.

Fig. 4.26 shows the nondimensional temperature profile inside the substrate material for three different heating periods for single and two successive step input pulses. The differences in temperature profiles appear in the surface region where the magnitude of temperature is high

Table 4.7 Laser Pulse Properties Used in the Simulations

Dimensionless pulse length ($\tau\alpha\delta^2$, τ is the pulse length)	0.00006
Dimensionless intensity ($I_o/k\delta$, I_o laser peak power intensity)	1

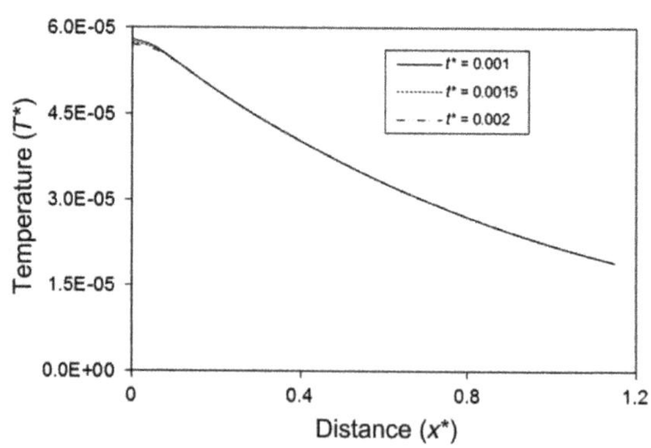

■ **FIGURE 4.26** Dimensionless temperature profiles inside the substrate material obtained at three different dimensionless heating periods for one step input pulse [44].

and the temperature gradient is small. The small temperature gradient in the surface region results from the energy gain in the substrate material from the irradiated field. It should be noted that the amount of the absorbed laser power varies exponentially with the depth inside the substrate material (Lambert's law [47]); i.e., it is maximum at the surface and decays exponentially with increasing depth. Consequently, the energy gain from the irradiated field dominates the diffusional energy transport from the surface vicinity to the solid bulk. It should be noted that the diffusional energy transport enhances as the temperature gradient increases. Moreover, as the depth below the surface increases, the temperature gradient becomes large due to a sharp change in temperature in this region and a smaller amount of absorbed energy from the irradiated field with increasing depth. Therefore, diffusional energy transport dominates internal energy gain of the substrate material through absorption in this region. Since the time domain selected is close to each other, the temperature variation along the depth is almost similar, particularly at some depth below the surface. It should be noted that the selection of a close time domain is due to examination of stress wave propagation in the substrate material; i.e., a long time domain obscures the appearance of the stress wave within the surface region since it has a considerably high speed.

Fig. 4.27 shows temporal variation of stress waves at different depths inside the substrate material. The stress levels and time domain shown in Fig. 4.27 is dimensionless. The propagation of the stress wave is evident, since the peak amplitude of the wave occurs at different times as the depth below the surface changes. The speed of the wave propagation appears to be constant, as expected, since the wave speed depends on

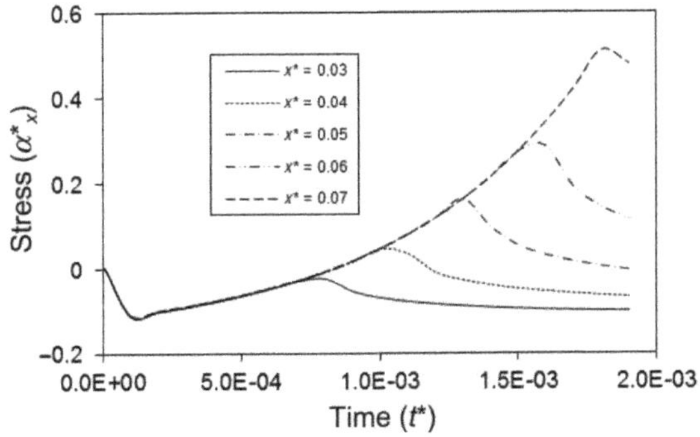

■ **FIGURE 4.27** Temporal variation of dimensionless thermal stress at different dimensionless depths inside the substrate material for a single pulse [44].

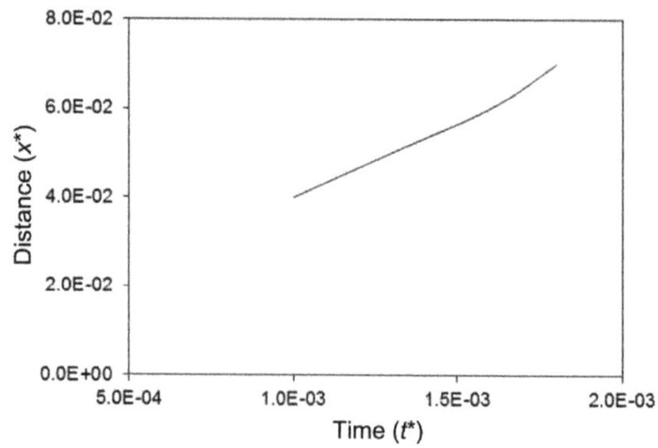

FIGURE 4.28 The magnitude of the stress wave peaks with time obtained for a single pulse [44].

Young's modulus and density of the substrate material ($c = \sqrt{\frac{E}{\rho}}$, where E and ρ are the Young's modulus and density of the substrate material, respectively), which is considered as constant in the analysis. This situation can also be seen in Fig. 4.28, in which the magnitude of the stress wave peaks is shown with time. The linear variation of the magnitude of the stress peaks with time is the indication of the constant propagation speed of the waves. Moreover, as time progresses, the stress wave becomes tensile with a compressive front. Since a stress free condition is considered at the surface, the stress magnitude is zero at the surface and the magnitude of stress is low in the surface vicinity. In the early heating period, thermal stress is compressive, particularly at some depth below the surface ($x^* = 0.07$). This is because of the thermal strain developed in this region—some depth below the surface material cannot expand freely and the magnitude of the thermal strain increases, which in turn generates compressive stress levels in this region.

4.6 SCHRÖDINGER EQUATION

4.6.1 Characteristics of Waves

Waves are generated from the source of a vibration or oscillation. It transfers the energy from one point to another without physical transfer of any material between the two points. The basic properties of waves are amplitude, wavelength, frequency, and speed. Fig. 4.29 represents the wave length and amplitude of wave. Amplitude is the maximum height of the wave from their rest position. The wavelength (λ) is the distance between the two consecutive crests. Frequency represents the number of

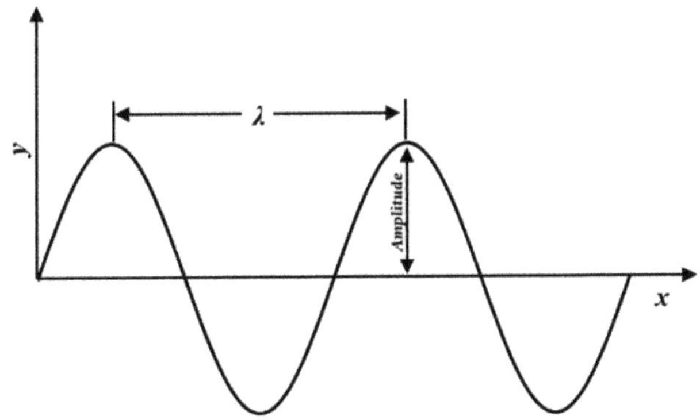

■ **FIGURE 4.29** Wavelength and amplitude of wave.

complete waves that pass from the given point in a certain amount of time, which can be given as

$$v = \frac{V}{\lambda} \tag{4.399}$$

where V is the velocity and v is the frequency.

The one-dimensional harmonic wave can be represented as

$$y(x, t) = A \sin(\omega t - kx) \tag{4.400}$$

where A is the amplitude, ω is the angular frequency ($\omega = 2\pi v$), t is the time, k is the wavenumber, and x is the distance in the x direction.

The wave can have two types of periodicity: one in time and one in space. The time periodicity is characterized by the frequency v, which equals the inverse of the period in time (T); the space periodicity is characterized by the wavelength (λ). Eq. (4.400) represents the wave that propagates in the x direction while it is vibrating along the y direction. There are two types of waves: transverse and longitudinal. In transverse waves, the particle displacement is perpendicular to the direction of wave propagation. In longitudinal waves, particle displacement is parallel to the direction of wave propagation.

4.6.2 Matter Waves

In 1924, the wave particle duality of light motivated de Broglie [29] to proposed the existence of matter waves. In the hypothesis de Broglie suggests that particles can behave like the wave. As the light wave represents the motion of photon, similarly motion of the material particle is governed by matter waves.

According to de Broglie the total energy E is related to the frequency v of the wave, given by

$$E = hv \tag{4.401}$$

and the momentum p of the entity is related to the wavelength λ of the associated wave given by the equation

$$p = h/\lambda \tag{4.402}$$

Here energy E and momentum p of the particles are connected to wave concepts by Planck's constant h. Eq. (4.402) can be written as

$$\lambda = h/p \tag{4.403}$$

This equation is known as the *de Broglie relation.* It predicts the de Broglie wavelength λ of a matter wave associated with the motion of a material particle having a momentum p.

Example 4.1

What is the de Broglie wavelength of a snooker ball (mass = 0.15 kg) moving at a speed of $V = 2\ \text{ms}^{-1}$?

Solution

$$m = 0.15\ \text{kg}$$

$$V = 2\ \text{ms}^{-1}$$

$$p = mV = (0.16)(2) = 0.32\ \text{kg ms}^{-1}$$

Using Eq. (4.403),

$$\lambda = \frac{6.6 \times 10^{-34}}{0.32} = 2.0625 \times 10^{-33}\ \text{m}$$

4.6.3 Schrödinger Equation

In 1925, Schrödinger introduced the equation in the form of the wave functions that govern the motion of small particles and specify how these waves are altered by external influence. Schrödinger states that the wave function of any material obeys the following [48]:

$$i\hbar\frac{\partial\Psi}{\partial t} = -\frac{\hbar^2}{2m}\nabla^2\Psi + V\Psi \tag{4.404}$$

where $\hbar(=h/2\pi)$ is the reduced Plank's constant, Ψ is the wave function, t is the time, m is the mass, and V is the potential energy.

Schrödinger himself was unable to give an explanation of the wave function. Later, in 1926, Born defined the wave function [26]. He stated that the product of wave function and its complex conjugate is the probability density function:

$$\int_{-\infty}^{\infty} \Psi(x,t)\Psi*(x,t)dx = 1 \tag{4.405}$$

where $\Psi*$ is the complex conjugate of the wave function.

The wave function can be used to calculate the momentum and energy of the system. Since Eq. (4.405) represents the statistical probability, the quantities of the system like momentum and energy can be represented by average or expected values. The expected value of any quantity can be calculated as

$$\langle Q\rangle = \int \Psi * Q\Psi dx \tag{4.406}$$

where $\langle Q\rangle$ represents the expected value and Q is the operator for this quantity.

The expected value of the momentum $\langle p\rangle$ can be calculated while using the momentum operator [49] in Eq. (4.406), which gives

$$\langle p\rangle = \int \Psi * \left(\frac{\hbar}{i}\frac{\partial}{\partial x}\right)\Psi dx \tag{4.407}$$

where $(\hbar/i)(\partial/\partial x)$ is known as momentum operator.

The kinetic energy (T) can be given as

$$T = \frac{1}{2}mv^2 = \frac{p^2}{2m} \tag{4.408}$$

Using the momentum operator $p\rightarrow(\hbar/i)(\partial/\partial x)$, the value of expected value of kinetic energy (T) can be given as

$$\langle T\rangle = \frac{1}{2m}\int \Psi * \left(\frac{\hbar}{i}\frac{\partial}{\partial x}\right)\left(\frac{\hbar}{i}\frac{\partial}{\partial x}\right)\Psi dx \tag{4.409}$$

$$\Rightarrow \langle T\rangle = -\frac{\hbar^2}{2m}\int \Psi*\frac{\partial^2\Psi}{\partial x^2}dx \tag{4.410}$$

4.6.4 Solution of the Schrödinger Equation

The one-dimensional Schrödinger equation is given as [49]

$$i\hbar\frac{\partial\Psi(x,t)}{\partial t} = -\frac{\hbar^2}{2m}\frac{\partial^2\Psi(x,t)}{\partial x^2} + V\Psi(x,t) \tag{4.411}$$

where x represents the x axis.

The Schrödinger equation can be solved by using the separation of variables. Assume the following:

$$\Psi(x,t) = \psi(x)\varphi(t) \tag{4.412}$$

where ψ is a function of x only and φ is a function of t alone.

Therefore,

$$\frac{\partial\Psi}{\partial t} = \psi\frac{\partial\varphi}{\partial t} \tag{4.413}$$

and

$$\frac{\partial^2\Psi}{\partial x^2} = \varphi\frac{\partial^2\Psi}{\partial x^2} \tag{4.414}$$

Now Eq. (4.411) becomes

$$i\hbar\psi\frac{\partial\varphi}{\partial t} = -\varphi\frac{\hbar^2}{2m}\frac{\partial^2\psi}{\partial x^2} + V\psi\varphi \tag{4.415}$$

After simplification we get the following:

$$i\hbar\frac{1}{\varphi}\frac{\partial\varphi}{\partial t} = -\frac{1}{\psi}\frac{\hbar^2}{2m}\frac{\partial^2\psi}{\partial x^2} + V \tag{4.416}$$

In this equation, the left-hand side is the function of t alone and the right-hand side is only the function of x. Assume E as the separation constant, then

$$i\hbar\frac{1}{\varphi}\frac{\partial\varphi}{\partial t} = E \tag{4.417}$$

and

$$-\frac{1}{\psi}\frac{\hbar^2}{2m}\frac{\partial^2\psi}{\partial x^2} + V = E \tag{4.418}$$

Eq. (4.418) can be written as

$$-\frac{\hbar^2}{2m}\frac{\partial^2\psi}{\partial x^2} + V\psi = E\psi \tag{4.419}$$

This equation is the time-independent Schrödinger equation.

The Hamiltonian operator, which represents the total energy (summation of kinetic and potential), can be given as

$$H(x,p) = \frac{p^2}{2m} + V(x) \tag{4.420}$$

Utilizing the momentum operator $p \to (\hbar/i)(\partial/\partial x)$, the Hamiltonian operator can be written as

$$H = -\frac{\hbar^2}{2m}\frac{\partial^2}{\partial x^2} + V(x) \tag{4.421}$$

Therefore, the time-independent Schrödinger equation (Eq. 4.419) can be written as

$$H\psi = E\psi \tag{4.422}$$

Now, consider Eq. (4.417); the solution for $\varphi(t)$ can be given as

$$\varphi(t) = e^{-iEt/\hbar} \tag{4.423}$$

Eq. (4.412) can be written as

$$\Psi(x,t) = \psi(x)e^{-iEt/\hbar} \tag{4.424}$$

Eq. (4.424) represents the time-dependent solution of the Schrödinger equation.

The value of the separation of variables is different for each energy level $(E_1, E_2, E_3, \ldots.)$, therefore each energy level has a different wave function $(\Psi_1, \Psi_2, \Psi_3, \ldots.)$. The linear combination of separable solutions (wave functions $(\Psi_1, \Psi_2, \Psi_3, \ldots.)$) represent the general solution, which can be given as

$$\Psi(x,t) = \sum_{n=1}^{\infty} c_n\psi_n(x)e^{-iE_nt/\hbar} \tag{4.425}$$

where c_n represents the constants.

4.6.5 Examples of the Schrödinger Equation

4.6.5.1 *The Infinite Square Well*

Consider a particle of mass m confined in an infinitely deep potential energy well as shown in Fig. 4.30. Because the particle is inside the well, the probability of finding the particle outside the well is zero $(\psi(x) = 0)$.

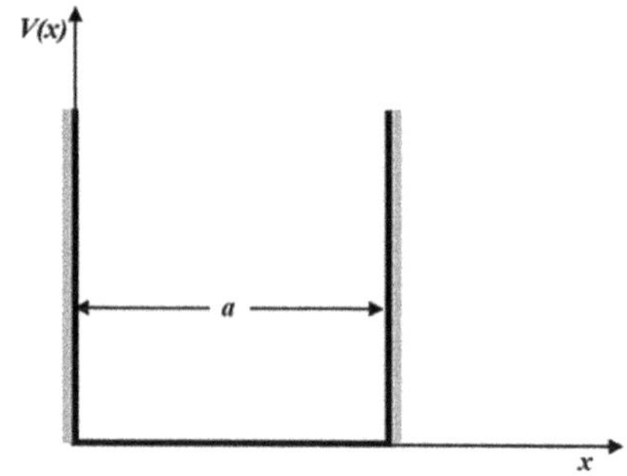

■ **FIGURE 4.30** The infinite square well potential.

Suppose the potential is given as

$$V(x) = \begin{cases} 0, & \text{if } 0 \le x \le a \\ \infty, & \text{otherwise} \end{cases} \tag{4.426}$$

Now the time-independent Schrödinger equation (Eq. 4.419) inside the well can be written as

$$-\frac{\hbar^2}{2m}\frac{\partial^2 \psi}{\partial x^2} = E\psi \tag{4.427}$$

or

$$\frac{\partial^2 \psi}{\partial x^2} = -k^2\psi \tag{4.428}$$

where

$$k = \sqrt{2mE}/\hbar \tag{4.429}$$

The general solution of Eq. (4.428) is

$$\psi(x) = A \sin kx + B \cos kx \tag{4.430}$$

where A and B are the constants. Using the boundary condition when $x = 0$, $\psi = 0$ into Eq. (4.430) gives

$$0 = A \sin 0 + B \cos 0 \tag{4.431}$$

$$\Rightarrow B = 0 \tag{4.432}$$

Therefore,

$$\psi(x) = A \sin kx \tag{4.433}$$

Now, using the boundary condition, $x = a$; $\psi = 0$ into Eq. (4.433):

$$0 = A \sin ka \tag{4.434}$$

For nontrivial solutions $A \neq 0$, therefore

$$\sin ka = 0 \tag{4.435}$$

$$\Rightarrow ka = 0,\ \pm\pi,\ \pm 2\pi,\ \pm 3\pi, \ldots. \tag{4.436}$$

Since $k=0$ makes $\psi = 0$ (from Eq. 4.428), it is not a possible solution. Since $\sin(-\theta) = -\sin(\theta)$, the negative sign can be absorbed by the constant A. Therefore, the solution is

$$k_n = \frac{n\pi}{a} \quad \text{with} \quad n = 1, 2, 3, \ldots. \tag{4.437}$$

Using Eq. (4.429) the energy E is

$$E_n = \frac{\hbar^2 k_n^2}{2m} = \frac{n^2\pi^2\hbar^2}{2ma^2} \tag{4.438}$$

$$E_n = E_1 n^2 \tag{4.439}$$

The wave function is

$$\psi_n(x) = A \sin\left(\frac{n\pi}{a}x\right) \tag{4.440}$$

Using the property of normalization,

$$\int_0^a \psi_n \psi_n^* dx = \int_0^a |A|^2 \sin^2(kx) dx \tag{4.441}$$

$$\Rightarrow |A| = \sqrt{2/a} \tag{4.442}$$

The wave function can be written as

$$\psi_n(x) = \sqrt{\frac{2}{a}} \sin\left(\frac{n\pi}{a}x\right) \tag{4.443}$$

Eq. (4.443) represents the wave function for the free particle confined in the infinite well. The lowest energy state for $n=1$ is called the ground state. The increment in energy is proportional to n^2 as given in Eq. (4.439); the higher energy states, $n \geq 2$, are the called the excited state. The first few stationary states are plotted in Fig. 4.31.

Eq. (4.443), which represents the time-independent solution for an infinite square well, has the following properties:

1. With respect to the center of well, the functions are odd and even, respectively.
2. The number of nodes in the wave function increases with an successive increment in energy state.

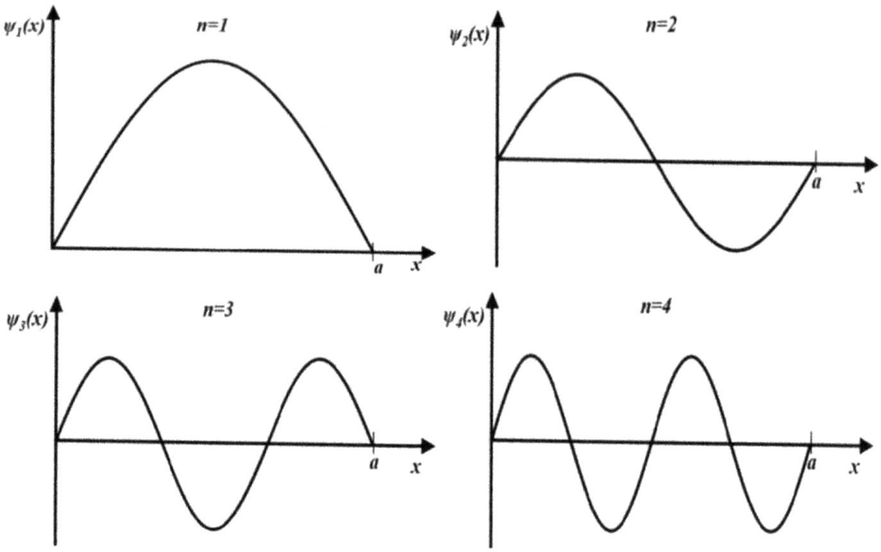

■ **FIGURE 4.31** The wave function for first four levels.

3. The states are mutually orthogonal (orthonormal); i.e.,

$$\int_{-\infty}^{\infty} \psi_m(x) * \psi_n(x)dx = \begin{cases} 0, & \text{if } m \neq n \\ 1, & \text{if } m = n \end{cases} \tag{4.444}$$

4. The wave functions are complete such that any function $f(x)$ can be represented by the linear combination of them, which can be given as

$$f(x) = \sum_{n=1}^{\infty} c_n \psi_n(x) = \sqrt{\frac{2}{a}} \sum_{n=1}^{\infty} c_n \sin\left(\frac{n\pi}{a}x\right) \tag{4.445}$$

where c_n represent the constants, using the orthonormal nature of the wave functions, which can calculated as [49]

$$c_n = \int \psi_n(x) * f(x)dx \tag{4.446}$$

The time-dependent wave function can be written as

$$\Psi(x,t) = \sum_{n=1}^{\infty} c_n \sqrt{\frac{2}{a}} \sin\left(\frac{n\pi}{a}x\right) e^{-i(n^2\pi^2\hbar/2ma^2)t} \tag{4.447}$$

Example 4.2

The initial wave function for a particle in an infinite square well is given as

$$\Psi(x,0) = \begin{cases} A(L^2 - x^2), & 0 \le x \le L \\ 0, & \text{otherwise} \end{cases}$$

where L is the length of the well.

Find out the wave function $\Psi(x,t)$ for a particle.

Solution

Fig. 4.32 represents the initial wave function for the particle.

In order to find out the value of A use the normalization (Eq. 4.405):

$$\int_0^L \Psi(x,0)\Psi*(x,0)dx = \int_0^L A^2(L^2-x^2)^2 dx = 1$$

Solving this gives

$$A = \sqrt{\frac{15}{8L^5}} \tag{4.448}$$

Now to solve for c_n, Eq. (4.446) is used with

$$f(x) = \sum_{n=1}^{\infty} c_n \psi_n(x) = \Psi(x,0)$$

Therefore,

$$\begin{aligned} c_n &= \int \psi_n(x) * \Psi(x,0)dx \\ c_n &= \sqrt{\frac{2}{L}} \int_0^L \sin\left(\frac{n\pi}{L}x\right)\sqrt{\frac{15}{8L^5}}(L^2 - x^2)dx \\ c_n &= \sqrt{\frac{30}{8L^6}}\left[L^2 \int_0^L \sin\left(\frac{n\pi}{L}x\right)dx - \int_0^L x^2 \sin\left(\frac{n\pi}{L}x\right)dx\right] \\ c_n &= \frac{\sqrt{15}}{2L^3}[(\text{int}-1) - (\text{int}-2)] \end{aligned} \tag{4.449}$$

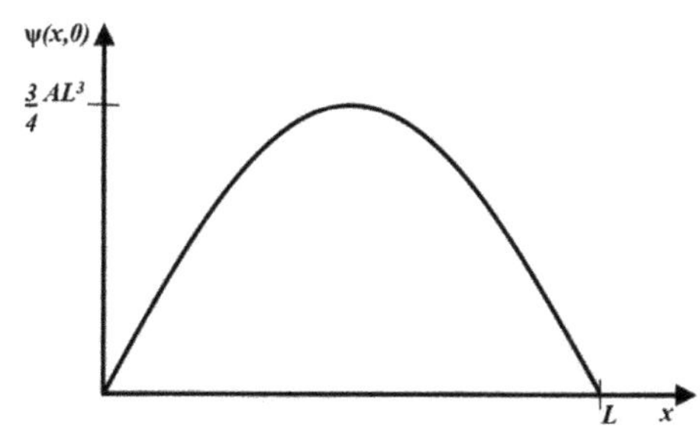

■ **FIGURE 4.32** The initial wave function for the particle in the infinite square well.

$$\text{int}-1 = L^2\int_0^L \sin\left(\frac{n\pi}{L}x\right)dx = -\frac{L^3}{n\pi}\left[\cos\left(\frac{n\pi}{L}x\right)\right]_0^L$$
$$\text{int}-1 = -\frac{L^3}{n\pi}[\cos(n\pi)-1] \tag{4.450}$$
$$\text{int}-2 = \int_0^L x^2\sin\left(\frac{n\pi}{L}x\right)dx$$

Using integration by parts $\int udv = uv - \int vdu$

$$\text{int}-2 = \int_0^L x^2\sin\left(\frac{n\pi}{L}x\right)dx = -\left[\frac{L}{n\pi}x^2\cos\left(\frac{n\pi}{L}x\right)\right]_0^L + 2\frac{L}{n\pi}\int_0^L x\cos\left(\frac{n\pi}{L}x\right)dx$$
$$\text{int}-2 = -\left[\frac{L}{n\pi}x^2\cos\left(\frac{n\pi}{L}x\right)\right]_0^L + 2\frac{L}{n\pi}\left\{\left[\frac{L}{n\pi}x\sin\left(\frac{n\pi}{L}x\right)\right]_0^L - \frac{L}{n\pi}\int_0^L \sin\left(\frac{n\pi}{L}x\right)dx\right\}$$
$$\text{int}-2 = -\left[\frac{L}{n\pi}x^2\cos\left(\frac{n\pi}{L}x\right)\right]_0^L + 2\frac{L}{n\pi}\left\{\left[\frac{L}{n\pi}x\sin\left(\frac{n\pi}{L}x\right)\right]_0^L + \left[\left(\frac{L}{n\pi}\right)^2\cos\left(\frac{n\pi}{L}x\right)\right]_0^L\right\}$$
$$\text{int}-2 = 2\left(\frac{L}{n\pi}\right)^2\left[x\sin\left(\frac{n\pi}{L}x\right)\right]_0^L - \left[\frac{L}{n\pi}x^2\cos\left(\frac{n\pi}{L}x\right)\right]_0^L + 2\left[\left(\frac{L}{n\pi}\right)^3\cos\left(\frac{n\pi}{L}x\right)\right]_0^L$$
$$\text{int}-2 = -\left[\frac{L^3}{n\pi}\cos(n\pi)\right] + 2\left(\frac{L}{n\pi}\right)^3[\cos(n\pi)-1] \tag{4.451}$$

Now, substituting the int-1 and int-2 from Eqs. (4.450) and (4.451) into Eq. (4.449) gives

$$c_n = \frac{\sqrt{15}}{2L^3}\left[-\frac{L^3}{n\pi}[\cos(n\pi)-1] + \left[\frac{L^3}{n\pi}\cos(n\pi)\right] - 2\left(\frac{L}{n\pi}\right)^3[\cos(n\pi)-1]\right]$$
$$c_n = \frac{\sqrt{15}}{2}\left[\frac{1}{n\pi} + 2\left(\frac{1}{n\pi}\right)^3[1-\cos(n\pi)]\right] \tag{4.452}$$

Using Eq. (4.447) gives

$$\Psi(x,t) = \frac{1}{2}\sqrt{\frac{30}{L}}\sum_{n=1}^{\infty}\left[\frac{1}{n\pi} + 2\left(\frac{1}{n\pi}\right)^3[1-\cos(n\pi)]\right]\sin\left(\frac{n\pi}{L}x\right)e^{-i(n^2\pi^2\hbar/2mL^2)t} \tag{4.453}$$

4.7 CLOSING REMARKS

Energy transport in micro/nanoscale structures, in general, involve nonequilibrium energy transport situations due to the length scale or short transport durations. In this case, the formulation of thermal transport in terms of the Fourier heating law fails to predict the correct temperature increase in the micro/nanostructures. The efforts to formulate the micro/nanoscale thermal energy transport led to the development of hyperbolic equations, incorporating various physical analysis, as well as the EPRT. One of the examples for the hyperbolic formulation of the problem is the Catteneo equation. The other examples extend toward the formulation of the heating problem by using two-equation and electron kinetic theory approaches. On the other hand, thermal stress becomes critical across the micro/nanoscale structures because of the large temperature gradients developed across these structures. The formulation of the problem usually incorporates the classical approach such as momentum equation. Nevertheless, the solution of such approaches still provide useful information and the strain and stress formations in the structures. Consequently, in the present chapter, formulation of microscopic energy transport in thin metallic and dielectric thin films is formulated. Analytical solution of the hyperbolic equations adopting various analytical methods is introduced. The case studies are introduced and the findings are discussed accordingly. The details of the formulation of the electron kinetic theory pertinent to laser short-pulse heating is also presented and the numerical solution of temperature field is presented due to laser short-pulse heating of metallic substrate. In later sections, phonon transport is formulated after considering the phonon radiative transport. The Casimir limit is also presented to demonstrate the thermal energy transport for the structures smaller (or comparable) than the phonon mean free path. TBR is also discussed and possible approaches formulating the problem are introduced and formulations are provided. Analytical solution for the thermal stress field is also introduced and the formulation of the stress field is included.

REFERENCES

[1] H. Al-Qahtani, B.S. Yilbas, Temperature and stress fields for short pulse heating of solids, J. Thermophys. Heat Transfer 25 (1) (2011) 173–176.

[2] B.S. Yilbas, M. Pakdemirli, Analytical solution for temperature field in electron and lattice sub-systems during heating of solid film, Phys. B 382 (1–2) (2006) 213–219.

[3] B.A. Boley, J.H. Weiner, Theory of Thermal Stresses, Robert E. Krieger Publishing Company, Malabar, FL, 1985.

[4] T. Darabseh, M. Naji, M. Al-Nimr, Transient thermal stresses in an orthotropic cylinder under the hyperbolic heat conduction model, Heat Transfer Eng. 29 (7) (2008) 632–642.

[5] J.K. Chen, J.E. Beraun, L.E. Grimes, D.Y. Tzou, Short-time thermal effects on thermomechanical response caused by pulsed lasers, J. Thermophys. Heat Transfer 17 (1) (2003) 35–42.

[6] P.H. Tehrani, M.R. Eslami, Boundary element analysis of coupled thermoelasticity with relaxation times in finite domain, AIAA J. 38 (3) (2000) 534–541.

[7] G.L. Eesley, Generation of nonequilibrium electron and lattice temperatures in copper by picosecond laser pulses, Phys. Rev. B 33 (1986) 2144.

[8] A.H. Nayfeh, Introduction to Perturbation Techniques, first ed., John Wiley and Sons, New York, 1981.

[9] B.S. Yilbas, Improved formulation of electron kinetic theory approach for laser ultra-short-pulse heating, Int. J. Heat Mass Transfer 49 (13–14) (2006) 2227–2238.

[10] B.S. Yilbas, S.Z. Shuja, Laser short-pulse heating of surfaces, J. Phys. D. Appl. Phys. 32 (16) (1999) 1947–1954.

[11] C.L. Tien, J.H. Lienhard, Statistical Thermodynamics, Hemisphere, Washington DC, 1979.

[12] B.S. Yilbas, A.F.M. Arif, Material response to thermal loading due to short pulse laser heating, Int. J. Heat Mass Transfer 44 (20) (2001) 3787–3798.

[13] H.E. Elsayed-Ali, T.B. Norris, M.A. Pessot, G.A. Mourou, Time-resolved observation of electron-phonon relaxation in copper, Phys. Rev. Lett. 58 (12) (1987) 1212.

[14] G. Chen, Ballistic-diffusive heat-conduction equations, Phys. Rev. Lett. 86 (11) (2001) 2297–2300.

[15] L.B. Loeb, The Kinetic Theory of Gases, Dover, New York, NY, 1961.

[16] M.I. Kaganov, I.M. Lifshiftz, L.V. Tanatarov, Relaxation between electrons and crystalline lattice, JETP 4 (1957) 173–178.

[17] S.D. Brorson, A. Kazeroonian, J.S. Moodera, D.W. Face, T.K. Cheng, E.P. Ippen, et al., Femtosecond room-temperature measurement of the electron-phonon coupling constant in metallic superconductors, Phys. Rev. Lett. 64 (18) (1990) 2172.

[18] M. Honner, J. Kunes, On the wave diffusion and parallel nonequilibrium heat conduction, J. Heat Transfer 121 (3) (1999) 702–707.

[19] T.-W. Tsai, Y.-M. Lee, Y.-H. Liao, Analysis of microscale heat transfer and ultrafast thermoelasticity in a multi-layered metal film, in: Proceedings of the ASME Summer Heat Transfer Conference 2009, HT2009, 2009, vol. 2, pp. 87–95.

[20] S.D. Brorson, J.G. Fujimoto, E.P. Ippen, Femtosecond electronic heat-transport dynamics in thin gold films, Phys. Rev. Lett. 59 (17) (1987) 1962–1965.

[21] B.S. Yilbas, M. Kassas, Investigation into the seebeck coefficient in a two-layer assembly during laser short-pulse heating, Japanese J. Appl. Physics, Part 1 Regul. Pap. Short Notes Rev. Pap 44 (8) (2005) 6068–6074.

[22] A. Majumdar, Microscale heat conduction in dielectric thin films, J. Heat Transfer 115 (1) (1993) 7–16.

[23] H.B.G. Casimir, Note on the conduction of heat in crystals, Physica 5 (6) (1938) 495–500.

[24] M.F. Modest, Radiative Heat Transfer, McGraw Hill, New York, 1993.

[25] S. Bin Mansoor, Absorption of Light Waves and Energy Transport in Thin Silicon Film in Relation to Photovoltaic Applications, Ph.D Thesis, KFUPM, 2010.
[26] G. Chen, Nanoscale Energy Transport and Conversion, Oxford University Press, 2005.
[27] J.M. Ziman, Electrons and Phonons: The Theory of Transport Phenomena in Solids, Oxford University Press, London, 1960.
[28] R.L. Liboff, Kinetic Theory, second ed., Willey, New York, 1998.
[29] L. De Broglie, Recherches sur la Theorie des Quanta, Ann. Phys. (Paris) 3 (10) (1925) 22.
[30] B.S. Yilbas, S.Bin Mansoor, Transient effects of phonon transport in two-dimensional silicon film, Numer. Heat Transfer, Part A 62 (9) (2012) 742–760.
[31] B.S. Yilbas, S.B. Mansoor, Phonon transport in two-dimensional silicon–diamond film pair, J. Thermophys. Heat Transfer 27 (3) (2013) 465–473.
[32] B.N. Brockhouse, Lattice vibrations in silicon and germanium, Phys. Rev. Lett. 2 (6) (1959) 256–258.
[33] A. Ward, D. Broido, D. Stewart, G. Deinzer, Ab initio theory of the lattice thermal conductivity in diamond, Phys. Rev. B 80 (12) (2009) 125203.
[34] B.S. Yilbas, S.B. Mansoor, Phonon transport and equivalent equilibrium temperature in thin silicon films, J. Non-Equilib. Thermodyn. 38 (2) (2013) 153–174.
[35] M. Planck, M. Masius, The Theory of Heat Radiation, P. Blakiston's Son & Co., 1914.
[36] A. Majumdar, P. Reddy, Role of electron-phonon coupling in thermal conductance of metal-nonmetal interfaces, Appl. Phys. Lett. 84 (23) (2004) 4768–4770.
[37] R. Stedman, G. Nilsson, Dispersion relations for phonons in aluminum at 80 and 300K, Phys. Rev. 145 (2) (1966) 492–500.
[38] A.C. Bouley, N.S. Mohan, D.H. Damon, The lattice thermal conductivity of copper and aluminum alloys at low temperatures, Therm. Conduct. 14 (1976) 81–88.
[39] C. Ni, J.Y. Murthy, Improved phonon transport modeling using boltzmann transport equation with anisotropic relaxation times, in Proceedings of the ASME InterPack Conference 2009, IPACK2009, 2010, vol. 2, pp. 99–109.
[40] A. Bejan, Advanced Engineering Thermodynamics, McGraw Hill, New York, NY, 1988.
[41] G. Chen, Thermal conductivity and ballistic-phonon transport in the cross-plane direction of superlattices, Phys. Rev. B 57 (23) (1998) 14958–14973.
[42] E.T. Swartz, R.O. Pohl, Thermal boundary resistance, Rev. Mod. Phys. 61 (3) (1989) 605–668.
[43] A.J. Minnich, G. Chen, S. Mansoor, B.S. Yilbas, Quasiballistic heat transfer studied using the frequency-dependent Boltzmann transport equation, Phys. Rev. B 84 (23) (2011) 235207.
[44] B.S. Yilbas, N. Ageeli, Thermal stress development due to laser step input pulse heating, J. Therm. Stress. 29 (8) (2006) 721–751.
[45] B.S. Yilbas, M. Kalyon, Formulation of laser pulse heating: a closed form solution including heating and cooling cycles with pulse parameter variation, Lasers Eng. 14 (3–4) (2004) 213–228.
[46] A.D. Kovalenko, Thermoelaticity: Basic Theory and Applications, Wolters-Noordhoff Publishing, Groningen, The Netherlands, 1969.

[47] C.F. Bohren, D.R. Huffman, Absorption and Scattering of Light by Small Particles, Wiley-Interscience publication, New York, 1998.
[48] E. Schrödinger, Quantisation as a problem of characteristic values, Ann. Phys. 79 (1926), 361-376-529.
[49] D.J. Griffiths, Introduction to Quantum Mechanics, second ed., Pearson Prentice Hall, 2004.

Chapter 5

Analytical Treatment of Phonon Transport in Thin Films

5.1 INTRODUCTION

The analytical solution of the equation for phonon radiative transport is presented for thin films. In the first case, the reduction of the equation for phonon radiative transport to hyperbolic form is presented. The analytical solution of the phonon radiative transfer equation for steady-state analysis under temperature disturbance is introduced as a second case. The solution for the dielectric and the metallic films are included separately. The two-equation model solution of temperature distribution is also accounted for the comparison of the analytical solution of the phonon radiative transport equation. It should be noted that the analytical solution of the hyperbolic equation due to electron kinetic theory model is presented last. The solution for the hyperbolic heat conduction for thin films was introduced previously [1]; however, the resulting equation is limited with the space and time domain due to the solution technique introduced in the perturbation method. As an extension of previous study [1], the hyperbolic heat conduction equation can be derived from the Boltzmann equation. In order to demonstrate the general solution of the resulting equations, a volumetric heat source resembling the unidirectional beam irradiation, such as laser beam, is incorporated in the equations. The resulting equations are simulated for the practical cases to demonstrate temperature variation in thin films. The mathematical analyses are given in line with the previous studies [2–5].

5.2 ANALYTICAL SOLUTION OF REDUCED EQUATION FOR PHONON RADIATIVE TRANSPORT

Consider the Boltzmann transport equation (BTE) in reduced form of the phonon radiative transfer in a substrate material [6]:

$$\frac{1}{v}\frac{\partial I_\omega}{\partial t} + \mu\frac{\partial I_\omega}{\partial x} = \frac{\left(\frac{1}{2}\int_{-1}^{1} I_\omega d\mu\right) - I_\omega}{v\tau(\omega, T)} \tag{5.1}$$

Heat Transport in Micro- and Nanoscale Thin Films.
DOI: http://dx.doi.org/10.1016/B978-0-323-42979-5.00005-0

where the phonon intensity $I_\omega = I_\omega(x, \mu, \omega, t)$ can be defined as

$$I_\omega(x, \mu, \omega, t) = \sum_{p=1}^{3} v f_\omega(x, t)\hbar\omega D(\omega) \tag{5.2}$$

In this expression, v is the speed of the phonons; $f_\omega(x, t)$ is the density of the phonon ensemble in phase space at point x and time t; $\hbar$ is Planck's constant divided by 2π; ω is the angular frequency of the phonons, and ranges from 0 to ω_D; τ is the phonon relaxation time, which is temperature dependent; $D(\omega)$ is the density of states; $\mu = \cos\theta$; θ is the polar angle; and p is the polarization index. There are three polarizations of the phonons: two transverse (mutually orthogonal) and one longitudinal.

The heat flux in the x direction is given as

$$\begin{aligned} q_x(x,t) &= \sum_p \int_{4\pi} \int_0^{\omega_D} v_x f_\omega(x,t)\hbar\omega D(\omega)d\omega d\Omega \\ &= 2\pi \sum_p \int_{-1}^{1} \int_0^{\omega_D} v\mu f_\omega(x,t)\hbar\omega D(\omega)d\omega d\mu \\ &= 2\pi \int_{-1}^{1} \int_0^{\omega_D} \mu I_\omega(x,\mu,\omega,t)d\omega d\mu \end{aligned} \tag{5.3}$$

where $v_x = v\mu$ ($\mu = \cos\theta$). In the following derivation we will assume that the relaxation time is independent of the phonon frequency ω. Multiplying Eq. (5.1) by $2\pi\mu$ and integrating it with respect to μ from -1 to 1 and with respect to ω from 0 to ω_D yields

$$\begin{aligned} &\frac{2\pi}{v}\frac{\partial}{\partial t}\int_0^{\omega_D}\int_{-1}^{1} \mu I_\omega d\mu d\omega + 2\pi \int_0^{\omega_D}\int_{-1}^{1} \mu^2 \frac{\partial I_\omega}{\partial x} d\mu d\omega \\ &= \frac{2\pi \int_0^{\omega_D}\int_{-1}^{1} \mu\left(\frac{1}{2}\int_{-1}^{1} I_\omega d\mu\right) d\mu d\omega - 2\pi \int_0^{\omega_D}\int_{-1}^{1} \mu I_\omega d\mu d\omega}{v\tau(T)} \end{aligned} \tag{5.4}$$

The first term on the left-hand side of Eq. (5.4) can be simplified to

$$\frac{2\pi}{v}\frac{\partial}{\partial t}\int_0^{\omega_D}\int_{-1}^{1} \mu I_\omega d\mu d\omega = \frac{1}{v}\frac{\partial q_x(x,t)}{\partial t} \tag{5.5}$$

The last term on the right-hand side of Eq. (5.4) may be simplified by using the definition of the heat flux as

$$\frac{2\pi \int_0^{\omega_D}\int_{-1}^{1} \mu I_\omega d\mu d\omega}{v\tau(T)} = \frac{q_x(x,t)}{v\tau(T)} \tag{5.6}$$

The first term on the right-hand side is identically zero because it was integrated two times with respect to μ. After the first integration the integrand contains only the first power of μ and a second integration will make the expression 0:

$$\frac{2\pi\int_0^{\omega_D}\int_{-1}^{1}\mu\left(\frac{1}{2}\int_{-1}^{1}I_\omega d\mu\right)d\mu d\omega}{v\tau(T)}=0 \tag{5.7}$$

The second term on the left-hand side of Eq. (5.4) may be simplified as follows:

$$2\pi\int_0^{\omega_D}\int_{-1}^{1}\mu^2\frac{\partial I_\omega}{\partial x}d\mu d\omega=2\pi\sum_p\int_0^{\omega_D}\int_{-1}^{1}\mu^2 v\frac{\partial f_\omega(x,t)}{\partial x}\hbar\omega D(\omega)d\mu d\omega \tag{5.8}$$

If the phonon gas is not very far from the equilibrium condition then we may write [7]

$$f_\omega(x,t)\approx f_\omega^o(T(x,t)) \tag{5.9}$$

where $f_\omega^o(T(x,t))$ is the equilibrium density of phonon ensemble at a point x and time t corresponding to the temperature $T(x,t)$ at that point and time. For phonons this is given by the Bose–Einstein distribution function. This allows the following approximations to be made [7]:

$$\frac{\partial f_\omega(x,t)}{\partial x}\approx\frac{\partial f_\omega^o(T(x,t))}{\partial x}=\frac{\partial f_\omega^o(T(x,t))}{\partial T}\frac{dT}{dx} \tag{5.10}$$

Substituting Eq. (5.10) into Eq. (5.8) yields

$$\begin{aligned}2\pi\int_0^{\omega_D}\int_{-1}^{1}\mu^2\frac{\partial I_\omega}{\partial x}d\mu d\omega&=2\pi\sum_p\int_0^{\omega_D}\int_{-1}^{1}\mu^2 v\frac{df_\omega^o}{dT}\frac{dT}{dx}\hbar\omega D(\omega)d\mu d\omega\\&=\frac{1}{v\tau}\frac{dT}{dx}2\pi\sum_p\int_0^{\omega_D}\int_{-1}^{1}\tau\mu^2v^2\frac{df_\omega^o}{dT}\hbar\omega D(\omega)d\mu d\omega\\&=\frac{1}{v\tau}\frac{dT}{dx}2\pi\sum_p\int_0^{\omega_D}\int_{-1}^{1}\tau v_x^2\frac{df_\omega^o}{dT}\hbar\omega D(\omega)d\mu d\omega\\&=\frac{1}{v\tau}\frac{dT}{dx}k\end{aligned} \tag{5.11}$$

In this expression, the thermal conductivity k is defined as [6]

$$\begin{aligned}k&=\frac{2\pi}{3}\sum_p\int_0^{\omega_D}\int_{-1}^{1}\tau v^2\frac{df_\omega^o}{dT}\hbar\omega D(\omega)d\mu d\omega\\&=2\pi\sum_p\int_0^{\omega_D}\int_{-1}^{1}\tau v_x^2\frac{df_\omega^o}{dT}\hbar\omega D(\omega)d\mu d\omega\end{aligned} \tag{5.12}$$

The last expression follows from the fact that the medium is isotropic, therefore, the average speed of the phonons in the three-coordinate direction is approximately the same, i.e., $v^2 = v_x^2 + v_y^2 + v_z^2 \approx 3v_x^2$. Substituting Eqs. (5.5), (5.6), (5.7), and (5.11) in Eq. (5.4) yields

$$\frac{1}{v}\frac{\partial q_x}{\partial t} + \frac{k}{v\tau(T)}\frac{dT}{dx} = \frac{0 - q_x}{v\tau(T)} \tag{5.13}$$

or

$$\tau\frac{\partial q_x}{\partial t} + q_x = -k\frac{dT}{dx} \tag{5.14}$$

The energy conservation relation is given as

$$C\frac{\partial T}{\partial t} + \nabla \cdot \mathbf{q} = 0 \tag{5.15}$$

where C is the volumetric specific heat capacity and $\mathbf{q}$ is the heat flux vector. For one-dimensional heat transfer, the previous relation may be written as

$$C\frac{\partial T}{\partial t} + \frac{\partial q_x}{\partial x} = 0 \tag{5.16}$$

Differentiating Eq. (5.14) with respect to x and assuming constant thermal conductivity k yields

$$\tau\frac{\partial^2 q_x}{\partial x \partial t} + \frac{\partial q_x}{\partial x} = -k\frac{d^2T}{dx^2} \Rightarrow \frac{\partial q_x}{\partial x} = -k\frac{d^2T}{dx^2} - \tau\frac{\partial^2 q_x}{\partial x \partial t} \tag{5.17}$$

Substituting the last expression in Eq. (5.16) yields

$$C\frac{\partial T}{\partial t} - k\frac{d^2T}{dx^2} - \tau\frac{\partial^2 q_x}{\partial x \partial t} = 0 \tag{5.18}$$

Differentiating Eq. (5.16) with respect to t and rearranging terms yields

$$\frac{\partial^2 q_x}{\partial t \partial x} = -C\frac{\partial^2 T}{\partial t^2} \tag{5.19}$$

Substituting Eq. (5.19) into Eq. (5.18) and rearranging terms finally yields the equation similar to the Cattaneo equation:

$$\tau C\frac{\partial^2 T}{\partial t^2} + C\frac{\partial T}{\partial t} = k\frac{d^2T}{dx^2} \tag{5.20}$$

From the kinetic theory it is known that $k = \frac{Cv^2\tau}{3}$, which may be rearranged to yield

$$\frac{3k}{v^2} = C\tau \Rightarrow \frac{k}{v_x^2} = C\tau \tag{5.21}$$

Substituting the last expression in Eq. (5.20) and simplifying yields

$$\frac{1}{v_x^2}\frac{\partial^2 T}{\partial t^2} + \frac{1}{\alpha}\frac{\partial T}{\partial t} = \frac{d^2 T}{dx^2} \tag{5.22}$$

where α is the thermal diffusivity.

In order to generalize Eq. (5.22), consider the laser short-pulse heating situation where the volumetric source term is added to Eq. (5.22). The source term resembles the absorption of the laser heating source. Since the laser pulse intensity varies with time, exponential function can be introduced to account for the time decay of the laser pulse [8]. Eq. (5.22) with the volumetric source term can be written as

$$A\frac{\partial^2}{\partial t^2}T(x,t) + B\frac{\partial}{\partial t}T(x,t) = \frac{\partial^2}{\partial x^2}T(x,t) + I_0\delta e^{-\delta x}e^{-\beta t} \tag{5.23}$$

where $I_0 = \frac{I'_0\delta}{k}$ (I'_0 is the peak power intensity of the laser pulse, δ is the absorption coefficient, and k is the thermal conductivity), β is the laser pulse parameter A ($= \frac{\rho C_p}{k\tau}$, ρ is the density, C_p is the specific heat, τ is the relaxation time), and B $\left(= \frac{\rho C_p}{k}\right)$ are the coefficients. Temporal variation of I'_0 is shown in Fig. 5.1. The initial and boundary conditions appropriate for the laser heating situation are given by

$$\begin{aligned} &T(x,0) = 0 \\ &\frac{\partial T}{\partial t}(x,0) = 0 \end{aligned} \tag{5.24}$$

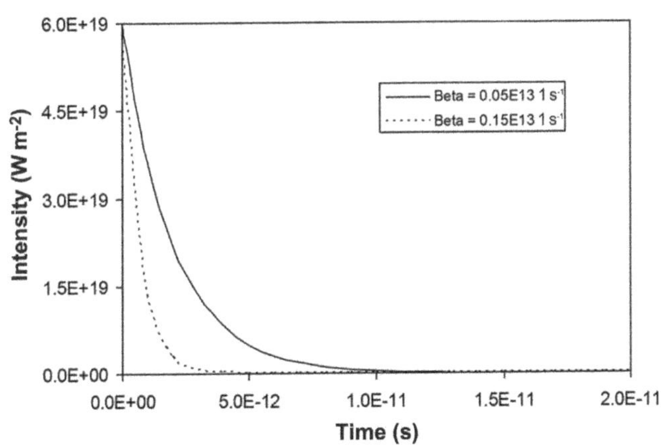

FIGURE 5.1 Temporal variation of laser pulse intensity for two laser pulse parameter (β) values [2].

In Eq. (5.24), initially, it is assumed that temperature is zero inside the substrate material and time derivative of temperature is also zero.

$$T(\infty, t) = 0$$
$$\frac{\partial T}{\partial x}(0, t) = 0 \tag{5.25}$$

In Eq. (5.25), it is assumed that at large depth below the surface substrate material remains at the initial temperature. In addition, no convection boundary is assumed at the surface during the short heating period. Applying the Fouriercos transform with respect to x, by requiring that, for a physically meaningful system, $T(x,t)$ and its first derivatives in x vanish as $x \to \infty$, it results in

$$A\frac{\partial^2}{\partial t^2}U(s,t) + B\frac{\partial}{\partial t}T(s,t) = -s^2U(s,t) - \sqrt{\frac{2}{\pi}}\frac{\partial U}{\partial x}(0,t) + \sqrt{\frac{2}{\pi}}\frac{I_0\delta e^{-\beta t}}{\delta^2+s^2} \tag{5.26}$$

where s is the Fouriercos variable and $U(s,t)$ is the Fouriercos of $T(x,t)$ with respect to x, which is defined by $U(s,t) = \sqrt{\frac{2}{\pi}}\int_0^\infty T(x,t)\cos(sx)dx$ and the inverse Fouriercos of $U(s,t)$ is given by $T(x,t) = \sqrt{\frac{2}{\pi}}\int_0^\infty U(s,t)\cos(sx)ds$.

By applying the boundary conditions for Eq. (5.25), we can get

$$A\frac{\partial^2}{\partial t^2}U(s,t) + B\frac{\partial}{\partial t}U(s,t) = -s^2U(s,t) + \sqrt{\frac{2}{\pi}}\frac{I_0\delta e^{-\beta t}}{\delta^2+s^2} \tag{5.27}$$

Now, Eq. (5.27) can be solved through applying the Laplace transform with respect to t,

$$\begin{aligned} &Ap^2\,\overline{U}(s,p) + Bp\,\overline{U}(s,p) - A\frac{\partial U}{\partial t}(s,0) - U(s,0)(Ap+B) \\ &= -s^2\,\overline{U}(s,p) + \sqrt{\frac{2}{\pi}}\frac{I_0\delta}{(\delta^2+s^2)(p+\beta)} \end{aligned} \tag{5.28}$$

where p is the Laplace variable and $\overline{U}(s,p)$ is the Laplace of $U(s,t)$ with respect to t.

However, applying the initial conditions by using the formula $U(s,t) = \sqrt{\frac{2}{\pi}}\int_0^\infty T(x,t)\cos(sx)dx$ gives

$$U(s,0) = 0$$
$$\frac{\partial U}{\partial t}(s,0) = 0 \tag{5.29}$$

So, Eq. (5.28) will reduce to

$$Ap^2\overline{U}(s,p) + Bp\overline{U}(s,p) = -s^2\overline{U}(s,p) + \sqrt{\frac{2}{\pi}}\frac{I_0\delta}{(\delta^2+s^2)(p+\beta)} \tag{5.30}$$

Then $\overline{U}(s,p)$ can be given as:

$$\overline{U}(s,p) = \sqrt{\frac{2}{\pi}}\frac{I_0\delta}{(\delta^2+s^2)(p+\beta)(Ap^2+Bp+s^2)} \tag{5.31}$$

The temperature equation in the Fouriercos domain in Eq. (5.28) can be given by using the following formula [9]:

$$F_c\left\{\frac{\partial^2}{\partial x^2}u(x,t)\right\} = -s^2F_c\{u(x,t)\} - \sqrt{\frac{2}{\pi}}\frac{\partial}{\partial x}u(0,t). \tag{5.32}$$

$$F_c\{e^{-ax}\} = \sqrt{\frac{2}{\pi}}\frac{a}{a^2+s^2}, \quad (a>0). \tag{5.33}$$

The solution of the temperature equation in the Fouriercos–Laplace domain (Eq. 5.28) can be obtained by using

$$\begin{aligned}\overline{U}(s,p) &= \sqrt{\frac{2}{\pi}}\frac{I_0\delta}{\delta^2+s^2}\frac{1}{(p+\beta)(Ap^2+Bp+s^2)}\\ &= \sqrt{\frac{2}{\pi}}\frac{I_0\delta}{\delta^2+s^2}\left(\frac{l_1}{p+\beta}+\frac{l_2}{p-\omega_1}+\frac{l_3}{p-\omega_2}\right)\end{aligned} \tag{5.34}$$

where $\omega_1 = \frac{-B+\sqrt{B^2-4As^2}}{2A}$, $\omega_2 = \frac{-B-\sqrt{B^2-4As^2}}{2A}$.

Then the Laplace inversion of the previous solution (Eq. 5.34) can be given as

$$\begin{aligned}T(x,t) &= \sqrt{\frac{2}{\pi}}\frac{I_0\delta}{\delta^2+s^2}L^{-1}\left\{\frac{l_1}{p+\beta}+\frac{l_2}{p-\omega_1}+\frac{l_3}{p-\omega_2}\right\}\\ &= \sqrt{\frac{2}{\pi}}\frac{I_0\delta}{\delta^2+s^2}\left(l_1e^{-\beta t}+l_2e^{\omega_1 t}+l_3e^{\omega_2 t}\right)\end{aligned} \tag{5.35}$$

where l_1, l_2, and l_3 are given as follows:

$$l_1 = \lim_{p\to-\beta}(p+\beta)\frac{1}{(p+\beta)(Ap^2+Bp+s^2)} = \frac{1}{(\beta+\omega_1)(\beta+\omega_2)} = \frac{1}{A\beta^2-B\beta+s^2},$$

$$l_2 = \lim_{p\to\omega_1}(p-\omega_1)\frac{1}{(p+\beta)(Ap^2+Bp+s^2)}$$

or

$$l_2 = \lim_{p\to\omega_1}\frac{1}{(p+\beta)(p-\omega_2)} = \frac{1}{(\omega_1+\beta)(\omega_1-\omega_2)}$$

or

$$l_2 = \frac{(\beta+\omega_2)}{(\beta+\omega_2)(\omega_1+\beta)(\omega_1-\omega_2)} = \frac{1}{A\beta^2-B\beta+s^2}\frac{(\beta+\omega_2)}{(\omega_1-\omega_2)}$$

or

$$l_2 = \frac{1}{A\beta^2-B\beta+s^2}\frac{2A\beta-B+\sqrt{B^2-4As^2}}{2\sqrt{B^2-4As^2}},$$

$$l_3 = \lim_{p\to\omega_2}(p-\omega_2)\frac{1}{(p+\beta)(Ap^2+Bp+s^2)}$$

or

$$l_3 = \lim_{p\to\omega_2}\frac{1}{(p+\beta)(p-\omega_1)} = \frac{1}{(\omega_2+\beta)(\omega_2-\omega_1)}$$

or

$$l_3 = \frac{(\beta+\omega_1)}{(\beta+\omega_1)(\omega_2+\beta)(\omega_2-\omega_1)} = -\frac{1}{A\beta^2-B\beta+s^2}\frac{(\beta+\omega_1)}{(\omega_1-\omega_2)}$$

or

$$l_3 = -\frac{1}{A\beta^2-B\beta+s^2}\frac{2A\beta-B-\sqrt{B^2-4As^2}}{2\sqrt{B^2-4As^2}}. \tag{5.36}$$

Hence the Laplace inversion is

$$T(x,t) = \sqrt{\frac{2}{\pi}}\frac{I_0\delta}{\delta^2+s^2}\left(l_1e^{-\beta t}+l_2e^{\omega_1 t}+l_3e^{\omega_2 t}\right) = \sqrt{\frac{2}{\pi}}\frac{I_0\delta}{(\delta^2+s^2)(A\beta^2-B\beta+s^2)}$$

$$\times\begin{pmatrix} e^{-\beta t}+e^{-\frac{B}{2A}t}\dfrac{2A\beta-B+\sqrt{B^2-4As^2}}{2\sqrt{B^2-4As^2}}e^{\frac{\sqrt{B^2-4As^2}}{2A}t} \\ \\ -e^{-\frac{B}{2A}t}\dfrac{2A\beta-B-\sqrt{B^2-4As^2}}{2\sqrt{B^2-4As^2}}e^{-\frac{\sqrt{B^2-4As^2}}{2A}t} \end{pmatrix}$$

or

$$= \sqrt{\frac{2}{\pi}}\frac{I_0\delta}{(\delta^2+s^2)(A\beta^2-B\beta+s^2)}\begin{pmatrix} e^{-\beta t}-e^{-\frac{Bt}{2A}}\cos h\left(\dfrac{t\sqrt{B^2-4As^2}}{2A}\right) \\ \\ +\dfrac{2A\beta-B}{\sqrt{B^2-4As^2}}\mathrm{e}^{-\frac{Bt}{2A}}\sin h\left(\dfrac{t\sqrt{B^2-4As^2}}{2A}\right) \end{pmatrix} \tag{5.37}$$

After the Laplace inversion of Eq. (5.31), we can have

$$U(s,t)=\sqrt{\frac{2}{\pi}}\frac{I_0\delta}{(\delta^2+s^2)(A\beta^2-B\beta+s^2)} \times\left(\begin{array}{l} e^{-\beta t}-e^{-\frac{Bt}{2A}}\cos\text{h}\left(\dfrac{t\sqrt{B^2-4As^2}}{2A}\right) \\ +\dfrac{2A\beta-B}{\sqrt{B^2-4As^2}}e^{-\frac{Bt}{2A}}\sin\text{h}\left(\dfrac{t\sqrt{B^2-4As^2}}{2A}\right)\end{array}\right) \tag{5.38}$$

Then the final solution is the inverse Fouriercos transform to $U(s,t)$; i.e.,

$$T(x,t)=\sqrt{\frac{2}{\pi}}\int_0^\infty U(s,t)\cos(sx)ds. \tag{5.39}$$

There are two positive singularities for $U(s,t)$, which are given as

$$sg_1=\sqrt{B\beta-A\beta^2} \quad \text{and} \quad sg_2=\frac{B}{2\sqrt{A}} \tag{5.40}$$

Hence, the final solution can be given by the closed form,

$$T(x,t)=\sqrt{\frac{2}{\pi}}\left(\int_0^{sg_1}U(s,t)\cos(sx)ds+\int_{sg_1}^{sg_2}U(s,t)\cos(sx)ds+\int_{sg_2}^{\infty}U(s,t)\cos(sx)ds\right) \tag{5.41}$$

where $U(s,t)$ is given by Eq. (5.38). This integral can be evaluated numerically to obtain the value of $T(x,t)$ at given (x,t). To do so, note that although the range of integration is from 0 to ∞, $U(s,t)$ generally decays with s, so we can truncate the integral at a certain finite (but large enough) value of s.

5.3 ANALYTICAL SOLUTION OF STEADY-STATE EQUATION FOR PHONON RADIATIVE TRANSPORT (EPRT) DUE TO TEMPERATURE DISTURBANCE ACROSS A THIN FILM

Although the numerical solution of the Boltzmann equation is possible, the simulations require excessive computer power and computational efforts. An analytical solution to the Boltzmann equation reduces the computational efforts and provides the analytical relations between the film properties and the phonon intensity distribution. Two types of films can be considered for the analytical solution of the equation for the phonon radiative transport, namely metallic and dielectric films.

The phonon transport is the governing mechanism for energy transfer in the dielectric thin films. However, separation of the electron and the lattice subsystems, due to nonequilibrium heating situations, gives rise to consideration of energy transport in the electron subsystem as well as phonon transport in the lattice subsystem. Therefore, these considerations are introduced in the following subsections in light of the previous studies [3,4].

5.3.1 Consideration of Dielectric Thin Film

In thin dielectric films, in some cases, the film thickness is comparable to the mean free path of the phonons and it is necessary to solve the BTE to determine the transport properties. The BTE is derived from Liouville's equation, which governs the time evolution of the N particle distribution function of a collection of particles in $6N$ dimensional phase space. However, the collision integral is a complicated expression accounting for the effects of collisions between the particles. It may be replaced by a simpler expression by assuming the relaxation time approximation [6]. For a thin film, which has a small thickness as compared to its other two dimensions, we may use the one-dimensional BTE to describe the dynamics of the dilute phonon gas. However, the phonons can be considered as moving between the two surfaces of the thin film, across the thickness, which is similar to the photons when they are transferring thermal energy between two surfaces across a vacuum or across a participating medium. Therefore, the equation of radiative transfer for phonons can be used to describe the energy transport across the thin film; i.e., [6]

$$\frac{1}{v}\frac{\partial I_\omega}{\partial t} + \mu\frac{\partial I_\omega}{\partial x} = -\frac{(I_\omega - I_\omega^o)}{v\tau} \tag{5.42}$$

where I_ω is the phonon intensity in the gray film, x is the coordinate variable along the film thickness, μ is the cosine of the polar angle, t is the time variable, and Λ ($\Lambda = v\tau$) is the mean free path of the phonons. For steady-state phonon transfer, $\partial I/\partial t = 0$ in Eq. (5.42) and the resulting equation becomes

$$\mu\frac{\partial I_\omega}{\partial x} = -\frac{(I_\omega - I_\omega^o)}{\Lambda} \tag{5.43}$$

The polar angle changes during the radiative transport, which in turn leads two identical radiative transport equations as shown:

$$\mu\frac{\partial I_\omega^+}{\partial x} = \frac{I_\omega^* - I_\omega^+}{\Lambda}, \quad \text{when} \quad 0 < \mu < 1 \tag{5.44}$$

and

$$\mu \frac{\partial I_\omega^-}{\partial x} = \frac{I_\omega^* - I_\omega^-}{\Lambda}, \quad \text{when} \quad -1 < \mu < 0 \tag{5.45}$$

where I_ω^+ represents the emitted intensity from the high temperature edge, I_ω^- corresponds to the emitted intensity from the low temperature edge, and $I_\omega^*(x) = \frac{1}{2}\int_{-1}^{1} I_\omega(x,\mu)d\mu$.

Since the film edges are at different temperatures across the film thickness (L), the corresponding boundary conditions become

$$\begin{aligned} I_\omega^+(0,\mu) &= \frac{CvT_{\text{left}}}{4\pi} = a \\ I_\omega^-(L,\mu) &= \frac{CvT_{\text{back}}}{4\pi} = b \end{aligned} \tag{5.46}$$

where C is the specific heat and v is the acoustic speed in the film. In order to introduce the analytical solution to Eqs. (5.44) and (5.45), the following procedure is adopted. In this case, multiplying both Eqs. (5.44) and (5.45) with the multiplier $\exp\left(\frac{x}{\mu\Lambda}\right)$ gives

$$\begin{aligned} \frac{\partial}{\partial x}\left(\exp\left(\frac{x}{\mu\Lambda}\right)I_\omega^+\right) &= \exp\left(\frac{x}{\mu\Lambda}\right)\frac{I_\omega^*}{\mu\Lambda}, \quad \text{when} \quad 0 < \mu < 1 \\ \frac{\partial}{\partial x}\left(\exp\left(\frac{x}{\mu\Lambda}\right)I_\omega^-\right) &= \exp\left(\frac{x}{\mu\Lambda}\right)\frac{I_\omega^*}{\mu\Lambda}, \quad \text{when} \quad -1 < \mu < 0 \end{aligned} \tag{5.47}$$

By integrating both sides with respect to x and applying the boundary conditions (Eq. 5.47), the solution of Eqs. (5.44) and (5.45) can be expressed as follows:

$$\begin{aligned} I_\omega^+(x,\mu) &= \exp\left(\frac{-x}{\mu\Lambda}\right)\left(\int_0^x \exp\left(\frac{\xi}{\mu\Lambda}\right)\frac{I_\omega^*(\xi)}{\mu\Lambda}d\xi + I_\omega^+(0,\mu)\right), \quad \text{when} \quad 0 < \mu < 1 \\ I_\omega^-(x,\mu) &= \exp\left(\frac{-x}{\mu\Lambda}\right)\left(\int_L^x \exp\left(\frac{\xi}{\mu\Lambda}\right)\frac{I_\omega^*(\xi)}{\mu\Lambda}d\xi + \exp\left(\frac{L}{\mu\Lambda}\right)I_\omega^-(L,\mu)\right), \\ &\text{when} \quad -1 < \mu < 0 \end{aligned} \tag{5.48}$$

Eq. (5.48) is also presented in the open literature [10].

However, the term $I_\omega^*(x) = \frac{1}{2}\int_{-1}^{1} I_\omega(x,\mu)d\mu = \frac{1}{2}\left(\int_{-1}^{0} I_\omega^-(x,\mu)d\mu + \int_0^1 I_\omega^+(x,\mu)d\mu\right)$ then yields

$$\begin{aligned} &\int_{-1}^{0}\left(\int_L^x \exp\left(\frac{\xi - x}{\mu\Lambda}\right)\frac{I_\omega^*(\xi)}{\mu\Lambda}d\xi + b\exp\left(\frac{L-x}{\mu\Lambda}\right)\right)d\mu \\ &+ \int_0^1\left(\int_0^x \exp\left(\frac{\xi - x}{\mu\Lambda}\right)\frac{I_\omega^*(\xi)}{\mu\Lambda}d\xi + a\exp\left(\frac{-x}{\mu\Lambda}\right)\right)d\mu = 2I_\omega^*(x) \end{aligned} \tag{5.49}$$

The integral equation (Eq. 5.49) can be written as the Fredholm integral equation of the second kind. In this case,

$$I_\omega^*(x) - \int_0^L K(x,\xi) I_\omega^*(\xi) d\xi = f(x) \tag{5.50}$$

where the kernel $K(x,\xi) = \frac{1}{2\Lambda} Ei\left(1, \frac{|x-\xi|}{\Lambda}\right)$ and $f(x) = \frac{1}{2}\left(aEi\left(2, \frac{x}{\Lambda}\right) + bEi\left(2, \frac{L-x}{\Lambda}\right)\right)$.

It is clear that the kernel is weakly singular, so in order to remove this singularity at the diagonal, $s = t$, we can consider the partition $\{x_0, x_1, \ldots, x_N\}$ of $[0,L]$ and then approximate the $I_\omega^*(x)$ on small intervals $[x_j, x_{j+1}]$ by $\frac{I_\omega^*(x_j) + I_\omega^*(x_{j+1})}{2}$. Finally, the singularity disappears by integrating the kernel. Here are the details:

Since Eq. (5.49) is true for each x_i, then

$$\int_{-1}^{0}\left(\int_L^{x_i} \exp\left(\frac{\xi - x_i}{\mu\Lambda}\right)\frac{I_\omega^*(\xi)}{\mu\Lambda} d\xi + b\exp\left(\frac{L-x_i}{\mu\Lambda}\right)\right) d\mu + \int_0^1\left(\int_0^{x_i} \exp\left(\frac{\xi - x_i}{\mu\Lambda}\right)\frac{I_\omega^*(\xi)}{\mu\Lambda} d\xi + a\exp\left(\frac{-x_i}{\mu\Lambda}\right)\right) d\mu = 2I_\omega^*(x_i), \quad i = 0\ldots N. \tag{5.51}$$

Distributing the integral over the partition gives

$$\int_{-1}^{0}\left(-\sum_{j=i}^{N-1}\int_{x_j}^{x_{j+1}} \exp\left(\frac{\xi - x_i}{\mu\Lambda}\right)\frac{I_\omega^*(\xi)}{\mu\Lambda} d\xi + b\exp\left(\frac{L-x_i}{\mu\Lambda}\right)\right) d\mu + \int_0^1\left(\sum_{j=0}^{i-1}\int_{x_j}^{x_{j+1}} \exp\left(\frac{\xi - x_i}{\mu\Lambda}\right)\frac{I_\omega^*(\xi)}{\mu\Lambda} d\xi + a\exp\left(\frac{-x_i}{\mu\Lambda}\right)\right) d\mu = 2I_\omega^*(x_i), \quad i = 0\ldots N. \tag{5.52}$$

Since $I_\omega^*(x) \approx \frac{I_\omega^*(x_j) + I_\omega^*(x_{j+1})}{2}$ on each interval $[x_j, x_{j+1}]$, then

$$\int_{-1}^{0}\left(-\sum_{j=i}^{N-1}\left(\frac{I_\omega^*(x_j) + I_\omega^*(x_{j+1})}{2}\right)\int_{x_j}^{x_{j+1}} \exp\left(\frac{\xi - x_i}{\mu\Lambda}\right)\frac{1}{\mu\Lambda} d\xi + b\exp\left(\frac{L-x_i}{\mu\Lambda}\right)\right) d\mu + \int_0^1\left(\sum_{j=0}^{i-1}\left(\frac{I_\omega^*(x_j) + I_\omega^*(x_{j+1})}{2}\right)\int_{x_j}^{x_{j+1}} \exp\left(\frac{\xi - x_i}{\mu\Lambda}\right)\frac{1}{\mu\Lambda} d\xi + a\exp\left(\frac{-x_i}{\mu\Lambda}\right)\right) d\mu = 2I_\omega^*(x_i),$$
$$i = 0\ldots N. \tag{5.53}$$

Integrating the kernel gives

$$\int_{-1}^{0}\left(-\sum_{j=i}^{N-1}\left(I_\omega^*(x_j) + I_\omega^*(x_{j+1})\right)\left(\exp\left(\frac{x_{j+1} - x_i}{\mu\Lambda}\right) - \exp\left(\frac{x_j - x_i}{\mu\Lambda}\right)\right) + 2b\exp\left(\frac{L-x_i}{\mu\Lambda}\right)\right) d\mu + \int_0^1\left(\sum_{j=0}^{i-1}\left(I_\omega^*(x_j) + I_\omega^*(x_{j+1})\right)\left(\exp\left(\frac{x_{j+1} - x_i}{\mu\Lambda}\right) - \exp\left(\frac{x_j - x_i}{\mu\Lambda}\right)\right) + 2a\exp\left(\frac{-x_i}{\mu\Lambda}\right)\right) d\mu$$
$$= 4I_\omega^*(x_i), \quad i = 0\ldots N. \tag{5.54}$$

Therefore, distributing the integral with respect to μ gives the linear system:

$$
\begin{aligned}
&4I_\omega^*(x_i) + \sum_{j=i}^{N-1}(I_\omega^*(x_j) + I_\omega^*(x_{j+1}))d(i,j) - \sum_{j=0}^{i-1}(I_\omega^*(x_j) + I_\omega^*(x_{j+1}))d(i,j) \\
&= 2aEi\left(2,\frac{x_i}{\Lambda}\right) + 2bEi\left(2,\frac{L-x_i}{\Lambda}\right), \quad i = 0\ldots N.
\end{aligned}
\tag{5.55}
$$

where $Ei(2,z) = \int_0^1 \exp\left(-\frac{z}{\mu}\right)d\mu$ and $d(i,j) = Ei\left(2,\frac{|x_{j+1}-x_i|}{\Lambda}\right) - Ei\left(2,\frac{|x_j - x_i|}{\Lambda}\right)$.

After shifting the index, Eq. (5.55) can be rewritten as

$$
\begin{aligned}
&\sum_{j=i+1}^{N}(d(i,j) + d(i,j-1))I_\omega^*(x_j) - \sum_{j=0}^{i-1}(d(i,j) + d(i,j-1))I_\omega^*(x_j) \\
&+ (4 + d(i,i) - d(i,i-1))I_\omega^*(x_i) = 2aEi\left(2,\frac{x_i}{\Lambda}\right) + 2bEi\left(2,\frac{L-x_i}{\Lambda}\right), \quad i = 0\ldots N.
\end{aligned}
\tag{5.56}
$$

where $d_{i,N} = d_{i,-1} = 0, \quad i = 0\ldots N.$

Now, defining

$$
A_{ij} = \begin{cases}
-d(i,0) & j = 0, \quad 1 \le i \le N \\
-d(i,j) - d(i,j-1) & 1 \le j \le i-1, \quad 2 \le i \le N \\
4 + d(i,i) - d(i,i-1) & j = i, \quad 0 \le i \le N \\
d(i,j) + d(i,j-1) & i+1 \le j \le N-1, \quad 0 \le i \le N-2 \\
d(i,N-1) & j = N, \quad 0 \le i \le N-1
\end{cases}
\tag{5.57}
$$

$X_i = I_\omega^*(x_i)$ and $B_i = 2aEi\left(2,\frac{x_i}{\Lambda}\right) + 2bEi\left(2,\frac{L-x_i}{\Lambda}\right)$, enable us to rewrite Eq. (5.56) as

$$
\sum_{j=0}^{N} A_{ij}X_j = B_i, \quad i = 0\ldots N \tag{5.58}
$$

Writing this in matrix form gives

$$
AX = B \tag{5.59}
$$

Finally, the $I_\omega^*(x_j)$ can be estimated by solving the matrix form in Eq. (5.59), and the corresponding temperature can be given by

$$
T(x_j) = \frac{4\pi I_\omega^*(x_j)}{Cv} \tag{5.60}
$$

where C is the specific heat and v is the acoustic speed in thin film.

5.3.2 Consideration of Thin Metallic Film

The analytical solution presented in the previous study [11] is limited with the emitted and the reflected phonon intensities across the film, and the solution for the equilibrium intensity is not provided. The analytical solution of the Boltzmann equation pertinent to metallic thin film is presented. Since the thermal separation takes place between the electron and the lattice subsystem, the solution covers the radiative energy transfer in both subsystems. The results are extended to include the effects of metallic film thickness on the phonon intensity distribution.

The BTE can be reduced to a phonon radiative transport equation, which can be written as [6]

$$\frac{1}{v}\frac{\partial I}{\partial t} + \mu\frac{\partial I}{\partial x} = \frac{\frac{1}{2}\int_{-1}^{1} I d\mu - I}{\Lambda} \tag{5.61}$$

In Eq. (5.61), $I = I\,(x, \mu, t)$ is the phonon intensity in units of Wm^{-2}, x is the distance along the film thickness, $\mu = \cos(\theta)$, θ is the angle of the velocity (momentum) vector of a phonon from the x axis, and t is the time. v is the phonon speed and Λ is the phonon mean free path. However, for the metal thin films, thermal separation of electron and phonon subsystems takes place during the heating process and restructuring of the energy equation becomes necessary to account for the energy transport in each subsystem. Therefore, in each subsystem, a separate equation for radiative phonon transport should be written in line with the Boltzmann equation. The thermal communication between the electron and the lattice subsystems takes place through the electron–phonon coupling process. In order to account for the thermal communication of the subsystems, electron–phonon coupling needs to be incorporated in the governing equation of radiative phonon transport.

5.3.3 The Coupled System of Equations of Phonon Radiative Transfer in Lattice Subsystems of Aluminum Film

Energy transport in the aluminum thin films is considered to be governed by the radiative transport. In order to analyze the energy transport in the lattice subsystem, the modified Boltzmann equation satisfying the conservation of energy can be written as

$$\frac{1}{v_p}\frac{\partial I_p}{\partial t} + \mu\frac{\partial I_p}{\partial x} = \frac{\frac{1}{2}\int_{-1}^{1} I_p d\mu - I_p}{\Lambda_p} - \frac{G}{2}\left(\frac{1}{C_p v_p}\int_{-1}^{1} I_p d\mu - \frac{1}{C_e v_e}\int_{-1}^{1} I_e d\mu\right) \tag{5.62}$$

In Eq. (5.62), I_p, v_p, Λ_p, C_p, and G are the phonon intensity, speed, mean free path, and volumetric specific heat in the lattice subsystem and the electron–phonon coupling constant, respectively.

In the case of the electron subsystem, the proposed equation for the energy transport satisfying the conservation of energy can be written as

$$\frac{1}{v_e}\frac{\partial I_e}{\partial t} + \mu\frac{\partial I_e}{\partial x} = \frac{\frac{1}{2}\int_{-1}^{1} I_e d\mu - I_e}{\Lambda_e} - \frac{G}{2}\left(\frac{1}{C_e v_e}\int_{-1}^{1} I_e d\mu - \frac{1}{C_p v_p}\int_{-1}^{1} I_p d\mu\right) \quad (5.63)$$

In Eq. (5.63), I_e, v_e, Λ_e, and C_e are the intensity, speed, mean free path, and volumetric specific heat in the electron subsystem, respectively.

In the case of the steady transport, Eq. (5.62) satisfies the conservation of energy and it reduces to the following equations for emitted and reflected phonon intensities (Fig. 5.2).

Emitted phonon intensity from the high temperature edge is

$$\frac{\partial I_p^+}{\partial x} + \frac{1}{\mu\Lambda_p} I_p^+ = \left(\frac{1}{\mu\Lambda_p} - \frac{G}{\mu C_p v_p}\right) I_p^* + \frac{G}{\mu C_e v_e} I_e^*, \quad \text{when} \quad 0 < \mu < 1 \quad (5.64)$$

Reflected phonon intensity from the low temperature edge is

$$\frac{\partial I_p^-}{\partial x} + \frac{1}{\mu\Lambda_p} I_p^- = \left(\frac{1}{\mu\Lambda_p} - \frac{G}{\mu C_p v_p}\right) I_p^* + \frac{G}{\mu C_e v_e} I_e^*, \quad \text{when} \quad -1 < \mu < 0 \quad (5.65)$$

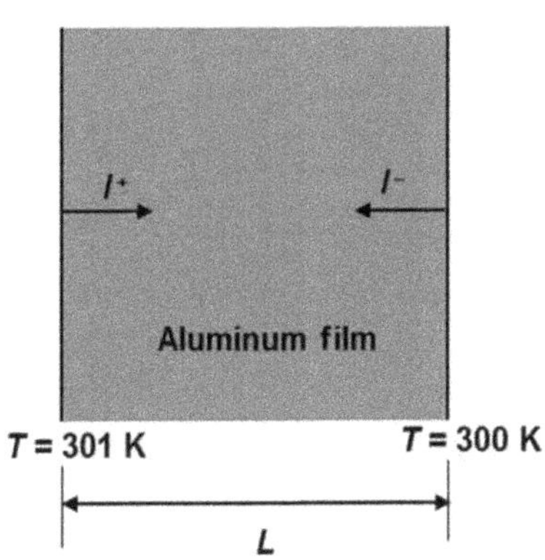

■ **FIGURE 5.2** A schematic view of aluminum film and constant temperature boundaries at the film edge [3].

It should be noted that Eq. (5.63), which is presented for the energy transport in the electron subsystem, can be simplified to steady transport, which also satisfies the conservation of energy.

Emitted electron intensity from the high temperature edge is

$$\frac{\partial I_e^+}{\partial x} + \frac{1}{\mu\Lambda_e} I_e^+ = \left(\frac{1}{\mu\Lambda_e} - \frac{G}{\mu C_e v_e}\right) I_e^* + \frac{G}{\mu C_p v_p} I_p^*, \text{ when } 0 < \mu < 1 \quad (5.66)$$

Reflected electron intensity from the low temperature edge is

$$\frac{\partial I_e^-}{\partial x} + \frac{1}{\mu\Lambda_e} I_e^- = \left(\frac{1}{\mu\Lambda_e} - \frac{G}{\mu C_e v_e}\right) I_e^* + \frac{G}{\mu C_p v_p} I_p^*, \text{ when } -1 < \mu < 0 \quad (5.67)$$

where the equilibrium intensities are

$$I_p^*(x) = \frac{1}{2}\int_{-1}^{1} I_p(x,\mu)d\mu = \frac{1}{2}\left(\int_{-1}^{0} I_p^-(x,\mu)d\mu + \int_{0}^{1} I_p^+(x,\mu)d\mu\right) \quad (5.68)$$

and

$$I_e^*(x) = \frac{1}{2}\int_{-1}^{1} I_e(x,\mu)d\mu = \frac{1}{2}\left(\int_{-1}^{0} I_e^-(x,\mu)d\mu + \int_{0}^{1} I_e^+(x,\mu)d\mu\right) \quad (5.69)$$

The heating situation is considered to take place between two constant temperatures across the film; in which case, temperature at each film edge is considered to be constant and its value differs at each edge. The relevant boundary conditions at the film edges can be expressed as the left and the right boundary conditions, accordingly.

5.3.3.1 *Left Boundary Condition*

For the lattice site, constant temperature (301K) is assumed at the left boundary; therefore, the left boundary condition for the lattice subsystem is

$$I_p^+(0,\mu) = \frac{C_p v_p \times 301}{4\pi} = a \quad (5.70)$$

The thermionic electron emission is not considered from the edges of the film; therefore, electrons do not escape from the film edges. Hence, an insulated boundary condition is assumed at the left boundary for the electron subsystem.

$$I_e^+(0,\mu) = -2\int_{-1}^{0} \mu I_e^-(0,\mu)d\mu = c \quad \text{(insulated boundary)} \quad (5.71)$$

5.3.3.2 Right Boundary Condition

For the lattice site, a constant temperature (300K) is assumed at the right boundary; therefore, the right boundary condition for the lattice subsystem is

$$I_p^-(L,\mu) = \frac{C_p v_p \times 300}{4\pi} = b \tag{5.72}$$

The insulated boundary condition at the right boundary of the electron subsystem is

$$I_e^-(L,\mu) = 2\int_0^1 \mu I_e^+(L,\mu)d\mu = d \quad \text{(insulated boundary)} \tag{5.73}$$

In order to introduce the semianalytical solution to Eqs. (5.64) to (5.67), the following procedure is adopted. In this case, multiplying both Eqs. (5.64) to (5.67) with the multipliers $\exp\left(\frac{x}{\mu\Lambda_p}\right)$ and $\exp\left(\frac{x}{\mu\Lambda_e}\right)$, respectively, gives

$$\begin{aligned}
&\frac{\partial}{\partial x}\left(\exp\left(\frac{x}{\mu\Lambda_p}\right)I_p^+\right) = \left(\frac{1}{\mu\Lambda_p} - \frac{G}{\mu C_p v_p}\right)\exp\left(\frac{x}{\mu\Lambda_p}\right)I_p^* + \frac{G}{\mu C_e v_e}\exp\left(\frac{x}{\mu\Lambda_p}\right)I_e^*,\\
&\quad \text{when} \quad 0<\mu<1\\
&\frac{\partial}{\partial x}\left(\exp\left(\frac{x}{\mu\Lambda_p}\right)I_p^-\right) = \left(\frac{1}{\mu\Lambda_p} - \frac{G}{\mu C_p v_p}\right)\exp\left(\frac{x}{\mu\Lambda_p}\right)I_p^* + \frac{G}{\mu C_e v_e}\exp\left(\frac{x}{\mu\Lambda_p}\right)I_e^*,\\
&\quad \text{when} \quad -1<\mu<0
\end{aligned} \tag{5.74}$$

$$\begin{aligned}
&\frac{\partial}{\partial x}\left(\exp\left(\frac{x}{\mu\Lambda_e}\right)I_e^+\right) = \left(\frac{1}{\mu\Lambda_e} - \frac{G}{\mu C_e v_e}\right)\exp\left(\frac{x}{\mu\Lambda_e}\right)I_e^* + \frac{G}{\mu C_p v_p}\exp\left(\frac{x}{\mu\Lambda_e}\right)I_p^*,\\
&\quad \text{when} \quad 0<\mu<1\\
&\frac{\partial}{\partial x}\left(\exp\left(\frac{x}{\mu\Lambda_e}\right)I_e^-\right) = \left(\frac{1}{\mu\Lambda_e} - \frac{G}{\mu C_e v_e}\right)\exp\left(\frac{x}{\mu\Lambda_e}\right)I_e^* + \frac{G}{\mu C_p v_p}\exp\left(\frac{x}{\mu\Lambda_e}\right)I_p^*,\\
&\quad \text{when} \quad -1<\mu<0
\end{aligned} \tag{5.75}$$

By integrating both sides with respect to x and applying the boundary conditions, the solutions of Eqs. (5.65) to (5.67) can be expressed as follows:

$$\begin{aligned}
&I_p^+(x,\mu) = \int_0^x \frac{1}{\mu}\exp\left(\frac{\xi - x}{\mu\Lambda_p}\right)(\lambda_p I_p^*(\xi) + \gamma_e I_e^*(\xi))d\xi + a\exp\left(\frac{-x}{\mu\Lambda_p}\right),\\
&\quad \text{when} \quad 0<\mu<1\\
&I_p^-(x,\mu) = \int_L^x \frac{1}{\mu}\exp\left(\frac{\xi - x}{\mu\Lambda_p}\right)(\lambda_p I_p^*(\xi) + \gamma_e I_e^*(\xi))d\xi + b\exp\left(\frac{L-x}{\mu\Lambda_p}\right),\\
&\quad \text{when} \quad -1<\mu<0
\end{aligned} \tag{5.76}$$

$$
\begin{aligned}
I_e^+(x,\mu) &= \int_0^x \frac{1}{\mu}\exp\left(\frac{\xi-x}{\mu\Lambda_e}\right)(\lambda_e I_e^*(\xi)+\gamma_p I_p^*(\xi))d\xi + c\exp\left(\frac{-x}{\mu\Lambda_e}\right), \\
&\text{when} \quad 0<\mu<1 \\
I_e^-(x,\mu) &= \int_L^x \frac{1}{\mu}\exp\left(\frac{\xi-x}{\mu\Lambda_e}\right)(\lambda_e I_e^*(\xi)+\gamma_p I_p^*(\xi))d\xi + d\exp\left(\frac{L-x}{\mu\Lambda_e}\right), \\
&\text{when} \quad -1<\mu<0
\end{aligned}
\tag{5.77}
$$

where $\lambda_p = \frac{1}{\Lambda_p} - \frac{G}{C_p v_p}$, $\lambda_e = \frac{1}{\Lambda_e} - \frac{G}{C_e v_e}$, $\gamma_p = \frac{G}{C_p v_p}$, and $\gamma_e = \frac{G}{C_e v_e}$.

Using the boundary conditions $\int_0^1 \mu I_e^+(L,\mu)d\mu = \frac{d}{2}$, $\int_0^1 \mu I_e^-(0,-\mu)d\mu = \frac{c}{2}$ and Eq. (5.75), we can express the values of c and d in terms of $I_p^*(x)$ and $I_e^*(x)$ as follows:

$$
\begin{aligned}
2\int_0^L Ei\left(2,\frac{L-\xi}{\Lambda_e}\right)(\lambda_e I_e^*(\xi)+\gamma_p I_p^*(\xi))d\xi + 2\,c\,Ei\left(3,\frac{L}{\Lambda_e}\right) &= d \\
2\int_0^L Ei\left(2,\frac{\xi}{\Lambda_e}\right)(\lambda_e I_e^*(\xi)+\gamma_p I_p^*(\xi))d\xi + 2\,d\,Ei\left(3,\frac{L}{\Lambda_e}\right) &= c
\end{aligned}
\tag{5.78}
$$

where the generalized exponential integral $Ei(n,z) = \int_1^\infty \frac{e^{-zt}}{t^n}dt = \int_0^1 \mu^{n-2}\exp\left(-\frac{z}{\mu}\right)d\mu$.

Solving Eq. (5.78) gives

$$
\begin{aligned}
c &= 2\int_0^L \psi(\xi)(\lambda_e I_e^*(\xi)+\gamma_p I_p^*(\xi))d\xi \\
d &= 2\int_0^L \chi(\xi)(\lambda_e I_e^*(\xi)+\gamma_p I_p^*(\xi))d\xi
\end{aligned}
\tag{5.79}
$$

where

$$
\begin{aligned}
\psi(\xi) &= \frac{Ei\left(2,\frac{\xi}{\Lambda_e}\right) + 2Ei\left(3,\frac{L}{\Lambda_e}\right)Ei\left(2,\frac{L-\xi}{\Lambda_e}\right)}{\left(1-4Ei\left(3,\frac{L}{\Lambda_e}\right)^2\right)}, \\
\chi(\xi) &= \frac{Ei\left(2,\frac{L-\xi}{\Lambda_e}\right) + 2Ei\left(3,\frac{L}{\Lambda_e}\right)Ei\left(2,\frac{\xi}{\Lambda_e}\right)}{\left(1-4Ei\left(3,\frac{L}{\Lambda_e}\right)^2\right)}.
\end{aligned}
\tag{5.80}
$$

However, $I_p^*(x) = \frac{1}{2}\int_{-1}^1 I_p(x,\mu)d\mu = \frac{1}{2}\left(\int_{-1}^0 I_p^-(x,\mu)d\mu + \int_0^1 I_p^+(x,\mu)d\mu\right)$ and $I_e^*(x) = \frac{1}{2}\int_{-1}^1 I_e(x,\mu)d\mu = \frac{1}{2}\left(\int_{-1}^0 I_e^-(x,\mu)d\mu + \int_0^1 I_e^+(x,\mu)d\mu\right)$; then it yields the following system of integral equations in $I_p^*(x)$ and $I_e^*(x)$:

$$2I_p^*(x) = \left[\begin{array}{l} \int_{-1}^{0}\left(\int_{L}^{x}\frac{1}{\mu}\exp\left(\frac{\xi-x}{\mu\Lambda_p}\right)(\lambda_p I_p^*(\xi)+\gamma_e I_e^*(\xi))d\xi + b\exp\left(\frac{L-x}{\mu\Lambda_p}\right)\right)d\mu \\ +\int_{0}^{1}\left(\int_{0}^{x}\frac{1}{\mu}\exp\left(\frac{\xi-x}{\mu\Lambda_p}\right)(\lambda_p I_p^*(\xi)+\gamma_e I_e^*(\xi))d\xi + a\exp\left(\frac{-x}{\mu\Lambda_p}\right)\right)d\mu \end{array}\right]$$

$$2I_e^*(x) = \left[\begin{array}{l} \int_{-1}^{0}\left(\int_{L}^{x}\frac{1}{\mu}\exp\left(\frac{\xi-x}{\mu\Lambda_e}\right)(\lambda_e I_e^*(\xi)+\gamma_p I_p^*(\xi))d\xi + d\exp\left(\frac{L-x}{\mu\Lambda_e}\right)\right)d\mu \\ +\int_{0}^{1}\left(\int_{0}^{x}\frac{1}{\mu}\exp\left(\frac{\xi-x}{\mu\Lambda_e}\right)(\lambda_e I_e^*(\xi)+\gamma_p I_p^*(\xi))d\xi + c\exp\left(\frac{-x}{\mu\Lambda_e}\right)\right)d\mu \end{array}\right] \quad (5.81)$$

The system of integral Eq. (5.81) can be written as the following system of Fredholm integral equation of the second kind; i.e.,

$$\begin{aligned} 2I_p^*(x) - \int_0^L K_p(x,\xi)(\lambda_p I_p^*(\xi)+\gamma_e I_e^*(\xi))d\xi &= f(x) \\ 2I_e^*(x) - \int_0^L K_e(x,\xi)(\lambda_e I_e^*(\xi)+\gamma_p I_p^*(\xi))d\xi &= g(x) \end{aligned} \quad (5.82)$$

where the K_e and K_p are

$$K_p(x,\xi) = Ei\left(1,\frac{|x-\xi|}{\Lambda_p}\right) \quad (5.83)$$

and

$$K_e(x,\xi) = Ei\left(1,\frac{|x-\xi|}{\Lambda_e}\right) \quad (5.84)$$

and the sources are

$$f(x) = a\,Ei\left(2,\frac{x}{\Lambda_p}\right) + b\,Ei\left(2,\frac{L-x}{\Lambda_p}\right) \quad (5.85)$$

and

$$g(x) = c\,Ei\left(2,\frac{x}{\Lambda_e}\right) + d\,Ei\left(2,\frac{L-x}{\Lambda_e}\right) \quad (5.86)$$

Since the values of c and d are given using Eq. (5.79), then $g(x)$ can be rewritten as

$$g(x) = \int_0^L K_g(x,\xi)(\lambda_e I_e^*(\xi)+\gamma_p I_p^*(\xi))d\xi \quad (5.87)$$

where

$$K_g(x,\xi) = 2Ei\left(2,\frac{x}{\Lambda_e}\right)\psi(\xi) + 2Ei\left(2,\frac{L-x}{\Lambda_e}\right)\chi(\xi) \tag{5.88}$$

So, the system of integral Eq. (5.82) can be written as

$$\begin{aligned} &2I_p^*(x) - \int_0^L K_p(x,\xi)(\lambda_p I_p^*(\xi) + \gamma_e I_e^*(\xi))d\xi = f(x) \\ &2I_e^*(x) - \int_0^L (K_e(x,\xi) + K_g(x,\xi))(\lambda_e I_e^*(\xi) + \gamma_p I_p^*(\xi))d\xi = 0 \end{aligned} \tag{5.89}$$

In order to solve this system, let's consider the partition $\{x_0, x_1, \ldots, x_N\}$ of $[0, L]$; then we get the following discretization of the system (Eq. 5.89):

$$\begin{aligned} &2I_p^*(x_i) - \int_0^L K_p(x_i,\xi)(\lambda_p I_p^*(\xi) + \gamma_e I_e^*(\xi))d\xi = f(x_i), \\ &2I_e^*(x_i) - \int_0^L (K_e(x_i,\xi) + K_g(x_i,\xi))(\lambda_e I_e^*(\xi) + \gamma_p I_p^*(\xi))d\xi = 0, \quad i = 0\ldots N. \end{aligned} \tag{5.90}$$

Now, distributing the integral over the partition gives

$$\begin{aligned} &2I_p^*(x_i) - \sum_{j=0}^{N-1}\int_{x_j}^{x_{j+1}} K_p(x_i,\xi)(\lambda_p I_p^*(\xi) + \gamma_e I_e^*(\xi))d\xi = f(x_i), \\ &2I_e^*(x_i) - \sum_{j=0}^{N-1}\int_{x_j}^{x_{j+1}} (K_e(x_i,\xi) + K_g(x_i,\xi))(\lambda_e I_e^*(\xi) + \gamma_p I_p^*(\xi))d\xi = 0, \quad i = 0\ldots N. \end{aligned} \tag{5.91}$$

Approximating the values of $I_p^*(x)$ and $I_e^*(x)$ on small intervals $[x_j, x_{j+1}]$ by $\frac{1}{2}(I_p^*(x_j) + I_p^*(x_{j+1}))$ and $\frac{1}{2}(I_e^*(x_j) + I_e^*(x_{j+1}))$, respectively, gives the linear system

$$\begin{aligned} &2I_p^*(x_i) - \frac{1}{2}\sum_{j=0}^{N-1}(\lambda_p(I_p^*(x_j) + I_p^*(x_{j+1})) + \gamma_e(I_e^*(x_j) + I_e^*(x_{j+1})))d(i,j) = f(x_i), \\ &2I_e^*(x_i) - \frac{1}{2}\sum_{j=0}^{N-1}(\lambda_e(I_e^*(x_j) + I_e^*(x_{j+1})) + \gamma_p(I_p^*(x_j) + I_p^*(x_{j+1})))s(i,j) = 0, \quad i = 0\ldots N. \end{aligned} \tag{5.92}$$

where

$$f(x_i) = a\,Ei\left(2,\frac{x_i}{\Lambda_p}\right) + b\,Ei\left(2,\frac{L-x_i}{\Lambda_p}\right), \quad d(i,j) = \int_{x_j}^{x_{j+1}} K_p(x_i,\xi)d\xi \tag{5.93}$$

and

$$s(i,j) = \int_{x_j}^{x_{j+1}} (K_e(x_i,\xi) + K_g(x_i,\xi))d\xi. \tag{5.94}$$

Since $\int_{x_j}^{x_{j+1}} Ei(n,z)dz = Ei(n+1,x_j) - Ei(n+1,x_{j+1})$, then

$$d(i,j) = \Lambda_p \left| Ei\left(2, \frac{|x_j - x_i|}{\Lambda_p}\right) - Ei\left(2, \frac{|x_{j+1} - x_i|}{\Lambda_p}\right) \right| \tag{5.95}$$

and

$$\begin{aligned} s(i,j) = {} & \Lambda_e \left| Ei\left(2, \frac{|x_j - x_i|}{\Lambda_e}\right) - Ei\left(2, \frac{|x_{j+1} - x_i|}{\Lambda_e}\right) \right| \\ & + 2Ei\left(2, \frac{x_i}{\Lambda_e}\right) \int_{x_j}^{x_{j+1}} \psi(\xi)d\xi + 2Ei\left(2, \frac{L - x_i}{\Lambda_e}\right) \int_{x_j}^{x_{j+1}} \chi(\xi)d\xi. \end{aligned} \tag{5.96}$$

where

$$\begin{aligned} & \int_{x_j}^{x_{j+1}} \psi(\xi)d\xi \\ & = \Lambda_e \left(\frac{Ei\left(3, \frac{x_j}{\Lambda_e}\right) - Ei\left(3, \frac{x_{j+1}}{\Lambda_e}\right) + 2Ei\left(3, \frac{L}{\Lambda_e}\right)\left(Ei\left(3, \frac{L - x_{j+1}}{\Lambda_e}\right) - Ei\left(3, \frac{L - x_j}{\Lambda_e}\right)\right)}{\left(1 - 4Ei\left(3, \frac{L}{\Lambda_e}\right)^2\right)} \right) \end{aligned} \tag{5.97}$$

and

$$\begin{aligned} & \int_{x_j}^{x_{j+1}} \chi(\xi)d\xi \\ & = \Lambda_e \left(\frac{Ei\left(3, \frac{L - x_{j+1}}{\Lambda_e}\right) - Ei\left(3, \frac{L - x_j}{\Lambda_e}\right) + 2Ei\left(3, \frac{L}{\Lambda_e}\right)\left(Ei\left(3, \frac{x_j}{\Lambda_e}\right) - Ei\left(3, \frac{x_{j+1}}{\Lambda_e}\right)\right)}{\left(1 - 4Ei\left(3, \frac{L}{\Lambda_e}\right)^2\right)} \right). \end{aligned} \tag{5.98}$$

Finally, equilibrium intensities in electron and lattice subsystems, $I_p^*(x_j)$ and $I_e^*(x_j)$, can be determined by solving the linear algebraic system (Eq. 5.92), and the corresponding temperatures can be given by

$$\begin{aligned} T_p(x_j) &= \frac{4\pi I_p^*(x_j)}{C_p v_p}, \\ T_e(x_j) &= \frac{4\pi I_e^*(x_j)}{C_e v_e}. \end{aligned} \tag{5.99}$$

5.3.4 Two-Equation Model

The two-equation model associated with the electron and lattice subsystems was presented in the previous study [12]. A brief description of the equations is given next, in line with the radiative transport in the electron and lattice subsystems. Since the steady radiative transport is considered to solve the governing equations in the aluminum film, a similar situation is considered for the two-equation model.

5.3.4.1 *Phonon Subsystem*

The energy equation in the phonon subsystem for the steady transport can be written as

$$k_p \frac{\partial^2 T_p}{\partial x^2} - G(T_p - T_e) = 0 \tag{5.100}$$

5.3.4.2 *Electron Subsystem*

The equation for the steady energy transport in the electron subsystem can be written as

$$k_e \frac{\partial^2 T_e}{\partial x^2} - G(T_e - T_p) = 0 \tag{5.101}$$

In order to compare the results of the two-equation model with the findings of the phonon radiative transport in the aluminum film, similar boundary conditions are introduced for those of the phonon radiative transport equations.

5.3.4.3 *Left Boundary Condition*

In the lattice subsystem, a constant temperature boundary is assumed, in which case, the left boundary is considered to be at 301K; i.e.,

$$T_p(0) = 301\,\text{K} \tag{5.102}$$

However, in the electron subsystem, an insulated boundary is considered to account for the nonthermionic boundary at the film edge; i.e.,

$$\left.\frac{dT_e}{dx}\right|_{x=0} = 0 \quad \text{(insulated boundary)} \tag{5.103}$$

Similar to the phonon transport equations, the right boundary condition takes the following forms.

5.3.4.4 *Right Boundary Condition*

In the lattice subsystem, a constant temperature boundary is assumed at the right edge of the aluminum film; therefore, the right boundary of the aluminum film is considered to be at 300K; i.e.,

$$T_p(L) = 300\text{K} \tag{5.104}$$

Similarly, an insulated boundary is assumed at the left edge of the aluminum film for the electron subsystem:

$$\left.\frac{dT_e}{dx}\right|_{x=L} = 0 \quad \text{(insulated boundary)} \tag{5.105}$$

The general solutions of Eqs. (5.100) and (5.101) can be given as

$$T_p(x) = C_1 + C_2 x - \frac{k_e}{k_p} C_3 \exp\left(\sqrt{\frac{G(k_e + k_p)}{k_e k_p}} x\right) - \frac{k_e}{k_p} C_4 \exp\left(-\sqrt{\frac{G(k_e + k_p)}{k_e k_p}} x\right)$$

$$T_e(x) = C_1 + C_2 x + C_3 \exp\left(\sqrt{\frac{G(k_e + k_p)}{k_e k_p}} x\right) + C_4 \exp\left(-\sqrt{\frac{G(k_e + k_p)}{k_e k_p}} x\right) \tag{5.106}$$

Now, involving the boundary conditions gives the values of the four arbitrary constants as follows:

$$C_1 = \frac{\left(\begin{array}{l}\left(601\, k_e^{3/2} + 301\, F\right)\exp\left(\sqrt{\frac{G(k_e + k_p)}{k_e k_p}} L\right) \\ \quad + \left(601\, k_e^{3/2} - 301\, F\right)\exp\left(-\sqrt{\frac{G(k_e + k_p)}{k_e k_p}} L\right) - 1202\, k_e^{3/2}\end{array}\right)}{R}$$

$$C_2 = \frac{F}{LR}\left(\exp\left(-\sqrt{\frac{G(k_e + k_p)}{k_e k_p}} L\right) - \exp\left(\sqrt{\frac{G(k_e + k_p)}{k_e k_p}} L\right)\right)$$

$$C_3 = \frac{\sqrt{k_e k_p}}{R}\left(1 - \exp\left(-\sqrt{\frac{G(k_e + k_p)}{k_e k_p}} L\right)\right)$$

$$C_4 = \frac{\sqrt{k_e k_p}}{R}\left(1 - \exp\left(\sqrt{\frac{G(k_e + k_p)}{k_e k_p}} L\right)\right) \tag{5.107}$$

where $F = \sqrt{G\, k_p(k_p + k_e)} L$

and $R = (2\, k_e^{3/2} + F)\exp\left(\sqrt{\frac{G(k_e + k_p)}{k_e k_p}} L\right) + (2\, k_e^{3/2} - F)\exp\left(-\sqrt{\frac{G(k_e + k_p)}{k_e k_p}} L\right) - 4\, k_e^{3/2}$.

A computer program can be developed to solve Eqs. (5.98) and (5.106) to obtain temperature variation in the electron and lattice subsystems for various values of the metallic film thickness.

5.4 ANALYTICAL SOLUTION TO FORMULATION OF ELECTRON KINETIC THEORY

Improved electron kinetic theory approach yields the hyperbolic equation for the temperature inside the films. However, the analytical solutions for hyperbolic heat conduction equation were presented earlier [13–15]; the main feature was the solution of the Cattaneo equation with the appropriate boundary conditions. However, short-pulse heating of metallic surfaces is involved with nonequilibrium energy transfer in the lattice and the electron subsystems. This requires the consideration the electron and the lattice subsystems in the analysis. The approximate solution of the two-equation model was presented by Yilbas et al. [16] using the perturbation method. However the analytical solution for electron and lattice temperatures provided reasonable accurate solutions for the limited time and space scales. This was because of the approximations made for the first and zeroth order solutions. Therefore, extension of the previous study [16] is necessary for extended time and space scales. In order to generalize the solution, the volumetric heat source is introduced, which can resemble the laser short-pulse heating. However, the surface heat source is also considered, which can resemble the short thermal contact of the film surface. The analytical solution for the governing equations can be obtained incorporating the Lie point symmetries and similarity solutions. This provides the closed form solution for electron and lattice temperature distribution in the heat transferring region. The mathematical analyses are given next, in line with the previous study [5].

The improved electron kinetic theory formulation for volumetric heat source resembles the short-pulse laser heating process, yielding the energy transport equation in the lattice site, which is [17]:

$$\left[\left(1+\tau_s\frac{\partial}{\partial t^*}\right)-\frac{\lambda^2}{f}\frac{\partial^2}{\partial x^{*2}}\right]C_L\frac{\partial T_L}{\partial t^*}=k\frac{\partial^2 T_L}{\partial x^{*2}}+\tau_p\frac{\partial}{\partial t^*}\left(k\frac{\partial^2 T_L}{\partial x^{*2}}\right)+\left(1+\tau_p\frac{\partial}{\partial t^*}\right)(f\delta I(t^*)\exp(-\delta|x^*|)) \tag{5.108}$$

where τ_s is the electron–phonon characteristic time ($\tau_s = C_E/G$); G is the electron–phonon coupling factor; λ is the mean free path of the electrons; f is the fraction of excess energy change; C_L and C_E are the lattice and electron heat capacities, respectively; k is the thermal conductivity;

τ_p is the electron mean free time between electron–phonon coupling; $I(t) = I_\circ \exp(-dt^*)$, where $I_\circ$ is the laser peak power intensity, $\exp(-dt^*)$ is the temporal distribution function of laser pulse, d is the temporal distribution parameter, δ is the absorption coefficient; x^* is the lattice depth; and t^* is the time variable. T_L and T_E are the lattice site and electron temperatures, respectively. Introducing the following equalities and dimensionless variables

$$\begin{aligned} &\lambda^2 = \frac{fk}{G}, \quad \tau_s = \frac{C_E}{G} \\ &\theta_E = \frac{T_E}{T_o}, \quad \theta_L = \frac{T_L}{T_o}, \quad x = x^*\delta, \quad t = \frac{t^*}{C_E/G}, \quad \alpha = \frac{k\delta^2}{G}, \quad \epsilon = \frac{C_E}{C_L} \end{aligned} \tag{5.109}$$

to Eq. (5.108) yields, finally,

$$\alpha(1+\mu)\frac{\partial^3\theta_L}{\partial x^2 \partial t} + \epsilon\alpha\frac{\partial^2\theta_L}{\partial x^2} - \frac{\partial^2\theta_L}{\partial t^2} - \frac{\partial\theta_L}{\partial t} - \beta\exp(-x)\exp(-\gamma t) = 0 \tag{5.110}$$

where

$$\mu = \frac{\tau_p G}{C_L}, \quad \beta = \frac{f\, I_o \epsilon\delta(\gamma - 1)}{T_o G}, \quad \gamma = \frac{dC_E}{G}. \tag{5.111}$$

This model is the improved energy transport equation including ballistic effects with the source in dimensionless form. Once the lattice site temperature is determined, the electron temperature can be found from

$$\frac{\partial\theta_L}{\partial t} = \epsilon(\theta_E - \theta_L) \tag{5.112}$$

So, our goal will be to find the exact solution of the improved model:

$$\begin{aligned} &\alpha(1+\mu)\frac{\partial^3\theta_L}{\partial x^2 \partial t} + \epsilon\,\alpha\,\frac{\partial^2\theta_L}{\partial x^2} - \frac{\partial^2\theta_L}{\partial t^2} - \frac{\partial\theta_L}{\partial t} - \beta\exp(-x)\exp(-\gamma t) = 0, \\ &\frac{\partial\theta_L}{\partial t} = \epsilon(\theta_E - \theta_L) \end{aligned} \tag{5.113}$$

By using the MAPLE program, we can find that the improved model system (Eq. 5.113) admits the following two Lie point symmetries:

$$\begin{aligned} X_1 &= \frac{\partial}{\partial x} + (\epsilon - \gamma)He^{-x-\gamma t}\frac{\partial}{\partial\theta_E} + \epsilon He^{-x-\gamma t}\frac{\partial}{\partial\theta_L} \\ X_2 &= \frac{\partial}{\partial t} + (\epsilon - \gamma)\gamma He^{-x-\gamma t}\frac{\partial}{\partial\theta_E} + \epsilon H\gamma e^{-x-\gamma t}\frac{\partial}{\partial\theta_L} \end{aligned} \tag{5.114}$$

where

$$H = \frac{\beta}{\epsilon(\gamma^2 + \gamma(\alpha(\mu + 1) - 1) - \alpha\epsilon)} \tag{5.115}$$

So we can get a reduced system by using the symmetry generator $Y = X_1 + bX_2$, where the generator Y has a canonical form $Y = \frac{\partial}{\partial s}$ when

$$\frac{dx}{1} = \frac{dt}{b} = \frac{d\theta_E}{(1 + b\gamma)(\epsilon - \gamma)He^{-x-\gamma t}} = \frac{d\theta_L}{(1 + b\gamma)\epsilon He^{-x-\gamma t}} = \frac{d\xi}{0} = \frac{ds}{1} = \frac{dv}{0} = \frac{dw}{0} \tag{5.116}$$

Then the canonical coordinates can be given as follows:

$$\xi = t - bx, \quad s = x, \quad v(\xi) = \theta_E(x, t) + (\epsilon - \gamma)He^{-x-\gamma t}, \quad w(\xi) = \theta_L(x, t) + \epsilon He^{-x-\gamma t} \tag{5.117}$$

or

$$\begin{aligned} &t = \xi + bs, \quad x = s, \quad \theta_E(x, t) = v(\xi) - (\epsilon - \gamma)He^{-(1+b\gamma)s-\gamma\xi}, \\ &\theta_L(x, t) = w(\xi) - \epsilon He^{-(1+b\gamma)s-\gamma\xi} \end{aligned} \tag{5.118}$$

Substituting the new coordinates, Eq. (5.118), into the system, Eq. (5.113) yields

$$\begin{aligned} &\alpha b^2(1 + \mu)\frac{d^3w(\xi)}{d\xi^3} + (\alpha\epsilon b^2 - 1)\frac{d^2w(\xi)}{d\xi^2} - \frac{dw(\xi)}{d\xi} = 0, \\ &\frac{dw(\xi)}{d\xi} - \epsilon v(\xi) + \epsilon w(\xi) = 0. \end{aligned} \tag{5.119}$$

Solving Eq. (5.119) (second equation) for $v(\xi)$ yields

$$v(\xi) = \frac{1}{\epsilon}\frac{dw(\xi)}{d\xi} + w(\xi) \tag{5.120}$$

Substituting $\eta(\xi) = \frac{dw(\xi)}{d\xi}$ into Eq. (5.119) (first equation) yields

$$\alpha b^2(1 + \mu)\frac{d^2\eta(\xi)}{d\xi^2} + (\alpha b^2\epsilon - 1)\frac{d\eta(\xi)}{d\xi} - \eta(\xi) = 0. \tag{5.121}$$

The solution of Eq. (5.121) can easily be presented as

$$\eta(\xi) = c_1 e^{\lambda_1\xi} + c_2 e^{\lambda_2\xi} \tag{5.122}$$

where

$$\lambda_1 = \frac{\sigma + \sqrt{\sigma^2 + 4\alpha b^2(1 + \mu)}}{2\alpha b^2(1 + \mu)}, \quad \lambda_2 = \frac{\sigma - \sqrt{\sigma^2 + 4\alpha b^2(1 + \mu)}}{2\alpha b^2(1 + \mu)}, \quad \sigma = 1 - \alpha b^2\varepsilon \tag{5.123}$$

Now since $\eta(\xi) = \frac{dw(\xi)}{d\xi}$, $w(\xi)$ can be given as

$$w(\xi) = C_1 + C_2 e^{\lambda_1 \xi} + C_3 e^{\lambda_2 \xi} \tag{5.124}$$

Substituting Eq. (5.124) into Eq. (5.120) yields

$$v(\xi) = C_1 + C_2\left(\frac{\lambda_1}{\varepsilon} + 1\right)e^{\lambda_1 \xi} + C_3\left(\frac{\lambda_2}{\varepsilon} + 1\right)e^{\lambda_2 \xi} \tag{5.125}$$

Substituting back the canonical coordinates of Eq. (5.117) into Eqs. (5.124) and (5.125) yields the solution of the original system (Eq. 5.111) as

$$\theta_L(x,t) = C_1 + C_2 e^{\lambda_1 (t-bx)} + C_3 e^{\lambda_2 (t-bx)} - \varepsilon H e^{-x-\gamma t} \tag{5.126}$$

and

$$\theta_E(x,t) = C_1 + C_2\left(\frac{\lambda_1}{\epsilon} + 1\right)e^{\lambda_1 (t-bx)} + C_3\left(\frac{\lambda_2}{\epsilon} + 1\right)e^{\lambda_2 (t-bx)} - (\varepsilon - \gamma) H e^{-x-\gamma t} \tag{5.127}$$

5.4.1 Consideration of Volumetric Heat Source (Case 1)

Assume a semi-infinite substrate material heated. The boundary conditions for the problem can be written as follows:

$$\begin{aligned} &\frac{\partial \theta_E}{\partial x}(0,t) = 0, \quad \frac{\partial \theta_E}{\partial x}(\infty,t) = 0 \\ &\theta_E(x,\infty) = \theta_o, \quad \theta_L(x,\infty) = \theta_o \end{aligned} \tag{5.128}$$

Using the boundary conditions $\theta_L(x,\infty) = \theta_o$, $\theta_E(x,\infty) = \theta_o$ and since $\lambda_1 > 0$ and $\lambda_2 < 0$, Eq. (5.123), C_2 vanishes ($C_2 = 0$) and $C_1 = \theta_o$.

Introducing the boundary condition $\frac{\partial \theta_E}{\partial x}(\infty,t) = 0$ and since $\lambda_2 < 0$, shows that $b < 0$.

Finally, using the boundary condition $\frac{\partial \theta_E}{\partial x}(0,t) = 0$, the two equations result:

$$\begin{aligned} &0 = b\lambda_2 C_3\left(\frac{\lambda_2}{\epsilon} + 1\right) - H(\epsilon - \gamma) \\ &\lambda_2 = -\gamma \end{aligned} \tag{5.129}$$

Solving Eq. (5.129) by using Eq. (5.123), the parameter b and the constant C_3 can be obtained as

$$b^2 = \frac{1-\gamma}{\alpha\gamma(\gamma(1+\mu)-\epsilon)}$$
$$C_3 = \frac{H(\epsilon-\gamma)}{b\gamma\left(\frac{\gamma}{\epsilon}-1\right)} \tag{5.130}$$

Since $b<0$ is known from the boundary condition, then

$$b = -\sqrt{\frac{1-\gamma}{\alpha\gamma(\gamma(1+\mu)-\epsilon)}}, \text{ where}\left\{\gamma>1\ \&\ \gamma<\frac{\epsilon}{1+\mu}\right\} \text{ or } \left\{\gamma<1\ \&\ \gamma>\frac{\epsilon}{1+\mu}\right\} \tag{5.131}$$

And so

$$C_3 = \frac{H(\epsilon-\gamma)}{\gamma\left(1-\frac{\gamma}{\epsilon}\right)}\sqrt{\frac{\alpha\gamma(\gamma(1+\mu)-\epsilon)}{1-\gamma}} \tag{5.132}$$

Then solution of the boundary value problem can be given, when $\left\{\gamma>1\ \&\ \gamma<\frac{\epsilon}{1+\mu}\right\}$ or $\left\{\gamma<1\ \&\ \gamma>\frac{\epsilon}{1+\mu}\right\}$ as follows:

$$\begin{aligned}\theta_{\mathrm{L}}(x,t) &= \theta_o + \frac{\epsilon}{\gamma}H\sqrt{\frac{\alpha\gamma(\gamma(1+\mu)-\epsilon)}{1-\gamma}}\exp\left(-\gamma\left(t+\sqrt{\frac{1-\gamma}{\alpha\gamma(\gamma(1+\mu)-\epsilon)}}x\right)\right)\\ &\quad -\epsilon He^{-x-\gamma t}\\ \theta_{\mathrm{E}}(x,t) &= \theta_o + \frac{(\epsilon-\gamma)}{\gamma}H\sqrt{\frac{\alpha\gamma(\gamma(1+\mu)-\epsilon)}{1-\gamma}}\exp\left(-\gamma\left(t+\sqrt{\frac{1-\gamma}{\alpha\gamma(\gamma(1+\mu)-\epsilon)}}x\right)\right)\\ &\quad -(\epsilon-\gamma)He^{-x-\gamma t}\end{aligned} \tag{5.133}$$

It should be noted that the solution for $\theta_{\mathrm{L}}(x,t)$ and $\theta_{\mathrm{E}}(x,t)$ in Eq. (5.133) is only valid for $\left\{\gamma>1\ \&\ \gamma<\frac{\epsilon}{1+\mu}\right\}$ or $\left\{\gamma<1\ \&\ \gamma>\frac{\epsilon}{1+\mu}\right\}$, in which case, ε is of order 10^{-2} for metals and $\gamma<1$ for short-pulse heating situations.

5.4.2 Consideration of Surface Heat Source (Case 2)

In the absence of the volumetric source, $f=0$, we get that $\beta=0$. So the system (Eq. 5.113) will reduce to the following system, which is solved approximately in [18].

$$\begin{aligned}&\alpha(1+\mu)\frac{\partial^3\theta_{\mathrm{L}}}{\partial x^2\partial t}+\epsilon\alpha\frac{\partial^2\theta_{\mathrm{L}}}{\partial x^2}-\frac{\partial^2\theta_{\mathrm{L}}}{\partial t^2}-\frac{\partial\theta_{\mathrm{L}}}{\partial t}=0,\\ &\frac{\partial\theta_{\mathrm{L}}}{\partial t}=\epsilon(\theta_{\mathrm{E}}-\theta_{\mathrm{L}})\end{aligned} \tag{5.134}$$

Solution of the system (Eq. 5.134) can be given by substituting for $H=0$ in Eqs. (5.126) and (5.127) as

$$\theta_{\mathrm{L}}(x,t)=C_1+C_2e^{\lambda_1(t-bx)}+C_3e^{\lambda_2(t-bx)} \tag{5.135}$$

and

$$\theta_{\mathrm{E}}(x,t)=C_1+C_2\left(\frac{\lambda_1}{\epsilon}+1\right)e^{\lambda_1(t-bx)}+C_3\left(\frac{\lambda_2}{\epsilon}+1\right)e^{\lambda_2(t-bx)} \tag{5.136}$$

Now, assume a semi-infinite substrate material heated with a time decaying source from the surface. The boundary conditions for the problem can be written as follows:

$$\begin{aligned}&\frac{\partial\theta_{\mathrm{E}}}{\partial x}(0,t)=-\bar{I}_{\mathrm{o}}e^{-k_1t}, \quad &&\frac{\partial\theta_{\mathrm{E}}}{\partial x}(\infty,t)=0\\ &\theta_{\mathrm{E}}(x,\infty)=\theta_{\mathrm{o}}, &&\theta_{\mathrm{L}}(x,\infty)=\theta_{\mathrm{o}}\end{aligned} \tag{5.137}$$

The dimensionless source amplitude $\bar{I}_{\mathrm{o}}$ is related to the dimensional one I_{o} through the relation $\bar{I}_{\mathrm{o}}=\frac{I_{\mathrm{o}}}{k\delta T_{\mathrm{o}}}$, and the dimensionless power k_1 is related to the dimensional one d through the relation $k_1=\frac{dC_{\mathrm{E}}}{G}=\gamma$.

Using the boundary conditions $\theta_{\mathrm{L}}(x,\infty)=\theta_{\mathrm{o}}$, $\theta_{\mathrm{E}}(x,\infty)=\theta_{\mathrm{o}}$ and since $\lambda_1>0$ and $\lambda_2<0$, Eq. (5.123), C_2 vanishes ($C_2=0$) and $C_1=\theta_{\mathrm{o}}$.

Introducing the boundary condition $\frac{\partial\theta_{\mathrm{E}}}{\partial x}(\infty,t)=0$, since $\lambda_2<0$, it shows that $b<0$.

Finally, using the boundary condition $\frac{\partial\theta_{\mathrm{E}}}{\partial x}(0,t)=-\bar{I}_{\mathrm{o}}e^{-k_1t}$, the two equations result:

$$\begin{aligned}&\bar{I}_{\mathrm{o}}=b\lambda_2C_3\left(\frac{\lambda_2}{\epsilon}+1\right)\\ &\lambda_2=-k_1\end{aligned} \tag{5.138}$$

Solving Eq. (5.138) by using Eq. (5.123), the parameter b and the constant C_3 can be obtained as

$$\begin{aligned}b^2&=\frac{1-k_1}{\alpha k_1(k_1(1+\mu)-\epsilon)}\\ C_3&=\frac{\bar{I}_{\mathrm{o}}}{bk_1\left(\frac{k_1}{\epsilon}-1\right)}\end{aligned} \tag{5.139}$$

Since $b<0$ is known from the boundary condition, then

$$b=-\sqrt{\frac{1-k_1}{\alpha k_1(k_1(1+\mu)-\epsilon)}},\ \text{where}\left\{k_1>1\ \&\ k_1<\frac{\epsilon}{1+\mu}\right\}\ \text{or}\ \left\{k_1<1\ \&\ k_1>\frac{\epsilon}{1+\mu}\right\} \tag{5.140}$$

And so

$$C_3 = \frac{\bar{I}_o}{k_1\left(1 - \frac{k_1}{\epsilon}\right)}\sqrt{\frac{\alpha k_1(k_1(1+\mu)-\epsilon)}{1-k_1}} \tag{5.141}$$

Then solution of the boundary value problem can be given, when $\left\{k_1 > 1 \,\&\, k_1 < \frac{\epsilon}{1+\mu}\right\}$ or $\left\{k_1 < 1 \,\&\, k_1 > \frac{\epsilon}{1+\mu}\right\}$ as follows:

$$\theta_L(x,t) = \theta_o + \frac{\epsilon \bar{I}_o}{k_1(\epsilon - k_1)}\sqrt{\frac{\alpha k_1(k_1(1+\mu)-\epsilon)}{1-k_1}}\exp\left(-k_1\left(t + \sqrt{\frac{1-k_1}{\alpha k_1(k_1(1+\mu)-\epsilon)}}x\right)\right)$$

$$\theta_E(x,t) = \theta_o + \frac{\bar{I}_o}{k_1}\sqrt{\frac{\alpha k_1(k_1(1+\mu)-\epsilon)}{1-k_1}}\exp\left(-k_1\left(t + \sqrt{\frac{1-k_1}{\alpha k_1(k_1(1+\mu)-\epsilon)}}x\right)\right) \tag{5.142}$$

It should be noted that the solution for $\theta_L(x,t)$ and $\theta_E(x,t)$ in Eq. (5.142) is only valid for $\left\{k_1 > 1 \,\&\, k_1 < \frac{\epsilon}{1+\mu}\right\}$ or $\left\{k_1 < 1 \,\&\, k_1 > \frac{\epsilon}{1+\mu}\right\}$, in which case, ε is of order 10^{-2} for metals and $k_1 < 1$ for short-pulse heating situations.

The exact values and the graphs of the solution for $\theta_L(x,t)$ and $\theta_E(x,t)$ in Eqs. (5.133) and (5.142) can be given by using the MATLAB or MAPLE programs directly.

5.5 FINDINGS AND DISCUSSIONS

The findings from the analytical treatment of equations for phonon radiative transport are presented for the following cases, in line with the findings of the previous studies [2–5].

5.5.1 Reduced Form of Equation for Phonon Radiative Transport

To adopt the analytical solution for the reduced form of the phonon radiative transport equation (Eq. 5.41), the laser short-pulse heating of solid surface is considered. Since the reduced equation for phonon transport becomes similar to the Cattaneo equation, the analytical solution for the Cattaneo equation with appropriate boundary conditions is presented. In this case, the time exponentially decaying pulse is incorporated as a volumetric source in the Cattaneo equation. However, there are two positive singularities for $U(s,t)$ in Eq. (5.41), so we can rewrite the integral as a sum of three improper integrals and then evaluate them numerically by

Table 5.1 Properties Used in the Simulations

α (m^2s^{-1})	β (1 s^{-1})	δ (1 m^{-1})	I_o (W m^{-2})	A (s^2 m^{-2})	B (s m^{-2})
0.227×10^{-4}	0.05×10^{13} & 0.15×10^{13}	6.16×10^{7}	0.6×10^{20}	2.1×10^{-10}	44,052.86

using adaptive Gauss–Kronrod quadrature [18]. This method is included in MATLAB, where the function **quadgk(fun,a,b,'RelTol',1e-6,'AbsTol',1e-6)** attempts to approximate the integral of a scalar-valued function *fun* from *a* to *b* using high-order global adaptive quadrature and take both relative error tolerance and absolute error tolerance as 10^{-6}. Moreover, the function of quadgk attempts to satisfy that "error bound <= max(AbsTol,RelTol*|Q|)" so that our results have excellent accuracy. The physical properties used in the simulations are given in Table 5.1.

Figs. 5.3 and 5.4 show temperature distribution inside the substrate material at different heating periods and for the laser pulse parameters $\left(\beta = 0.05 \times 10^{13}\ s^{-1}\right)$ and $\left(\beta = 0.15 \times 10^{13}\ s^{-1}\right)$. Temperature decays gradually in the surface region and it becomes sharp as the distance from the surface increases toward the solid bulk. The gradual decay of the temperature in the surface region is attributed to the volumetric heat source in the hyperbolic conduction equation. In this case, energy absorbed from the irradiated field increases the internal energy gain of the substrate material in the surface region. As the distance from the surface increases beyond the absorption depth, energy gain by the substrate material is governed by the wave form of the heat conduction, since absorbed energy from the irradiated laser energy almost diminishes at this depth. However, temperature decay becomes sharp at some depth below the surface as the heating period progresses. This is attributed to less energy gain from the irradiated field with progressing time in the surface region. Consequently, energy transfer through conduction from the surface vicinity to the solid bulk increases with progressing time. In addition, the rate of energy gain from the irradiated field is larger than the energy transfer through conduction from the surface region to the solid bulk. This is more pronounced in the early heating period when the laser beam intensity reaches its peak, which in turn, results in increasing temperature difference between the surface region and the solid bulk. Since the heat conduction is not only governed by the classical diffusion, the wave nature of the heat transfer, temperature decay becomes sharp inside the substrate material during the long heating periods. This is true for all the pulses incorporated in the simulation.

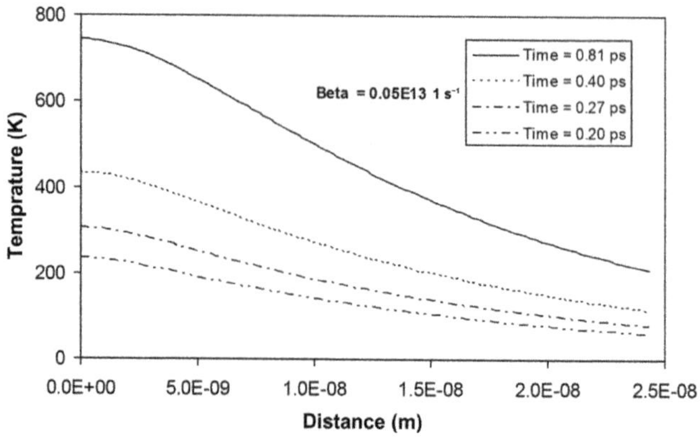

■ **FIGURE 5.3** Temperature variation inside the substrate material for different heating periods and for pulse parameter $\beta = 0.05 \times 10^{13}\ 1\ s^{-1}$ [2].

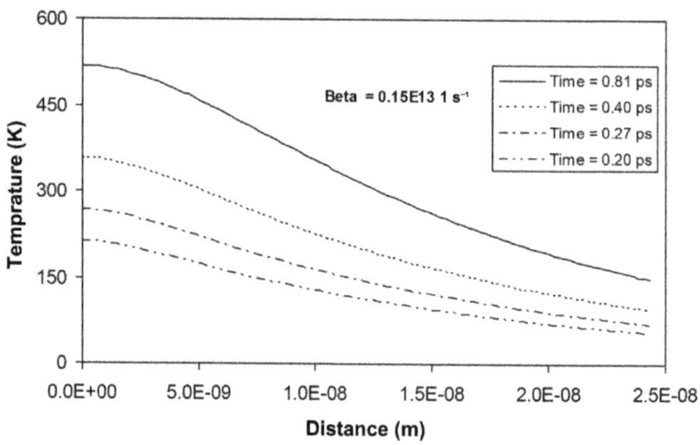

■ **FIGURE 5.4** Temperature variation inside the substrate material for different heating periods and for pulse parameter $\beta = 0.15 \times 10^{13}\ 1\ s^{-1}$ [2].

5.5.2 Analytical Solution of Equation for Phonon Radiative Transport for the Steady-State Temperature Disturbance Across the Thin Film

The simulation of the analytical solution of the phonon radiative transport equation is presented according to the film material whether dielectric or metallic. The findings are presented in line with the previous studies [3–5].

5.5.2.1 *Dielectric Thin Film (Silicon and Diamond Films)*

The analytical solution for the radiative transport equation in thin films is presented by incorporating the Fredholm integral equation of the second kind. The resulting equation for intensity distribution is reformulated in the temperature equation by using Eq. (5.60). A computer program is developed to predict dimensionless temperature distribution for silicon and diamond thin films of various thicknesses. The dimensionless temperature is defined through the ratio of relative temperature rise in the film to temperature difference across the film edges, i.e., $\frac{(T - T_{\text{right}})}{(T_{\text{left}} - T_{\text{right}})}$, where T_{right} is the temperature at the right edge of the film and T_{left} is the temperature at the left edge of the film, provided that temperature at the film edges is kept constant as $T_{\text{right}} = 301\text{K}$ and $T_{\text{left}} = 300\text{K}$. Table 5.2 gives the properties used in the simulations.

Fig. 5.5 shows the dimensionless temperature $\left(\frac{(T - T_{\text{right}})}{(T_{\text{left}} - T_{\text{right}})}\right)$ distribution in the diamond film for various thicknesses. The presence of a temperature jump at both edges ($x/L = 0$ and $x/L = 1$) is evident. This is attributed to the emitted phonon intensity, which attains low values at the edge interface. However, temperature jump increases at the edges as the film thickness reduces. This is associated with the phonon transport; in this case, the mean free path becomes comparable to the film thickness and the phonons emitted at the high temperature edge do not undergo much scattering in the film. This, in turn, reduces the thermal resistance to the phonons in the film. This situation is more pronounced at the edges of the film. Since the steady-state radiative phonon transport equation is solved analytically, temperature distribution follows the trend, which is almost linearly decay along the film thickness. However, in the vicinity of the edges, particularly at the low temperature edge, phonons emitted from this edge undergo interactions/superposition with those emitted from the high temperature edge. This modifies the temperature distribution in the neighborhood of the low temperature edge. However, as the

Table 5.2 Properties of Silicon and Diamond Films Used in the Simulations

	Volumetric Specific Heat (J m^{-3}K)	Phonon Speed (m s^{-1})	Mean Free Path (nm)
Silicon	1.653×10^6	8430	33.8
Diamond	1.815×10^6	12,288	447

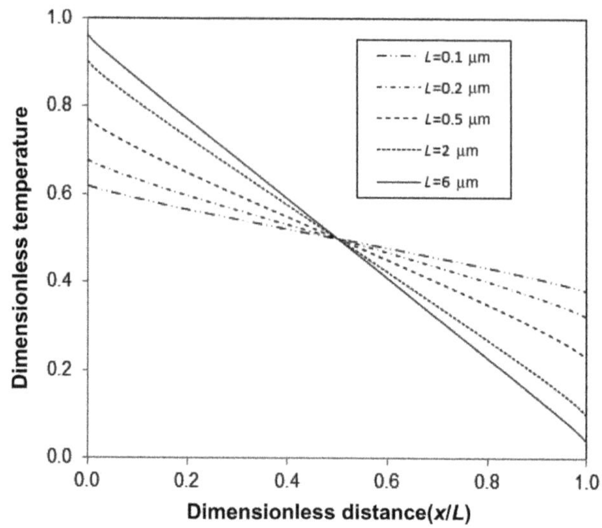

■ **FIGURE 5.5** Dimensionless temperature $\left(\frac{(T - T_{\text{right}})}{(T_{\text{left}} - T_{\text{right}})}\right)$ variation along the dimensionless distance (x/L) for various thicknesses of the diamond film (L) [4].

film thickness increases, this effect almost diminishes in this region because of the emitted phonons from the high temperature edge, which suffers from the large scattering before reaching the low temperature edge. Therefore, its effect on the temperature variation becomes less in this region as compared to the small thickness film.

Fig. 5.6 shows dimensionless temperature $\left(\frac{(T - T_{\text{right}})}{(T_{\text{left}} - T_{\text{right}})}\right)$ distribution along the dimensionless film length for different thicknesses of the silicon film. Temperature behavior along the film length is similar to those shown in Fig. 5.5, provided that the temperature jump at the film edges are smaller for the silicon film as compared to that of the diamond film. In addition, the influence of the film thickness on the temperature distribution is not the same as that of the diamond film. This behavior is associated with the properties of diamond and silicon, which are considerably different (Table 5.2). In this case, diamond has significantly high phonon speed and the mean free path as compared to those corresponding to the silicon film. The high value of the mean free path results in phonon behavior, which is similar to the ballistic phonon behavior in the thin films. In this case, phonons emitted from the high temperature edge do not suffer from a large amount of scattering before reaching the low temperature edge.

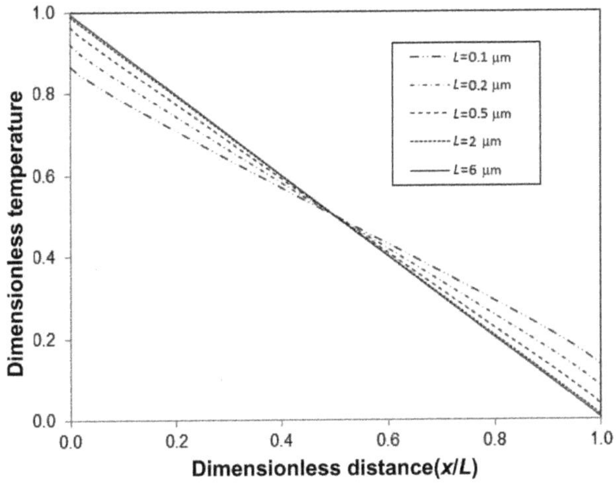

■ **FIGURE 5.6** Dimensionless temperature $\left(\frac{(T - T_{right})}{(T_{left} - T_{right})}\right)$ variation along the dimensionless distance (x/L) for various thicknesses of the silicon film (L) [4].

This lowers the thermal resistance to the phonon transport. In the case of the silicon film, this situation does not take place due to the small mean free path; therefore, the phonons emitted from the high temperature edge suffer significant large scattering and phonon intensity reduces prior to reaching the low temperature edge. In addition, since the phonon speed is much lower for silicon film than that of the diamond film, the rate of scattering in the film attains higher values for the silicon film while causing considerable reduction in the phonon intensity in the film. The film thickness considered in the present study is significantly larger than the mean free path of the silicon film; however, it is comparable to the mean free path of the diamond film.

5.5.2.2 *Metallic Thin Film (Aluminum Film)*

The analytical solution obtained for the equation of phonon radiative energy transport (Eq. 5.81) is simulated for the thin aluminum film. It should be noted that the system of integral equations (Eq. 5.81) is presented in the form of the Fredholm integral equation of the second kind and they are solved analytically through the integral transformations and discretization techniques. In the simulations, the constant temperature boundary conditions are assumed at the edges of the aluminum film. To compare the results of the analytical solution for the radiative transport, the two-equation model is used and the solution for temperature distributions in the electron and lattice subsystems are obtained for the similar boundary conditions, which are used for the radiative transport equations. The resulting temperature distribution in the film is simulated for various values of the aluminum film thickness. Table 5.3 gives aluminum film properties used in the simulations.

Fig. 5.7 shows the comparison of the phonon temperature obtained from the analytical solution and predicted from the numerical simulation.

Table 5.3 Properties of Aluminum Film Used in the Simulations

	Aluminum, Phonons	Aluminum, Electrons
Thermal conductivity ($Wm^{-1}K^{-1}$)	$11.3 = .05 \times k_e$	225.7
Volumetric specific heat ($Jm^{-3}K^{-1}$)	2,294,396.6	23,659
Speed (ms^{-1})	6420	2.03×10^6
Mean free path, $\Lambda = 3k/Cv$, (m)	2.3×10^{-9}	14.1×10^{-9}
Relaxation time, $\tau = \Lambda/v$, (s)	3.58×10^{-13}	6.945×10^{-15}
Electron–Phonon coupling constant in aluminum (Wm^{-3})		2.4642×10^{17}

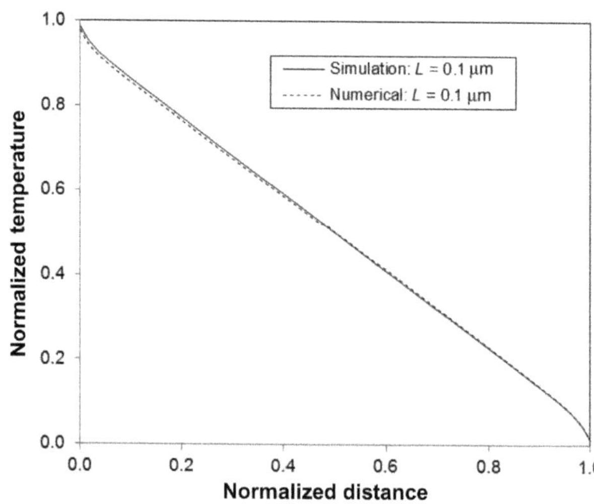

■ **FIGURE 5.7** Normalized temperature in the lattice system due to analytical and numerical solutions of radiative transport equation. The normalized temperature corresponds to $(T_p - T_{min})/(T_{max} - T_{min})$, where T_p is the phonon temperature, T_{max} is the high temperature at the film edge (301K), and T_{min} is the low temperature at the film edge (300K) [3].

Temperature profiles corresponding to both results are in good agreement. The small discrepancies are because of the round-off errors resulting from the numerical simulations; nevertheless, there are negligibly small.

Fig. 5.8 shows the normalized temperature distribution in the lattice subsystem, which is obtained from the analytical solution of the radiative transport equation, for different film thicknesses. The distance is normalized according to each film thickness (x/L) and temperature is normalized according to $(T_p - T_{min})/(T_{max} - T_{min})$, where T_p is the phonon temperature, T_{max} is the high temperature at the film edge (301K), and T_{min} is the low temperature at the film edge (300K). Temperature jump is observed at the edges of the film, which is associated with the emitted phonons from the edges. The phonons emitted from the edges do not undergo substantial scattering at the vicinity of the edges while resulting low thermal resistance in this region. In the case of the thin film (0.1 μm thickness), the phonon scattering results in almost linear variation of temperature in the lattice site. This is mainly associated with acoustic velocity of phonons; in this case, as the film thickness reduces, the time taken for the emitted phonons from the high temperature edge to reach at the low temperature edge becomes shorter. This, in turn, creates a ballistic behavior of phonons while modifying the scattering rate and temperature increase

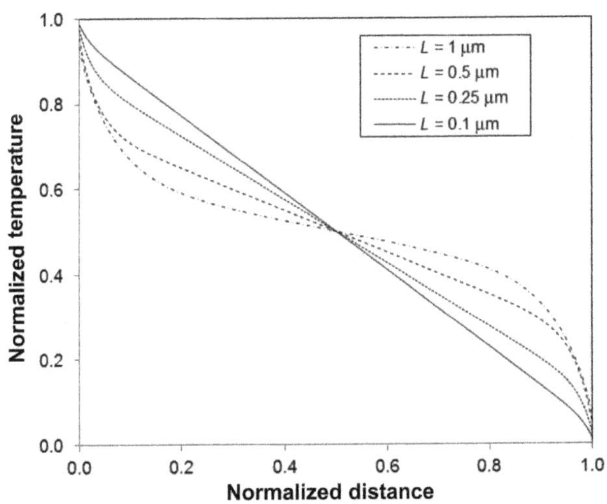

■ **FIGURE 5.8** Normalized temperature in the lattice subsystem along the normalized film thickness (x/L) for various aluminum film thicknesses. The normalized temperature corresponds to $(T_p - T_{min})/(T_{max} - T_{min})$, where T_p is the phonon temperature, T_{max} is the high temperature at the film edge (301K) and T_{min} is the low temperature at the film edge (300K) [3].

in the film. As the film thickness increases, phonons undergo large scattering in the film, causing different behavior of temperature distribution across the film. It should be noted that radiative transfer is governed by the phonons emitted from the film edges due to constant temperature heat sources at the film edges. The phonon intensity emitted from the high temperature edge can reflect from the low temperature edge (I^-) upon reaching the low temperature edge. This, in turn, modifies temperature behavior in the close region of the edges. Consequently, temperature variation becomes curvier in the edge regions of the film, which is more pronounced for the thick films.

Fig. 5.9 shows the normalized electron temperature $(T_e-T_{min})/(T_{max}-T_{min})$, where T_e is the electron temperature, T_{max} is the high temperature at the film edge (301K), and T_{min} is the low temperature at the film edge (300K) distribution along the normalized film length (x/L) for various film thicknesses. Electron temperature distribution shows similar behavior to those obtained for the lattice subsystem, provided that temperature jump at the edges are higher than that of the lattice site phonons. In addition, emitted phonons from the high temperature edge transfer some of their energy to electrons through electron–phonon coupling. This, in turn, results in electron excess energy and temperature increase in the electron subsystem. Therefore, electron excess energy gain depends on phonon energy distribution in the

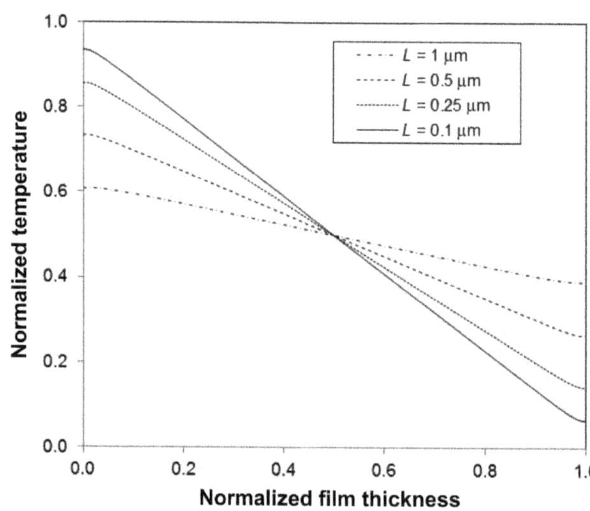

■ **FIGURE 5.9** Normalized temperature in the electron subsystem along the normalized film thickness (x/L) for various aluminum film thicknesses. The normalized temperature corresponds to $(T_e - T_{min})/(T_{max} - T_{min})$, where T_e is the electron temperature, T_{max} is the high temperature at the film edge (301K), and T_{min} is the low temperature at the film edge (300K) [3].

film. In the vicinity of the edges, emitted phonons do not undergo substantial scattering and electron–phonon coupling results in a small increase in electron temperature in this region, which in turn appears as a larger temperature jump than that of phonons in this region. Since the emitted phonons are dissipated through scattering in the film and at the same time phonons lose some fraction of their intensity to the electron subsystem through the electron–phonon coupling, electron temperature decays differently from lattice site temperature in the film. In this case, electron temperature decay demonstrates rather linear variation in the film that is not observed for lattice site temperature. It should be noted that thermionic emission of electrons at the film edges are omitted because of the low temperature range (301–300K) in the film and the insulated boundary is assumed at the edges of the film for the electron subsystem. This also influences electron temperature distribution in the film, which is particularly true in the regions close to the edges. However, in the midsection of the film ($x/L = 0.5$), lattice site energy transfer to the electron subsystem is not as high as that which takes place in the regions close to the edges. This can be observed from the comparison of electron and lattice site temperatures in the film.

Fig. 5.10 shows the comparison of normalized electron temperature distributions obtained from the radiative transport and the two-equation

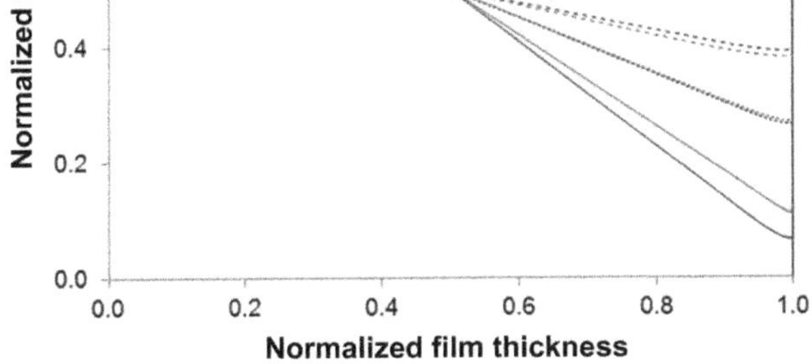

■ **FIGURE 5.10** Comparison of normalized temperature in the electron subsystem, due to radiative transport and the two-equation model, along the normalized film thickness (x/L) for various aluminum film thicknesses. The normalized temperature corresponds to $(T_e - T_{min})/(T_{max} - T_{min})$, where T_e is the electron temperature, T_{max} is the high temperature at the film edge (301K), and T_{min} is the low temperature at the film edge (300K) [3].

model along the normalized film length for three different film thicknesses. The differences between the results of both models appear to be significant for the film thickness of 0.1 μm. This is attributed to phonon intensity distribution and ballistic behavior of phonons for small film thicknesses. In this case, phonons do not contribute to scattering in the film for thinner film (0.1 μm) and, therefore, phonon energy transfer to electron subsystem through electron–phonon coupling becomes small in the film. This lowers electron temperature for the radiative transport model than that of the results of the two-equation model.

5.5.3 Analytical Solution of Hyperbolic Equation Pertinent to Electron Kinetic Theory Approach

To simulate the solution presented in Eq. (5.142), a short-pulse heating of a gold surface is considered after incorporating the ballistic contribution of electrons to nonequilibrium energy transfer during the short-pulse heating process. It should be noted that the finite Lie point symmetry is used to obtain the closed-form solutions for electron and lattice site temperature distributions. Table 5.4 gives the data used in the simulations.

Fig. 5.11 shows dimensionless electron temperature distribution in the surface region obtained from the volumetric and surface heat sources for different time periods. Electron temperature attains higher values in the surface region. This is attributed to electron excess energy gain in the surface region, which is more pronounced during the early heating period where the pulse peak intensity is high (Eq. 5.108). Electron temperature

Table 5.4 The Properties Used in the Simulations

Property	Values
C_L	2.8×10^6 $Jm^{-3}K^{-1}$
C_E	2.1×10^4 $Jm^{-3}K^{-1}$
G	2.6×10^{16} $Wm^{-3}K^{-1}$
k	315 $Wm^{-1}K^{-1}$
δ	10^9 m^{-1}
d	0.03×10^{13} s^{-1}
τ_p	0.024×10^{-12} s
T_o	300K
θ_o	1
I_o	10^{13} W m^{-2}
f	1

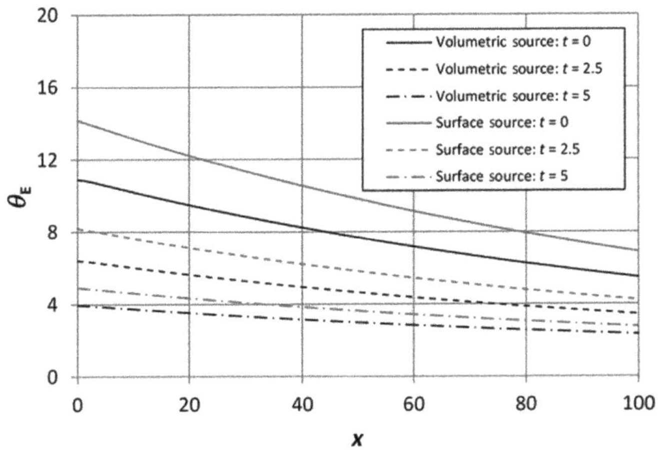

■ **FIGURE 5.11** Dimensionless electron temperature distributions along the dimensionless depth obtained from closed form solutions for volumetric and surface sources at different heating periods [4].

reduces sharply in the surface region and it decays gradually at some depth below the surface. Electrons in the surface region undergo collisions with the lattice site while transferring some fraction of their excess energy to the lattice site. Since the number of collisions is less in the early heating period, electron temperature remains high during this period. As the depth below the surface increases, electrons lose some of their excess energy to the lattice site, resulting in reduced electron temperature in this region. It should be noted that the amount of electron excess energy transfer through the collisional process is associated with the same amount of energy gain by the lattice site. Consequently, with progressing time, electrons further lose their excess energy to the lattice site and, at the same, energy gain by the lattice site increases lattice site temperature. This reduces further electron temperatures. The energy transfer between electrons and lattice site continues until thermal equilibrium is obtained between the electron and the lattice subsystems. Electron temperature attains higher values for surface heat source than that corresponding to the volumetric source. Although power intensity at the surface for both cases is the same, the boundary condition associated with electron temperature gradient at the surface is responsible for the attainment of high electron temperature for the surface heat source model. Temperature decays relatively gradually at some distance below the surface for the surface source case than that of the volumetric source case. This is related to the local excess energy gain of electrons from the irradiated field through the absorption for the volumetric source case.

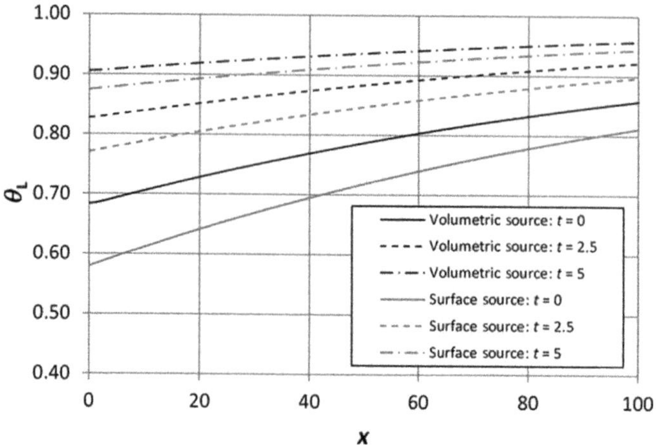

■ **FIGURE 5.12** Dimensionless lattice site temperature distributions along the dimensionless depth obtained from closed form solutions for volumetric and surface sources at different heating periods [4].

Fig. 5.12 shows the dimensionless lattice site temperature distribution in the surface region for the volumetric and surface sources for different heating periods. Lattice site temperature is higher at the surface for the surface heat source than that of the volumetric heat source. This is attributed to the high electron excess energy due to the high electron temperature gradient for the surface heat source, which is transferred to the lattice site through the collisional process. As the distance from the surface increases, lattice site temperature increases due to the energy transfer from electrons to the lattice site through the collisional process. In this case, electrons lose some of their excess energy to the lattice site at the surface region; therefore, the large fraction of remaining electron excess energy transfers to the lattice site through the collisional process as the distance from the surface increases. Moreover, lattice site temperature increases at long heating periods. This is associated with the number of collisions between electrons and the lattice site, which increases with progressing time while giving rise to increased lattice site temperature. This is true for surface and volumetric sources. The differences in lattice site temperature increase inside the substrate material due to volume and surface sources are associated with the local electron excess energy gain from the irradiated field through the absorption while modifying the amount of electron excess energy transfer to the lattice site.

5.6 CLOSING REMARKS

Since the formulation of energy transport across the micro/nanoscale structures are provided in Chapter 4, Analysis of Energy Transport Equations at Micro/Nanoscales, the analytical approaches to such formulations are provided in this chapter. The computational power for the solution of the thermal energy transport equation is expensive and requires excessive programing. In order to avoid such efforts and introduce the functional relation between the temperature and stress fields with the heating and transport parameters, analytical solutions to the thermal energy transport problems are presented. The equation for phonon radiative transport is an integral differential equation and the analytical solution to the transport problem becomes challenging. The analytical solution for the equation for phonon transport is introduced for the steady heating situation of dielectric thin film and findings are discussed for the conditions of temperature disturbance across the thin film. Application of electron kinetic theory is introduced and analytical solutions for volumetric and surface source heating situations are provided. The case study is introduced and the findings of the analytical solution are discussed for the resulting temperature and stress fields. Analytical treatment of coupling of electron and lattice subsystems is considered and the close form solution for temperature field in the metallic film is presented. The application of the analytical formulation of the temperature and stress fields is presented for laser short-pulse heating of the metallic thin film.

REFERENCES

[1] B.S. Yilbas, M. Pakdemirli, Analytical solution for temperature field in electron and lattice sub-systems during heating of solid film, Phys. B 382 (1–2) (2006) 213–219.

[2] B.S. Yilbas, A.Y. Al-Dweik, S. Bin Mansour, Analytical solution of hyperbolic heat conduction equation in relation to laser short-pulse heating, Phys. B 406 (8) (2011) 1550–1555.

[3] B.S. Yilbas, A.Y. Al-Dweik, S. Bin Mansoor, Non-equilibrium energy transport in a thin metallic film: Analytical solution for radiative transport equation, Phys. B 454 (2014) 15–22.

[4] B.S. Yilbas, A.Y. Al-Dweik, S. Bin Mansoor, Analytical solution for phonon transport across thin films, J. Non-Equilib. Thermodyn. 38 (4) (2013) 377–390.

[5] B.S. Yilbas, A.Y. Al-Dweik, Exact solution for temperature field due to non-equilibrium heating of solid substrate, Phys. B 406 (23) (2011) 4523–4528.

[6] A. Majumdar, Microscale heat conduction in dielectric thin films, J. Heat Transfer 115 (1) (1993) 7–16.

[7] C. Kittel, Introduction to Solid State Physics, sixth ed., Willey, 1986.

[8] B.S. Yilbas, A closed form solution for temperature rise inside solid substrate due to time exponentially varying pulse, Int. J. Heat Mass Transfer 45 (9) (2002) 1993–2000.
[9] L. Debnath, D. Bhatta, Integral Transforms and Their Applications, second ed., Chapman & Hall/CRC, New York, 2007.
[10] Z.M. Zhang, Nano Microscale Heat Transfer, McGraw-Hill, New York, 2007.
[11] G. Chen, Nanoscale Energy Transport and Conversion, Oxford University Press, 2005.
[12] G.L. Eesley, Generation of nonequilibrium electron and lattice temperatures in copper by picosecond laser pulses, Phys. Rev. B (1986).
[13] M. Pakdemirli, I.T. Dolapçi, B.S. Yilbaş, Symmetries and approximate solution of energy transfer equations in short pulse laser heating, Int. J. Therm. Sci. 46 (9) (2007) 908–913.
[14] H.M. Al-Theeb, B.S. Yilbas, Analytical solution for electron and lattice site temperatures due to laser-induced non-equilibrium energy transport in metals, Can. J. Phys. 88 (7) (2010) 479–491.
[15] H. Al-Qahtani, B.S. Yilbas, The closed form solutions for cattaneo and stress equations due to step input pulse heating, Phys. B 405 (18) (2010) 3869–3874.
[16] B.S. Yilbas, I.T. Dolapçi, M. Pakdemirli, Analytical solution for non-equilibrium energy transfer in gold: influence of ballistic contribution of electrons on energy transfer, Int. J. Therm. Sci. 48 (2) (2009) 383–390.
[17] B.S. Yilbas, Improved formulation of electron kinetic theory approach for laser ultra-short-pulse heating, Int. J. Heat Mass Transfer 49 (13–14) (2006) 2227–2238.
[18] L.F. Shampine, Vectorized adaptive quadrature in MATLAB, J. Comput. Appl. Math. 211 (2) (2008) 131–140.

Chapter 6

Heat Transfer Applications in One- and Two-Dimensional Thin Films

6.1 INTRODUCTION

Phonon transport in one- and two-dimensional films involves the solution of the Boltzmann transport equation (BTE). In the case of the dielectric thin films, the application of the BTE becomes straightforward, particularly for the frequency-independent consideration of the transport. Similar arguments can apply to the two-dimensional thin dielectric films when the frequency-independent solution of the BTE is considered. However, in the case of metallic thin films, a coupling of electron and lattice subsystems needs to be considered because of the thermal communication of both systems. In this case, electron–phonon coupling plays an important role to couple the energy transport equations of the electron and lattice subsystems. This situation can be simplified by introducing two phonon radiative transport equations for both subsystems and coupling them via the electron–phonon coupling parameter. On the other hand, in the case of the frequency-dependent solution of the BTE, the dispersion relation of the film becomes important to account for the transport in the branch of the phonon modes. This argument is true for one- and two-dimensional thin films. In the case of the metallic thin films, the same analysis is required to determine the transport characteristics in the electron and lattice subsystems, provided that electron–phonon coupling should be introduced to couple radiative transport equations for the electron and lattice subsystems.

The analytical solution of the BTE requires significant efforts, and in some cases, such as two-dimensional films, the analytical solution of the BTE becomes extremely difficult. Therefore, the numerical solution of the governing equation of the transport becomes unavoidable. Since the phonon transport involves the wave nature of radiative transport, care must be taken when solving the radiative transport equation numerically. In this case, the transport equation is treated as the radiative equation and the discrete ordinate method, in general, is adopted to discretize the

Heat Transport in Micro- and Nanoscale Thin Films.
DOI: http://dx.doi.org/10.1016/B978-0-323-42979-5.00006-2

transport equation. In addition the boundary conditions need to be incorporated in the numerical solution of the transport equation. In this chapter, numerical treatment of the radiative transport equation pertinent to the thin dielectric and metallic thin films is presented in light of the previous studies [1–6]. The appropriate case studies are also included to demonstrate the physical response of the film to the thermal disturbances according to the identified boundary conditions.

6.2 TRANSIENT ANALYSIS AND CASE STUDIES

The phonon transport in thin films is, in general, transient and the radiative transport equation is considered to be transient. In this section the transient analysis of the phonon transport in one- and two-dimensional films is presented in line with previous studies [1–4]. In addition, a frequency-dependent solution of the radiative phonon transport equation is also provided in relation to the previous studies [1–4].

6.2.1 One-Dimensional Frequency-Dependent Analysis of Dielectric Films and Applications

The analysis is involved with phonon transport characteristics in a thin silicon film and it is presented in the light of previous work [1]. Thermal energy transport in dielectric thin films involves diffusive or a mixture of diffusive and ballistic transport depending on the film size. The consideration of Casimir and Fourier heating models to formulate such thermal energy transport involve errors due to the internal and boundary scatterings of the films. Phonons are the main energy carriers in the thin films, particularly in the dielectric films. Phonons emanating from the film edges, due to temperature disturbances, undergo diffusive or diffusive/ballistic behaviors depending on the film size and the phonon mean free path. Determining the diffusive or ballistic contribution of phonon transport is critical for the assessment of the film's thermal properties such as thermal conductivity. Consequently, the ballistic and quasiballistic behavior of phonons alters the transport characteristics, and the size effect based on the phonon mean free path influences significantly the thermal resistance of the film. The phonon intensity distribution, due to the varying phonon mean free path, is characterized by the intensity ratio of acoustic phonon branches ($\frac{I}{I^o_{\max}}$, where $I^o_{\max}$ is the maximum equilibrium intensity, I is the intensity of an acoustic phonon branch at any location and time in the film) and the phonon transport characteristics are assessed by the wavelength ratio ($\eta_k = \frac{k_{\max} - k_{\min}}{k_{\max} - k_{\text{cut-off}}}$, where k is the wavenumber, $k = 2\pi/\lambda$, where λ is the phonon wavelength, and $k_{\text{cut-off}}$ is the cut off wavenumber based on the film thickness, $k_{\text{cut-off}} = \frac{2\pi}{L}$). The wavelength ratio (η_k) provides information about the film size effect on the transport characteristics, since $k_{\text{cut-off}}$ is based on the film thickness. In this section phonon transport characteristics for one-dimensional silicon film are presented in light of the previous study [1]. The frequency-dependent solution of the phonon radiative transfer equation is solved in line with the boundary conditions.

In the silicon layer, the equation of phonon radiative transfer (EPRT) is incorporated in the analysis, details of which are given in Chapter 4, Analysis of Energy Transport Equations at Micro/Nanoscales, Section 4.4.5. In silicon, there are in general two acoustic-transverse phonon branches, one acoustic-longitudinal branch, two acoustic-optical branches, and one optical-longitudinal branch. Therefore, a separate EPRT must be solved for each phonon branch and for each wavenumber. The heat capacity and group velocity of phonons is calculated from the dispersion relation for the [001] symmetry direction. Due to the symmetry of the [001] direction the two acoustic-transverse branches are degenerate; that is, it is not possible to distinguish between them and consequently they have the same dispersion relation (same wavevector and frequency ranges). Similarly, the two optical-transverse branches are also degenerate and have the same dispersion relation. This results in a total of four phonon branches. These equations are as follows.

Transverse Acoustic

$$\frac{1}{v_{\text{TA,k}}}\frac{\partial I_{\text{TA,k}}}{\partial t} + \mu\frac{\partial I_{\text{TA,k}}}{\partial x} = \frac{\frac{1}{2}\int_{-1}^{1} I_{\text{TA,k}}d\mu - I_{\text{TA,k}}}{v_{\text{TA,k}}\tau_{\text{TA,k}}} \quad \begin{matrix} t>0, & 0<x<1 \\ -1<\mu<1, & k_{\min}<k<k_{\max} \end{matrix} \tag{6.1}$$

Longitudinal Acoustic

$$\frac{1}{v_{\text{LA,k}}}\frac{\partial I_{\text{LA,k}}}{\partial t} + \mu\frac{\partial I_{\text{LA,k}}}{\partial x} = \frac{\frac{1}{2}\int_{-1}^{1} I_{\text{LA,k}}d\mu - I_{\text{LA,k}}}{v_{\text{LA,k}}\tau_{\text{LA,k}}} \quad \begin{matrix} t>0, & 0<x<1 \\ -1<\mu<1, & k_{\min}<k<k_{\max} \end{matrix} \tag{6.2}$$

Transverse Optical

$$\frac{1}{v_{\text{TO,k}}}\frac{\partial I_{\text{TO,k}}}{\partial t} + \mu\frac{\partial I_{\text{TO,k}}}{\partial x} = \frac{\frac{1}{2}\int_{-1}^{1} I_{\text{TO,k}}d\mu - I_{\text{TO,k}}}{v_{\text{TO,k}}\tau_{\text{TO,k}}} \quad \begin{matrix} t>0, & 0<x<1 \\ -1<\mu<1, & k_{\min}<k<k_{\max} \end{matrix} \tag{6.3}$$

Longitudinal Optical

$$\frac{1}{v_{\mathrm{LO,k}}}\frac{\partial I_{\mathrm{LO,k}}}{\partial t}+\mu\frac{\partial I_{\mathrm{LO,k}}}{\partial x}=\frac{\frac{1}{2}\int_{-1}^{1}I_{\mathrm{LO,k}}d\mu-I_{\mathrm{LO,k}}}{v_{\mathrm{LO,k}}\tau_{\mathrm{LO,k}}}\quad \begin{array}{l}t>0,\quad 0<x<1,\\ -1<\mu<1,\quad k_{\min}<k<k_{\max}\end{array} \tag{6.4}$$

It is important to note that in the previous equations, the phonon intensity I depends not only on the variables t, x, μ, but also on the wavenumber k as well as the particular phonon branch, TA, LA, TO, or LO. Similarly, the parameters used in the previous equations such as the group velocity v and the relaxation time τ are dependent on the wavenumber and the particular phonon branch. The volumetric heat capacity and phonon group velocity at each wavenumber and for each phonon branch is calculated from the dispersion relation according to the formulas,

$$C_{\mathrm{p,k}}(T)=k_B\left(\frac{\hbar\omega_{\mathrm{p}}(k)}{k_BT}\right)^2\frac{\exp^{\frac{\hbar\omega_{\mathrm{p}}(k)}{k_BT}}}{\left(\exp^{\frac{\hbar\omega_{\mathrm{p}}(k)}{k_BT}}-1\right)^2}\frac{k^2}{2\pi^2},\quad \text{and}\quad v_{\mathrm{p,k}}=\frac{\partial\omega}{\partial k} \tag{6.5}$$

In these formulas, $\omega_{\mathrm{p}}(k)$ is the wavenumber dependent frequency that is to be determined from the experimentally obtained dispersion relation [7]. The subscript p represents a particular phonon branch. The formulas for the relaxation time $\tau_{\mathrm{p,k}}$ are given as

$$\begin{aligned}\tau_{\mathrm{TA,k}}^{(N)}&=\frac{2.3\times10^{24}}{253,322\times\omega^2T\left[1-\exp\left(-3T/\theta_{\mathrm{D,TA}}\right)\right]}\\ \tau_{\mathrm{TA,k}}^{(U)}&=\frac{5.3\times10^{48}}{2012\times\omega^4T\left[1-\exp\left(-3T/\theta_{\mathrm{D,TA}}\right)\right]}\end{aligned}\qquad \frac{1}{\tau_{\mathrm{TA,k}}}=\frac{1}{\tau_{\mathrm{TA,k}}^{(N)}}+\frac{1}{\tau_{\mathrm{TA,k}}^{(U)}} \tag{6.6}$$

$$\begin{aligned}\tau_{\mathrm{LA}}^{(N)}&=\frac{2.3\times10^{24}}{163,921\times\omega^2T\left[1-\exp\left(-3T/\theta_{\mathrm{D,LA}}\right)\right]}\\ \tau_{\mathrm{LA}}^{(U)}&=\frac{5.3\times10^{48}}{507\times\omega^4T\left[1-\exp\left(-3T/\theta_{\mathrm{D,LA}}\right)\right]}\end{aligned}\qquad \frac{1}{\tau_{\mathrm{LA,k}}}=\frac{1}{\tau_{\mathrm{LA,k}}^{(N)}}+\frac{1}{\tau_{\mathrm{LA,k}}^{(U)}} \tag{6.7}$$

In these formulas the superscripts (N) and (U) represent the normal and Umklapp phonon scattering processes. These relaxation times were obtained from literature [8]. The Debye temperatures are given as $\theta_{\mathrm{D,TA}}=\hbar\omega_{\mathrm{max,TA}}/k_B$ and $\theta_{\mathrm{D,LA}}=\hbar\omega_{\mathrm{max,LA}}/k_B$. The authors couldn't find similar formulas for the optical phonon branches; however, they did find molecular dynamics simulations for calculating the optical phonon relaxation times for the transverse as well as for the longitudinal polarizations.

The curves between the wavenumber and relaxation times were found to be similar to those for the acoustic phonons. Therefore, the same formulas as those for the acoustic phonons were used. However, the coefficients were adjusted so that the calculated, temperature dependent, thermal conductivity ($K(T)$) matches the experimentally determined. The formulas are

$$\begin{aligned} \tau_{\text{TO,k}}^{(N)} &= \frac{2.3\times10^{24}}{2533.22\times\omega^2 T\left[1-\exp\left(-3T/\theta_{\text{D,TO}}\right)\right]} \\ \tau_{\text{TO,k}}^{(U)} &= \frac{5.3\times10^{48}}{20.12\times\omega^4 T\left[1-\exp\left(-3T/\theta_{\text{D,TO}}\right)\right]} \end{aligned} \quad \frac{1}{\tau_{\text{TO,k}}} = \frac{1}{\tau_{\text{TO,k}}^{(N)}} + \frac{1}{\tau_{\text{TO,k}}^{(U)}} \tag{6.8}$$

$$\begin{aligned} \tau_{\text{LO}}^{(N)} &= \frac{2.3\times10^{24}}{16,392.1\times\omega^2 T\left[1-\exp\left(-3T/\theta_{\text{D,LO}}\right)\right]} \\ \tau_{\text{LO}}^{(U)} &= \frac{5.3\times10^{48}}{50.7\times\omega^4 T\left[1-\exp\left(-3T/\theta_{\text{D,LO}}\right)\right]} \end{aligned} \quad \frac{1}{\tau_{\text{LO,k}}} = \frac{1}{\tau_{\text{LO,k}}^{(N)}} + \frac{1}{\tau_{\text{LO,k}}^{(U)}} \tag{6.9}$$

$$K(T) = \sum_{\text{p}=1}^{6} \int_0^{k_{\max}} \frac{C_{\text{p,k}} v_{\text{p,k}}^2 \tau_{\text{p,k}}}{3}\, dk \tag{6.10}$$

The Debye temperatures are $\theta_{\text{D,TO}} = \theta_{\text{D,LO}} = 690$ (K) [9]. In the previous equations the symbol k represents both the wavenumber as well as the thermal conductivity. However, in the case of a frequency-independent solution of the BTE, the phonon group velocity, phonon mean free path, and phonon relaxation time remain constant. The typical values for silicon are phonon group velocity $v = 8430$ m s^{-1}, phonon mean free path $\Lambda = 33.8$ nm, phonon relaxation time, $\tau = 4$ ps.

6.2.1.1 ***A Case Study and Application***

Consider the thin silicon film and transient thermal transport takes place due to temperature disturbances across the film. In this case the initial and boundary conditions are given in the following subsections.

6.2.1.1.1 **Initial Condition**

In accordance with the case study, the thin silicon film is considered to be at uniform temperature. Therefore, initially, it is considered that the silicon film is at a constant ambient temperature of $T_o = 300$K. This means that the intensity in the silicon film is constant everywhere and corresponds to the temperature T_o. The initial intensity may be calculated from the following formulas:

$$I_{TA,k}(x,\mu,0) = \frac{C_{TA,k} v_{TA,k} T_o}{4\pi} \quad 0<x<1, \quad -1<\mu<1, \quad k_{min}<k<k_{max} \tag{6.11}$$

$$I_{LA,k}(x,\mu,0) = \frac{C_{LA,k} v_{LA,k} T_o}{4\pi} \quad 0<x<1, \quad -1<\mu<1, \quad k_{min}<k<k_{max} \tag{6.12}$$

$$I_{TO,k}(x,\mu,0) = \frac{C_{TO,k} v_{TO,k} T_o}{4\pi} \quad 0<x<1, \quad -1<\mu<1, \quad k_{min}<k<k_{max} \tag{6.13}$$

$$I_{LO,k}(x,\mu,0) = \frac{C_{LO,k} v_{LO,k} T_o}{4\pi} \quad 0<x<1, \quad -1<\mu<1, \quad k_{min}<k<k_{max} \tag{6.14}$$

In the preceding formulas, $C_{TA,k}$ is the volumetric heat capacity of the silicon layer per unit wavenumber at the particular wavenumber k, and for the TA phonon branch, at temperature T_o. Similar heat capacities are defined for the other phonon branches. The total volumetric heat capacity for the silicon layer may be calculated as

$$\begin{aligned} C(T) &= 2C_{TA} + C_{LA} + 2C_{TO} + C_{LO} \\ &= 2\int_{k_{min}}^{k_{max}} C_{TA,k}dk + \int_{k_{min}}^{k_{max}} C_{LA,k}dk + 2\int_{k_{min}}^{k_{max}} C_{TO,k}dk + \int_{k_{min}}^{k_{max}} C_{LO,k}dk \\ &= \int_{k_{min}}^{k_{max}} \left(2C_{TA,k} + C_{LA,k} + 2C_{TO,k} + C_{LO,k}\right)dk \end{aligned} \tag{6.15}$$

6.2.1.1.2 **Boundary Conditions**

In accordance with the case study, temperature disturbance is introduced at the edges of the film. The left boundary condition corresponds to the front surface of the silicon film as depicted in Fig. 6.1. At all times $t>0$ it is considered that the front surface of the silicon film is maintained at a constant temperature of $T_{left} = 301\text{K}$. This gives the following boundary conditions:

$$I_{TA,k}(0,\mu,t) = \frac{C_{TA,k} v_{TA,k} T_{left}}{4\pi} \quad t>0, \quad -1<\mu<1, \quad k_{min}<k<k_{max} \tag{6.16}$$

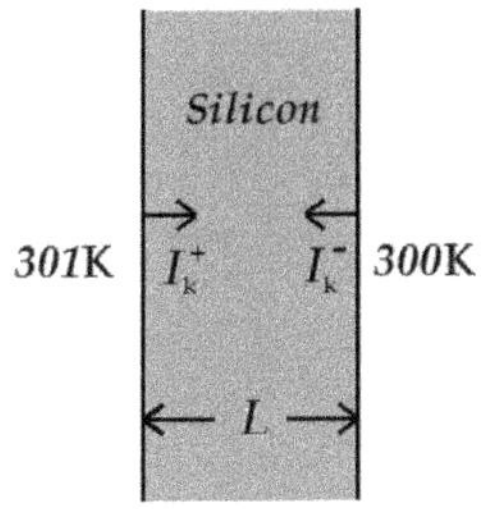

■ **FIGURE 6.1** A schematic view of a silicon film and temperature disturbance at the film edges [1].

$$I_{LA,k}(0,\mu,t)=\frac{C_{LA,k}v_{LA,k}T_{left}}{4\pi}\qquad t>0,\quad -1<\mu<1,\quad k_{min}<k<k_{max} \tag{6.17}$$

$$I_{TO,k}(0,\mu,t)=\frac{C_{TO,k}v_{TO,k}T_{left}}{4\pi}\qquad t>0,\quad -1<\mu<1,\quad k_{min}<k<k_{max} \tag{6.18}$$

$$I_{LO,k}(0,\mu,t)=\frac{C_{LO,k}v_{LO,k}T_{left}}{4\pi}\qquad t>0,\quad -1<\mu<1,\quad k_{min}<k<k_{max} \tag{6.19}$$

The right boundary condition corresponds to the back surface of the silicon film as depicted in Fig. 6.1. At all times $t>0$ it is considered that the back surface of the silicon film is maintained at a constant temperature of $T_{right}=$ 300K. The boundary conditions at this face can be defined similarly. A temperature difference of 1K between the two film surfaces is imposed mainly because of (1) the dispersion relation for silicon available is at room temperature, which is approximately 300K; (2) due to a film thickness in micrometers a temperature difference more than 1K may induce very large heat flux and stress fields, which may damage the film; and (3) the equation for radiative phonon transport is based on the linearized BTE and therefore is valid only for small temperature differences.

6.2.1.1.3 Limitations on the Phonon Wavenumber

The phonon wavenumber k is an independent variable in the EPRT. Its usual range is from *0* to k_{max}, corresponding to a phonon wavelength λ of λ_{max} to λ_{min}. This is because the wavenumber is inversely proportional to the wavelength, $k=2\pi/\lambda$. In a silicon film of thickness of the order of micrometers, the size effects start to appear. One of these size effects originates from the fact that a silicon film cannot sustain a lattice wave, along its thickness, of wavelength greater than its thickness. This means that only those lattice waves are allowed whose wavelength falls between the range $\lambda_{\max}$ to $\lambda_{\min}$. This corresponds to a phonon wavenumber range of $k_{\min}$ to $k_{\max}$. For a silicon film of thickness L we may define a cut-off wavenumber as

$$k_{\text{cut-off}} = \frac{2\pi}{L} \tag{6.20}$$

It should be noted that changing the film thickness changes the $k_{\text{cut-off}}$ as defined in Eq. (6.20). Strictly speaking only those wavenumbers should be considered that fall in the range $k_{\text{cut-off}}$ to $k_{\max}$. The maximum wavenumber $k_{\max}$ corresponds to the minimum allowed wavelength in the silicon crystal, and this is equal to twice the distance between two neighboring silicon atoms. However the study is extended to include or exclude some fraction of the allowed wavenumbers from this range as presented next. Let the wavenumber ratio η_k be a number defined as

$$\eta_{\text{k}} = \frac{k_{\max} - k_{\min}}{k_{\max} - k_{\text{cut-off}}} \quad \rightarrow \quad k_{\min} = k_{\max} - \eta_{\text{k}}(k_{\max} - k_{\text{cut-off}}) \tag{6.21}$$

$k_{\max}$ and $k_{\text{cut-off}}$ are constants for a known film thickness. The variable η_{k} takes on the values $\{1,0.9,0.8,0.7,0.6,0.5\}$. $k_{\min}$, which correspond to the minimum value of the wavenumber in any particular simulation, and it is varied in such a way that η_{k} takes the previous values. Setting η_{k} values and using Eq. (6.1), $k_{\min}$ can be evaluated. As a comparison a simulation is also run, which corresponds to the range 0 to $k_{\max}$.

6.2.1.1.4 **Definition of Temperatures and Heat Flux**

Equivalent equilibrium phonon temperature defined anywhere in the thin silicon layer is the analog of the usual temperature defined in the diffusive limit in any medium. It represents the average energy of all phonons around a local point and it is equivalent to the equilibrium temperature of phonons when they redistribute adiabatically to an equilibrium state. It is defined as

$$T(x,t) = 2\pi \frac{\int_{k_{\min}}^{k_{\max}} \int_{-1}^{1} (2I_{\text{TA,k}} + I_{\text{LA,k}} + 2I_{\text{TO,k}} + I_{\text{LO,k}})d\mu\, dk}{\int_{k_{\min}}^{k_{\max}} (2C_{\text{TA,k}}v_{\text{TA,k}} + C_{\text{LA,k}}v_{\text{LA,k}} + 2C_{\text{TO,k}}v_{\text{TO,k}} + C_{\text{LO,k}}v_{\text{LO,k}})dk} \tag{6.22}$$

The heat flux is defined at all points inside the layer and at all times. It is given as

$$q''(x,t)_x = 2\pi \int_{k_{\min}}^{k_{\max}} \int_{-1}^{1} \mu(2I_{\text{TA,k}} + I_{\text{LA,k}} + 2I_{\text{TO,k}} + I_{\text{LO,k}})d\mu\, dk \tag{6.23}$$

In these equations $C_{\text{TA},k}$, is the wavenumber-dependent volumetric heat capacity at the nominal temperature of $T_o = 300\text{K}$, and $v_{i,k}$ is the wavenumber-dependent phonon group velocity for the transverse acoustic branch for the [001] crystal plane in the silicon. Note that cubic cleavage occurs on the [001] planes in silicon, which is parallel to the faces of a cube for a crystal with cubic symmetry. Moreover, similar definitions can apply for the longitudinal acoustic, transverse optical, and longitudinal optical branches of phonons.

6.2.1.2 ***Numerical Method***

The EPRT in the silicon film with the associated initial and boundary conditions is solved numerically using the finite-difference method. This is an integrodifferential equation and may be solved using an iterative technique. Since the equation is time-dependent, the implicit scheme is used to solve it. Consider the EPRT for a particular phonon branch, in the silicon film:

$$\frac{1}{v_k}\frac{\partial I_k}{\partial t} + \mu\frac{\partial I_k}{\partial x} = \frac{\int_{-1}^{1} I_k d\mu - I_k}{\Lambda_k} \tag{6.24}$$

The variable μ may vary between $-1 \le \mu \le 1$. To facilitate the solution of the EPRT it is customary to break the intensity into parts; when $0 \le \mu \le 1$, $I_k = I_k^-(x,\mu,t)$, and when $0 \le \mu \le 1$, $I_k = I_k^+(x,\mu,t)$. Therefore, we may write separate equations for the two intensities. These are:

$$\frac{1}{v_k}\frac{\partial I_k^-}{\partial t} + \mu\frac{\partial I_k^-}{\partial x} = \frac{\int_{-1}^{0} I_k^- d\mu + \int_{0}^{1} I_k^+ d\mu - I_k^-}{\Lambda_k} \qquad -1 \le \mu \le 0 \tag{6.25}$$

and

$$\frac{1}{v_k}\frac{\partial I_k^+}{\partial t} + \mu\frac{\partial I_k^+}{\partial x} = \frac{\int_{-1}^{0} I_k^- d\mu + \int_{0}^{1} I_k^+ d\mu - I_k^+}{\Lambda_k} \qquad 0 \le \mu \le 1 \tag{6.26}$$

To solve these equations an implicit method is used for the time discretization and a discrete ordinates method is used for the discretization of the angular term. The time-derivative in each equation is approximated by means of the first-order backward-difference. The partial derivative with respect to the coordinate variable x is descritized using a forward-difference for the first equation in which $-1 \le \mu \le 0$ and is descritized using a backward-difference for the second equation in which $0 \leq \mu \leq 1$. This procedure is important to obtain stable discretizations. The integrals are calculated by first discretizing the cosine of the azimuthal angle μ into a number of divisions and then using the Simpson's rule to evaluate the integrals numerically for each time-step and each coordinate location. It is important to keep in mind that the intensities I_k^- and I_k^+ are two separate variables and each needs an initial condition and a single boundary condition. For the intensity I_k^-, the initial condition is as given earlier, and the boundary condition corresponds to the conditions at the back face. For the intensity I_k^+, the initial condition is again the same and the boundary condition is the left boundary condition for the silicon layer as discussed earlier.

Starting from the initial conditions in the film, at each time-step the EPRTs are solved in the silicon layer by means of an iterative procedure. In each iteration, a solution is obtained, which gives the approximate intensity distribution in the silicon film. From this solution the integral corresponding to the equilibrium intensity is calculated and the next iteration is performed. After convergence the computations proceed to the next time-step. Once the simulation is complete, the equivalent equilibrium phonon temperature may be evaluated at each time-step and at each location inside the silicon film.

6.2.1.3 *Findings of the Case Study*

The phonon intensity ratio is introduced, which is the ratio of intensity of phonons in a certain branch (longitudinal and transverse), according to the dispersion relations, to total phonon intensity determined over the phonon branches and modes (acoustic and optical). In addition, the wavenumber ratio ($\eta_k = \frac{k_{max} - k_{min}}{k_{max} - k_{cut\text{-}off}}$, where k is the wavenumber) is introduced to assess the influence of quasiballistic and ballistic phonons on the transport characteristics. Note that varying $k_{cut\text{-}off}$ corresponds to varying the film thickness in such a way that reduced $k_{cut\text{-}off}$ corresponds to increased film thickness (Eq. 6.20).

Fig. 6.2 shows phonon intensity ratio ($\frac{I_{LA}}{I^o_{max}}$, where I_{LA} is the phonon intensity due to longitudinal acoustic branch and I^o_{max} is the maximum equilibrium intensity in the film) along the film thickness for two heating periods and steady state (S/S) solution. I^o_{max} is defined as $I^o_{max} = \frac{T_o}{4\pi}\int_{k_{min}}^{k_{max}}(2C_{TA,k}v_{TA,k} + C_{LA,k}v_{LA,k} + 2C_{TO,k}v_{TO,k} + C_{LO,k}v_{LO,k})dk$, and I_{LA} is defined as $I_{LA} = \frac{1}{2}\int_{k_{min}}^{k_{max}}\int_{-1}^{1}(2I_{TA,k} + I_{LA,k} + 2I_{TO,k} + I_{LO,k})\,d\mu\,dk$. The phonon intensity ratios due to longitudinal acoustic and transverse acoustic branches are demonstrated separately for the wavenumber ratio of

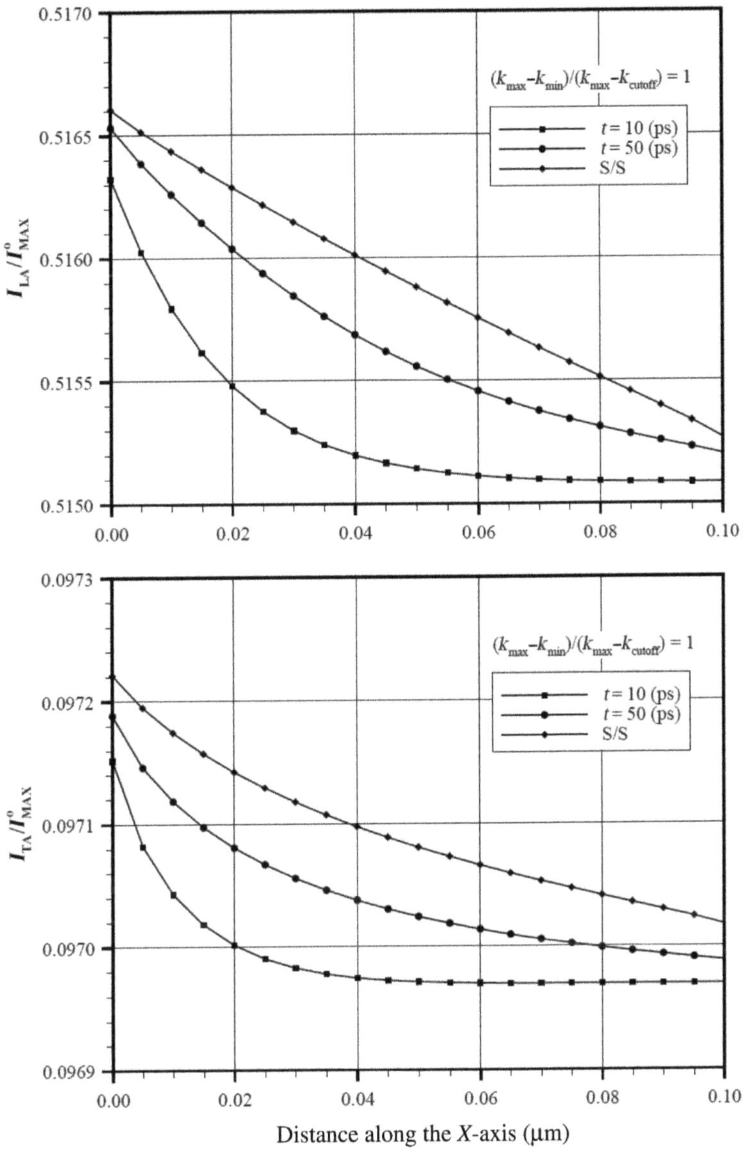

■ **FIGURE 6.2** Phonon intensity ratio for acoustic longitudinal and acoustic transverse branches along the film thickness for wavenumber ratio $\eta_k = 1$ [1].

$\eta_k = 1$. Note that $\eta_k = 1$ corresponds to the case in which all phonons do not suffer from cut-off, since all the phonon wavenumbers are incorporated in the analysis. However, the value of $\eta_k < 1$ allows only phonons with the wavenumbers equal or less than the cut-off wavenumbers contributing to the film resistance. In this case, phonons having wavelengths greater than the film thickness act like ballistic phonons but do not contribute to the film resistance. Intensity ratio remains high in the region close to the high-temperature edge and decreases sharply as the distance increases toward the low-temperature edge. The sharp decay in phonon intensity is more pronounced in the early heating periods. Since the frequency-dependent solution of the EPRT is considered including all modes and branches of the phonons, temporal response of the longitudinal acoustic branch differs from transverse acoustic. In addition, the intensity ratio shows that the longitudinal acoustic branch has a larger contribution than the transverse acoustic to the energy transport in the thin film. The time to reach a quasisteady state transport is earlier for the longitudinal acoustic branch. Consequently, longitudinal acoustic phonons are the main energy carriers during the energy transfer and can transport to reach quasisteady heating in the early periods.

However, boundary scattering of the emitted phonons in the vicinity of the high-temperature edge causes temperature jump in the film side of the high-temperature edge. The similar arguments are observed at the low-temperature film edge; in this case, reflected phonons from the low-temperature edge also contribute to temperature jump at the film edge. Fig. 6.3 shows the intensity ratio along the film thickness for the same conditions shown in Fig. 6.2, provided that the wavenumber ratio is $\eta_k = 0.5$. Since the wavenumber is inversely proportional to the wavelengths of the phonons ($k = 2\pi/\lambda$) and the cut-off wavenumber ($k_{\text{cut-off}}$) is related to the thickness of the film ($k_{\text{cut-off}} = \frac{2\pi}{L}$, where L is the film thickness), reducing the value of $k_{\text{cut-off}}$ corresponds to the thicker film. Consequently, the effect of ballistic and quasiballistic phonons on the phonon transport characteristics changes with altering the $k_{\text{cut-off}}$. In addition, increasing the artificially minimum value of the phonon wavenumber ($k_{\min}$) allows the large wavelengths phonon to be considered in the radiative transport analysis. The parameter η_k takes on the values $\{1,0.9,0.8,0.7,0.6,0.5\}$ and $k_{\min}$ is calculated based on these values. Although the behavior of phonon intensity ratios corresponding to $\eta_k = 0.5$ and $\eta_k = 1$ are similar, their corresponding values differ considerably. In this case, $\eta_k = 0.5$ results in a high ratio of phonon intensity for both longitudinal and transverse acoustic branches. The attainment of a large value of phonon intensity ratio for $\eta_k = 0.5$ is attributed to the increased film thickness. Therefore, the effect of ballistic and quasiballistic phonons on the film resistance becomes less as compared to the thin films. This in turn enhances the phonon intensity

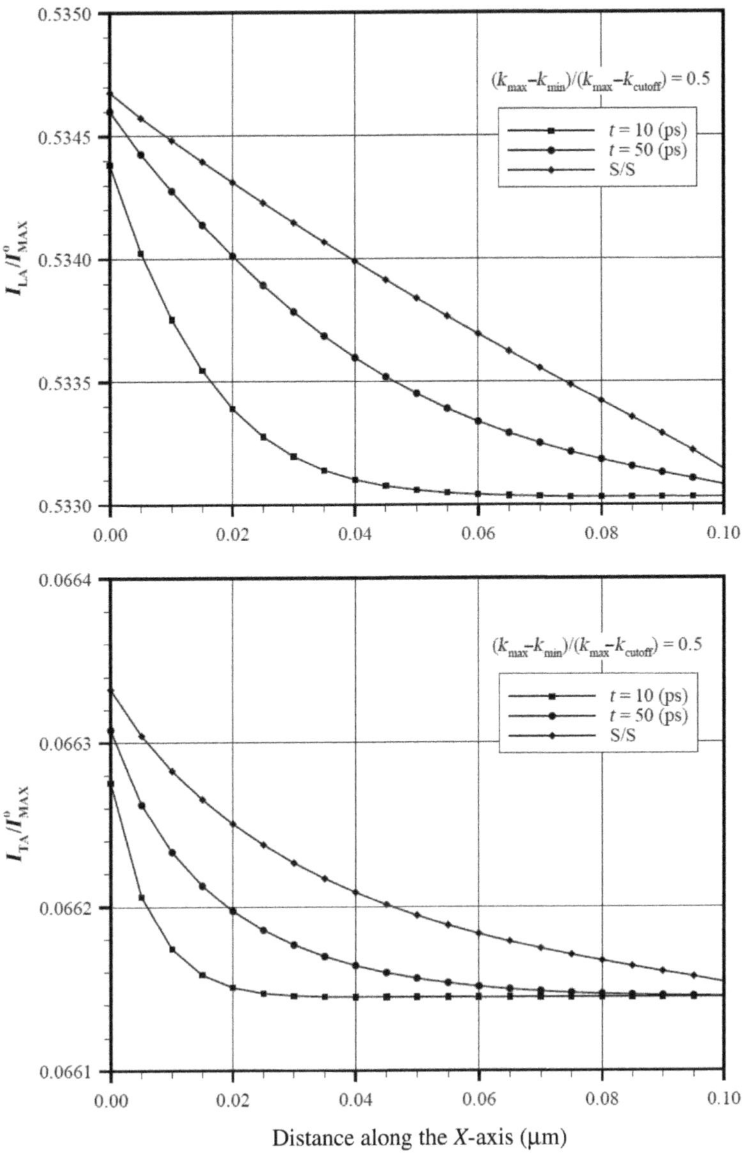

FIGURE 6.3 Phonon intensity ratio for acoustic longitudinal and acoustic transverse branches along the film thickness for wavenumber ratio $\eta_k = 0.5$ [1].

through scattering of phonons with a large spectrum of wavelengths in the film. However, only phonons with their wavelengths comparable to the film thickness can transfer energy without going through the scattering in the film.

Fig. 6.4 shows thermal conductivity variation with wavenumber ratio. Decreasing wavenumber ratio reduces thermal conductivity of the film. In this case, increasing k_{min} ($k = 2\pi/\lambda$) is responsible for the reduction of thermal conductivity. Therefore, increasing k_{min} limits the certain range of wavelength of phonons contributing to the radiative transport, because the wavenumber is inversely proportional to the wavelength of the phonons emitted from the high-temperature edge. Therefore, limiting the minimum value K in the simulations forces some of the phonons having a wavelength greater than $2\pi/K_{min}$ not to be emitted from the high-temperature film edge. Consequently, those phonons do not contribute to the film resistance through scattering. Since ballistic and quasiballistic phonons do not contribute to the scattering in the film, they have an adverse effect on the thermal conductivity [5]. However, for wavelength ratio of unity, thermal conductivity reaches to the bulk thermal conductivity of silicon [5]. Hence, increasing k_{min} by 20% shows the effect of ballistic phonons on the thermal conductivity of the film. Further increasing k_{min} beyond 20%, thermal conductivity reduces significantly. Consequently, phonons having wavelengths less than k_{min} do not contribute to thermal resistance of the film, which in turn, generates a size effect on the thermal properties of the film. In this case, 20% of the phonons

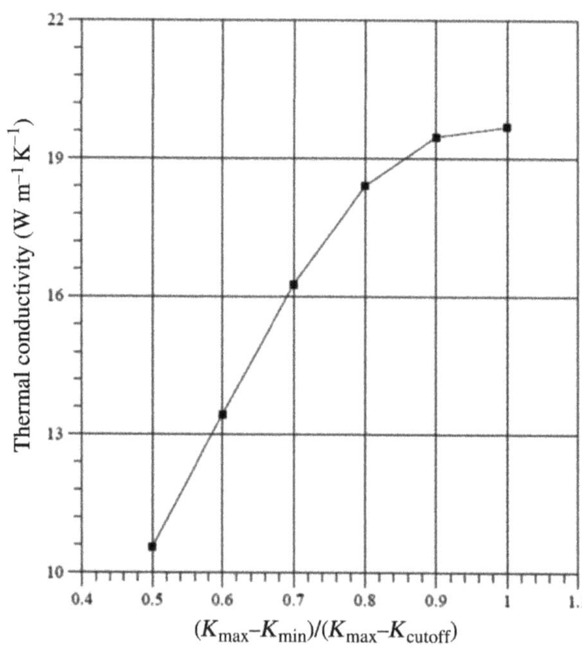

■ **FIGURE 6.4** Thermal conductivity of the film with wavenumber ratio [1].

having the large wavelengths act like the quasiballistic phonons in the silicon film and create the size effect on the transport properties while suppressing thermal conductivity and specific heat capacity of the silicon thin film.

6.2.2 One-Dimensional Frequency-Independent Analysis of Metallic Thin Films and Applications

Thermal separation of electron and lattice subsystems takes place in the metallic thin films when subjected to a thermal disturbance across film edges. This causes nonequilibrium energy transfer taking place in the metal substrate when the heating duration becomes comparable to electron relaxation time and the space is on the order of phonon mean free path. Consequently, nonequilibrium behavior results in thermal separation of electron and lattice subsystems in the heated region. The classical Fourier heating law fails to predict correct temperature rise in the substrate material during the heating process. Although heating models incorporating the hyperbolic heat equation predict temperature field in the substrate material, the findings are limited with the consideration of collisional process in which electron–phonon coupling plays an important role. However, energy transport in the lattice subsystem governs by phonon transport, which may be formulated using the BTE. Consequently, a comparative study incorporating the BTE and improved two-equation model for electron and lattice temperature fields is presented in this section. The formulation, analysis, and discussion of the problem related to the thin metallic film are given in light of the previous study [2].

Nonequilibrium heating in metals results in thermal separation of electron and phonon subsystems. In order to formulate energy transport in each subsystem using the BTE, we can write a separate EPRT for each subsystem. Since both subsystems thermally communicate during the energy transport, the electron–phonon coupling term needs to be incorporated in the EPRTs for each subsystem. The functional form of the coupling term enables the EPRTs to reduce to the standard two-equation model formulation in the diffusive limit. The EPRTs are incorporated into the solution; the details are given in Section 4.4.6.4 of Chapter 4, Analysis of Energy Transport Equations at Micro/Nanoscales.

Phonon Subsystem

$$\mu\frac{\partial I_p}{\partial x} = \frac{\frac{1}{2}\int_{-1}^{1} I_p d\mu - I_p}{\Lambda_p} + \frac{3G}{4\pi\Lambda_p}\mu\int_{o}^{x}(T_p - T_e)dy \qquad (6.27)$$

Electron Subsystem

$$\mu\frac{\partial I_e}{\partial x} = \frac{\frac{1}{2}\int_{-1}^{1} I_e d\mu - I_e}{\Lambda_e} + \frac{3G}{4\pi\Lambda_e}\mu\int_o^x (T_e - T_p)dy \tag{6.28}$$

Eqs. (6.27) and (6.28) can be reduced into the two-equation model; details are given in Section 4.4.6.4 of Chapter 4, Analysis of Energy Transport Equations at Micro/Nanoscales, which can be given as

$$C_p\frac{\partial T_p}{\partial t} = k_p\frac{\partial^2 T_p}{\partial x^2} - G(T_p - T_e) \tag{6.29}$$

and

$$C_e\frac{\partial T_e}{\partial t} = k_e\frac{\partial^2 T_e}{\partial x^2} - G(T_e - T_p) \tag{6.30}$$

6.2.2.1 ***A Case Study and Application***

One-dimensional aluminum thin film with a thermal disturbance due to temperature difference across the film edges is considered as a case study. The properties of aluminum used in the simulation are given in Table 6.1. The boundary conditions are provided according to the case study adopted.

6.2.2.1.1 **Boundary Conditions**

A constant temperature boundary is considered at both edges of the film. In this case, the left edge of the film is considered at 301K and the right edge of the film is at 300K. The phonon radiative intensity from the edges are I_p^+ and I_p^-. It is important to keep in mind that the intensities I_p^+ and I_p^- are two separate variables and each of them needs a single boundary condition.

Table 6.1 Properties Used in the Simulations [10–12]

	Phonons	Electrons
Thermal conductivity, ($Wm^{-1}K^{-1}$)	$11.3 = .05 \times k_e$ [11]	225.7 [10]
Volumetric heat capacity, ($Jm^{-3}K^{-1}$)	2,294,396.6 [12]	23,659 [10]
Speed, (ms^{-1})	6420	2.03×10^6
Mean free path, $\Lambda = 3k/Cv$, (m)	2.3×10^{-9}	14.1×10^{-9}
Relaxation time $\tau = \Lambda/v$, (s)	3.58×10^{-13}	6.945×10^{-15}
Electron–phonon coupling constant in aluminum, G, (Wm^{-3})	2.4642×10^{17}	

$$\text{Left Boundary Condition, } (x=0): \quad \begin{aligned} I_{\mathrm{p}}^{+}(0,\mu) &= C_{\mathrm{p}} v_{\mathrm{p}} T_{\mathrm{left}}/4\pi \\ I_{\mathrm{e}}^{+}(0,\mu) &= C_{\mathrm{e}} v_{\mathrm{e}} T_{\mathrm{left}}/4\pi \end{aligned} \tag{6.31}$$

$$\text{Right Boundary Condition, } (x=0): \quad \begin{aligned} I_{\mathrm{p}}^{-}(L,\mu) &= C_{\mathrm{p}} v_{\mathrm{p}} T_{\mathrm{right}}/4\pi \\ I_{\mathrm{e}}^{-}(L,\mu) &= C_{\mathrm{e}} v_{\mathrm{e}} T_{\mathrm{right}}/4\pi \end{aligned} \tag{6.32}$$

6.2.2.2 ***Numerical Method***

EPRTs in the aluminum film with the associated boundary conditions are solved numerically for the phonon and electron subsystems using the finite-difference method for the steady-state case. The governing equations (Eqs. 6.27 and 6.28) are in integrodifferential form and they can be solved using an iterative technique. For the numerical solution of the EPRT, it is convenient to use the implicit forms of the EPRTs. Consider the EPRT for the phonons (Eq. 6.27):

$$\mu \frac{\partial I_{\mathrm{p}}}{\partial x} = \frac{\frac{1}{2}\int_{-1}^{1} I_{\mathrm{p}} d\mu - I_{\mathrm{p}}}{\Lambda_{\mathrm{p}}} + \frac{3\,G}{4\pi\Lambda_{\mathrm{p}}}\mu \int_{o}^{x} (T_{\mathrm{p}} - T_{\mathrm{e}})\, dy$$

The variable μ varies between $-1 \le \mu \le 1$. To facilitate the solution of the EPRT it is customary to break the intensity into parts. When $-1 \le \mu \le 0$, $I_{\mathrm{p}} = I_{\mathrm{p}}^{-}(x,\mu)$ and when $0 \le \mu \le 1$, $I_{\mathrm{p}} = I_{\mathrm{p}}^{+}(x,\mu)$. Therefore, we may write separate equations for the two intensities:

$$\mu \frac{\partial I_{\mathrm{p}}^{+}}{\partial x} = \frac{\frac{1}{2}\left(\int_{-1}^{0} I_{\mathrm{p}}^{-} d\mu + \int_{0}^{1} I_{\mathrm{p}}^{+} d\mu\right) - I_{\mathrm{p}}^{+}}{\Lambda_{\mathrm{p}}} + \frac{3\,G}{4\pi\Lambda_{\mathrm{p}}}\mu \int_{o}^{x} (T_{\mathrm{p}} - T_{\mathrm{e}})\, dy, \quad 0 \le \mu \le 1 \tag{6.33}$$

and

$$\mu \frac{\partial I_{\mathrm{p}}^{-}}{\partial x} = \frac{\frac{1}{2}\left(\int_{-1}^{0} I_{\mathrm{p}}^{-} d\mu + \int_{0}^{1} I_{\mathrm{p}}^{+} d\mu\right) - I_{\mathrm{p}}^{-}}{\Lambda_{\mathrm{p}}} + \frac{3\,G}{4\pi\Lambda_{\mathrm{p}}}\mu \int_{o}^{x} (T_{\mathrm{p}} - T_{\mathrm{e}})\, dy, \quad -1 \le \mu \le 0 \tag{6.34}$$

To solve these equations the discrete ordinates method is used for the discretization of the angular term. The partial derivative with respect to the coordinate variable x is discretized using a backward-difference for the first equation in which $0 \le \mu \le 1$ and is discretized using a forward-difference for the second equation in which $-1 \le \mu \le 0$. This procedure is important to obtain stable discretization. The coordinate x is discretized into 40 divisions. The integrals with respect to the variable μ are calculated by first discretizing the variable μ into 80 divisions and then using

Simpson's rule to evaluate the integrals numerically for each coordinate location x. The integral with regard to the variable y should be carefully evaluated. To properly evaluate it, the x coordinate is temporarily discretized into 10 times the number of original divisions. Simpson's rule is then used to evaluate the integral at each original x coordinate location and each value of the coordinate μ. After the numerical integration the integral term is evaluated at each original, discretized x coordinate location. Exactly similar considerations apply to the intensity calculations for the electron subsystem. The EPRT is written in terms of two intensities:

$$\mu \frac{\partial I_e^+}{\partial x} = \frac{\frac{1}{2}\left(\int_{-1}^{0} I_e^- \, d\mu + \int_{0}^{1} I_e^+ \, d\mu\right) - I_e^+}{\Lambda_e} + \frac{3\,G}{4\pi\Lambda_e}\mu \int_{o}^{x} (T_e - T_p)\, dy, \quad 0 \le \mu \le 1 \tag{6.35}$$

and

$$\mu \frac{\partial I_e^-}{\partial x} = \frac{\frac{1}{2}\left(\int_{-1}^{0} I_e^- \, d\mu + \int_{0}^{1} I_e^+ \, d\mu\right) - I_e^-}{\Lambda_e} + \frac{3\,G}{4\pi\Lambda_e}\mu \int_{o}^{x} (T_e - T_p)\, dy, \quad -1 \le \mu \le 0 \tag{6.36}$$

6.2.2.2.1 **Solution Algorithm**

After discretizing Eq. (6.27) and (6.28), we can obtain a set of linear simultaneous equations. This system of equations has to be solved iteratively. The algorithm is given next.

1. Guess an intensity distribution at all values of the discretized variables x and μ, $I_p^+(x_i, \mu_j)$, $I_p^-(x_i, \mu_j)$, $I_e^+(x_i, \mu_j)$ & $I_e^-(x_i, \mu_j)$.
2. Impose the boundary conditions, $I_p^+(x_1, \mu_j) = C_p v_p T_{left}/4\pi$ and $I_p^-(x_{nx+1}, \mu_j) = C_p v_p T_{right}/4\pi$, $I_e^+(x_1, \mu_j) = C_e v_e T_{left}/4\pi$ and $I_e^-(x_{nx+1}, \mu_j) = C_e v_e T_{left}/4\pi$
3. Evaluate the integrals, $\int_{-1}^{1} I_p \, d\mu = \int_{-1}^{0} I_p^- \, d\mu + \int_0^1 I_p^+ \, d\mu$ and $\int_{-1}^{1} I_e \, d\mu = \int_{-1}^{0} I_e^- \, d\mu + \int_0^1 I_e^+ \, d\mu$.
4. Calculate the equivalent equilibrium temperatures, $C_p v_p T_p/4\pi = \frac{1}{2}\int_{-1}^{1} I_p d\mu$ and $C_e v_e T_e/4\pi = \frac{1}{2}\int_{-1}^{1} I_e d\mu$.
5. Solve the system of discretized equations, keeping the integrals, $\int_{-1}^{1} I_p \, d\mu$, $\int_{-1}^{1} I_e \, d\mu$ and the temperatures T_p and T_e as constants.
6. Repeat steps 3–5 until a converged solution is obtained.
7. Calculate the heat flux at each x location, $q_p'' = 2\pi \int_{-1}^{1} \mu I_p d\mu$ & $q_e'' = 2\pi \int_{-1}^{1} \mu I_e d\mu$.

Note that to quantify the phonon intensity distribution inside the film, equivalent equilibrium temperature is used as described earlier. However, equivalent equilibrium temperature is defined anywhere in the thin aluminum layer and is the analog of the usual temperature defined in the diffusive limit in any medium. It represents the average energy of all phonons around a local point and it is equivalent to the equilibrium temperature of phonons when they redistribute adiabatically to an equilibrium state.

6.2.2.3 *Findings of the Case Study*

Findings of equivalent equilibrium temperature inside the film are discussed in line with the previous study [2]. Fig. 6.5 shows phonon temperature difference due to the solutions of the BTE and the two-equation model for different film thicknesses. Temperature difference is high for small film thickness (0.01 μm), which is more pronounced in the region of a high-temperature edge. In this case, the two-equation model predicts

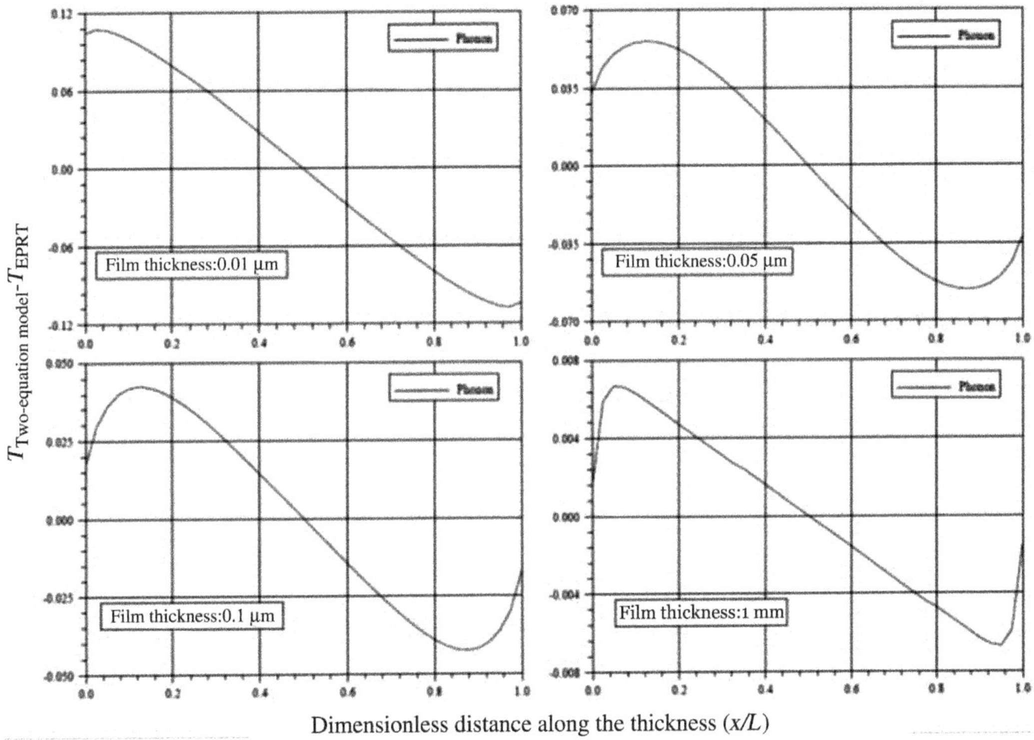

■ **FIGURE 6.5** Dimensionless equivalent equilibrium temperature difference due to two-equation model and the Boltzmann equation solution along the normalized film thickness for phonon subsystems for different film thicknesses [2].

higher temperature than the solution of the BTE. This is associated with the consideration of the presence of ballistic phonons in the film. Therefore, the presence of ballistic phonons does not contribute to phonon scattering in the region of the high-temperature edge. This yields low phonon temperature in this region. Moreover, reduced temperature difference in the vicinity of the cold temperature edge indicates that phonons reflected from the cold temperature edge contribute to phonon scattering while increasing phonon temperature obtained from the solution of the BTE. However, the low-temperature edge acts like a phonon sink while influencing phonon temperature increase in this region, which is particularly true for that obtained from the two-equation model. In this case, phonon temperature obtained from the two-equation model becomes less than that corresponding to the Boltzmann solution. This, in turn, results in negative temperature difference in the neighborhood of the cold edge of the film. As the film thickness increases, temperature difference reduces significantly and the location of maximum temperature difference moves away from the high-temperature edge. The ballistic phonons resulted from the solution of the BTE contributes to the attainment of the low-temperature difference at the high-temperature edge of the film. In addition, electron–phonon coupling resistance due to collisional energy transfer enhances the phonon temperature obtained from the two-equation model in this region. In the case of a low-temperature edge, phonons reflected from the cold edge undergo scattering in the vicinity of the low-temperature edge. This enhances phonon temperature as obtained from the BTE. Consequently, temperature difference becomes negative in this region.

Fig. 6.6 shows an electron temperature difference due to the solutions of the two-equation model and the BTE for different film thicknesses. Temperature difference is larger than phonon temperature difference as shown in Fig. 6.5, which is particularly true for small film thicknesses. This indicates that the two-equation model predicts high electron temperature in the high-temperature region. Since the film width is small, the number of electron–phonon collisions taking place in the film is less. This allows electron excess energy to remain high rather than transferring some fraction of their excess energy to phonons through the collisional process. In addition, ballistic phonons suppress phonon temperature rise in the high-temperature region of the film. Consequently, temperature difference remains high in this region. However, electron temperature obtained from the BTE becomes higher than its counterpart obtained from the two-equation model. The reflected electrons from the low-temperature edge contribute to electron temperature enhancement in this

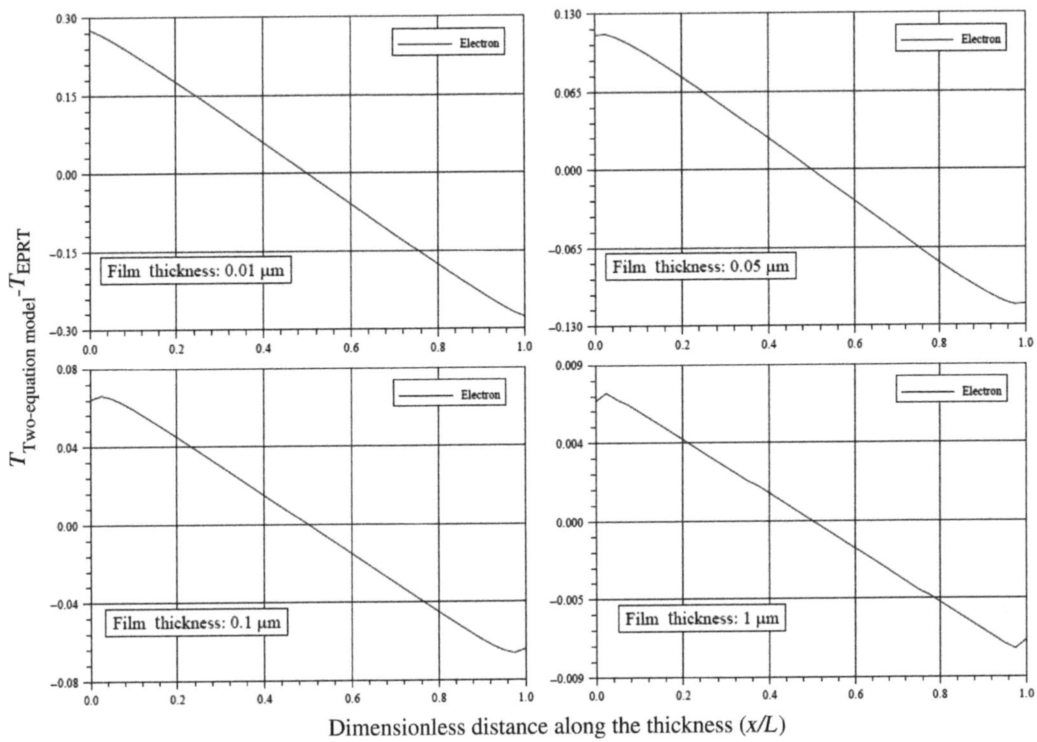

■ **FIGURE 6.6** Dimensionless equivalent equilibrium temperature difference due to the two-equation model and the Boltzmann equation solution along the normalized film thickness for electron subsystems for different film thicknesses [2].

region. Moreover, low-temperature edge acts like an energy sink for electrons, which in turn, lower electron temperature obtained from the two-equation model while resulting in a high-temperature difference in this region.

6.2.3 Two-Dimensional Frequency-Dependent Dielectric Thin Films and Applications

The molecular dynamics and BTE are generally used to address the phonon transport in two-dimensional thin films. Since the phonon transport can be simplified in the form of the radiative transport, the ballistic contribution of the phonons cannot be avoided and the frequency-dependent solution of the problem should be incorporated in the analysis. Although the frequency-dependent solution of the radiative transport equation includes the ballistic effects of the phonons, the solution depends on the boundary conditions and the thickness of the film. Frequency-dependent

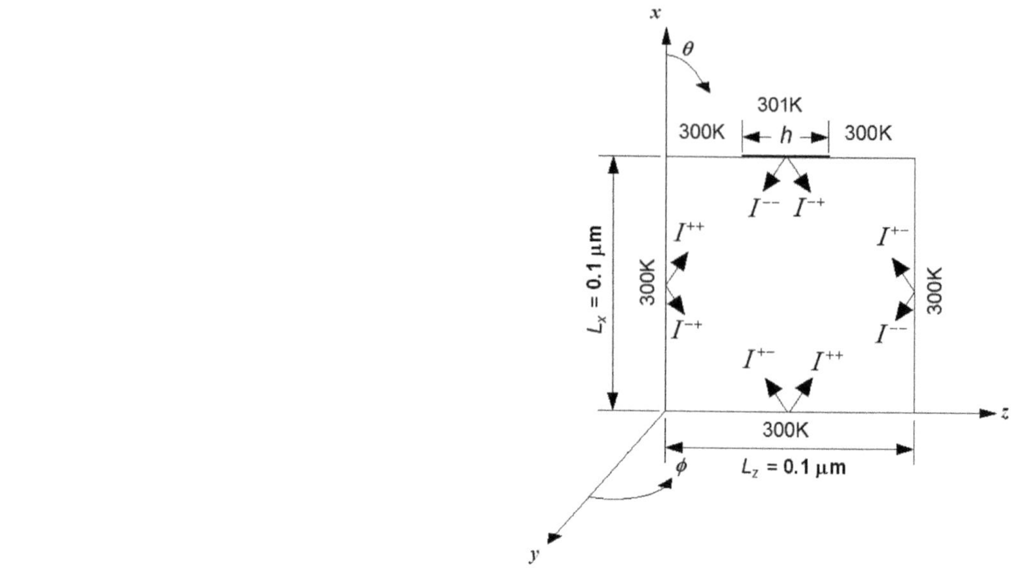

■ **FIGURE 6.7** Schematic of the film and the coordinate system. Phonon intensity emitted from the film edge is shown as I^+, and reflected from the edge is shown as I^-. The size of the heat source is denoted as h [3].

phonon transport in two-dimensional thin silicon film is presented in light of the previous study [3].

The solution of the BTE can be simplified after incorporating the relaxation-time approximation. This provides numerical ease in handling the phonon transport in the two-dimensional film. Fig. 6.7 shows the coordinate system used in the analysis. The transient two-dimensional and frequency (wavenumber)-dependent EPRT is written as

$$\frac{1}{v_k}\frac{\partial I_k}{\partial t} + \cos\theta\frac{\partial I_k}{\partial x} + \sin\theta\sin\phi\frac{\partial I_k}{\partial z} = \frac{\frac{1}{4\pi}\int_0^{2\pi}\int_0^{\pi} I_k \sin\theta\, d\theta\, d\phi - I_k}{\Lambda_k} \tag{6.37}$$

$$t>0,\quad 0<x<L_x,\quad 0<z<L_z,\quad 0<\theta<\pi,\quad 0<\phi<2\pi$$

In Eq. (6.37), the phonon intensity I_k is a function of x, z, t, θ, ϕ and k; i.e., $I_k = I_k(x, z, t, \theta, \phi)$. The details of EPRT are given Section 4.4.4 of Chapter 4, Analysis of Energy Transport Equations at Micro/Nanoscales.

The dispersion relations for the [001] direction in silicon is incorporated into the simulation, details of which are given in Section 6.2.1. There are four phonon branches: longitudinal acoustic, longitudinal optical, transverse acoustic, and transverse optical. Eq. (6.37) should be applied to each of these branches individually. These EPRTs are:

$$\frac{1}{v_{\mathrm{TA,k}}}\frac{\partial I_{\mathrm{TA,k}}}{\partial t}+\cos\theta\frac{\partial I_{\mathrm{TA,k}}}{\partial x}+\sin\theta\ \sin\phi\frac{\partial I_{\mathrm{TA,k}}}{\partial z}=\frac{\frac{1}{4\pi}\int_0^{2\pi}\int_0^{\pi}I_{\mathrm{TA,k}}\sin\theta\, d\theta\, d\phi-I_{\mathrm{TA,k}}}{\Lambda_{\mathrm{TA,k}}} \tag{6.38}$$

$$\frac{1}{v_{\mathrm{LA,k}}}\frac{\partial I_{\mathrm{LA,k}}}{\partial t}+\cos\theta\frac{\partial I_{\mathrm{LA,k}}}{\partial x}+\sin\theta\ \sin\phi\frac{\partial I_{\mathrm{LA,k}}}{\partial z}=\frac{\frac{1}{4\pi}\int_0^{2\pi}\int_0^{\pi}I_{\mathrm{LA,k}}\sin\theta\, d\theta\, d\phi-I_{\mathrm{LA,k}}}{\Lambda_{\mathrm{LA,k}}} \tag{6.39}$$

$$\frac{1}{v_{\mathrm{TO,k}}}\frac{\partial I_{\mathrm{TO,k}}}{\partial t}+\cos\theta\frac{\partial I_{\mathrm{TO,k}}}{\partial x}+\sin\theta\ \sin\phi\frac{\partial I_{\mathrm{TO,k}}}{\partial z}=\frac{\frac{1}{4\pi}\int_0^{2\pi}\int_0^{\pi}I_{\mathrm{TO,k}}\sin\theta\, d\theta\, d\phi-I_{\mathrm{TO,k}}}{\Lambda_{\mathrm{TO,k}}} \tag{6.40}$$

$$\frac{1}{v_{\mathrm{LO,k}}}\frac{\partial I_{\mathrm{LO,k}}}{\partial t}+\cos\theta\frac{\partial I_{\mathrm{LO,k}}}{\partial x}+\sin\theta\ \sin\phi\frac{\partial I_{\mathrm{LO,k}}}{\partial z}=\frac{\frac{1}{4\pi}\int_0^{2\pi}\int_0^{\pi}I_{\mathrm{LO,k}}\sin\theta\, d\theta\, d\phi-I_{\mathrm{LO,k}}}{\Lambda_{\mathrm{LO,k}}} \tag{6.41}$$

In Eqs. (6.38)–(6.41), T and L stand for transverse and longitudinal, respectively, and A and O stand for acoustic and optical, respectively. Note that the thermophysical properties of silicon are to be calculated from the dispersion relation.

In order to examine the phonon intensity behavior in the film, equivalent equilibrium temperature is introduced. However, the heat flux vector is defined at all points inside the layer, which is given as

$$q''_x(x,z,t)=\int_0^{k_{\max}}\int_0^{2\pi}\int_0^{\pi}\left(2I_{\mathrm{TA,k}}+I_{\mathrm{LA,k}}+2I_{\mathrm{TO,k}}+I_{\mathrm{LO,k}}\right)\cos\theta\ \sin\theta\ d\theta\, d\phi\, dk \tag{6.42}$$

and

$$q''_z(x,z,t)=\int_0^{k_{\max}}\int_0^{2\pi}\int_0^{\pi}\left(2I_{\mathrm{TA,k}}+I_{\mathrm{LA,k}}+2I_{\mathrm{TO,k}}+I_{\mathrm{LO,k}}\right)\sin\phi\ \sin^2\theta\ d\theta\, d\phi\, dk \tag{6.43}$$

$$\mathbf{q}''=q''_x\mathbf{i}+q''_z\mathbf{k} \tag{6.44}$$

6.2.3.1 *A Case Study and Application*

Two-dimensional silicon film is considered and temperature disturbance at one edge of the film is introduced. In addition, the effect of heat source size at the film edge on the transport characteristics is presented. The initial and boundary conditions are set in line with the previous study [3].

6.2.3.1.1 Initial and Boundary Conditions

There are four boundaries for silicon film, bottom, top, left, and right (Fig. 6.7). In addition to the EPRT, the phonon intensity specification at the boundaries and the initial condition should be prescribed to complete the problem specification. The geometry of the physical domain and the boundary conditions remain unchanged with time. Initially, the silicon layer is assumed to be at a uniform temperature of 300K. Then, after that for all times the following boundary conditions are imposed. The bottom, left, and right boundaries of the silicon layer remain at the initial temperature of 300K. The temperature at the top boundary also remains constant at 300K except on a small region around the center of the top boundary, where it is equal to 301K. These initial and boundary conditions are described next.

6.2.3.1.1.1 *Initial Condition*

$$
\begin{aligned}
I_{\mathrm{TA,k}}(x,z,t=0,\theta,\phi) &= \frac{C_{\mathrm{TA,k}}v_{\mathrm{TA,k}}T_{\mathrm{o}}}{4\pi}, \qquad 0<x<L_x,\quad 0<z<L,\quad 0<\theta<\pi,\quad 0<\phi<2\pi \\
I_{\mathrm{LA,k}}(x,z,t=0,\theta,\phi) &= \frac{C_{\mathrm{LA,k}}v_{\mathrm{LA,k}}T_{\mathrm{o}}}{4\pi} \\
I_{\mathrm{TO,k}}(x,z,t=0,\theta,\phi) &= \frac{C_{\mathrm{TO,k}}v_{\mathrm{TO,k}}T_{\mathrm{o}}}{4\pi} \\
I_{\mathrm{LO,k}}(x,z,t=0,\theta,\phi) &= \frac{C_{\mathrm{LO,k}}v_{\mathrm{LO,k}}T_{\mathrm{o}}}{4\pi}
\end{aligned}
\tag{6.45}
$$

6.2.3.1.1.2 *Top Boundary Condition*

$$
\begin{array}{l|l}
I_{\mathrm{TA,k}}(x=L_x,z,t,\theta,\phi)=\dfrac{C_{\mathrm{TA,k}}v_{\mathrm{TA,k}}T_{\mathrm{o}}}{4\pi} & I_{\mathrm{TA,k}}(x=L_x,z,t,\theta,\phi)=\dfrac{C_{\mathrm{TA,k}}v_{\mathrm{TA,k}}(T_{\mathrm{o}}+1)}{4\pi} \\
I_{\mathrm{LA,k}}(x=L_x,z,t,\theta,\phi)=\dfrac{C_{\mathrm{LA,k}}v_{\mathrm{LA,k}}T_{\mathrm{o}}}{4\pi} & I_{\mathrm{LA,k}}(x=L_x,z,t,\theta,\phi)=\dfrac{C_{\mathrm{LA,k}}v_{\mathrm{LA,k}}(T_{\mathrm{o}}+1)}{4\pi} \\
I_{\mathrm{TO,k}}(x=L_x,z,t,\theta,\phi)=\dfrac{C_{\mathrm{TO,k}}v_{\mathrm{TO,k}}T_{\mathrm{o}}}{4\pi} & I_{\mathrm{TO,k}}(x=L_x,z,t,\theta,\phi)=\dfrac{C_{\mathrm{TO,k}}v_{\mathrm{TO,k}}(T_{\mathrm{o}}+1)}{4\pi} \\
I_{\mathrm{LO,k}}(x=L_x,z,t,\theta,\phi)=\dfrac{C_{\mathrm{LO,k}}v_{\mathrm{LO,k}}T_{\mathrm{o}}}{4\pi} & I_{\mathrm{LO,k}}(x=L_x,z,t,\theta,\phi)=\dfrac{C_{\mathrm{LO,k}}v_{\mathrm{LO,k}}(T_{\mathrm{o}}+1)}{4\pi} \\
t>0,\ 0<z<L_z\ \&\ |z-L_z/2|>h/2 & t>0,\ 0<z<L_z\ \&\ |z-L_z/2|<h/2 \\
\pi/2<\theta<\pi,\ 0<\phi<2\pi & \pi/2<\theta<\pi,\ 0<\phi<2\pi
\end{array}
\tag{6.46}
$$

6.2.3.1.1.3 *Bottom Boundary Condition*

$$
\begin{aligned}
I_{\mathrm{TA,k}}(x=0,z,t,\theta,\phi) &= \frac{C_{\mathrm{TA,k}}v_{\mathrm{TA,k}}T_{\mathrm{o}}}{4\pi} \qquad t>0,\quad 0<z<L_z,\quad 0<\theta<\pi/2,\quad 0<\phi<2\pi \\
I_{\mathrm{LA,k}}(x=0,z,t,\theta,\phi) &= \frac{C_{\mathrm{LA,k}}v_{\mathrm{LA,k}}T_{\mathrm{o}}}{4\pi} \\
I_{\mathrm{TO,k}}(x=0,z,t,\theta,\phi) &= \frac{C_{\mathrm{TO,k}}v_{\mathrm{TO,k}}T_{\mathrm{o}}}{4\pi} \\
I_{\mathrm{LO,k}}(x=0,z,t,\theta,\phi) &= \frac{C_{\mathrm{LO,k}}v_{\mathrm{LO,k}}T_{\mathrm{o}}}{4\pi}
\end{aligned} \tag{6.47}
$$

6.2.3.1.1.4 *Left Boundary Condition*

$$
\begin{aligned}
I_{\mathrm{TA,k}}(x,z=0,t,\theta,\phi) &= \frac{C_{\mathrm{TA,k}}v_{\mathrm{TA,k}}T_{\mathrm{o}}}{4\pi} \qquad t>0,\quad 0<x<L_x,\quad 0<\theta<\pi,\quad 0<\phi<\pi \\
I_{\mathrm{LA,k}}(x,z=0,t,\theta,\phi) &= \frac{C_{\mathrm{LA,k}}v_{\mathrm{LA,k}}T_{o}}{4\pi} \\
I_{\mathrm{TO,k}}(x,z=0,t,\theta,\phi) &= \frac{C_{\mathrm{TO,k}}v_{\mathrm{TO,k}}T_{\mathrm{o}}}{4\pi} \\
I_{\mathrm{LO,k}}(x,z=0,t,\theta,\phi) &= \frac{C_{\mathrm{LO,k}}v_{\mathrm{LO,k}}T_{\mathrm{o}}}{4\pi}
\end{aligned} \tag{6.48}
$$

In Eqs. (6.45)–(6.48), $C_{\mathrm{TA,k}}$ is the wavenumber-dependent volumetric heat capacity at the nominal temperature of $T_{\mathrm{o}}=300\mathrm{K}$, and $v_{\mathrm{TA,k}}$ is the wavenumber-dependent phonon group velocity for the transverse acoustic branch for the [001] direction in silicon. Similar definitions apply for the longitudinal acoustic, transverse optical, and longitudinal optical branches of phonons.

The expression for the effective thermal conductivity of the film in terms of the wavenumber (k) can be written as [13]

$$
k(T) = \sum_{q=1}^{N}\sum_{p=1}^{3}\int_{0}^{k_{\max}} \frac{C_{\mathrm{k}}v_{\mathrm{k}}^{2}\tau_{\mathrm{k}}}{3}\,dk \tag{6.49}
$$

6.2.3.2 ***Numerical Method***

The EPRT in the silicon film with the associated initial and boundary conditions are solved numerically using the finite-difference method, which is the so-called discrete ordinates method. The EPRT is an integro-differential equation and may be solved using an iterative technique. We use the implicit scheme for temporal discretization. Consider the two-dimensional EPRT in the silicon film for any phonon branch,

$$\frac{1}{v_k}\frac{\partial I_k}{\partial t} + \cos\theta\frac{\partial I_k}{\partial x} + \sin\theta\,\sin\phi\frac{\partial I_k}{\partial z} = \frac{\frac{1}{4\pi}\int_0^{2\pi}\int_0^{\pi} I_k\,\sin\theta d\theta d\phi - I_k}{\Lambda_k} \tag{6.50}$$

A straightforward discretization strategy would be to replace the two, first derivatives on the left-hand side of the previous equation by forward-difference formulas and the time derivative term by a backward-difference formula. Doing so yields the following equations:

$$\frac{1}{v_k}\frac{I^p_{k,i,j,m,n} - I^{p-1}_{k,i,j,m,n}}{\Delta t} + \cos\theta_m \frac{I^p_{k,i,j,m,n} - I^p_{k,i-1,j,m,n}}{\Delta x} + \sin\theta_m \sin\phi_n \frac{I^p_{k,i,j,m,n} - I^p_{k,i,j-1,m,n}}{\Delta z} = \frac{(\text{integral})^p_{k,i,j} - I^p_{k,i,j,m,n}}{\Lambda_k} \tag{6.51}$$

$$I^p_{k,i,j,m,n}\left[\frac{1}{v_k\Delta t} + \frac{\cos\theta_m}{\Delta x} + \frac{\sin\theta_m\,\sin\phi_n}{\Delta z} + \frac{1}{\Lambda_k}\right] = I^p_{k,i-1,j,m,n}\left[\frac{\cos\theta_m}{\Delta x}\right] + I^p_{k,i,j-1,m,n}\left[\frac{\sin\theta_m\,\sin\phi_n}{\Delta z}\right] + \frac{(\text{integral})^p_{k,i,j}}{\Lambda_k} + \frac{I^{p-1}_{k,i,j,m,n}}{v_k\Delta t} \tag{6.52}$$

where $(\text{integral})^p_{k,i,j} = \frac{1}{4\pi}\int_0^{2\pi}\int_0^{\pi} I_k\,\sin\theta d\theta d\phi$.

It is well known that the finite difference discretization results in a stable numerical scheme, only if the coefficients of all the unknowns in the discretized equation are positive. Now, it may be observed in the previous equation that the coefficients are not always positive and their sign depends on the values of the angles θ and ϕ. When the coefficients are negative, it may be possible to use a backward difference and once again obtain a stable discretization. It turns out that to obtain stable discretizations for all values of θ and ϕ, we should partition the complete 4π solid angle into four quadrants. In the first quadrant, $0<\theta<\pi/2$, $0<\phi<\pi$, and we will denote the phonon intensity as $I^{p++}_{k,i,j,m,n}$; in the second quadrant, $\pi/2<\theta<\pi$, $0<\phi<\pi$, and we will denote the phonon intensity as $I^{p-+}_{k,i,j,m,n}$; in the third quadrant, $0<\theta<\pi/2$, $\pi<\phi<2\pi$, and we will denote the phonon intensity as $I^{p+-}_{k,i,j,m,n}$; finally, in the fourth quadrant, $\pi/2<\theta<\pi$, $\pi<\phi<2\pi$, and we will denote the phonon intensity as $I^{p--}_{k,i,j,m,n}$.

In the first quadrant, $\cos\theta$ and $\sin\theta\sin\phi$ are both positive, therefore the forward-difference formula for both $\partial I_k/\partial x$ and $\partial I_k/\partial z$ are to be used. In the second quadrant, only $\cos\theta$ is negative, so a backward difference has to be used for $\partial I_k/\partial x$. In the third quadrant, only $\sin\theta\,\sin\phi$ is negative, so a backward difference has to be used for $\partial I_k/\partial z$. In the fourth quadrant, both $\cos\theta$ and $\sin\theta\sin\phi$ are negative, therefore a backward-difference formula for both $\partial I_k/\partial x$ and $\partial I_k/\partial z$ are to be used. The discretized equations for each quadrant are as follows.

First quadrant: $0<\theta<\pi/2,\ \ 0<\phi<\pi$

$$I^{p++}_{k,i,j,m,n}\left[\frac{1}{v_k\Delta t}+\frac{\cos\theta_m}{\Delta x}+\frac{\sin\theta_m\sin\phi_n}{\Delta z}+\frac{1}{\Lambda_k}\right] = I^{p++}_{k,i-1,j,m,n}\left[\frac{\cos\theta_m}{\Delta x}\right]+I^{p++}_{k,i,j-1,m,n}\left[\frac{\sin\theta_m\sin\phi_n}{\Delta z}\right]+\frac{(\text{integral})^{p}_{k,i,j}}{\Lambda_k}+\frac{I^{p-1++}_{k,i,j,m,n}}{v_k\Delta t} \tag{6.53}$$

Second quadrant: $\pi/2<\theta<\pi,\ \ 0<\phi<\pi$

$$I^{p-+}_{k,i,j,m,n}\left[\frac{1}{v_k\Delta t}-\frac{\cos\theta_m}{\Delta x}+\frac{\sin\theta_m\sin\phi_n}{\Delta z}+\frac{1}{\Lambda_k}\right] = I^{p\ -+}_{k,i+1,j,m,n}\left[-\frac{\cos\theta_m}{\Delta x}\right]+I^{p-+}_{k,i,j-1,m,n}\left[\frac{\sin\theta_m\sin\phi_n}{\Delta z}\right]+\frac{(\text{integral})^{p}_{k,i,j}}{\Lambda_k}+\frac{I^{p-1-+}_{k,i,j,m,n}}{v_k\Delta t} \tag{6.54}$$

Third quadrant: $0<\theta<\pi/2,\ \ \pi<\phi<2\pi$

$$I^{p+-}_{k,i,j,m,n}\left[\frac{1}{v_k\Delta t}+\frac{\cos\theta_m}{\Delta x}-\frac{\sin\theta_m\sin\phi_n}{\Delta z}+\frac{1}{\Lambda_k}\right] = I^{p+-}_{k,i-1,j,m,n}\left[\frac{\cos\theta_m}{\Delta x}\right]+I^{p+-}_{k,i,j+1,m,n}\left[-\frac{\sin\theta_m\sin\phi_n}{\Delta z}\right]+\frac{(\text{integral})^{p}_{k,i,j}}{\Lambda_k}+\frac{I^{p-1+-}_{k,i,j,m,n}}{v_k\Delta t} \tag{6.55}$$

Fourth quadrant: $\pi/2<\theta<\pi,\ \ \pi<\phi<2\pi$

$$I^{p--}_{k,i,j,m,n}\left[\frac{1}{v_k\Delta t}-\frac{\cos\theta_m}{\Delta x}-\frac{\sin\theta_m\sin\phi_n}{\Delta z}+\frac{1}{\Lambda_k}\right] = I^{p--}_{k,i+1,j,m,n}\left[-\frac{\cos\theta_m}{\Delta x}\right]+I^{p--}_{k,i,j+1,m,n}\left[-\frac{\sin\theta_m\sin\phi_n}{\Delta z}\right]+\frac{(\text{integral})^{p}_{k,i,j}}{\Lambda_k}+\frac{I^{p-1--}_{k,i,j,m,n}}{v_k\Delta t} \tag{6.56}$$

It is clear that the previous discretizations are unconditionally stable for the range of angles in which they are to be used. A second consequence of these discretizations is that the boundary conditions may now be implemented at all the boundaries; otherwise on the face of it, the original EPRT would accommodate boundary conditions at only two boundaries. To explain the implementation of the boundary conditions, refer to Fig. 6.7. As can be seen, each of these four different equations needs two boundary conditions for its solution. Consider the intensity $I^{p++}_{k,i,j,m,n}$. The two boundary conditions for this intensity are provided by the equilibrium intensities at the left and bottom boundaries, at which the intensity $I^{p++}_{k,i,j,m,n}$ is defined. These boundaries are specified in Fig. 6.7. Hence starting from these boundary intensities, the equation for $I^{\mathrm{p}++}_{\mathrm{k},i,j,m,n}$ is solved, until the intensity is obtained for all the nodes denoted by $(x_i, z_j, t^{\mathrm{p}}, \theta_m, \phi_n)$, within the range defined for them, for the intensity $I^{\mathrm{p}++}_{\mathrm{k},i,j,m,n}$. Similar considerations apply for the intensities $I^{\mathrm{p}-+}_{\mathrm{k},i,j,m,n}$, $I^{\mathrm{p}+-}_{\mathrm{k},i,j,m,n}$, and $I^{\mathrm{p}--}_{\mathrm{k},i,j,m,n}$. In the discretized equations discussed earlier, the angles θ and ϕ also assume discrete values. The angle θ ranges from 0 to π, which is divided into 60 intervals. The angle ϕ ranges from 0 to 2π and this range is divided into 24 intervals. The integral over the solid angle, $\frac{1}{4\pi}\int_0^{2\pi}\int_0^{\pi} I_{\mathrm{k}}\sin\theta \, d\theta \, d\phi$, is evaluated using the Simpson's rule.

The numerical algorithm to solve the EPRT would start by first initializing the intensities at each of the grid points $(x_i, z_j, \theta_m, \phi_n)$ to the initial condition, as well as by initializing the unknown $(\text{integral})_{i,j}$ at each of the grid points (x_i, z_j) to the corresponding initial value as described in the section on initial condition. Then at each time-step, starting from the appropriate boundary equilibrium intensities, and after assuming all unknown variables at the current time-step to be equal to their corresponding values in the previous time-step, each of these four equations are solved to obtain better estimates of the intensities inside the domain. Subsequently, a better value of the integral is evaluated from these new estimates of the intensities at each spatial grid point. This whole procedure is to be repeated for each discrete wavenumber and for each of the four phonon branches. Once this is done, the equivalent equilibrium temperature is calculated at each grid point inside the domain. This whole procedure now constitutes a single iteration. Several iterations are performed until a converged solution is obtained for the current time-step. The criterion for convergence at each time-step is that the maximum absolute difference everywhere inside the domain, between the equivalent equilibrium temperature calculated at the current iteration and that at the previous iteration should become smaller than a preassigned small positive number. Mathematically, $\max\left| (T^{\mathrm{p}}_{i,j})^{\mathrm{iter}} - (T^{\mathrm{p}}_{i,j})^{\mathrm{iter}-1} \right| < \varepsilon$, where iter is the iteration index and $\varepsilon = 10^{-6}$. Once a converged solution is obtained, the algorithm moves onto the next time-step. Before moving on to the next time-step, the transient heat flux at each point may be calculated by the formulas,

$$q''_x(x,z,t) = \int_0^{k_{\max}}\int_0^{2\pi}\int_0^{\pi} (2I_{\mathrm{TA,k}} + I_{\mathrm{LA,k}} + 2I_{\mathrm{TO,k}} + I_{\mathrm{LO,k}})\cos\theta \sin\theta \, d\theta \, d\phi \, dk \tag{6.57}$$

$$q''_z(x,z,t) = \int_0^{k_{\max}}\int_0^{2\pi}\int_0^{\pi} (2I_{\mathrm{TA,k}} + I_{\mathrm{LA,k}} + 2I_{\mathrm{TO,k}} + I_{\mathrm{LO,k}})\sin\phi \sin^2\theta \, d\theta \, d\phi \, dk \tag{6.58}$$

6.2.3.3 Findings of the Case Study

The equivalent equilibrium temperature is used to assess the phonon intensity distribution at each location in the film. The findings are discussed in line with the previous study [3]. Fig. 6.8 shows dimensionless

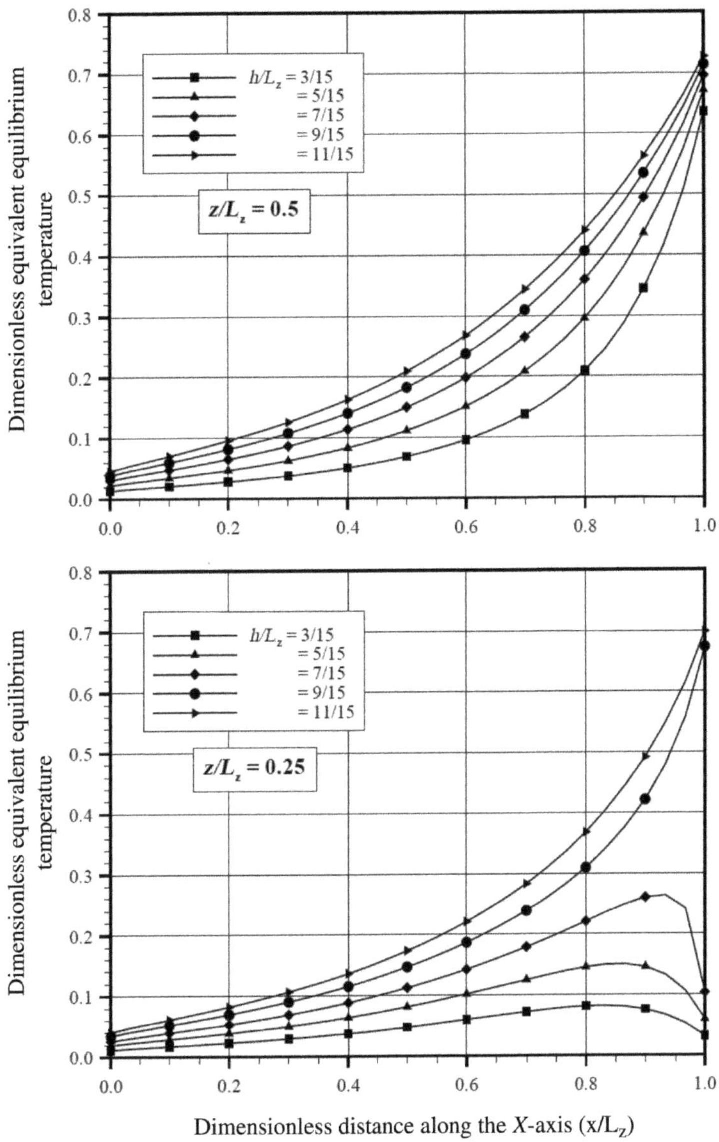

■ **FIGURE 6.8** Dimensionless temperature variation along the *x* axis for different heat source size and two locations in the film [3].

equivalent equilibrium temperature distribution along the normalized x axis (x/L_z) for different z-axis locations and different heat source sizes. It should be noted that the heat source size is normalized by the film width (L_z) and the curves are plotted at heating time of 12 ns. Since the heat source is located at the edge where $x = L_x$, equivalent equilibrium temperature remains high in the vicinity of this edge. Increasing distance along the film width results in attainment of low values of equivalent equilibrium temperature. This is associated with the low temperature of the film edge at $L_z = 0$. The influence of heat source size on equilibrium equivalent temperature is notable in the film. In this case lowering the heat source size lowers equivalent equilibrium temperature. This is attributed to the ballistic phonon behavior in the film. The ballistic phonons have high frequencies and do not undergo scattering in the film. This, in turn, lowers the resistance to the transport in the film. Equivalent equilibrium temperature reduces for reduced film thickness. Temperature jump at the interface of the heat source edge becomes larger for small-size heat source as compared to that of large-size heat source. This is again attributed to the presence of the ballistic phonons, which do not contribute to the scattering and dispersion in the film. However, short wavelength phonons undergo scattering in this region of the interface while contributing to the equilibrium equivalent temperature increase. Moreover, the equivalent equilibrium temperature difference, due to a different heat source size, becomes large at the x-axis location $x/L_x = 0.8$. This corresponds to the region next to the high-temperature interface. Consequently, phonons' relatively low frequencies undergo scattering in this region and ballistic phonons and the phonons having large wavelengths pass over this region without suffering from the film resistance due to scattering. In addition, the reflected phonons from the low-temperature edge contribute to phonon intensity increase in this region. However, this contribution is not high enough to have a significant effect on equivalent equilibrium temperature increase in this region. In the case of z-axis location $z/L_z = 0.25$, equivalent equilibrium temperature behaves similarly to that corresponding to the z-axis location at $z/L_z = 0.5$, provided that at small heat source size, this behavior differs. This is attributed to heat source size, which is less than the space corresponding to $0 \leq \frac{z}{L_z} \leq 0.25$. Therefore, the rate of longitudinal phonon waves emitted from the heat source and reflected from the low-temperature edges reached at this section is small. Therefore, their contribution to the phonon intensity in this region becomes small. Consequently, temperature jump at the interface becomes significantly high in this region.

Fig. 6.9 shows effective thermal conductivity with dimensionless heat source size. Thermal conductivity reduces with reducing heat source size.

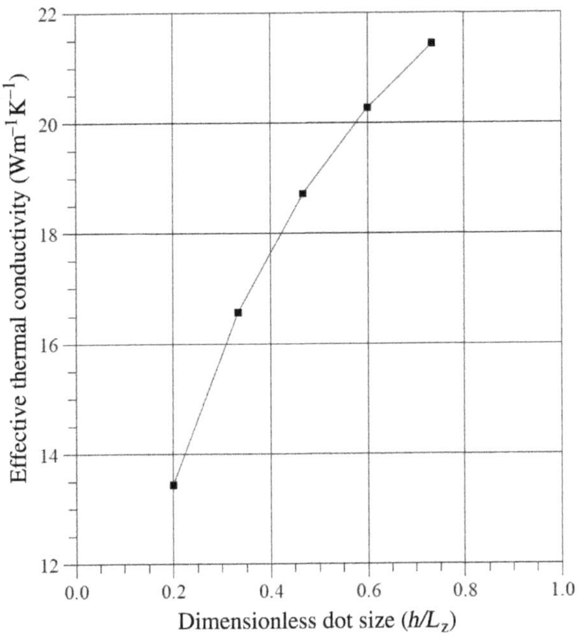

■ **FIGURE 6.9** Effective thermal conductivity with heat source size [4].

This behavior is attributed to the ballistic phonon behavior in the film, which does not contribute to the phonon intensity distribution in the film, which is more pronounced for the small-size heat source. The effective thermal conductivity is associated with the phonon scattering in the film while creating a thermal resistance for the phonon transport across the film. Therefore, small film thickness results in reduced thermal conductivity. This effect is also observed for small-size heat source, which is located at the film edge. In this case, ballistic phonons are responsible for the reduction of the effective thermal conductivity, since these phonons do not undergo scattering in the film while opposing the thermal resistance increase in the film. As the heat source size becomes comparable to the mean free path of the phonons, thermal conductivity reduces significantly, provided that the trend of the thermal conductivity decay is not in a linear form because of the complex nature of the phonon scattering in the film.

6.2.4 **Two-Dimensional Frequency-Independent Metallic Thin Films and Applications**

Nonequilibrium energy transfer in thin metallic films due to short duration of thermal excitation from an irradiated field is considered in this section in light of the previous study [4]. Nonequilibrium heating results in thermal separation of the electron and the lattice subsystems. Electrons absorb energy from the irradiated field and increase their excess energy in the electron subsystem. Through electron–phonon coupling, energy transfer occurs from the electron subsystem to the lattice subsystem. Since the radiation absorption takes place locally in the irradiated film, electron energy distribution in the electron subsystem becomes nonuniform. This in turn results in nonuniform energy transfer within the lattice subsystems. However, phonons are the main heat carriers in the lattice subsystem and the phonon scattering governs the thermal energy transport in the subsystem. To formulate the energy transport pertinent to the heating problem in a two-dimensional metallic film such as an aluminum film subjected to a laser irradiation, the BTE with the appropriate boundary conditions is considered. The equation of phonon radiative transport is given next, in line with the previous study [4].

The laser heating of the aluminum thin film should be modeled by means of a volumetric heat source term in the EPRT for the electron subsystem. The functional form of the electron–phonon coupling term as well as the volumetric heat source term should be incorporated in such a way that in the diffusive limit, the EPRT is expected to reduce the standard two-temperature model with a volumetric heat generation [14].

6.2.4.1 *EPRT for Phonon Subsystem*

The proposed modified BTE for the energy transport should satisfy the conservation of energy [14] in the lattice subsystem and can be written as

$$\frac{1}{v_p}\frac{\partial I_p}{\partial t} + \cos\theta\frac{\partial I_p}{\partial x} + \sin\theta\,\sin\phi\frac{\partial I_p}{\partial z} = \frac{\frac{1}{4\pi}\iint_{4\pi} I_p \sin\theta\, d\theta d\phi - I_p}{\Lambda_p} - \frac{G}{4\pi}\left(\frac{1}{C_p v_p}\iint_{4\pi} I_p \sin\theta\, d\theta d\phi - \frac{1}{C_e v_e}\iint_{4\pi} I_e \sin\theta\, d\theta d\phi\right) \tag{6.59}$$

I_p is the phonon intensity, ν_p is the phonon speed, Λ_p is the phonon mean free path, C_v is the phonon volumetric heat capacity, θ is the polar angle, ϕ is the azimuthal angle, and G is the electron–phonon coupling parameter.

6.2.4.2 *EPRT for Electron Subsystem*

The proposed equation for the energy transport in the electron subsystem should satisfy the conservation of energy [14] and can be written as

$$\frac{1}{v_e}\frac{\partial I_e}{\partial t}+\cos\theta\frac{\partial I_e}{\partial x}+\sin\theta\sin\phi\frac{\partial I_e}{\partial z}=\frac{\iint_{4\pi} I_e\sin\theta\,d\theta d\phi - I_e}{\Lambda_e}$$
$$-\frac{G}{4\pi}\left(\frac{1}{C_e v_e}\iint_{4\pi} I_e\sin\theta\,d\theta d\phi-\frac{1}{C_p v_p}\iint_{4\pi} I_p\sin\theta\,d\theta d\phi\right)+\frac{1}{4\pi}S(x,z,t) \tag{6.60}$$

where

$$S(x,z,t)=(1-R)\left(\frac{J}{t_p}\right)\alpha\exp\left[-\left(\frac{x-0.5L_x}{0.3L_x}\right)^2-\alpha z-2.77\left(t/t_p\right)^2\right] \tag{6.61}$$

In Eq. (6.60), ν_e is the electron speed, Λ_e is the electron mean free path, C_e is the electron volumetric heat capacity, θ is the polar angle, ϕ is the azimuthal angle, G is the electron–phonon coupling parameter, and S is the laser volumetric heat source term. R is the reflectance and α is the absorption coefficient. Eqs. (6.59) and (6.60) satisfy the law of conservation of energy, which can be verified by taking the zeroth moment of Eqs. (6.59) and (6.60). Multiply Eqs. (6.59) and (6.60) throughout by $\sin\theta d\theta d\phi$ and then integrate over the solid angle 4π, to get

$$\frac{1}{v_p}\frac{\partial}{\partial t}\iint_{4\pi} I_p\sin\theta\,d\theta\,d\phi+\frac{\partial}{\partial x}\iint_{4\pi} I_p\cos\theta\sin\theta\,d\theta\,d\phi+\frac{\partial}{\partial z}\iint_{4\pi} I_p\sin^2\theta\sin\phi\,d\theta d\phi$$
$$=\frac{\left(\frac{1}{4\pi}\iint_{4\pi} I_p\sin\theta\,d\theta\,d\phi\right)\iint_{4\pi}\sin\theta\,d\theta d\phi-\iint_{4\pi} I_p\sin\theta\,d\theta d\phi}{\Lambda_p}$$
$$-\frac{G}{4\pi}\left(\frac{1}{C_p v_p}\iint_{4\pi} I_p\sin\theta\,d\theta d\phi-\frac{1}{C_e v_e}\iint_{4\pi} I_e\sin\theta\,d\theta d\phi\right)\iint_{4\pi}\sin\theta d\theta d\phi \tag{6.62}$$

$$\frac{1}{v_e}\frac{\partial}{\partial t}\iint_{4\pi} I_e\sin\theta\,d\theta d\phi+\frac{\partial}{\partial x}\iint_{4\pi} I_e\cos\theta\sin\theta d\theta d\phi+\frac{\partial}{\partial z}\iint_{4\pi} I_e\sin^2\theta\sin\phi d\theta d\phi$$
$$=\frac{\left(\frac{1}{4\pi}\iint_{4\pi} I_e\sin\theta\,d\theta d\phi\right)\iint_{4\pi}\sin\theta\,d\theta d\phi-\iint_{4\pi} I_e\sin\theta\,d\theta d\phi}{\Lambda_p}$$
$$+\left[-\frac{G}{4\pi}\left(\frac{1}{C_e v_e}\iint_{4\pi} I_e\sin\theta d\theta d\phi-\frac{1}{C_p v_p}\iint_{4\pi} I_p\sin\theta d\theta d\phi\right)+\frac{1}{4\pi}S(x,z,t)\right]\iint_{4\pi}\sin\theta d\theta d\phi \tag{6.63}$$

In order to simplify these equations, we note the following definitions:

1. The phonon heat fluxes are defined as $q''_x=\iint_{4\pi} I_p\cos\theta\sin\theta d\theta d\phi$ and $q''_z=\iint_{4\pi} I_p\sin^2\theta\sin\phi d\theta d\phi$. (These definitions are to be used in the second and third terms on the left-hand side of Eqs. (6.62) and (6.63).)

2. The phonon and electron temperatures T_p and T_e are defined as $\frac{C_p v_p T_p}{4\pi} = \frac{1}{4\pi}\iint_{4\pi} I_p \sin\theta\, d\theta\, d\phi$ and $\frac{C_e v_e T_e}{4\pi} = \frac{1}{4\pi}\iint_{4\pi} I_e \sin\theta\, d\theta\, d\phi$.

Using the definitions in Eqs. (6.62) and (6.63), the following simplified forms are obtained:

$$C_p \frac{\partial T_p}{\partial t} + \nabla \cdot \mathbf{q}''_p = -G(T_p - T_e) \tag{6.64}$$

$$C_e \frac{\partial T_e}{\partial t} + \nabla \cdot \mathbf{q}''_e = -G(T_e - T_p) + S(x, z, t) \tag{6.65}$$

However, Eqs. (6.64) and (6.65) represent the modified two-temperature model, which satisfies the conservation of energy equations [15]. Therefore, Eqs. (6.59) and (6.60) intrinsically satisfy the principle of conservation of energy.

In the limit when the Knudsen number $\mathrm{Kn} = \Lambda/L \to 0$ (where Λ is the mean free path and L is the film thickness), Eqs. (6.59) and (6.60) should reduce to the two-temperature model. This can be verified as follows. Multiply Eq. (6.60) first by $\cos\theta \sin\theta d\theta d\phi$ and then by $\sin^2\theta \sin\phi d\theta d\phi$ and integrate over the solid angle 4π,

$$\begin{aligned}
&\frac{1}{v_e}\frac{\partial}{\partial t}\iint_{4\pi} I_e \cos\theta \sin\theta\, d\theta d\phi + \frac{\partial}{\partial x}\iint_{4\pi} I_e \cos^2\theta \sin\theta\, d\theta d\phi + \frac{\partial}{\partial z}\iint_{4\pi} I_e \cos\theta \sin^2\theta \sin\phi\, d\theta d\phi \\
&= \frac{\left(\frac{1}{4\pi}\iint_{4\pi} I_e \sin\theta\, d\theta d\phi\right)\iint_{4\pi} \cos\theta \sin\theta\, d\theta d\phi - \iint_{4\pi} I_e \cos\theta \sin\theta\, d\theta d\phi}{\Lambda_p} \\
&+ \left[-\frac{G}{4\pi}\left(\frac{1}{C_e v_e}\iint_{4\pi} I_e \sin\theta\, d\theta d\phi - \frac{1}{C_p v_p}\iint_{4\pi} I_p \sin\theta\, d\theta d\phi\right) \right. \\
&\left. + \frac{1}{4\pi} S(x,z,t)\right]\iint_{4\pi} \cos\theta \sin\theta\, d\theta d\phi
\end{aligned} \tag{6.66}$$

$$\begin{aligned}
&\frac{1}{v_e}\frac{\partial}{\partial t}\iint_{4\pi} I_e \sin^2\theta \sin\phi\, d\theta d\phi + \frac{\partial}{\partial x}\iint_{4\pi} I_e \cos\theta \sin^2\theta \sin\phi\, d\theta d\phi + \frac{\partial}{\partial z}\iint_{4\pi} I_e \sin^3\theta \sin^2\phi\, d\theta d\phi \\
&= \frac{\left(\frac{1}{4\pi}\iint_{4\pi} I_e \sin\theta\, d\theta d\phi\right)\iint_{4\pi} \sin^2\theta \sin\phi\, d\theta d\phi - \iint_{4\pi} I_e \sin^2\theta \sin\phi\, d\theta d\phi}{\Lambda_p} \\
&+ \left[-\frac{G}{4\pi}\left(\frac{1}{C_e v_e}\iint_{4\pi} I_e \sin\theta\, d\theta d\phi - \frac{1}{C_p v_p}\iint_{4\pi} I_p \sin\theta\, d\theta d\phi\right) \right. \\
&\left. + \frac{1}{4\pi} S(x,z,t)\right]\iint_{4\pi} \sin^2\theta \sin\phi\, d\theta d\phi
\end{aligned} \tag{6.67}$$

Using the definitions introduced before and carrying out the integrations, we get

$$\frac{1}{v_e}\frac{\partial q''_{x,e}}{\partial t} + \frac{\partial}{\partial x}\iint_{4\pi} I_e \cos^2\theta \sin\theta d\theta d\phi + \frac{\partial}{\partial z}\iint_{4\pi} I_e \cos\theta \sin^2\theta \sin\phi d\theta d\phi = -\frac{q''_{x,e}}{\Lambda_p} \tag{6.68}$$

$$\frac{1}{v_e}\frac{\partial q''_{z,e}}{\partial t} + \frac{\partial}{\partial x}\iint_{4\pi} I_e \cos\theta \sin^2\theta \sin\phi \, d\theta d\phi + \frac{\partial}{\partial z}\iint_{4\pi} I_e \sin^3\theta \sin^2\phi \, d\theta d\phi = -\frac{q''_{z,e}}{\Lambda_p} \tag{6.69}$$

where q is the heat flux. In the limit $\text{Kn} = \Lambda/L \to 0$, the intensity becomes isotropic and does not depend on the direction. This means that I_e ceases to be a function of θ and ϕ but is given as $I_e = I_e^o = \frac{C_e v_e T_e}{4\pi}$. These details can be used to carry out the integrations in the second and third terms of Eqs. (6.68) and (6.69).

$$\frac{1}{v_e}\frac{\partial q''_{x,e}}{\partial t} + \frac{4\pi}{3}\frac{C_e v_e}{4\pi}\frac{\partial T_e}{\partial x} = -\frac{q''_{x,e}}{\Lambda_p} \tag{6.70}$$

$$\frac{1}{v_e}\frac{\partial q''_{z,e}}{\partial t} + \frac{4\pi}{3}\frac{C_e v_e}{4\pi}\frac{\partial T_e}{\partial z} = -\frac{q''_{z,e}}{\Lambda_p} \tag{6.71}$$

T is the temperature.

The kinetic theory formula for the thermal conductivity is given as

$$k_p = \frac{C_p v_p \Lambda_p}{3} \quad k_e = \frac{C_e v_e \Lambda_e}{3} \tag{6.72}$$

Hence, Eqs. (6.70) and (6.71) can be transformed into

$$\frac{\Lambda_e}{v_e}\frac{\partial q''_{x,e}}{\partial t} + q''_{x,e} = -k_e\frac{\partial T_e}{\partial x} \tag{6.73}$$

$$\frac{\Lambda_e}{v_e}\frac{\partial q''_{z,e}}{\partial t} + q''_{z,e} = -k_e\frac{\partial T_e}{\partial z} \tag{6.74}$$

Multiplying Eq. (6.73) by $\hat{\mathbf{i}}$ and Eq. (6.74) by $\hat{\mathbf{j}}$ and then adding them together we get

$$\frac{\Lambda_e}{v_e}\frac{\partial \mathbf{q}''_e}{\partial t} + \mathbf{q}''_e = -k_e \nabla T_e \tag{6.75}$$

A similar equation is obtained for the phonon subsystem:

$$\frac{\Lambda_p}{v_p}\frac{\partial \mathbf{q}''_p}{\partial t} + \mathbf{q}''_p = -k_p \nabla T_p \tag{6.76}$$

The law of energy conservation is given by Eqs. (6.64) and (6.65). The heat flux vectors $\mathbf{q}''_p$ and $\mathbf{q}''_e$ can be eliminated between Eqs. (6.64) and (6.76) and between Eqs. (6.65) and (6.75). Also, noting that $\tau_p = \Lambda_p/v_p$ and $\tau_e = \Lambda_e/v_e$ are the relaxation times of the phonons and electrons, respectively, these equations may be written as

$$\tau_p \frac{\partial}{\partial t}\left[C_p \frac{\partial T_p}{\partial t} + G(T_p - T_e)\right] + C_p \frac{\partial T_p}{\partial t} = k_p \nabla^2 T_p - G(T_p - T_e) \quad (6.77)$$

$$\tau_e \frac{\partial}{\partial t}\left[C_e \frac{\partial T_e}{\partial t} + G(T_e - T_p) - S(x,z,t)\right] + C_e \frac{\partial T_e}{\partial t} = k_e \nabla^2 T_e - G(T_e - T_p) + S(x,z,t) \quad (6.78)$$

When the overall time scale of the heat transfer process is much larger than the relaxation times, Eqs. (6.77) and (6.78) can be further simplified to

$$C_p \frac{\partial T_p}{\partial t} = k_p \nabla^2 T_p - G(T_p - T_e) \quad (6.79)$$

$$C_e \frac{\partial T_e}{\partial t} = k_e \nabla^2 T_e - G(T_e - T_p) + S(x,z,t) \quad (6.80)$$

Eqs. (6.79) and (6.80) constitute the familiar two-temperature model [16].

6.2.4.3 *Integral Energy Balance*

All the boundaries of the film are considered to be defined through insulated boundary conditions; therefore, radiative intensities, which are obtained after the solution of the EPRT, do not have unique values. This is because a constant added to these variables can also satisfy the governing equations and the boundary conditions. To obtain a unique solution, the overall integral energy balance has to be satisfied. Physically it relates to the requirement that whatever thermal energy transferred by the laser pulse to the aluminum film up to a particular moment in time is stored within it and should be revealed in the overall temperature distribution. Mathematically we can write the heat source term:

$$\iiint S(x,z,t)\,dx\,dz\,dt = \iint \left[C_p(T_p - T_o) + C_e(T_e - T_o)\right] dx\,dz \quad (6.81)$$

During the solution of EPRT, this equation has to be satisfied for a particular time-step after a converged solution has been obtained for the intensities (I). If this equation is not satisfied, then all the intensity variables (I) are multiplied by a constant factor c, which can be determined for each time-step by the procedure given here:

$$c = \frac{\iiint S(x,z,t)\,dx\,dz\,dt}{\iint \left[C_p(T_p - T_o) + C_e(T_e - T_o)\right] dx\,dz} \quad (6.82)$$

The various variables are modified according to the procedure mentioned:

$$
\begin{aligned}
I_{\mathrm{p}}^{++} &= cI_{\mathrm{p}}^{++} + (1-c)\frac{C_{\mathrm{p}}v_{\mathrm{p}}T_{\mathrm{o}}}{4\pi} & I_{\mathrm{e}}^{++} &= cI_{\mathrm{e}}^{++} + (1-c)\frac{C_{\mathrm{e}}v_{\mathrm{e}}T_{\mathrm{o}}}{4\pi} \\
I_{\mathrm{p}}^{-+} &= cI_{\mathrm{p}}^{-+} + (1-c)\frac{C_{\mathrm{p}}v_{\mathrm{p}}T_{\mathrm{o}}}{4\pi} & I_{\mathrm{e}}^{-+} &= cI_{\mathrm{e}}^{-+} + (1-c)\frac{C_{\mathrm{e}}v_{\mathrm{e}}T_{\mathrm{o}}}{4\pi} \\
I_{\mathrm{p}}^{+-} &= cI_{\mathrm{p}}^{+-} + (1-c)\frac{C_{\mathrm{p}}v_{\mathrm{p}}T_{\mathrm{o}}}{4\pi} & I_{\mathrm{e}}^{+-} &= cI_{\mathrm{e}}^{+-} + (1-c)\frac{C_{\mathrm{e}}v_{\mathrm{e}}T_{\mathrm{o}}}{4\pi} \\
I_{\mathrm{p}}^{--} &= cI_{\mathrm{p}}^{--} + (1-c)\frac{C_{\mathrm{p}}v_{\mathrm{p}}T_{\mathrm{o}}}{4\pi} & I_{\mathrm{e}}^{--} &= cI_{\mathrm{e}}^{--} + (1-c)\frac{C_{\mathrm{e}}v_{\mathrm{e}}T_{\mathrm{o}}}{4\pi} \\
I_{\mathrm{p}}^{o} &= cI_{\mathrm{p}}^{o} + (1-c)\frac{C_{\mathrm{p}}v_{\mathrm{p}}T_{\mathrm{o}}}{4\pi} & I_{\mathrm{e}}^{o} &= cI_{\mathrm{e}}^{o} + (1-c)\frac{C_{\mathrm{e}}v_{\mathrm{e}}T_{\mathrm{o}}}{4\pi} \\
T_{\mathrm{p}} &= cT_{\mathrm{p}} + (1-c)T_{\mathrm{o}} & T_{\mathrm{e}} &= cT_{\mathrm{e}} + (1-c)T_{\mathrm{o}}
\end{aligned}
\tag{6.83}
$$

6.2.4.4 *Definition of Equivalent Equilibrium Temperature*

This temperature, described in the preceding section, defined anywhere in the thin aluminum film is the analog of the usual thermodynamic temperature defined in the diffusive limit in any medium. It represents the average energy of all phonons around a local point and is equivalent to the equilibrium temperature of phonons when they redistribute adiabatically to an equilibrium state. It is defined as

$$
T_{\mathrm{p}} = \frac{\iint_{4\pi} I_{\mathrm{p}}\sin\theta\, d\theta d\phi}{C_{\mathrm{p}}v_{\mathrm{p}}} \qquad T_{\mathrm{e}} = \frac{\iint_{4\pi} I_{\mathrm{e}}\sin\theta\, d\theta d\phi}{C_{\mathrm{e}}v_{\mathrm{e}}}
\tag{6.84}
$$

6.2.4.5 *Numerical Method*

The energy transport equation, in terms of the EPRT in the aluminum films, is solved numerically with the associated initial and boundary conditions using the finite-difference method, which is also referred to as the discrete ordinates method. Fig. 6.10 shows the aluminum film and a schematic view of laser irradiation. The governing equations are in the integrodifferential form and they may be solved using an iterative technique.

The governing equations for the transient case are presented in Eqs. (6.59) and (6.60); however, for a numerical solution of these equations it may be more convenient to write them in a slightly modified form, as shown:

$$
\frac{1}{v_{\mathrm{p}}}\frac{\partial I_{\mathrm{p}}}{\partial t} + \cos\theta\frac{\partial I_{\mathrm{p}}}{\partial x} + \sin\theta\sin\phi\frac{\partial I_{\mathrm{p}}}{\partial z} = \frac{\frac{1}{4\pi}\iint_{4\pi} I_{\mathrm{p}}\sin\theta\, d\theta\, d\phi - I_{\mathrm{p}}}{\Lambda_{\mathrm{p}}} - \frac{G}{4\pi}\left(T_{\mathrm{p}} - T_{\mathrm{e}}\right)
\tag{6.85}
$$

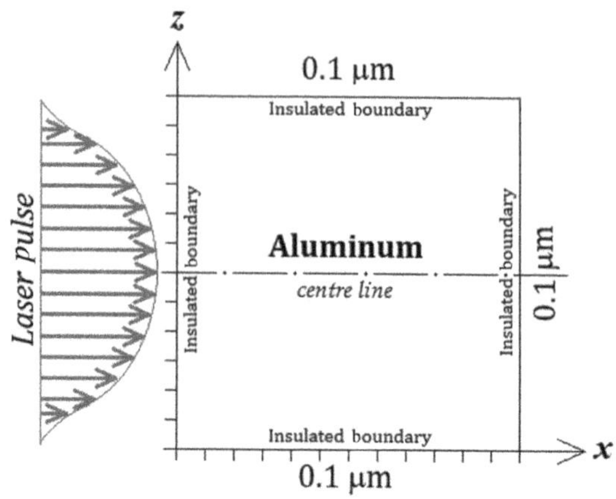

■ **FIGURE 6.10** A schematic view of aluminum film and laser irradiation [4].

and the governing equation of EPRT for the electron subsystem in the aluminum film is

$$\frac{1}{v_e}\frac{\partial I_e}{\partial t} + \cos\theta\frac{\partial I_e}{\partial x} + \sin\theta\sin\phi\frac{\partial I_e}{\partial z} = \frac{\iint_{4\pi} I_e \sin\theta d\theta d\phi - I_e}{\Lambda_e} + \frac{G(T_p - T_e) + S(x, z, t)}{4\pi} \tag{6.86}$$

We use the implicit scheme for temporal discretization. Consider the electron EPRT Eq. (6.86); a straightforward discretization strategy would be to replace the two, first derivatives on the left-hand side of the equation by forward-difference formulas and the time derivative term by a backward-difference formula. Doing so, we get the following equations:

$$\frac{1}{v_e}\frac{I^k_{e;i,j,m,n} - I^{k-1}_{e;i,j,m,n}}{\Delta t} + \cos\theta_m\frac{I^k_{e;i,j,m,n} - I^k_{e;i-1,j,m,n}}{\Delta x} + \sin\theta_m\sin\phi_n\frac{I^k_{e;i,j,m,n} - I^k_{e;i,j-1,m,n}}{\Delta z}$$
$$= \frac{(\text{integral})^k_{e;i,j} - I^k_{e;i,j,m,n}}{\Lambda_e} + \frac{G\left(T^k_{p;i,j} - T^k_{e;i,j}\right) + S^k_{i,j}}{4\pi} \tag{6.87}$$

and

$$I^k_{e;i,j,m,n}\left[\frac{1}{v_e\Delta t} + \frac{\cos\theta_m}{\Delta x} + \frac{\sin\theta_m\sin\phi_n}{\Delta z} + \frac{1}{\Lambda_e}\right] = I^k_{e;i-1,j,m,n}\left[\frac{\cos\theta_m}{\Delta x}\right]$$
$$+ I^k_{e;i,j-1,m,n}\left[\frac{\sin\theta_m\sin\phi_n}{\Delta z}\right] + \frac{(\text{integral})^k_{e;i,j}}{\Lambda_e} + \frac{G\left(T^k_{p;i,j} - T^k_{e;i,j}\right) + S^k_{i,j}}{4\pi} + \frac{I^{k-1}_{e;i,j,m,n}}{v_e\Delta t} \tag{6.88}$$

where $(\text{integral})^k_{e;i,j} = \frac{1}{4\pi}\iint_{4\pi} I^k_{e;i,j}\sin\theta d\theta d\phi$.

It is well known that the finite difference discretization results in a stable numerical scheme only if the coefficients of all the unknowns in the discretized equation are positive. It may be observed in the previous equation that the coefficients are not always positive and their sign depends on the values of the angles θ and ϕ. When the coefficients are negative, it may be possible to use a backward difference for discretization of the spatial derivatives and once again obtain a stable discretization. It turns out that to obtain stable discretization, we should partition the complete 4π solid angle into four quadrants. In the first quadrant, $0<\theta<\pi/2$, $0<\phi<\pi$ and we will denote the intensity as $I_{e;i,j,m,n}^{k++}$; in the second quadrant, $\pi/2<\theta<\pi$, $0<\phi<\pi$ and we will denote the phonon intensity as $I_{e;i,j,m,n}^{k-+}$; in the third quadrant, $0<\theta<\pi/2$, $\pi<\phi<2\pi$ and we will denote the phonon intensity as $I_{e;i,j,m,n}^{k+-}$; finally, in the fourth quadrant, $\pi/2<\theta<\pi$, $\pi<\phi<2\pi$ and we will denote the phonon intensity as $I_{e;i,j,m,n}^{k--}$.

In the first quadrant, $\cos\theta$ and $\sin\theta\sin\phi$ are both positive, and therefore the forward-difference formula for both $\partial I_e/\partial x$ and $\partial I_e/\partial z$ are to be used. In the second quadrant, only $\cos\theta$ is negative and therefore a backward-difference formula has to be used for $\partial I_e/\partial x$. In the third quadrant, only $\sin\theta\sin\phi$ is negative and therefore a backward-difference formula has to be used for $\partial I_e/\partial z$. In the fourth quadrant, both $\cos\theta$ and $\sin\theta\sin\phi$ are negative and therefore a backward-difference formula for both $\partial I_e/\partial x$ and $\partial I_e/\partial z$ are to be used.

It is clear that this discretization is unconditionally stable for the range of angles in which they are to be used. A second consequence of the discretization is that the boundary conditions may now be implemented at all the boundaries; otherwise on the face of it, the original EPRT would accommodate boundary conditions at only two boundaries. To explain the implementation of the boundary conditions we would refer to Fig. 6.11. As we can see, each of the four difference equations needs two boundary conditions for their solution. Consider the intensity $I_{e;i,j,m,n}^{k++}$. The two boundary conditions for this intensity are provided by the equilibrium intensities at the left and bottom boundaries, where the intensity $I_{e;i,j,m,n}^{k++}$ is defined. These boundaries are specified in Fig. 6.11. Hence starting from these boundary intensities, the equation for $I_{e;i,j,m,n}^{k++}$ is solved, until the intensity is obtained for all the nodes denoted by $(x_i, z_j, t^k, \theta_m, \phi_n)$, within the range defined for them, for the intensity $I_{e;i,j,m,n}^{k++}$. Similar considerations apply for the intensities $I_{e;i,j,m,n}^{k-+}$, $I_{e;i,j,m,n}^{k+-}$, and $I_{e;i,j,m,n}^{k--}$.

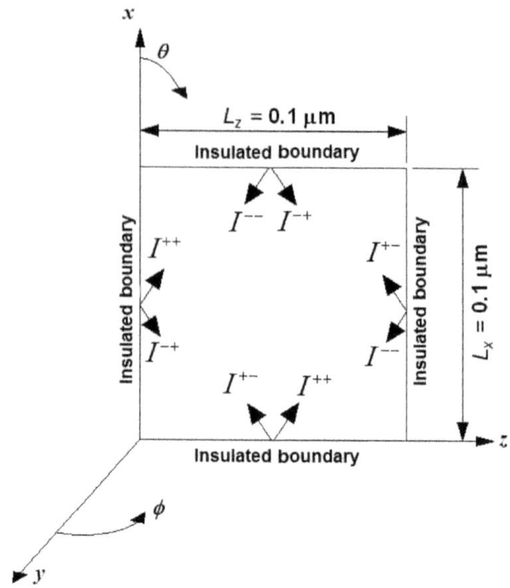

■ **FIGURE 6.11** Coordinate system and boundary conditions used in the simulations at the aluminum film [4].

The numerical algorithm to solve the EPRT would start by first initializing the intensities at each of the grid points $(x_i, z_j, \theta_m, \phi_n)$ to the initial condition, as well as by initializing the unknown $(\text{integral})_{i,j}$ at each of the grid points (x_i, z_j) to the corresponding initial value as described in the section on initial condition. Then at each time-step, starting from the appropriate boundary equilibrium intensities, and after assuming all unknown variables at the current time-step to be equal to their corresponding values in the previous time-step, each of the four equations are solved to obtain better estimates of the intensities inside the domain. Simultaneously the EPRT for phonons is also solved in exactly the same fashion. Subsequently, better values of the integrals are evaluated from these new estimates of the intensities at each spatial grid point. Once this is done, the equivalent equilibrium temperature is calculated at each grid point inside the domain. This whole procedure now constitutes a single iteration. Several iterations are performed until a converged solution is obtained for the current time-step. The criterion for convergence at each time-step is that the maximum absolute difference everywhere inside the domain, between the equivalent equilibrium temperature calculated at the current iteration and that calculated at the previous iteration, should become smaller than a preassigned small positive number. Mathematically, $\max\left|\,(T_{i,j}^{\mathrm{k}})^{\mathrm{iter}} - (T_{i,j}^{\mathrm{k}})^{\mathrm{iter}-1}\,\right| < \varepsilon$, where iter is the iteration index and $\varepsilon = 10^{-6}$. This criterion is to be satisfied independently by the phonon and electron temperatures. Once a converged solution is obtained, the algorithm moves onto the next time-step. Before moving on to the next time-step, the transient heat flux at each point may be calculated by the formulas described previously,

$$q''_x(x,z,t) = \int_0^{2\pi}\int_0^{\pi} I(x,z,t,\theta,\phi)\cos\theta\sin\theta d\theta\, d\phi \tag{6.89}$$

and

$$q''_z(x,z,t) = \int_0^{2\pi}\int_0^{\pi} I(x,z,t,\theta,\phi)\sin\phi\sin^2\theta d\theta\, d\phi \tag{6.90}$$

6.2.4.6 *A Case Study and Applications*

Laser short-pulse heating of two-dimensional aluminum thin film is considered in line with the previous study [4]. The laser irradiated energy is considered to be incident on one edge of the film. The appropriate initial and boundary conditions are provided next.

6.2.4.6.1 **Initial Condition**

At time $t = 0$, both subsystems are assumed to be at a uniform temperature of $T_o = 300\text{K}$. In terms of intensities the initial condition is written as

$$I_p^{++} = I_p^{-+} = I_p^{+-} = I_p^{--} = \frac{C_p v_p T_o}{4\pi} \quad I_e^{++} = I_e^{-+} = I_e^{+-} = I_e^{--} = \frac{C_p v_p T_o}{4\pi} \tag{6.91}$$

6.2.4.6.2 **Boundary Conditions**

It is important to keep in mind that the intensities I^{++}, I^{-+}, I^{+-}, and I^{--} are separate variables and each needs two boundary conditions. Also, the boundary condition are the same for the phonon as well as the electron subsystem, therefore the subscripts p and e will be dropped.

Left Boundary Condition: Insulated boundary

$$I^{++}(x,z=0,t,\theta,\phi) = I^{-+}(x,z=0,t,\theta,\phi)$$
$$= -\frac{1}{\pi}\left[\int_{\pi}^{2\pi}\int_0^{\pi/2} I^{+-}\sin\phi\sin^2\theta d\theta\, d\phi + \int_{\pi}^{2\pi}\int_{\pi/2}^{\pi} I^{--}\sin\phi\sin^2\theta d\theta\, d\phi\right] \tag{6.92}$$

Right Boundary Condition: Insulated boundary

$$I^{+-}(x, z = L_z, t, \theta, \phi) = I^{--}(x, z = L_z, t, \theta, \phi)$$
$$= \frac{1}{\pi}\left[\int_0^{\pi}\int_0^{\pi/2} I^{++}\sin\phi\sin^2\theta d\theta\, d\phi + \int_0^{\pi}\int_{\pi/2}^{\pi} I^{-+}\sin\phi\sin^2\theta d\theta\, d\phi\right] \tag{6.93}$$

Bottom Boundary Condition: Insulated boundary

$$I^{++}(x = 0, z, t, \theta, \phi) = I^{+-}(x = 0, z, t, \theta, \phi)$$
$$= -\frac{1}{\pi}\left[\int_0^{\pi}\int_{\pi/2}^{\pi} I^{-+}\cos\theta\sin\theta d\theta\, d\phi + \int_{\pi}^{2\pi}\int_{\pi/2}^{\pi} I^{--}\cos\theta\sin\theta d\theta\, d\phi\right] \tag{6.94}$$

Top Boundary Condition: Insulated boundary

$$I^{-+}(x = 0, z, t, \theta, \phi) = I^{--}(x = 0, z, t, \theta, \phi)$$
$$= -\frac{1}{\pi}\left[\int_0^{\pi}\int_0^{\pi/2} I^{++}\cos\theta\sin\theta d\theta\, d\phi + \int_0^{\pi}\int_{\pi/2}^{\pi} I^{+-}\cos\theta\sin\theta d\theta\, d\phi\right] \tag{6.95}$$

A numerical code is developed using MATLAB solver to solve the discretized phonon transport equations according to the initial and boundary conditions. Table 6.2 gives the properties used in the simulations.

6.2.4.7 *Findings of the Case Study*

The phonon radiate transport equation is incorporated in the analysis while introducing electron–phonon coupling across the electron and lattice subsystems. Equivalent equilibrium temperature is introduced to assess the thermal characteristics in the film. Since the insulated boundary conditions are incorporated in the analysis, the laser irradiated

Table 6.2 Properties Used in the Simulations [10–12]

	Phonons	Electrons
Thermal conductivity, ($Wm^{-1}K^{-1}$)	$11.3 = .05 \times k_e$ [11]	225.7 [10]
Volumetric heat capacity, ($Jm^{-3}K^{-1}$)	2,294,396.6 [12]	23,659 [10]
Speed, (ms^{-1})	6420	2.03×10^6
Mean free path, $\Lambda = 3k/Cv$, (m)	2.3×10^{-9}	14.1×10^{-9}
Relaxation time $\tau = \Lambda/v$, (s)	3.58×10^{-13}	6.945×10^{-15}
Film area, (μm^2)	0.1×0.1	
Electron–phonon coupling constant in aluminum (Wm^{-3})	2.4642×10^{17}	
Laser heat source parameters	$R = 0.5$; $\alpha = 1.35 \times 10^8$ ($1\ m^{-1}$); $t_p = 96 \times 10^{-15}$ (s); $J = 0.045$ ($J\ m^{-2}$)	

intensity absorbed by the film should satisfy the energy gain by the film during the heating period. Integral energy balance (Eq. 6.81) in the electron and lattice system reveals that the total energy absorbed from the irradiated field is equal to the energy gain by the film; the numerical simulations satisfy the energy balance requirement.

Fig. 6.12 shows equivalent equilibrium temperature along the z axis in the electron subsystem for different x-axis locations and various heating

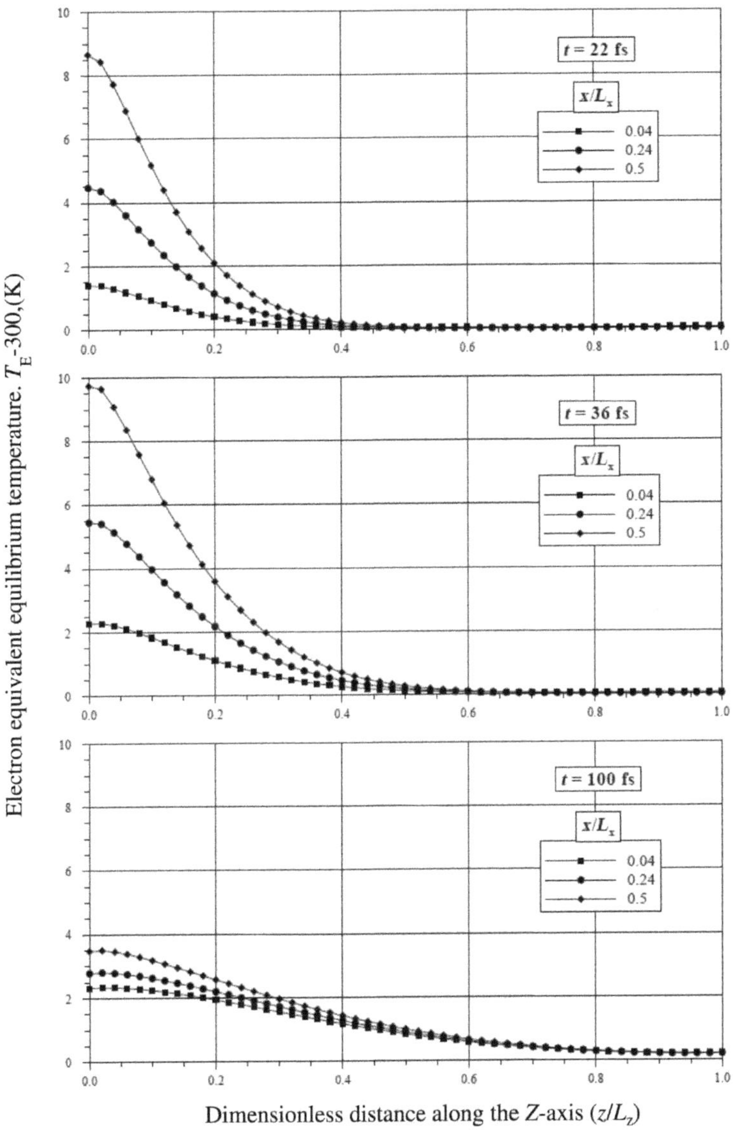

■ **FIGURE 6.12** Equivalent equilibrium temperature in the electron subsystem along the z axis [4].

periods. It should be noted that z- and x-axis locations are shown schematically in Fig. 6.10. Moreover, equivalent equilibrium temperature is introduced in the thin aluminum film in line with the analog of thermodynamic temperature defined in the diffusive limit in any medium. Equivalent equilibrium temperature corresponds to the average energy of all phonons around a local point when they redistribute adiabatically to an equilibrium state. Since the laser pulse intensity varies exponentially with time, energy absorbed by electrons from the irradiated filed also varies with time. This in turn modifies electron energy distribution in the electron subsystem with progressing time. Equivalent equilibrium temperature attains high values in the surface region along the z axis and as the depth below the surface increases, temperature decays sharply. This behavior is attributed to the absorption of the irradiated field in the aluminum film, which decreases exponentially according to the Beer–Lambert law. Note that laser irradiated energy is considered to be absorbed within the absorption depth according to the Beer–Lambert law in the analysis in accordance with the source term ($S(x, z, t)$) in Eq. (6.60). Since the laser power intensity distribution is Gaussian along the x axis, temperature varies at different locations along the x axis. In this case, the peak temperature occurs at the centerline ($x/L_x = 0.5$) along the z axis where the laser power intensity is the maximum. Equivalent equilibrium temperature follows almost laser pulse intensity distribution along the z axis, particularly along the centerline ($x/L_x = 0.5$). This is associated with the electron excess energy gain from the irradiated field, which decays exponentially with increasing distance along the z axis according to the Beer–Lambert law.

Fig. 6.13 shows equivalent equilibrium temperature along the z axis in the lattice subsystem for different x-axis locations and various heating periods. Equivalent equilibrium temperature for the lattice subsystem, where phonon transport takes place, follows almost electron temperature distribution along the film width, particularly in the early heating periods ($t \leq 167$ fs). This is attributed to the high electron excess energy gain from the irradiated field and excess electron energy transfer to the lattice site through electron–phonon coupling. It should be noted that laser pulse intensity decays exponentially with progressing time; therefore, energy gain by the lattice site from the electron subsystem through electron–phonon coupling remains high when electron excess energy is high. Since energy transport is governed by the phonon radiative transport in the lattice subsystem, phonon intensity distribution plays an important role in the rise of lattice-site temperature. Consequently, phonon scattering due to film resistance modifies phonon intensity in the film. This in turn

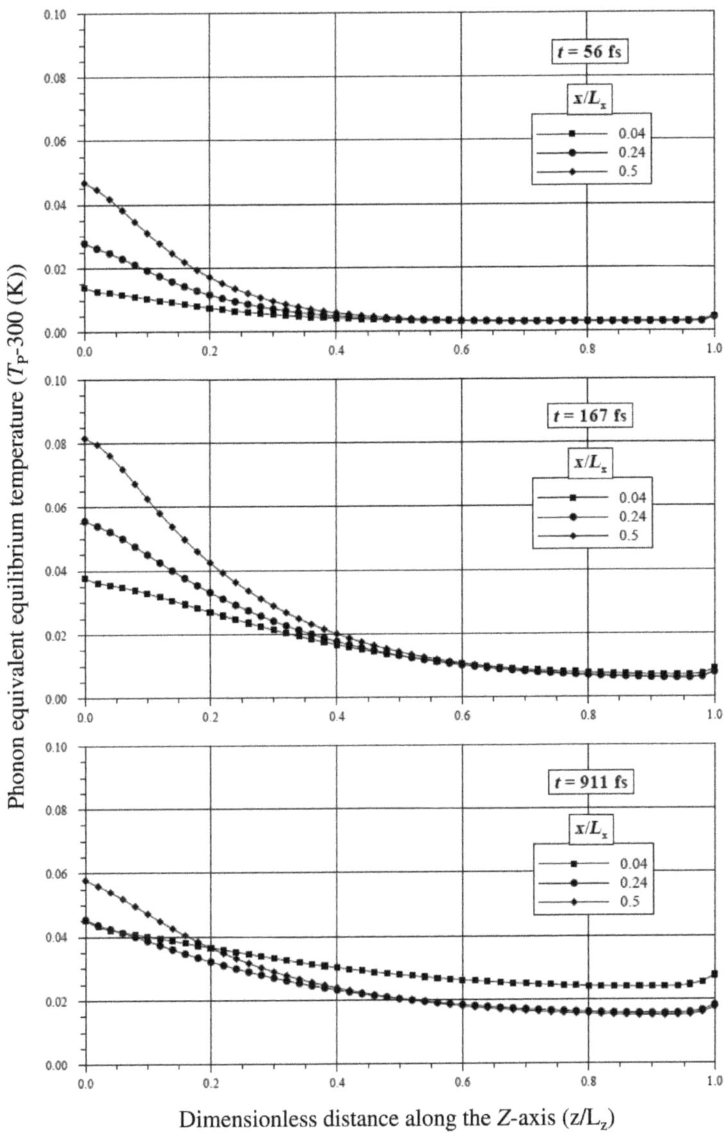

FIGURE 6.13 Equivalent equilibrium temperature in the lattice subsystem along the z axis [4].

alters equivalent equilibrium temperature distribution in the film; equivalent equilibrium temperature does not exactly follow electron temperature despite the constant electron–phonon coupling that is assumed during the energy exchange process. The emitted phonons in the irradiated region reflect from the insulated film boundaries. This raises the equivalent equilibrium temperature at the film boundaries. This situation can be observed from the three-dimensional view of equivalent equilibrium temperatures as well as equivalent equilibrium temperature distribution at the late heating period at the location near the film edge ($t = 911$ fs and $z/L_x \cong 1.$). Note that as the temperature increases at the film boundaries, phonon emission toward the irradiated region modifies temperature distribution in the film. This behavior can be observed in the late heating periods in the close region of the edges ($t = 911$ fs and $x/L_x = 0.04$).

6.3 CROSS-PLANE TRANSPORT AND CASE STUDIES

Some electronic devices involve multilayered thin films composed of dielectric–dielectric and dielectric–metallic thin film pairs. In this case, thermal management of the films pair becomes important to secure the high performance of the devices. Thermal energy transport takes place via phonon transport in thin films pair. However, interfacial resistance to thermal transport occurs because of the mismatch of the properties of both films. This requires the proper formulation of the transport equation together with the thermal boundary resistance at the interface. In this section, the cross-plane transport of the thermal energy is considered and case studies and applications are introduced in line with the previous studies [5,6].

6.3.1 Cross-Plane Transport Across One-Dimensional Dielectric and Metallic Thin Films and Applications

The micro/nanosized electronic devices can be composed of multithin films; cross-plane transport becomes important to resemble the heat conduction across the film pairs. Phonons are the main energy carriers in dielectric films, which must be incorporated in heat transport analysis for multifilms. On the other hand, in metallic films, electron and lattice subsystems thermally separate and electron–phonon coupling plays a major role in energy transfer between the subsystems. Since the model studies provide useful information on the physical processes taking place during the energy transport, establishing a sound modeling incorporating the physical aspects of the transport process is necessary. Modeling the phonon transport is possible using BTEs in dielectric thin films; however, in metallic films two-transport equations need to be introduced to account for the thermal separation of electron and lattice subsystems. This arrangement can predict transient transport characteristics in electron and lattice subsystems of the thin film. In this section, laser short-pulse heating of a composite of silicon-diamond-aluminum thin films is considered and the energy transport across the films is predicted using the first principle solutions. The BTE is introduced to account for the phonon transport in the silicon-diamond film pair while a modified version of the BTE is incorporated to model energy transport in the electron and lattice subsystems. Thermal boundary resistance is incorporated at the interfaces of the thin films. The transfer matrix method is adopted to determine the amount of laser beam energy transmitted, reflected, and absorbed in the silicon and diamond thin films in line with the previous study [5].

Fig. 6.14 shows the schematic view of the short-pulse laser heating of a silicon-diamond-aluminum layer. TBR stands for thermal boundary resistance; I^+ is the phonon intensity I for the phonons directed away from the left face and toward the right face within any film; I^- is the phonon intensity I for the phonons directed away from the right face and toward the left face within any film; L_1, L_2, and L_3 are the thicknesses of the silicon, diamond, and aluminum films, respectively. The subscripts p and e in the aluminum film denote the phonon and electron subsystems, respectively. Incident laser energy onto the silicon film can be partially reflected, absorbed, and transmitted into the diamond film. Similarly, laser beam in the diamond film can be partially reflected at the silicon-diamond interface, partially absorbed and transmitted to the aluminum

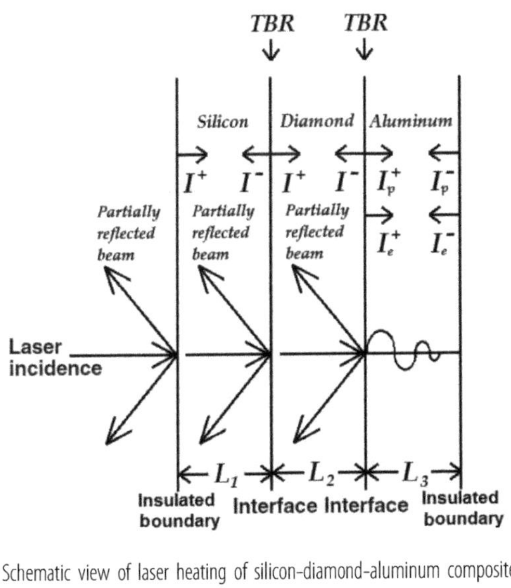

■ **FIGURE 6.14** Schematic view of laser heating of silicon-diamond-aluminum composite and boundary conditions [5].

film. Since aluminum film is opaque to the incident radiation, it partially reflects the incident beam at the aluminum-diamond interface and absorbs the remaining part of the laser irradiation. The beam intensity reflected, absorbed, and transmitted are determined by the transfer matrix method. However, energy transport in the films is formulated using the BTE. One of the approaches to solve the BTE in the silicon-diamond-aluminum thin films is to use the relaxation-time approximation and reduce the BTE to the EPRT. This makes the EPRT easier to solve numerically with the appropriate boundary conditions. Consequently, the transient, frequency-independent EPRT is formulated within the silicon-diamond-aluminum thin films. The boundaries of silicon and aluminum are considered to be insulated. At the interfaces of the silicon-diamond and diamond-aluminum films, appropriate interface conditions are introduced. The initial condition corresponds to uniform temperature in all three films. The EPRT and the associated boundary, interface, and initial conditions are described next.

6.3.1.1 *Equation of Phonon Radiative Transfer*

The one-dimensional, frequency-independent EPRT can be written as

$$\frac{1}{v}\frac{\partial I}{\partial t}+\mu\frac{\partial I}{\partial x}=\frac{\frac{1}{2}\int_{-1}^{1} I\,d\mu - I}{\Lambda} \quad t>0,\quad 0<x<L,\quad -1<\mu<1 \tag{6.96}$$

In this equation, the phonon intensity I is a function of x, μ, and t:

$$I = I(x, \mu, t) \tag{6.97}$$

where I is the phonon intensity, ν is the phonon speed, x is the distance along the film thickness, $\mu = \cos\theta$ and θ is the angle between the velocity vector and the x axis, t is the time, Λ is the phonon mean free path, and L is the film thickness. Eq. (6.96) is applicable to both silicon and diamond films with the appropriate parameter values being used in each film. To avoid overcomplication we have not considered the different phonon branches and polarizations in the silicon and diamond films, nor have we considered the dependence of the phonon intensity on the frequency.

In the aluminum film, there are two subsystems that can store and carry thermal energy independently. These are the phonon subsystem and the electron subsystem. As is the case with silicon and diamond films, the different phonon branches and their polarizations as well as the frequency dependence of the phonon intensity are not considered in the aluminum film. In this scenario we get two EPRTs, one for the phonon subsystem and the other for the electron subsystem. These two subsystems are coupled and can exchange thermal energy between them. This means that the two EPRTs are coupled, as shown:

$$\frac{1}{v_p}\frac{\partial I_p}{\partial t}+\mu\frac{\partial I_p}{\partial x}=\frac{\frac{1}{2}\int_{-1}^{1} I_p\,d\mu - I_p}{\Lambda_p}-\frac{G}{4\pi}(T_p-T_e)\quad t>0,\ 0<x<L,\ -1<\mu<1 \tag{6.98}$$

$$\frac{1}{v_e}\frac{\partial I_e}{\partial t}+\mu\frac{\partial I_e}{\partial x}=\frac{\frac{1}{2}\int_{-1}^{1} I_e\,d\mu - I_e}{\Lambda_e}-\frac{G}{4\pi}(T_e-T_p)+\frac{S}{4\pi},\quad t>0,\ 0<x<L,\ -1<\mu<1 \tag{6.99}$$

where G is the electron–phonon coupling parameter, S is the source term representing the absorption of the external irradiated field, T is the temperature, and the subscripts p and e stand for phonon and electron, respectively. The laser pulse intensity within a film is written as

$$I_{laser}(x,t)=\mathrm{Tr}\times(J/t_p)\exp\left[-\alpha x-2.77(t/t_p)^2\right] \tag{6.100}$$

It should be noted that the laser beam intensity is absorbed within the films according to the Beer–Lambert law and absorbed beam intensity is converted into a heat in the film. Therefore, the volumetric heat generation inside a film due to laser heating is then derived from the following equations:

$$S(x,t)=-\frac{\partial I_{laser}}{\partial x} \tag{6.101}$$

or

$$S(x,t)=\mathrm{Tr}\times(J/t_p)\alpha\exp\left[-\alpha x-2.77(t/t_p)^2\right] \tag{6.102}$$

The values of the parameters used in Eq. (6.102) are listed at the end of Table 6.3. In the heat source term, Tr is the transmittance (i.e., the fraction of the laser intensity incident on the aluminum film after passing through the silicon and diamond film), J/t_p is the original laser intensity, and α is the absorption coefficient in aluminum. Tr and α can be calculated through a transfer matrix algorithm, which is presented later in the section. The parameters used for this method are displayed in Table 6.4; the results for Tr and α as well as J and t_p are given in Table 6.3. Fig. 6.15 shows the temporal variation of laser pulse intensity outside the film ($I_{laser,o}/I_{laser}/\mathrm{Tr}$) for $x=0$ m, while Fig. 6.16 shows the spatial distribution of the absorbed laser intensity in the silicon-diamond-aluminum composite for $t=0$ s.

Table 6.3 Properties Used in the Simulations

	Silicon	Diamond	Aluminum, Phonons	Aluminum, Electrons
Thermal conductivity, k, ($Wm^{-1}K^{-1}$)	157	3320	$11.3 = .05 \times k_e$	$k_e = 225.7$
Volumetric heat capacity heat, C, ($Jm^{-3}K^{-1}$)	1.653×10^6	1.815×10^6	2,294,396.6	23,659
Speed, v (ms^{-1})	8430	12,288	6420	2.03×10^6
Mean free path, $\Lambda = 3k/Cv$, (m)	33.8×10^{-9}	446.6×10^{-9}	2.3×10^{-9}	14.1×10^{-9}
Film thickness, L (μm)	0.1	0.1	0.1	
Electron–phonon coupling constant in aluminum G, (Wm^{-3})				2.4642×10^{17}
Thermal boundary resistance at silicon-diamond interface, ($KW^{-1}m^2$)				1.7×10^{-8}
Thermal boundary resistance at diamond-aluminum interface, ($KW^{-1}m^2$)				2.6926×10^{-10}
Laser heat source parameters				Tr = 0.7507 $\alpha = 1.35 \times 10^8$ ($1\ m^{-1}$) $t_p = 96 \times 10^{-15}$ (s) $J = 0.02$ ($J\ m^{-2}$)

Table 6.4 Parameters Used in the Transfer Matrix Method

Permittivity of vacuum, ε_o, (Fm^{-1})	8.8542×10^{-12}
Permeability of vacuum, μ_o, (NA^{-2})	$4\pi \times 10^{-7}$
Speed of light, c, (ms^{-1})	3×10^8
Refractive index of air, n_o	1
Refractive index of substrate (aluminum), n_s	1.39
Wavelength of incident laser, λ_o, (nm)	630
Incident angle of the laser, θ_i	0
Refractive index of silicon, n_1	3.8780−0.0150i
Refractive index of diamond, n_2	2.4097 + 0i
Reflectance, Re	0.2161
Transmittance, Tr	0.7507
Absorptance, Ab	0.0332

6.3.1.2 *Initial and Boundary Conditions*

Since the phonons emitted from the high-temperature edge reflects from the low-temperature edge of the film, the emitted phonon intensity is denoted as I^+ while reflected phonon intensity is shown as I^- (Fig. 6.14).

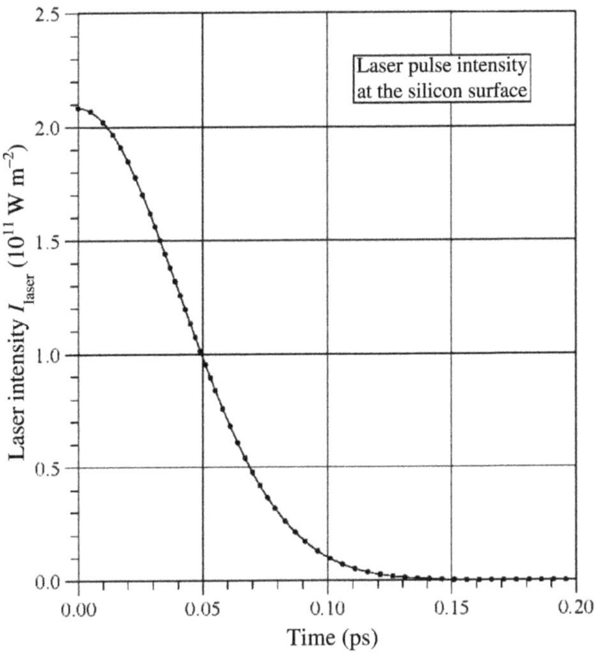

■ **FIGURE 6.15** Temporal variation of the laser pulse intensity at the silicon film surface [5].

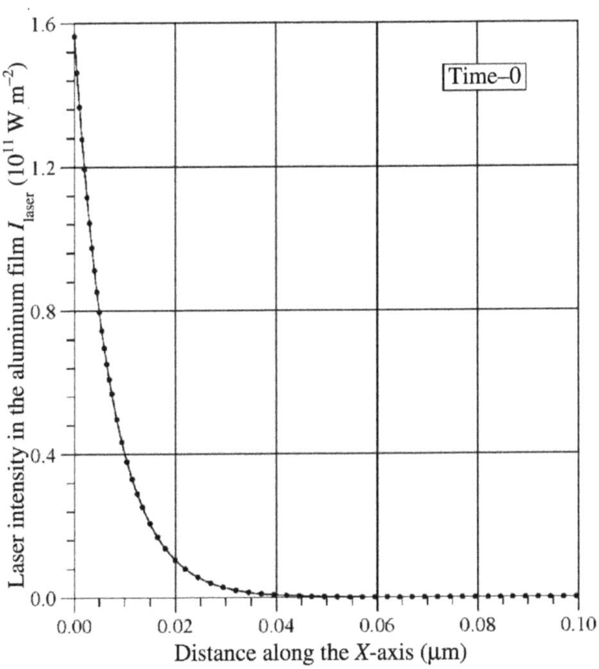

■ **FIGURE 6.16** Laser pulse intensity distribution along silicon-diamond-aluminum composite [5].

The boundary conditions for the composite films are given in the following subsections.

6.3.1.2.1 **Silicon Film**

6.3.1.2.1.1 ***Initial Condition*** Initially silicon film is considered to have a uniform temperature (T_{initial}), which is 300K,

$$I_{\text{Si}}(x_1, \mu, t=0) = \frac{C_{\text{Si}} v_{\text{Si}}}{4\pi} \times T_{\text{initial}} \tag{6.103}$$

where C is the volumetric heat capacity.

6.3.1.2.1.2 ***Left Boundary Condition*** Left boundary is considered to be insulated, which is

$$q''_{\text{Si}}(x_1 = 0, t) = 2\pi \int_{-1}^{1} \mu I_{\text{Si}}(x_1 = 0, \mu, t)\, d\mu = 0 \tag{6.104}$$

$$\int_{-1}^{0} \mu I^{-}_{\text{Si}}(0, \mu, t)\, d\mu + \int_{0}^{1} \mu I^{+}_{\text{Si}}(0, \mu, t)\, d\mu = 0 \tag{6.105}$$

At the boundary diffuse reflection of the phonons is assumed. This means that $I^{+}_{\text{Si}}(0, \mu, t)$ becomes independent of the variable μ at the boundary and, hence, it can be pulled out of the integral sign:

$$I^{+}_{\text{Si}}(0, \mu, t) \int_{0}^{1} \mu\, d\mu = -\int_{-1}^{0} \mu I^{-}_{\text{Si}}(0, \mu, t)\, d\mu \tag{6.106}$$

The emitted phonon intensity at the left boundary is then

$$I^{+}_{\text{Si}}(x_1 = 0, \mu, t) = -2 \int_{-1}^{0} \mu I^{-}_{\text{Si}}(x_1 = 0, \mu, t) d\mu \tag{6.107}$$

where q'' is the heat flux.

6.3.1.2.1.3 ***Right Boundary Condition*** Due to the mismatch between the properties of the silicon and diamond as well as diamond and aluminum films, a thermal boundary resistance occurs at the films interfaces. Therefore, the thermal boundary resistance is introduced at the right boundary of the silicon film (between the silicon and diamond film), which is

$$q''_{\text{Si}}(x_1 = L_1, t) = \frac{T_{\text{Si}}(x_1 = L_1, t) - T_{\text{Di}}(x_2 = 0, t)}{\text{TBR}_{\text{Si/Di}}} = q''_{\text{Di}}(x_2 = 0, t) \tag{6.108}$$

where TBR is the thermal boundary resistance.

6.3.1.2.2 **Diamond Film**

6.3.1.2.2.1 ***Initial Condition*** Initially diamond film is assumed to be at a uniform initial temperature (T_{initial}), which is 300K,

$$I_{\text{Di}}(x_2, \mu, t = 0) = \frac{C_{\text{Di}} v_{\text{Di}}}{4\pi} \times T_{\text{initial}} \tag{6.109}$$

6.3.1.2.2.2 ***Left Boundary Condition*** Thermal boundary resistance is introduced at the left boundary of the diamond film, which is

$$q''_{\text{Si}}(x_1 = L_1, t) = \frac{T_{\text{Si}}(x_1 = L_1, t) - T_{\text{Di}}(x_2 = 0, t)}{\text{TBR}_{\text{Si/Di}}} = q''_{\text{Di}}(x_2 = 0, t) \tag{6.110}$$

6.3.1.2.2.3 ***Right Boundary Condition*** Thermal boundary resistance is introduced at the right boundary of the diamond film (between diamond and aluminum film), which is

$$q''_{\text{Di}}(x_2 = L_2, t) = \frac{T_{\text{Di}}(x_2 = L_2, t) - T_{\text{Al,p}}(x_3 = 0, t)}{\text{TBR}_{\text{Di/Al}}} = q''_{\text{Al,p}}(x_3 = 0, t) \tag{6.111}$$

6.3.1.2.3 **Aluminum Film: Lattice (Phonon) Subsystem**

6.3.1.2.3.1 ***Initial Condition*** Initially, lattice subsystem in the aluminum film is assumed to be at a uniform initial temperature (T_{initial}), which is 300K,

$$I_{\text{p}}(x_3, \mu, t = 0) = \frac{C_{\text{p}} v_{\text{p}}}{4\pi} \times T_{\text{initial}} \tag{6.112}$$

6.3.1.2.3.2 ***Left Boundary Condition*** Thermal boundary resistance is introduced at the left boundary of the lattice subsystem in the aluminum film, which is 300K,

$$q''_{\text{Di}}(x_2 = L_2, t) = \frac{T_{\text{Di}}(x_2 = L_2, t) - T_{\text{Al,p}}(x_3 = 0, t)}{\text{TBR}_{\text{Di/Al}}} = q''_{\text{Al,p}}(x_3 = 0, t) \tag{6.113}$$

6.3.1.2.3.3 ***Right Boundary Condition*** Right boundary of the lattice subsystem in the aluminum film is considered to be insulated, which is

$$q''_{\text{Al,p}}(x_3 = L_3, t) = 2\pi \int_{-1}^{1} \mu I_{\text{p}}(x_3 = L_3, \mu, t)\, d\mu = 0 \tag{6.114}$$

$$\Rightarrow I_{\text{p}}^{-}(x_3 = L_3, \mu, t) = 2 \int_{0}^{1} \mu I_{\text{p}}^{+}(x_3 = L_3, \mu, t)\, d\mu \tag{6.115}$$

6.3.1.2.4 Aluminum Film: Electron Subsystem

6.3.1.2.4.1 ***Initial Condition*** Initially, the electron subsystem in the aluminum film is assumed to be at a uniform initial temperature (T_{initial}), which is 300K,

$$I_e(x_3, \mu, t=0) = \frac{C_e v_e}{4\pi} \times T_{\text{initial}} \tag{6.116}$$

6.3.1.2.4.2 ***Left Boundary Condition*** Left boundary of the electron subsystem in the aluminum film is considered to be insulated (no electron jump or diffusion to the diamond film), which is

$$q''_{\text{Al,e}}(x_3=0,t) = 2\pi \int_{-1}^{1} \mu I_e(x_3=0,\mu,t)\,d\mu = 0 \tag{6.117}$$

$$\Rightarrow I_e^+(x_3=0,\mu,t) = -2\int_{-1}^{0} \mu I_e^-(x_3=0,\mu,t)\,d\mu \tag{6.118}$$

6.3.1.2.4.3 ***Right Boundary Condition*** Right boundary of the electron subsystem in the aluminum film is considered to be insulated, which is

$$q''_{\text{Al,e}}(x_3=L_3,t) = 2\pi \int_{-1}^{1} \mu I_e(x_3=L_3,\mu,t)\,d\mu = 0 \tag{6.119}$$

$$\Rightarrow I_e^-(x_3=L_3,\mu,t) = 2\int_{0}^{1} \mu I_e^+(x_3=L_3,\mu,t)\,d\mu \tag{6.120}$$

In the previous equations C is the volumetric heat capacity at a temperature of $T_{\text{initial}} = 300\text{K}$ and v is the phonon group velocity.

6.3.1.3 *Interface Conditions*

At the silicon-diamond and diamond-aluminum interfaces, the heat flux in the left film should be equal to the heat flux in the right film. In addition, the equivalent equilibrium temperatures of the left and right films are related by the expression,

$$q''_{\text{left}} = \frac{T_{\text{left}} - T_{\text{right}}}{\text{TBR}} = q''_{\text{right}} \tag{6.121}$$

6.3.1.4 *Definition of Temperature and Heat Flux*

Equivalent equilibrium phonon temperature is defined as

$$T(x,t) = \frac{2\pi}{Cv}\int_{-1}^{1} I\, d\mu \tag{6.122}$$

The heat flux at any location inside the film and at any time is defined as

$$q''(x,t) = 2\pi \int_{-1}^{1} \mu I\, d\mu \tag{6.123}$$

6.3.1.5 *Transfer Matrix Approach for Multiple Thin Films*

Once the laser beam intensity reaches the silicon film, it undergoes reflection, transmission, and absorption. In order to determine the reflectance, transmittance, and absorption of the incident laser beam, a transfer matrix method is introduced, which provides the optical characteristics of the incident beam in each film. In addition, it allows determining the amount of laser energy transmitted to the aluminum film surface at the silicon-aluminum film interface and the absorption coefficient (α) of the incident laser beam across the film pair.

Optical path length for the ith film can be written as [17]

$$\delta_i = \left(\frac{2\pi}{\lambda_o}\right) n_i t_i \cos\theta_{ti} \tag{6.124}$$

where δ_i is the optical path length, λ_o is the wavelength in vacuum, n_i is the refractive index of the ith film, t_i is the thickness of the ith film, and θ_{ti} is the angle of refraction in the ith film.

Transfer matrix for the ith film is

$$M_i = \begin{bmatrix} m_{11} & m_{12} \\ m_{21} & m_{22} \end{bmatrix}_i = \begin{bmatrix} \cos\delta_i & \dfrac{j\sin\delta_i}{\gamma_i} \\ j\gamma_i \sin\delta_i & \cos\delta_i \end{bmatrix} \tag{6.125}$$

and

$$\gamma_o = n_o\sqrt{\frac{\varepsilon_o}{\mu_o}}\cos\theta_o, \quad \gamma_i = n_i\sqrt{\frac{\varepsilon_o}{\mu_o}}\cos\theta_{ti}, \quad \gamma_s = n_s\sqrt{\frac{\varepsilon_o}{\mu_o}}\cos\theta_{ts}, \quad j = \sqrt{-1} \tag{6.126}$$

and

$$n_o \sin\theta_o = \ldots = n_i \sin\theta_{ti} = \ldots = n_s \sin\theta_{ts} \tag{6.127}$$

where n_o is the refractive index of air, n_s is the refractive index of the substrate (the substrate is air itself), θ_o is the angle of incidence of the laser pulse, θ_{ts} is the angle of refraction in the substrate, ε_o is the permittivity of vacuum, and μ_o is the permeability of vacuum.

The transfer matrix for a stack of multiple layers consisting of different optical characteristics can be written as:

$$M = M_1 M_2 \cdots M_N = \begin{bmatrix} m_{11} & m_{12} \\ m_{21} & m_{22} \end{bmatrix}$$

The reflection coefficient and the transmittance coefficient for the multiple layers of different optical characteristics are

$$\text{Reflection Coefficient:} \quad r = \frac{\gamma_o m_{11} + \gamma_o \gamma_s m_{12} - m_{21} - \gamma_s m_{22}}{\gamma_o m_{11} + \gamma_o \gamma_s m_{12} + m_{21} + \gamma_s m_{22}} \tag{6.128}$$

$$\text{Transmission Coefficient:} \quad t = \frac{2\gamma_o}{\gamma_o m_{11} + \gamma_o \gamma_s m_{12} + m_{21} + \gamma_s m_{22}} \tag{6.129}$$

Reflectance over the stack of multiple layers is

$$\text{Re} = rr^* \tag{6.130}$$

Transmittance over the stack of multiple layers is

$$\text{Tr} = \left(\frac{n_s \cos\theta_{ts}}{n_0 \cos\theta_0}\right) tt^* \tag{6.131}$$

where the symbol $*$ denotes complex conjugate. All of these equations are valid for real as well as complex refractive indices of the films.

For real refractive indices, Re + Tr = 1 and Ab = 0, where Ab stands for absorptance, which is defined as the fraction of the incident power absorbed by the thin film.

For complex refractive indices, Ab + Re + Tr = 1.

6.3.1.6 *Refractive Index for a Dielectric Material*

The refractive index of each thin film dielectric is given by the following expression:

$$n^2 = (n_R + i n_I)^2 = 1 + \frac{Ne^2}{m\varepsilon_o}\left[\frac{1}{(\omega_o^2 - \omega^2) - i\omega\gamma}\right] \tag{6.132}$$

$$\Rightarrow n_R^2 - n_I^2 = 1 + \frac{Ne^2}{m\varepsilon_o}\left[\frac{\omega_o^2 - \omega^2}{\left(\omega_o^2 - \omega^2\right)^2 + \omega^2\gamma^2}\right] \ \& \ 2n_R n_I = \frac{Ne^2}{m\varepsilon_o}\left[\frac{\omega\gamma}{\left(\omega_o^2 - \omega^2\right)^2 + \omega^2\gamma^2}\right] \tag{6.133}$$

The preceding two equations can be solved simultaneously to get n_R and n_I. n_R is the real part of the refractive index and n_I is the imaginary part. It is related to the absorption in the thin film. ω_o is the resonance frequency of the electrons and changes from material to material. N is the number of atoms per unit volume (actually the number of elementary dipoles per unit volume). m is the mass of the electron and e is the electric charge of the electron. $\omega = 2\pi\upsilon$ and υ is the frequency of the monochromatic incident light. γ is the damping coefficient in the Drude model of the electron [17].

6.3.1.7 ***Extinction Coefficient and Absorption Coefficient***

The intensity of the laser beam reaching the silicon film surface can be written as

$$I_o(t) = (J/t_p)\exp\left[-2.77(t/t_p)^2\right] \tag{6.134}$$

Once the laser beam passes through the silicon and diamond films, it is partially reflected and partially transmitted from both films. The total intensity of the laser beam reaching at the aluminum surface is related to the total transmission of the laser beam reaching the aluminum film surface ($x = L_1 + L_2$). The total laser intensity transmitted to the aluminum film surface can be determined from the transmittance in Eq. (6.131). Therefore, the laser pulse intensity reaching the surface of aluminum film is

$$I_o(t) = \mathrm{Tr}(J/t_p)\exp\left[-2.77(t/t_p)^2\right] \tag{6.135}$$

Laser pulse intensity reaching the aluminum film surface is absorbed within the aluminum film according to the Beer–Lambert law. The Beer–Lambert law states that in an absorbing material, the attenuation of the transmitted power of the incident radiation is given as

$$I = I_o e^{-\alpha x} \tag{6.136}$$

I_o is the incident power, which is equal to $I_o(t)$, as given in Eq. (6.136). x is the distance along the thin film and α is the absorption coefficient. The imaginary part n_I of the complex refractive index is known as the extinction coefficient. The absorption coefficient is related to the extinction coefficient as

$$\alpha = 2\left(\frac{\omega}{c}\right)n_{\mathrm{I}} \tag{6.137}$$

where c is the speed of light in vacuum in (ms^{-1}). The absorption coefficient for the aluminum film is determined from this formula for the laser frequency considered ($\omega = 2\pi\upsilon$, where υ is the frequency of the incident laser radiation).

6.3.1.8 ***Numerical Method***

The EPRT in the silicon, diamond, and aluminum films with the associated initial and boundary conditions are solved numerically using the discrete ordinates method. Fig. 6.17 shows schematic view of the phonon intensities in the silicon-diamond-aluminum films composite. The EPRT is an integrodifferential equation and may be solved using an iterative technique, which was given in Section 6.2.1.

The phonon and electron EPRTs in the aluminum film are given as

$$\frac{1}{v_{\mathrm{p}}}\frac{\partial I_{\mathrm{p}}}{\partial t} + \mu\frac{\partial I_{\mathrm{p}}}{\partial x} = \frac{\frac{1}{2}\int_{-1}^{1} I_{\mathrm{p}}d\mu - I_{\mathrm{p}}}{\Lambda_{\mathrm{p}}} - \frac{G}{4\pi}(T_{\mathrm{p}} - T_{\mathrm{e}}) \tag{6.138}$$

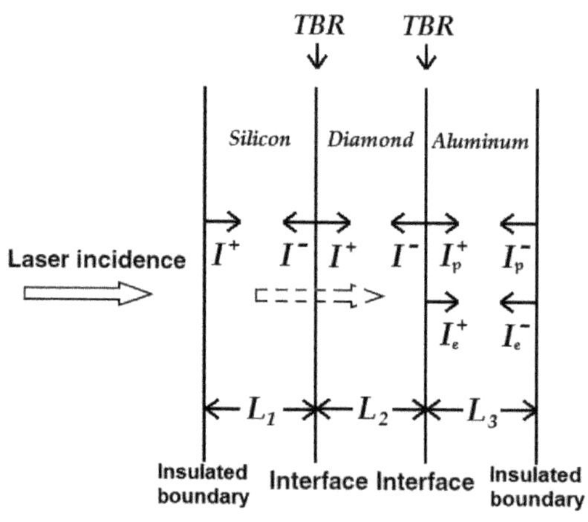

■ **FIGURE 6.17** Schematic view of phonon intensities in silicon-diamond-aluminum composite used in the numerical analysis [5].

$$\frac{1}{v_e}\frac{\partial I_e}{\partial t} + \mu\frac{\partial I_e}{\partial x} = \frac{\frac{1}{2}\int_{-1}^{1} I_e \, d\mu - I_e}{\Lambda_e} - \frac{G}{4\pi}(T_e - T_p) + \frac{S}{4\pi} \tag{6.139}$$

where

$$S(x,t) = \mathrm{Tr} \times (J/t_p)\alpha \exp\left[-\alpha x - 2.77(t/t_p)^2\right] \tag{6.140}$$

Each of these equations are first split into two equations corresponding to the intensities I_p^- and I_p^+ and I_e^- and I_e^+. These equations are

$$\frac{1}{v_p}\frac{\partial I_p^-}{\partial t} + \mu\frac{\partial I_p^-}{\partial x} = \frac{\int_{-1}^{0} I_p^- \, d\mu + \int_{0}^{1} I_p^+ \, d\mu - I_p^-}{\Lambda_p} - \frac{G}{4\pi}(T_p - T_e) \quad -1 \le \mu \le 0 \tag{6.141}$$

$$\frac{1}{v_p}\frac{\partial I_p^+}{\partial t} + \mu\frac{\partial I_p^+}{\partial x} = \frac{\int_{-1}^{0} I_p^- \, d\mu + \int_{0}^{1} I_p^+ \, d\mu - I_p^+}{\Lambda_p} - \frac{G}{4\pi}(T_p - T_e) \quad 0 \le \mu \le 1 \tag{6.142}$$

and

$$\frac{1}{v_e}\frac{\partial I_e^-}{\partial t} + \mu\frac{\partial I_e^-}{\partial x} = \frac{\int_{-1}^{0} I_e^- \, d\mu + \int_{0}^{1} I_e^+ \, d\mu - I_e^-}{\Lambda_e} - \frac{G}{4\pi}(T_e - T_p) + \frac{S}{4\pi} \quad -1 \le \mu \le 0 \tag{6.143}$$

$$\frac{1}{v_e}\frac{\partial I_e^+}{\partial t} + \mu\frac{\partial I_e^+}{\partial x} = \frac{\int_{-1}^{0} I_e^- \, d\mu + \int_{0}^{1} I_e^+ \, d\mu - I_e^+}{\Lambda_e} - \frac{G}{4\pi}(T_e - T_p) + \frac{S}{4\pi} \quad 0 \le \mu \le 1 \tag{6.144}$$

The time-derivative in these equations is discretized using the backward-difference formula corresponding to the implicit scheme. The partial derivative with respect to the coordinate variable x is discretized using a forward-difference for the first equation in which $-1 \le \mu \le 0$ and is discretized using a backward-difference for the second equation in which $0 \le \mu \le 1$. This procedure is important to obtain stable discretization. Integration is performed using Simpson's rule. The initial and boundary conditions are as described in the mathematics section. It should be observed that the electron EPRT has insulated boundary conditions at both ends of the aluminum film.

Starting from the initial conditions, at each time-step, EPRT is solved in the silicon and diamond films. Also, two coupled EPRTs are solved in the aluminum film corresponding to the phonon and electron subsystems in each film. To solve these equations in each time-step, iterations have to be performed. Starting from a guessed solution at the current time-step (which is usually the solution of the previous time-step), in each iteration, a solution is obtained that gives the approximate intensity distribution in the silicon, diamond, and aluminum films. From this solution the equilibrium intensities and then the interface intensities are calculated. These variables are then used in the next iteration. At each time-step the iterations are terminated once the equivalent equilibrium temperatures in each film change very little from the previous iteration. The convergence criteria at each time-step is $\max(|T_{iter} - T_{iter-1}|) \leq 10^{-7}$. This criterion has to be satisfied by the temperatures in all films. After convergence the computations proceed to the next time-step. The simulations are carried out to a time when almost steady-state conditions are reached.

6.3.1.9 *Findings of Laser Short-Pulse Heating of Three-Layer Thin Film Assembly*

Fig. 6.18 shows dimensionless equivalent equilibrium temperature $\left(\frac{T-300\text{K}}{301-300\text{K}}\right)\frac{T-300}{301-300}$ along the film composite for three heating periods.

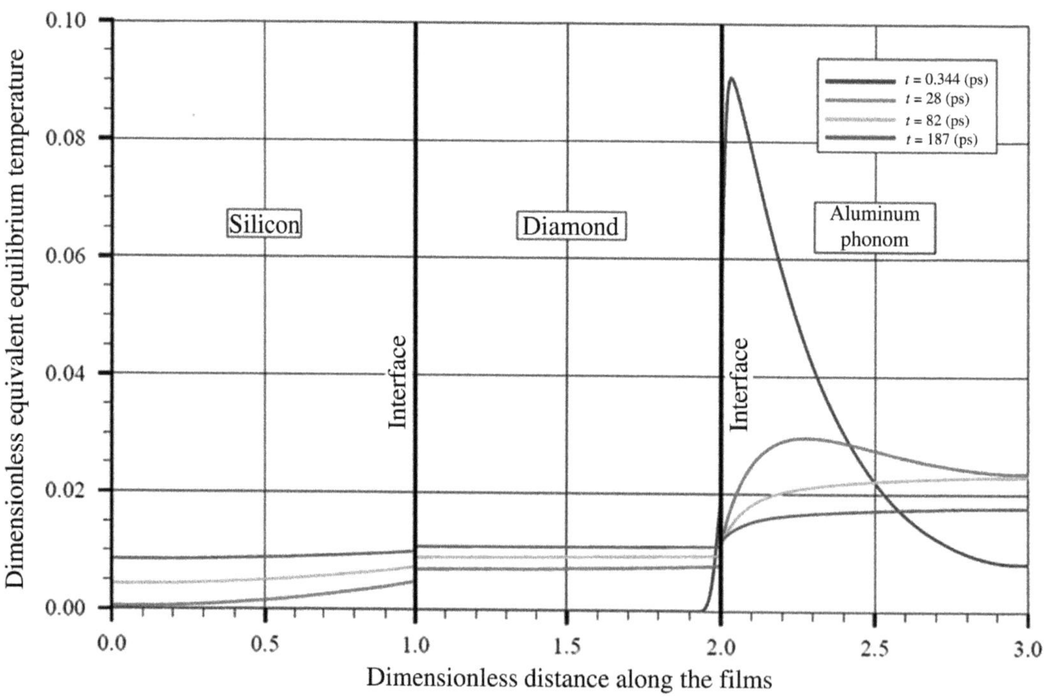

FIGURE 6.18 Dimensionless equivalent equilibrium temperature along silicon-diamond-aluminum composite for different heating durations [5].

This definition is chosen so as to ensure that the dimensionless equivalent equilibrium temperature for the phonons in any film remains between 0 and 1. It should be noted that initial temperature of the film composite (silicon-aluminum films pair) is considered to be 300K. The maximum phonon temperature, as obtained from the simulation data, can reach 301K in the film composite after the laser irradiation pulse. However, the absorption of the incident radiation mainly takes place in the aluminum film. This is because the absorption depths of silicon and diamond are very large and the film thickness is very small so that an incident beam can be transmitted to the aluminum film surface at the diamond-silicon interface. Therefore, the irradiated energy absorbed by electrons in the absorption depth of the substrate material (aluminum) causes electron excess energy gain in this region. This, in turn, results in thermal separation of electron and lattice subsystems in the film. However, electrons transfer some fraction of their excess energy to the lattice site through electron–phonon coupling. Therefore, lattice-site temperature increases sharply in the near vicinity of the aluminum-diamond interface. Since the laser pulse intensity decays exponentially with progressing time (Fig. 6.15), electron excess energy rise reduces with progressing time. Consequently, the amount of excess energy gain from the irradiated field by electrons becomes significant in the early heating period and the electron excess energy transfer to the lattice site by the collisional process substantiates during this period. This in turn increases lattice-site phonon temperature in the early heating period.

As the heating progresses, electron excess energy gain decreases because of reducing irradiated field intensity with progressing time. Hence, energy transferring to the lattice site reduces while lowering equilibrium equivalent temperature with progressing time. Since energy transfer takes place from aluminum film to the diamond film, lattice-site phonon temperature in the aluminum film reduces sharply. The presence of thermal boundary resistance between the diamond-silicon film interface causes temperature jump at the diamond film interface. Equivalent equilibrium temperature decays sharply in the diamond film in the near region of the interface. This is attributed to the scattering of the emitted phonons from the diamond film interface toward the diamond film. As the heating time progresses, energy transfer from the aluminum lattice site to diamond film across the interface enhances temperature rise in the diamond film. In this case, phonons emitted from the diamond film interface toward the diamond film increases with progressing time. Note that electron diffusion from the aluminum film to the diamond film through the interface is not considered. Therefore, an insulated boundary condition is incorporated for the electron subsystem at the diamond interface. In the case of the interface between the diamond-silicon films, temperature jump occurs because of the boundary scattering of the phonons transmitted from the diamond film to the silicon film. The phonons emitted from the silicon film interface toward the silicon film bulk undergo scattering in the film while contributing to the film resistance. In this case, equivalent equilibrium temperature reduces sharply in the early heating period; however, it remains the same along the silicon film with progressing time ($t = 187$ ps).

Fig. 6.19 shows the temporal behavior of dimensionless equivalent equilibrium temperature $\left(\frac{T - 300\text{K}}{301 - 300\text{K}}\right)$ at the silicon-diamond and the diamond-aluminum interface. In the case of aluminum film interface, phonon temperature (lattice site) remains high at the aluminum interface, which is also true for the diamond film interface in the early heating period. However, equivalent equilibrium temperature is higher at the aluminum film interface than that of the diamond film interface. Electron excess energy transfer in the aluminum lattice site increases phonon energy prior to transferring to diamond film across the interface. Since the thermal boundary resistance is introduced at the interface between the aluminum and diamond films, temperature mismatch, due to aluminum and diamond sites at the interface, occurs. However, boundary scattering of emitted phonons in the diamond film contributes to temperature jump at the diamond interface. As the time progresses, equivalent equilibrium temperature decays sharply up to 10 ps of the heating duration and temperature decay becomes gradual.

Sharp decay of equivalent equilibrium temperature in the diamond film interface is related to phonon intensity distribution in the interface region, which depends on the transmitted lattice-site phonons from the aluminum interface toward the diamond interface. However, temperature decay in the aluminum interface is associated with electron excess energy transfer from the electron subsystem to the lattice subsystem. In the case of silicon-diamond interface, interfacial thermal resistance introduced at the interface causes temperature jump across the interface. However, boundary scattering of the phonons in the diamond and silicon interfaces contribute to temperature jump at the interface. As the heating period increases equivalent equilibrium temperature at silicon and diamond films interface increases. The rate of this increase is significantly smaller than that of the rate of equivalent equilibrium temperature decrease at the aluminum-diamond films interface. The slight rate of increase in equivalent equilibrium temperature indicates that with progressing time phonons transmitted from the diamond film through the interface enhances equivalent equilibrium temperature in the silicon film.

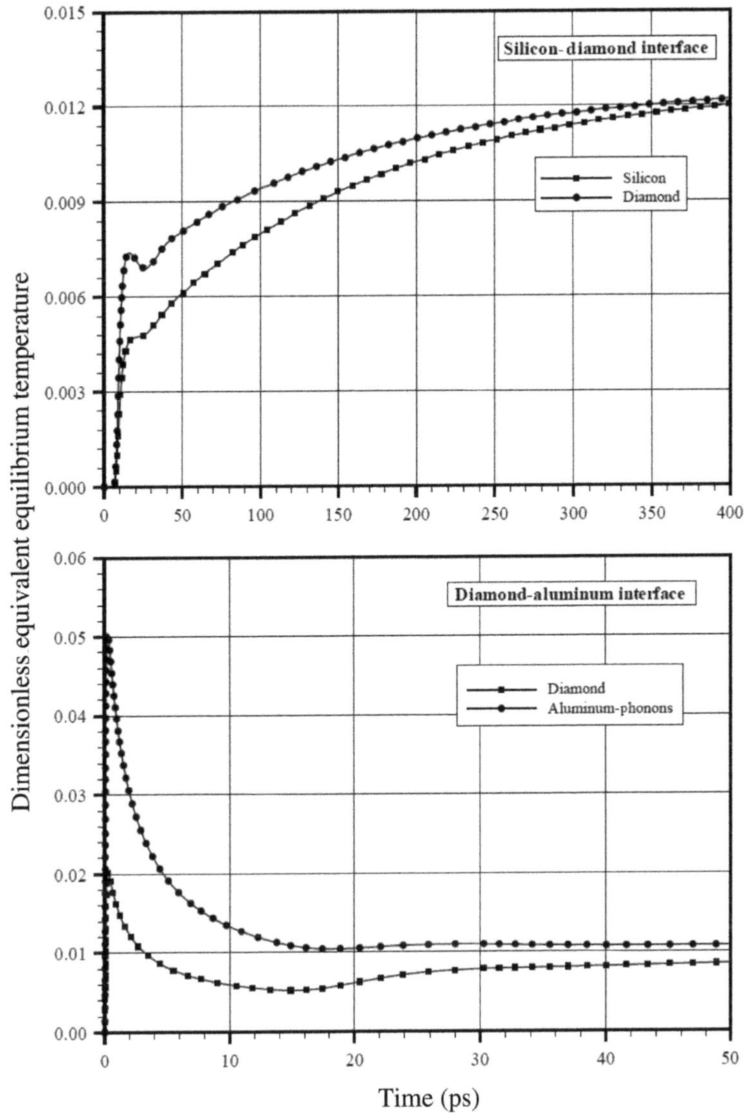

■ **FIGURE 6.19** Temporal variation of dimensionless equivalent equilibrium temperature at silicon-diamond and diamond-aluminum interfaces [5].

Fig. 6.20 shows dimensionless equivalent equilibrium temperature $\left(\frac{T-300\text{K}}{301-300\text{K}}\right)$ in the lattice and electron subsystems in the aluminum film for various heating periods. Electron temperature attains higher values than those of the lattice-site temperatures because of the excess energy gain from the irradiated field. Since electrons cannot travel into the

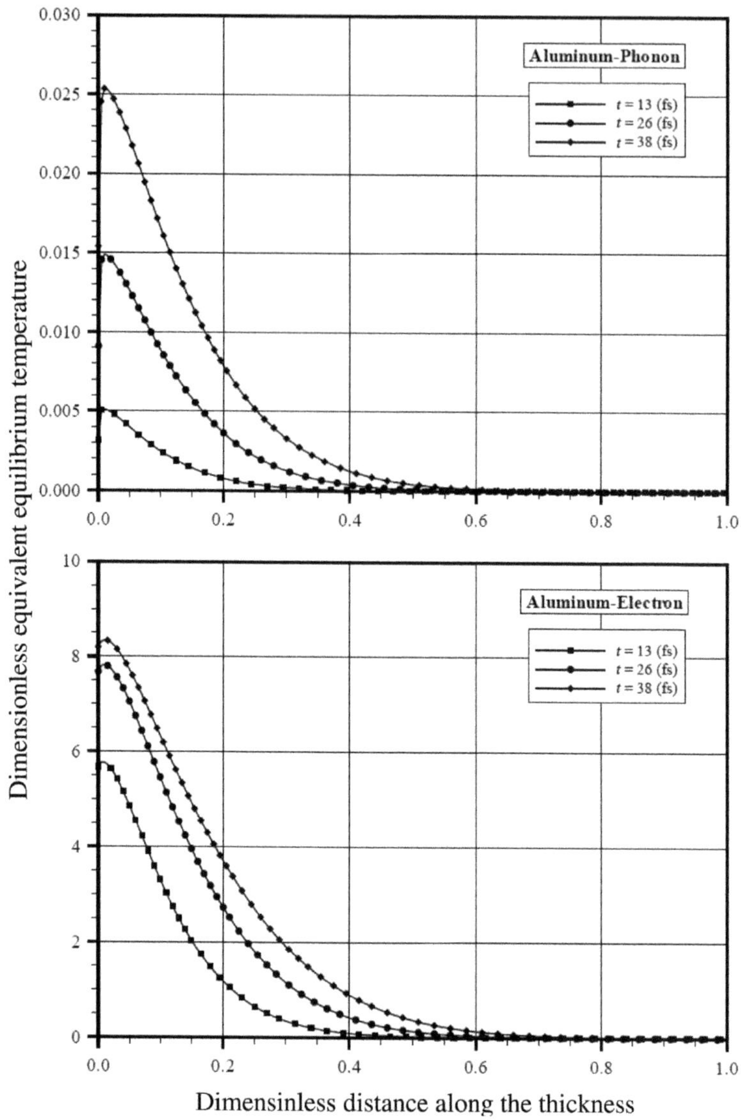

■ **FIGURE 6.20** Dimensionless equivalent equilibrium temperature in the lattice and electron subsystems along the aluminum film thickness for various heating periods [5].

diamond film through the interface, an insulated boundary condition is adopted for the electron subsystem at the interface. This results in gradual change of temperature in the close region of the interface of the electron subsystem. As the distance increases from the interface toward the aluminum film bulk, temperature decay becomes sharp. This is related to the energy absorption from the irradiated field, which follows the Beer–Lambert law; the absorbed laser beam intensity decays exponentially with increasing depth beyond the interface. Increasing heating duration reduces temperature in the electron subsystem. This is associated with the temporal behavior of the laser pulse intensity, which decreases with progressing time (Fig. 6.15). In the case of the lattice subsystem, phonon transport across the interface toward the diamond film lowers temperature at the interface ($x = 0$). Therefore, the maximum temperature moves away from the interface.

The presence of thermal boundary resistance at the aluminum-diamond interface results in lattice-site temperature decay at a small rate as compared to that corresponding to temperature decay in the electronic subsystem. In addition, some of the phonons reflected from the interface undergo scattering toward the aluminum film bulk while contributing to temperature increase in the aluminum film. Lattice site temperature decay is relatively gradual along the film thickness as compared to that of the electron subsystem. This is related to the electron–phonon coupling constant, which facilities excess energy transfer from the electron subsystem to the lattice subsystem. Equivalent equilibrium temperature distribution depends on the phonon intensity distribution in the lattice subsystem, which depends mainly on the phonon scattering in the film. However, in the electron subsystem the carrier mobility is high and the relaxation time is short, which in turn results in electron excess energy gain dependence of equivalent equilibrium temperature in the electron subsystem.

Fig. 6.21 shows temporal variation of dimensionless equivalent equilibrium temperature $\left(\frac{T-300\mathrm{K}}{301-300\mathrm{K}}\right)$ in the electron and the lattice subsystems for various locations in the aluminum thin film. Temperature in both subsystems remains high at the aluminum-diamond interface ($x/L = 0$). Temperature increases sharply to reach its peak in the early heating period due to the energy gain from the irradiated field. Since the irradiated energy reduces with time (Fig. 6.15), electron excess energy gain reduces with progressing time. This results in a decrease of temperature in the electron subsystem with progressing time. The time occurrence of the peak temperature changes in the electron subsystem. This is attributed to the electron excess energy distribution with time, which alters due to decay of the irradiated field with progressing time (Fig. 6.15). In the case of the lattice subsystem, temperature decay is fast at the aluminum film interface ($x/L = 0$) and becomes gradual with increasing time. This is associated with the energy transfer from the electron subsystem to the lattice subsystem via electron–phonon coupling in the early heating period and phonon transfer from aluminum film to the diamond film across the

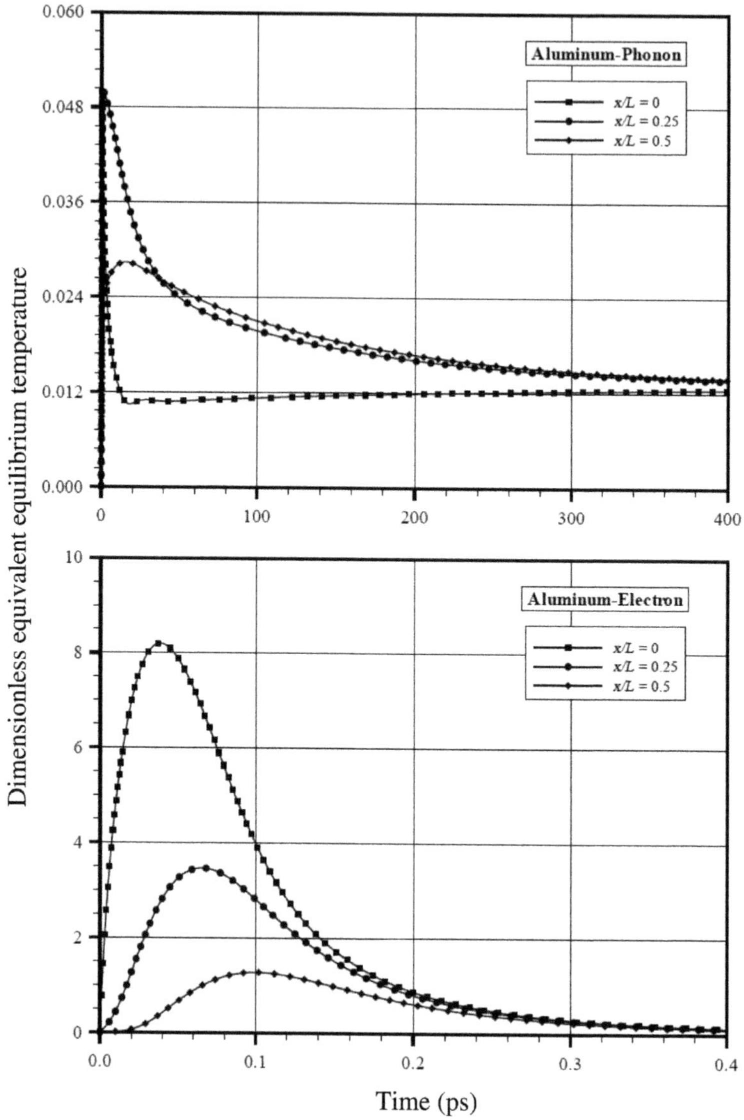

■ **FIGURE 6.21** Temporal variation of equivalent equilibrium temperature in lattice and electron subsystems in aluminum film at various locations [5].

interface. In this case, energy gain by the lattice subsystem is high in the early heating period because of high rate of electron excess energy transfer due to high electron excess energy gain from the irradiated field. As the time progresses, phonon transport from aluminum film toward the diamond film reduces equivalent equilibrium temperature in the lattice site at the interface. As the distance increases toward the aluminum bulk ($x/L = 0.5$), temporal behavior of temperature differs significantly from that at the interface in the lattice subsystem. In this case, lattice-site temperature reduces sharply and remains almost the same with progressing time.

Phonon scattering in the film is responsible for the gradual temperature increase in this region with progressing time. In this case, phonons emitted from the region close to the diamond film interface, where the absorption of irradiated field is large, undergo several scatterings prior to reaching the location $x/L = 0.5$. Phonon intensity is modified and reduced by the phonon scattering prior to reaching this location. Since equivalent equilibrium temperature represents the average energy of all phonons around a local point when they redistribute adiabatically to an equilibrium state, equivalent equilibrium temperature reduces in this region.

6.3.2 Cross-Plane Transport Across Two-Dimensional Dielectric Thin Films and Applications

Energy transport in thin films is important in electronic industry, since electronic device performance reduces notably with increasing temperature. The phonon transport mainly governs the energy transport in dielectric thin films and in some cases two thin films can be in thermal contact resulting in thermal boundary resistance at the contact interface due to mismatch of acoustic speeds and phonon frequencies. The model study provides physical insight into the energy transport process when two thin films are in thermal contact. Consequently, investigation into phonon transport in two adjacent thin films becomes essential. In the present section, phonon transport in thin films of silicon-diamond pair with the presence of thermal boundary resistance at the contact interface is considered. The transient analysis for two-dimensional phonon transport is considered and frequency-dependent BTE is solved numerically. The findings (equivalent equilibrium temperature) in both films are presented in line with the previous study [6].

In the analysis, transient, wavenumber (frequency)-dependent EPRT is formulated within the two-dimensional silicon-diamond films pair. The boundary conditions used are those of constant temperature (Fig. 6.22). At the interface of the silicon and diamond films, an appropriate interface condition including the thermal boundary resistance has to be imposed, which should satisfy the energy balance at the interface.

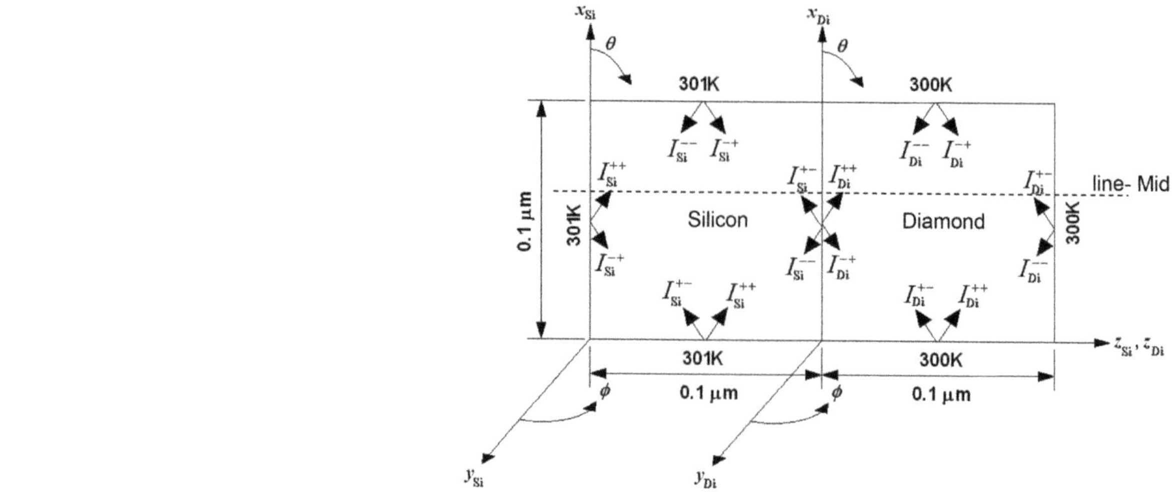

■ **FIGURE 6.22** A schematic view of silicon and diamond films pair, direction of emitted phonons, and the coordinate system [6].

Fig. 6.22 shows the films pair and the coordinate system used in the analysis. The transient two-dimensional and frequency (wavenumber)-dependent EPRT is incorporated into the analysis, which can be given as

$$\frac{1}{v_k}\frac{\partial I_k}{\partial t} + \cos\theta\frac{\partial I_k}{\partial x} + \sin\theta\,\sin\phi\frac{\partial I_k}{\partial z} = \frac{\frac{1}{4\pi}\int_0^{2\pi}\int_0^{\pi} I_k \sin\theta\, d\theta\, d\phi - I_k}{\Lambda_k} \tag{6.145}$$

$$t>0,\quad 0<x<L_x,\quad 0<z<L_z,\quad 0<\theta<\pi,\quad 0<\phi<2\pi$$

In Eq. (6.145), the phonon intensity I_k is a function of x, z, t, θ, ϕ and k: $I_k = I_k(x, z, t, \theta, \phi)$. The details of EPRT are given in Section 4.4.4 of Chapter 4, Analysis of Energy Transport Equations at Micro/Nanoscales.

The dispersion relations represent various polarizations and modes of vibrations of the silicon [8] and diamond [13]. Moreover, there are three different polarizations and two different modes of vibrations of the phonons. Waves can propagate in a dielectric crystal in three different polarizations, one longitudinal and two transverse polarizations. The dispersion relations presented in this study are for the [001] direction in silicon and diamond, which in reality is a symmetry direction. Along this direction the two transverse waves coincide and therefore we have only one dispersion curve for the two transverse waves for each mode of vibration. The other curve is for the longitudinal wave. There are two modes of longitudinal and transverse waves: acoustic waves and optical waves. Hence we have now six different types of waves, but since the transverse waves are identical, we finally end up with four different types of waves and hence four different curves in the dispersion relation. The EPRT given earlier has to be applied to each of these waves individually. These EPRTs are

$$\frac{1}{v_{TA,k}}\frac{\partial I_{TA,k}}{\partial t}+\cos\theta\frac{\partial I_{TA,k}}{\partial x}+\sin\theta\sin\phi\frac{\partial I_{TA,k}}{\partial z}=\frac{\frac{1}{4\pi}\int_0^{2\pi}\int_0^{\pi} I_{TA,k}\sin\theta\, d\theta\, d\phi - I_{TA,k}}{\Lambda_{TA,k}} \tag{6.146}$$

$$\frac{1}{v_{LA,k}}\frac{\partial I_{LA,k}}{\partial t}+\cos\theta\frac{\partial I_{LA,k}}{\partial x}+\sin\theta\sin\phi\frac{\partial I_{LA,k}}{\partial z}=\frac{\frac{1}{4\pi}\int_0^{2\pi}\int_0^{\pi} I_{LA,k}\sin\theta\, d\theta\, d\phi - I_{LA,k}}{\Lambda_{LA,k}} \tag{6.147}$$

$$\frac{1}{v_{TO,k}}\frac{\partial I_{TO,k}}{\partial t}+\cos\theta\frac{\partial I_{TO,k}}{\partial x}+\sin\theta\sin\phi\frac{\partial I_{TO,k}}{\partial z}=\frac{\frac{1}{4\pi}\int_0^{2\pi}\int_0^{\pi} I_{TO,k}\sin\theta\, d\theta\, d\phi - I_{TO,k}}{\Lambda_{TO,k}} \tag{6.148}$$

$$\frac{1}{v_{LO,k}}\frac{\partial I_{LO,k}}{\partial t}+\cos\theta\frac{\partial I_{LO,k}}{\partial x}+\sin\theta\sin\phi\frac{\partial I_{LO,k}}{\partial z}=\frac{\frac{1}{4\pi}\int_0^{2\pi}\int_0^{\pi} I_{LO,k}\sin\theta\, d\theta\, d\phi - I_{LO,k}}{\Lambda_{LO,k}} \tag{6.149}$$

In Eqs. (6.146)–(6.149), T and L stand for transverse and longitudinal, respectively, and A and O stand for acoustic and optical, respectively. This description is valid for both silicon and diamond films without any modification. However, the dispersion relation and hence the thermophysical properties of silicon and diamond are different.

Equivalent equilibrium phonon temperature is defined as

$$T(x,z,t)=\frac{\int_0^{k_{max}}\int_0^{2\pi}\int_0^{\pi} I_k\sin\theta\, d\theta\, d\phi\, dk}{\int_0^{k_{max}}\left(2\, C_{TA,k}v_{TA,k}+C_{LA,k}v_{LA,k}+2\, C_{TO,k}v_{TO,k}+C_{LO,k}v_{LO,k}\right)dk} \tag{6.150}$$

The heat flux vector is defined at all points inside the layer. It is given as

$$q_x''(x,z,t)=\int_0^{k_{max}}\int_0^{2\pi}\int_0^{\pi}(2I_{TA,k}+I_{LA,k}+2I_{TO,k}+I_{LO,k})\cos\theta\sin\theta d\theta\, d\phi\, dk \tag{6.151}$$

and

$$q_z''(x,z,t) = \int_0^{k_{\max}} \int_0^{2\pi} \int_0^{\pi} (2I_{\mathrm{TA,k}} + I_{\mathrm{LA,k}} + 2I_{\mathrm{TO,k}} + I_{\mathrm{LO,k}}) \sin\phi \sin^2\theta d\theta \, d\phi \, dk \tag{6.152}$$

and

$$\mathbf{q}'' = q_x''\mathbf{i} + q_z''\mathbf{k} \tag{6.153}$$

6.3.2.1 *Initial and Boundary Conditions*

There are four boundaries for both the silicon and diamond. These are the bottom, top, left, and right boundaries (Fig. 6.22). In addition, the right boundary for silicon and the left boundary for diamond make a single interface. At this interface an appropriate condition has to be defined, which is derived from the energy balance condition at the interface. The phonon intensity specification at the boundaries, the interface condition, and the initial condition should be prescribed to complete the problem specification. The geometry of the physical domain and the boundary conditions do not change with time. Initially, the silicon and diamond layers are assumed to be at a uniform temperature of 300K. Then, after that for all times the following boundary conditions are imposed. The left, bottom, and top boundaries of the silicon layer are assumed to be at a constant temperature of 301K. The top, right, and bottom boundaries of the diamond layer are assumed to be at a constant temperature of 300K. At the interface, the temperatures in the silicon and diamond layers are variable and they are determined by means of the interface condition. The initial, boundary, and interface conditions are described next.

6.3.2.1.1 **Silicon Film**

6.3.2.1.1.1 *Initial Condition*

$$\begin{aligned}
I_{\mathrm{Si,TA,k}}(x_{\mathrm{Si}}, z_{\mathrm{Si}}, t=0,\theta,\phi) &= \frac{C_{\mathrm{Si,TA,k}} v_{\mathrm{Si,TA,k}} T_{\mathrm{initial}}}{4\pi}, \quad \begin{array}{l} 0 < x_{\mathrm{Si}} < L_{x,\mathrm{Si}},\ 0 < z_{\mathrm{Si}} < L_{z,\mathrm{Si}}, \\ 0 < \theta < \pi, \quad 0 < \phi < 2\pi \end{array} \\
I_{\mathrm{Si,LA,k}}(x_{\mathrm{Si}}, z_{\mathrm{Si}}, t=0,\theta,\phi) &= \frac{C_{\mathrm{Si,LA,k}} v_{\mathrm{Si,LA,k}} T_{\mathrm{initial}}}{4\pi} \\
I_{\mathrm{Si,TO,k}}(x_{\mathrm{Si}}, z_{\mathrm{Si}}, t=0,\theta,\phi) &= \frac{C_{\mathrm{Si,TO,k}} v_{\mathrm{Si,TO,k}} T_{\mathrm{initial}}}{4\pi} \\
I_{\mathrm{Si,LO,k}}(x_{\mathrm{Si}}, z_{\mathrm{Si}}, t=0,\theta,\phi) &= \frac{C_{\mathrm{Si,LO,k}} v_{\mathrm{Si,LO,k}} T_{\mathrm{initial}}}{4\pi}
\end{aligned} \tag{6.154}$$

6.3.2.1.1.2 *Bottom Boundary Condition*

$$
\begin{aligned}
I_{\mathrm{Si,TA,k}}(x_{\mathrm{Si}}=0, z_{\mathrm{Si}}, t, \theta, \phi) &= \frac{C_{\mathrm{Si,TA,k}} v_{\mathrm{Si,TA,k}} T_{\mathrm{bottom}}}{4\pi}, \quad \begin{array}{l} t>0, \quad 0<z_{\mathrm{Si}}<L_{z,\mathrm{Si}}, \\ 0<\theta<\pi/2, \quad 0<\phi<2\pi \end{array} \\
I_{\mathrm{Si,LA,k}}(x_{\mathrm{Si}}=0, z_{\mathrm{Si}}, t, \theta, \phi) &= \frac{C_{\mathrm{Si,LA,k}} v_{\mathrm{Si,LA,k}} T_{\mathrm{bottom}}}{4\pi} \\
I_{\mathrm{Si,TO,k}}(x_{\mathrm{Si}}=0, z_{\mathrm{Si}}, t, \theta, \phi) &= \frac{C_{\mathrm{Si,TO,k}} v_{\mathrm{Si,TO,k}} T_{\mathrm{bottom}}}{4\pi} \\
I_{\mathrm{Si,LO,k}}(x_{\mathrm{Si}}=0, z_{\mathrm{Si}}, t, \theta, \phi) &= \frac{C_{\mathrm{Si,LO,k}} v_{\mathrm{Si,LO,k}} T_{\mathrm{bottom}}}{4\pi}
\end{aligned}
\tag{6.155}
$$

6.3.2.1.1.3 *Top Boundary Condition*

$$
\begin{aligned}
I_{\mathrm{Si,TA,k}}\left(x_{\mathrm{Si}}=L_{x,\mathrm{Si}}, z_{\mathrm{Si}}, t, \theta, \phi\right) &= \frac{C_{\mathrm{Si,TA,k}} v_{\mathrm{Si,TA,k}} T_{\mathrm{top}}}{4\pi}, \quad \begin{array}{l} t>0, \quad 0<z_{\mathrm{Si}}<L_{z,\mathrm{Si}}, \\ \pi/2<\theta<\pi, \quad 0<\phi<2\pi \end{array} \\
I_{\mathrm{Si,LA,k}}\left(x_{\mathrm{Si}}=L_{x,\mathrm{Si}}, z_{\mathrm{Si}}, t, \theta, \phi\right) &= \frac{C_{\mathrm{Si,LA,k}} v_{\mathrm{Si,LA,k}} T_{\mathrm{top}}}{4\pi} \\
I_{\mathrm{Si,TO,k}}\left(x_{\mathrm{Si}}=L_{x,\mathrm{Si}}, z_{\mathrm{Si}}, t, \theta, \phi\right) &= \frac{C_{\mathrm{Si,TO,k}} v_{\mathrm{Si,TO,k}} T_{\mathrm{top}}}{4\pi} \\
I_{\mathrm{Si,LO,k}}\left(x_{\mathrm{Si}}=L_{x,\mathrm{Si}}, z_{\mathrm{Si}}, t, \theta, \phi\right) &= \frac{C_{\mathrm{Si,LO},k} v_{\mathrm{Si,LO},k} T_{\mathrm{top}}}{4\pi}
\end{aligned}
\tag{6.156}
$$

6.3.2.1.1.4 *Left Boundary Condition*

$$
\begin{aligned}
I_{\mathrm{Si,TA,k}}(x_{\mathrm{Si}}, z_{\mathrm{Si}}=0, t, \theta, \phi) &= \frac{C_{\mathrm{Si,TA,k}} v_{\mathrm{Si,TA,k}} T_{\mathrm{left}}}{4\pi} \quad \begin{array}{l} t>0, \quad 0<x_{\mathrm{Si}}<L_{x,\mathrm{Si}}, \\ 0<\theta<\pi, \quad 0<\phi<\pi \end{array} \\
I_{\mathrm{Si,LA,k}}(x_{\mathrm{Si}}, z_{\mathrm{Si}}=0, t, \theta, \phi) &= \frac{C_{\mathrm{Si,LA,k}} v_{\mathrm{Si,LA,k}} T_{\mathrm{left}}}{4\pi} \\
I_{\mathrm{Si,TO,k}}(x_{\mathrm{Si}}, z_{\mathrm{Si}}=0, t, \theta, \phi) &= \frac{C_{\mathrm{Si,TO,k}} v_{\mathrm{Si,TO,k}} T_{\mathrm{left}}}{4\pi} \\
I_{\mathrm{Si,LO,k}}(x_{\mathrm{Si}}, z_{\mathrm{Si}}=0, t, \theta, \phi) &= \frac{C_{\mathrm{Si,LO,k}} v_{\mathrm{Si,LO,k}} T_{\mathrm{left}}}{4\pi}
\end{aligned}
\tag{6.157}
$$

In Eqs. (6.154)–(6.157), $C_{\mathrm{Si,TA,k}}$ is the wavenumber-dependent volumetric heat capacity at the nominal temperature of $T_o = 300\mathrm{K}$, and $v_{\mathrm{Si,TA,k}}$ is the wavenumber dependent phonon group velocity for the transverse acoustic branch for the [001] direction in silicon. Similar definitions apply for the longitudinal acoustic, transverse optical, and longitudinal optical branches of phonons.

6.3.2.2.2 Diamond Film

6.3.2.2.2.1 *Initial Condition*

$$
\begin{aligned}
&I_{\mathrm{Di,TA,k}}(x_{\mathrm{Di}}, z_{\mathrm{Di}}, t=0, \theta, \phi) = \frac{C_{\mathrm{Di,TA,k}} v_{\mathrm{Di,TA,k}} T_{\mathrm{initial}}}{4\pi}, \quad \begin{matrix} 0<x_{\mathrm{Di}}<L_{x,\mathrm{Di}}, \ 0<z_{\mathrm{Di}}<L_{z,\mathrm{Di}}, \\ 0<\theta<\pi, \ 0<\phi<2\pi \end{matrix} \\
&I_{\mathrm{Di,LA,k}}(x_{\mathrm{Di}}, z_{\mathrm{Di}}, t=0, \theta, \phi) = \frac{C_{\mathrm{Di,LA,k}} v_{\mathrm{Di,LA,k}} T_{\mathrm{initial}}}{4\pi} \\
&I_{\mathrm{Di,TO,k}}(x_{\mathrm{Di}}, z_{\mathrm{Di}}, t=0, \theta, \phi) = \frac{C_{\mathrm{Di,TO,k}} v_{\mathrm{Di,TO,k}} T_{\mathrm{initial}}}{4\pi} \\
&I_{\mathrm{Di,LO,k}}(x_{\mathrm{Di}}, z_{\mathrm{Di}}, t=0, \theta, \phi) = \frac{C_{\mathrm{Di,LO,k}} v_{\mathrm{Di,LO,k}} T_{\mathrm{initial}}}{4\pi}
\end{aligned}
\tag{6.158}
$$

6.3.2.2.2.2 *Bottom Boundary Condition*

$$
\begin{aligned}
&I_{\mathrm{Di,TA,k}}(x_{\mathrm{Di}}=0, z_{\mathrm{Di}}, t, \theta, \phi) = \frac{C_{\mathrm{Di,TA,k}} v_{\mathrm{Di,TA,k}} T_{\mathrm{bottom}}}{4\pi}, \quad \begin{matrix} t>0, \ 0<z_{\mathrm{Di}}<L_{z,\mathrm{Di}}, \\ 0<\theta<\pi/2, \ 0<\phi<2\pi \end{matrix} \\
&I_{\mathrm{Di,LA,k}}(x_{\mathrm{Di}}=0, z_{\mathrm{Di}}, t, \theta, \phi) = \frac{C_{\mathrm{Di,LA,k}} v_{\mathrm{Di,LA,k}} T_{\mathrm{bottom}}}{4\pi} \\
&I_{\mathrm{Di,TO,k}}(x_{\mathrm{Di}}=0, z_{\mathrm{Di}}, t, \theta, \phi) = \frac{C_{\mathrm{Di,TO,k}} v_{\mathrm{Di,TO,k}} T_{\mathrm{bottom}}}{4\pi} \\
&I_{\mathrm{Di,LO,k}}(x_{\mathrm{Di}}=0, z_{\mathrm{Di}}, t, \theta, \phi) = \frac{C_{\mathrm{Di,LO,k}} v_{\mathrm{Di,LO,k}} T_{\mathrm{bottom}}}{4\pi}
\end{aligned}
\tag{6.159}
$$

6.3.2.2.2.3 *Top Boundary Condition*

$$
\begin{aligned}
&I_{\mathrm{Di,TA,k}}\left(x_{\mathrm{Di}}=L_{x,\mathrm{Di}}, z_{\mathrm{Di}}, t, \theta, \phi\right) = \frac{C_{\mathrm{Di,TA,k}} v_{\mathrm{Di,TA,k}} T_{\mathrm{top}}}{4\pi}, \quad \begin{matrix} t>0, \ 0<z_{\mathrm{Di}}<L_{z,\mathrm{Di}}, \\ \pi/2<\theta<\pi, \ 0<\phi<2\pi \end{matrix} \\
&I_{\mathrm{Di,LA,k}}\left(x_{\mathrm{Di}}=L_{x,\mathrm{Di}}, z_{\mathrm{Di}}, t, \theta, \phi\right) = \frac{C_{\mathrm{Di,LA,k}} v_{\mathrm{Di,LA,k}} T_{\mathrm{top}}}{4\pi} \\
&I_{\mathrm{Di,TO,k}}\left(x_{\mathrm{Di}}=L_{x,\mathrm{Di}}, z_{\mathrm{Di}}, t, \theta, \phi\right) = \frac{C_{\mathrm{Di,TO,k}} v_{\mathrm{Di,TO,k}} T_{\mathrm{top}}}{4\pi} \\
&I_{\mathrm{Di,LO,k}}\left(x_{\mathrm{Di}}=L_{x,\mathrm{Di}}, z_{\mathrm{Di}}, t, \theta, \phi\right) = \frac{C_{\mathrm{Di,LO,k}} v_{\mathrm{Di,LO,k}} T_{\mathrm{top}}}{4\pi}
\end{aligned}
\tag{6.160}
$$

6.3.2.2.2.4 *Right Boundary Condition*

$$
\begin{aligned}
&I_{\mathrm{Di,TA,k}}\left(x_{\mathrm{Di}}, z_{\mathrm{Di}}=L_{z,\mathrm{Di}}, t, \theta, \phi\right) = \frac{C_{\mathrm{Di,TA,k}} v_{\mathrm{Di,TA,k}} T_{\mathrm{left}}}{4\pi}, \quad \begin{matrix} t>0, \ 0<x_{\mathrm{Di}}<L_{x,\mathrm{Di}}, \\ 0<\theta<\pi, \ \pi<\phi<2\pi \end{matrix} \\
&I_{\mathrm{Di,LA,k}}\left(x_{\mathrm{Di}}, z_{\mathrm{Di}}=L_{z,\mathrm{Di}}, t, \theta, \phi\right) = \frac{C_{\mathrm{Di,LA,k}} v_{\mathrm{Di,LA,k}} T_{\mathrm{left}}}{4\pi} \\
&I_{\mathrm{Di,TO,k}}\left(x_{\mathrm{Di}}, z_{\mathrm{Di}}=L_{z,\mathrm{Di}}, t, \theta, \phi\right) = \frac{C_{\mathrm{Di,TO,k}} v_{\mathrm{Di,TO,k}} T_{\mathrm{left}}}{4\pi} \\
&I_{\mathrm{Di,LO,k}}\left(x_{\mathrm{Di}}, z_{\mathrm{Di}}=L_{z,\mathrm{Di}}, t, \theta, \phi\right) = \frac{C_{\mathrm{Di,LO,k}} v_{\mathrm{Di,LO,k}} T_{\mathrm{left}}}{4\pi}
\end{aligned}
\tag{6.161}
$$

In Eqs. (6.158)–(6.161), $C_{\mathrm{Di,TA,k}}$ is the wavenumber-dependent volumetric heat capacity at the nominal temperature of $T_o = 300\mathrm{K}$, and $v_{\mathrm{Di,TA,k}}$ is the wavenumber-dependent phonon group velocity for the transverse acoustic branch for the [001] direction in diamond. Similar definitions apply for the longitudinal acoustic, transverse optical, and longitudinal optical branches of phonons.

6.3.2.3 ***Interface Condition***

At the interface, the following energy balance should be satisfied between the silicon and diamond films:

$$q''_{z,\mathrm{Si}}(x, z_{\mathrm{Si}} = L_{z,\mathrm{Si}}, t) = \frac{T_{\mathrm{Si}}(x, z_{\mathrm{Si}} = L_{z,\mathrm{Si}}, t) - T_{\mathrm{Di}}(x, z_{\mathrm{Di}} = 0, t)}{\mathrm{TBR}} = q''_{z,\mathrm{Di}}(x, z_{\mathrm{Di}} = 0, t) \tag{6.162}$$

In these equations, T_{Si} and T_{Di} are the equivalent equilibrium temperatures. Physically, this means that the heat flux exiting the silicon film in the z direction should be the same as the heat flux entering the diamond film in the z direction at the interface. Between the silicon and diamond films there exists a resistance to heat flow, called the TBR or Kapitsa resistance. This resistance will exist even if there is perfect contact between the films. TBR may be found theoretically by using the discrete mismatch model or the acoustic mismatch model. Or it may be found experimentally. A theoretically determined value of TBR is used, based on the discrete mismatch model.

The use of the interface condition needs further elaboration. At the interface, any unknown intensity I in the silicon is expressed in terms of a single unknown temperature $T_{\mathrm{gb,Si}}$. This is the gray-body temperature for silicon at the interface. Similarly, any unknown intensity at the interface in diamond is expressed in terms of the temperature $T_{\mathrm{gb,Di}}$. This is the gray-body temperature for diamond at the interface. The temperatures $T_{\mathrm{gb,Si}}$ and $T_{\mathrm{gb,Di}}$ will in general be a function of the x coordinate and the time t. The heat flux $q''_{z,\mathrm{Si}}$ and $q''_{z,\mathrm{Di}}$ at the interface may be represented in terms of the integrals of the known and unknown intensities and in turn, in terms of the integrals of the known intensities and the gray-body temperatures. Similarly the equivalent equilibrium temperatures at the interface for the silicon and diamond should also be represented in terms of the integrals of the known intensities and the gray-body temperatures $T_{\mathrm{gb,Si}}$ and $T_{\mathrm{gb,Di}}$. Substituting all these expressions in the interface condition yields two equations for the two unknowns $T_{\mathrm{gb,Si}}$ and $T_{\mathrm{gb,Di}}$, and hence these can be determined. Once these temperatures are known, the unknown intensities can be determined and the EPRTs can then be solved subject to appropriate boundary conditions. The unknown intensities in terms of the gray-body temperatures at the interface are listed as follows.

$$
\begin{aligned}
I_{\mathrm{Si,TA,k}}\left(x_{\mathrm{Si}}, z_{\mathrm{Si}} = L_{z,\mathrm{Si}}, t, \theta, \phi\right) &= \frac{C_{\mathrm{Si,TA,k}} v_{\mathrm{Si,TA,k}} T_{\mathrm{gb,Si}}}{4\pi}, \quad t>0, \quad 0<x_{\mathrm{Si}}<L_{x,\mathrm{Si}}, \quad 0<\theta<\pi, \quad \pi<\phi<2\pi \\
I_{\mathrm{Si,LA,k}}\left(x_{\mathrm{Si}}, z_{\mathrm{Si}} = L_{z,\mathrm{Si}}, t, \theta, \phi\right) &= \frac{C_{\mathrm{Si,LA,k}} v_{\mathrm{Si,LA,k}} T_{\mathrm{gb,Si}}}{4\pi} \\
I_{\mathrm{Si,TO,k}}\left(x_{\mathrm{Si}}, z_{\mathrm{Si}} = L_{z,\mathrm{Si}}, t, \theta, \phi\right) &= \frac{C_{\mathrm{Si,TO,k}} v_{\mathrm{Si,TO,k}} T_{\mathrm{gb,Si}}}{4\pi} \\
I_{\mathrm{Si,LO,k}}\left(x_{\mathrm{Si}}, z_{\mathrm{Si}} = L_{z,\mathrm{Si}}, t, \theta, \phi\right) &= \frac{C_{\mathrm{Si,LO,k}} v_{\mathrm{Si,LO,k}} T_{\mathrm{gb,Si}}}{4\pi}
\end{aligned}
\tag{6.163}
$$

$$
\begin{aligned}
I_{\mathrm{Di,TA,k}}(x_{\mathrm{Di}}, z_{\mathrm{Di}} = 0, t, \theta, \phi) &= \frac{C_{\mathrm{Di,TA,k}} v_{\mathrm{Di,TA,k}} T_{\mathrm{gb,Di}}}{4\pi} \quad t>0, \quad 0<x_{\mathrm{Di}}<L_{x,\mathrm{Di}}, \quad 0<\theta<\pi, \quad 0<\phi<\pi \\
I_{\mathrm{Di,LA,k}}(x_{\mathrm{Di}}, z_{\mathrm{Di}} = 0, t, \theta, \phi) &= \frac{C_{\mathrm{Di,LA,k}} v_{\mathrm{Di,LA,k}} T_{\mathrm{gb,Di}}}{4\pi} \\
I_{\mathrm{Di,TO,k}}(x_{\mathrm{Di}}, z_{\mathrm{Di}} = 0, t, \theta, \phi) &= \frac{C_{\mathrm{Di,TO,k}} v_{\mathrm{Di,TO,k}} T_{\mathrm{gb,Di}}}{4\pi} \\
I_{\mathrm{Di,LO,k}}(x_{\mathrm{Di}}, z_{\mathrm{Di}} = 0, t, \theta, \phi) &= \frac{C_{\mathrm{Di,LO,k}} v_{\mathrm{Di,LO,k}} T_{\mathrm{gb,Di}}}{4\pi}
\end{aligned}
\tag{6.164}
$$

6.3.2.4 ***Numerical Method***

The EPRTs in the silicon and diamond films with the associated initial and boundary conditions are solved numerically using the finite-difference method. These are integrodifferential equations and can be solved using an iterative technique, which is given in Section 6.2.9.

The numerical algorithm to solve the EPRT would start by first initializing the intensities at each of the grid points $\left(x_i, z_j, \theta_m, \phi_n\right)$ to the initial condition, as well as by initializing the unknown $(\text{integral})_{i,j}$ at each of the grid points (x_i, z_j) to the corresponding initial value as described in the section on initial condition. Also, at the interface, the gray-body temperatures will also be initialized. Then at each time-step, starting from the appropriate boundary equilibrium intensities and after assuming all unknown variables at the current time-step to be equal to their corresponding values in the previous time-step, each of the previous four equations are solved to obtain better estimates of the intensities inside the domain. Subsequently, a better value of the integral is evaluated from these new estimates of the intensities at each spatial grid point. This whole procedure is to be repeated for each discrete wavenumber and for each of the four phonon branches. Once this is done, the equivalent equilibrium temperature is calculated at each grid point inside the domain. Also, the gray-body temperatures $T_{gb,Si}$ and $T_{gb,Di}$ at each point along the interface are calculated from the interface condition. Using these temperatures, the unknown intensities at the interface are updated. This whole procedure now constitutes a single iteration. Several iterations are performed until a converged solution is obtained for the current time-step. The criterion for convergence is that the maximum absolute difference everywhere inside the domain, between the equivalent equilibrium temperature calculated at the current iteration and that at the previous iteration should become smaller than a preassigned small positive number. Mathematically, $\max\left|(T_{i,j}^{\mathrm{P}})^h - (T_{i,j}^{\mathrm{P}})^{h-1}\right| < \varepsilon$, where h is the iteration index and $\varepsilon = 10^{-6}$. Once a converged solution is obtained, the algorithm moves on to the next time-step. Once a converged solution is obtained for the current time-step, the transient heat flux at each point may be calculated by the formulas,

$$q_x''(x,z,t) = \int_0^{k_{\max}} \int_0^{2\pi} \int_0^{\pi} \left(2I_{\mathrm{TA,k}} + I_{\mathrm{LA,k}} + 2I_{\mathrm{TO,k}} + I_{\mathrm{LO,k}}\right)\cos\theta\sin\theta d\theta\, d\phi\, dk \tag{6.165}$$

$$q_z''(x,z,t) = \int_0^{k_{\max}} \int_0^{2\pi} \int_0^{\pi} \left(2I_{\mathrm{TA,k}} + I_{\mathrm{LA,k}} + 2I_{\mathrm{TO,k}} + I_{\mathrm{LO,k}}\right)\sin\phi\sin^2\theta d\theta\, d\phi\, dk \tag{6.166}$$

6.3.2.5 *Findings of Laser Short-Pulse Heating of Aluminum-Diamond Thin Films Pair*

Fig. 6.23 shows dimensionless equivalent equilibrium temperature along the dimensionless width of the films pair for different heating periods. The x-axis location is at the middle line (midline) as shown in Fig. 6.22. Temperature jump occurs at the silicon film front face, where the temperature is 301K, which is more pronounced in the early heating period. This is associated with the emitted phonons from the front surface and the reflected phonons from the interface reaching at the front surface of the silicon film. In this case, emitted phonons do not undergo substantial scattering and the amount of reflected phonons arriving at the front face is not significant in the early heating period. This, in turn, causes a lesser amount of scattering of reflected phonons in this region. As the time progresses, equivalent equilibrium temperature in the silicon film and temperature jump at the interface become small. In this case, emitted and reflected phonons undergo scattering in the film while causing increased

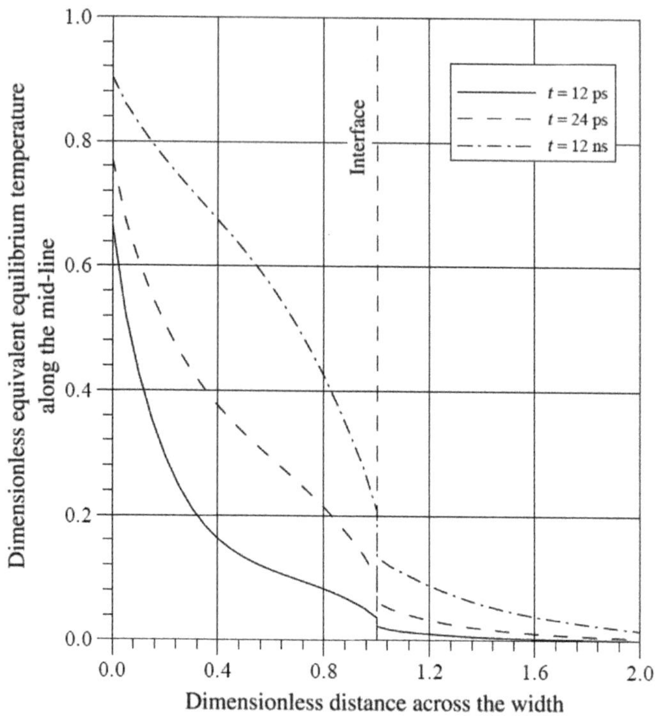

■ **FIGURE 6.23** Equivalent equilibrium temperature along the width of the films pair for different heating periods. *x*-axis location is midline ($x/L = 0.5$) [6].

equivalent equilibrium temperature. However, at the interface of the films pair, temperature jump occurs due to the presence of thermal boundary resistance. Consequently, the amount of phonons that are transported across the interface becomes less, giving rise to attainment of low equivalent equilibrium temperature at the diamond interface. As time progresses, equivalent equilibrium temperature increases at the diamond interface and in the diamond film.

Fig. 6.24 shows dimensionless equivalent equilibrium temperature along the thickness of the silicon and diamond films for two heating periods and at two z-axis locations. Note that thickness extends along the x axis (Fig. 6.22) and high temperature (301K) boundary conditions are located at the edges of the silicon film. Equivalent equilibrium temperature attains low values at mid-thickness, where $x/L_x = 0.5$, which is more pronounced during the early heating period. In this case, emitted phonons from the high-temperature edges do not undergo a large number of scatterings in the silicon film during the early heating periods. Temperature at locations $x/L_x = 0$ and $x/L_x = 1.0$ attains high values

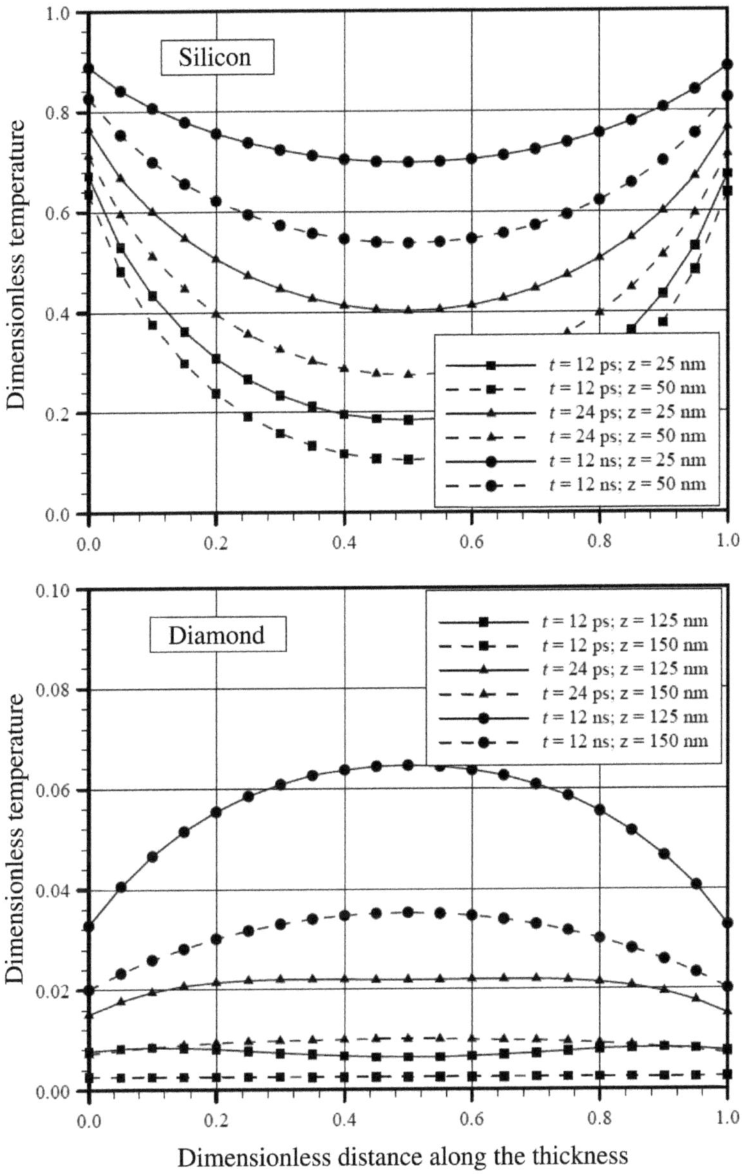

■ **FIGURE 6.24** Equivalent equilibrium temperature along the thickness of the films pair at different *z*-axis locations for different heating periods [6].

because of high-temperature edges, which are located at these edges. Temperature jump at high-temperature edges is large in the early heating periods and as the heating period progresses, temperature jump becomes small.

In the case of diamond film, the temperature is kept at 300K at all edges of the film, except the interface where thermal boundary resistance is present. Temperature increases in the diamond film with progressing time due to transmitted phonons from the interface. Equivalent equilibrium temperature attains significantly lower values in the diamond film as compared to that corresponding to the silicon film. This is attributed to the presence of thermal boundary resistance; not all phonons at all frequencies emitted from the silicon film can be transmitted to the diamond film due to frequency mismatch at the interface of the films. Temperature attains high values at location $x/L_x = 0.5$ unlike the silicon film. This is attributed to the presence of low-temperature edges (300K) around the diamond film. Consequently, low-temperature edges act like a phonon sink, lowering equivalent equilibrium temperature in the region of the edges. In addition, phonons emitted from the low-temperature edges do not contribute notably to equivalent equilibrium temperature increase in edge regions.

Fig. 6.25 shows dimensionless equivalent equilibrium temperature along the thickness for the silicon and diamond films at the interface for the silicon and diamond films for different heating periods. Equivalent equilibrium temperature remains low at late heating period ($t = 12$ ns) at location $x/L_x = 0.5$ due to high-temperature edges of the silicon film. However, emitted phonons from the high-temperature edges suffer a high degree of scattering while increasing equivalent equilibrium temperature in the edge regions. Phonon transport from the high-temperature edges results in parabolic-like equivalent equilibrium temperature distribution along the thickness at the interface. Since lesser amounts of phonons are transmitted from the silicon film to the diamond film in the early heating period, equivalent equilibrium temperature remains low in the early heating period, which is more pronounced at the edges of the diamond film ($x/L_x = 0$ and $x/L_x = 1$). The low-temperature edges are responsible for the attainment of low equivalent equilibrium temperature in this region. The amount of transmitted phonons from silicon film to diamond film remain low even for long heating periods due to the presence of thermal boundary resistance. This results in considerably lower temperature in the diamond film interface than that of the silicon film interface.

Fig. 6.26 shows dimensionless equivalent equilibrium temperature along the film width (z axis) for the heating period $t = 24$ ns. The frequency-independent results of equivalent equilibrium temperature are also shown for the comparison reason. It should be noted that the x-axis location is the half thickness ($x/L_x = 0.5$). Equivalent equilibrium temperature decays sharply in the region of high-temperature edge toward the silicon interface while temperature decay is gradual for the frequency-independent solution. This is attributed to the contribution of ballistic phonons to the

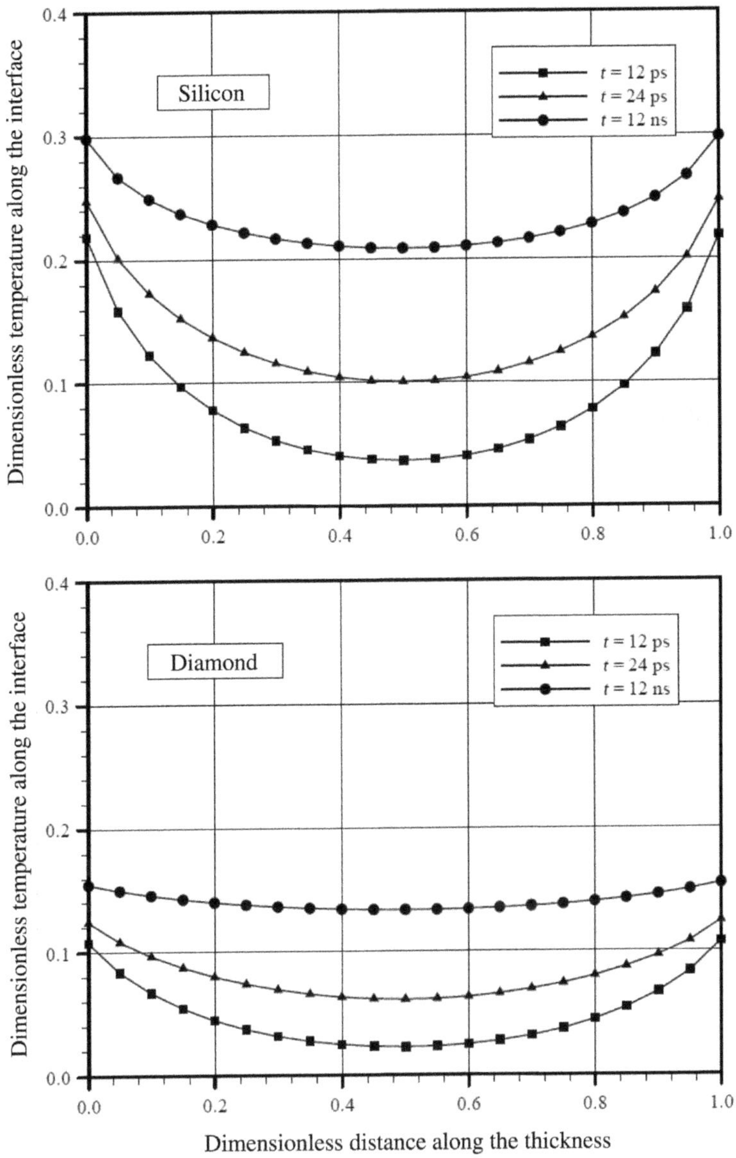

■ **FIGURE 6.25** Equivalent equilibrium temperature along the thickness of the films pair at the interface for different heating periods [6].

energy transport in the silicon film. In this case, ballistic phonons travel with less scattering within the silicon film while suppressing equivalent equilibrium temperature increase. In addition, the presence of ballistic phonons enhances temperature jump at the high-temperature edge of the

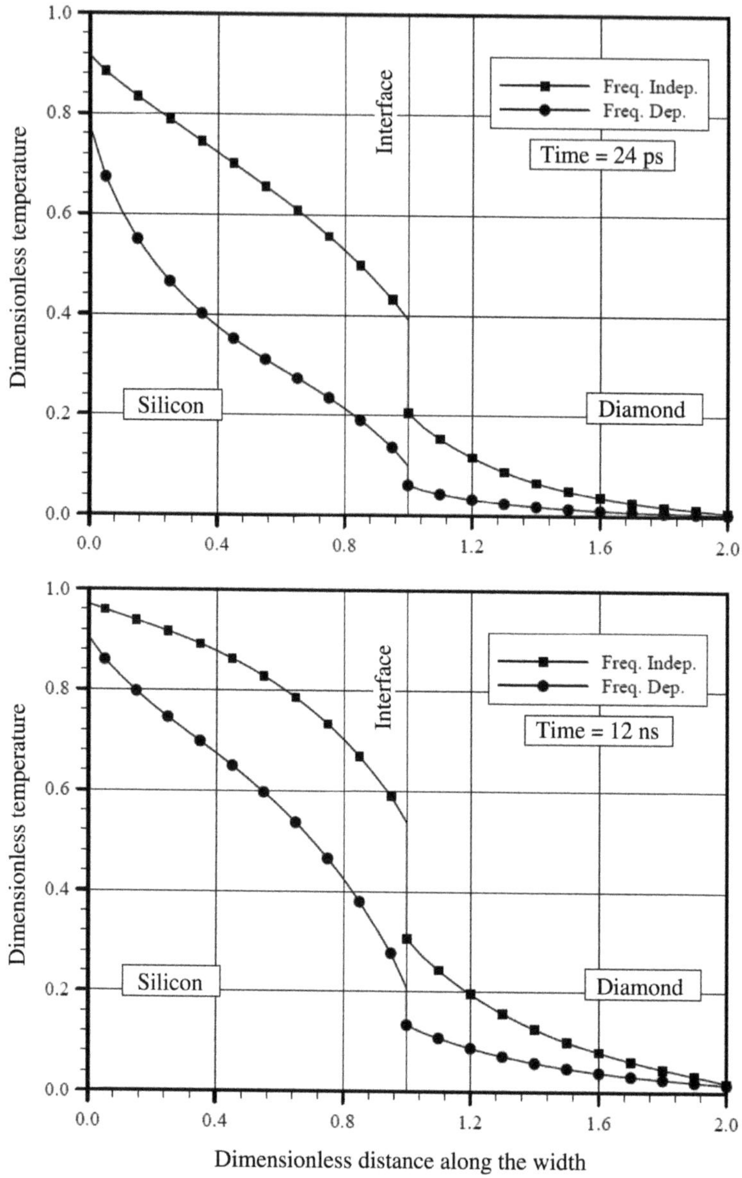

■ **FIGURE 6.26** Equivalent equilibrium temperature along the width of the films pair due to frequency-dependent and frequency-independent solutions of Boltzmann equation for two heating periods [6].

silicon film. Since equivalent equilibrium temperature is low at the silicon film interface for a frequency-dependent solution, temperature jump at the interface is lower for the frequency-dependent solution than for that of an independent solution. The presence of thermal boundary resistance at the interface lowers equivalent equilibrium temperature at the diamond interface. The ballistic phonons in the diamond film modifies temperature distribution in the diamond film; frequency-dependent results in different temperature distribution than the frequency-independent solution in the diamond film. Although equivalent equilibrium temperature is low at the diamond interface, temperature jump at the diamond interface is larger than that of the high-temperature edge of the silicon film. This is associated with the emitted phonons from the diamond interface toward the bulk of the diamond film. Note that all the edges of the diamond film are considered to be at 300K, except the interface. Consequently, low-temperature edges act like phonon sinks while lowering equivalent equilibrium temperature in the diamond film.

6.4 CLOSING REMARKS

Phonon transport in metallic and dielectric films with one-dimensional and two-dimensional configurations is considered. Frequency-dependent and -independent solutions of phonon radiative transport are introduced. The cross-plane transport of phonons is also incorporated via introducing the thermal boundary resistance at the interface of the thin films. The applications are extended to include the dielectric and metallic thin films. However, in metallic thin films, thermal separation takes place between the electron and lattice subsystems. Thermal communication between these subsystems determines the thermal states in the thin metallic film. The thermal communication of these systems is formulated through electron–phonon coupling; the radiative phonon transport equation is written for the individual system and they are, then, coupled through the electron–phonon coupling factor. In the case of the cross-plane transport, thermophysical properties of each film differ. This, in turn, causes thermal resistance across the interface between the films pair. This situation is formulated via thermal boundary resistance, which is introduced by considering acoustic and diffusive mismatch models. To assess the phonon intensity distribution in the film, equivalent equilibrium temperature is introduced. Equivalent equilibrium temperature is defined anywhere in the thin film, which is the analog of the usual thermodynamic temperature defined in the diffusive limit in any medium. It represents the average energy of all phonons around a local point and is equivalent to the equilibrium temperature of phonons when they redistribute adiabatically to an equilibrium state. Case studies are introduced to discuss the application of phonon radiative transport equation in different film configurations including metallic, dielectric, and films pairs. In addition, the frequency-dependent formulation and solution as well as frequency-independent solution of the radiative transport equation are incorporated in the case studies.

REFERENCES

[1] S.B. Mansoor, B.S. Yilbas, Phonon transport characteristics in a thin silicon film, J. Comput. Theor. Transp. 44 (3) (2015) 154–174.

[2] B.S.S. Yilbas, S. Bin Mansoor, Phonon and electron transport in aluminum thin film: Influence of film thickness on electron and lattice temperatures, Phys. B 407 (24) (2012) 4643–4648.

[3] B.S. Yilbas, S. Bin Mansoor, Influence of heat source size on phonon transport in thin silicon film, Transp. Theory. Stat. Phys. 42 (2–3) (2014) 65–84.

[4] S. Bin Mansoor, B. Sami Yilbas, Laser short-pulse heating of an aluminum thin film: energy transfer in electron and lattice sub-systems, Phys. B 470–471 (2015) 82–91.

[5] S.B. Mansoor, B.S. Yilbas, S. Bin Mansoor, B.S. Yilbas, Thermal transport across a thin film composite due to laser short-pulse heating, J. Non-Equilib. Thermodyn. 40 (2) (2015) 103–120.

[6] B.S. Yilbas, S.B. Mansoor, Phonon transport in two-dimensional silicon–diamond film pair, J. Thermophys. Heat Transfer 27 (3) (2013) 465–473.

[7] B.N. Brockhouse, Lattice vibrations in silicon and germanium, Phys. Rev. Lett. 2 (6) (1959) 256–258.

[8] A. Ward, D.A. Broido, Intrinsic phonon relaxation times from first-principles studies of the thermal conductivities of Si and Ge, Phys. Rev. B 81 (8) (2010) 85205.

[9] J.L. Moll, C.-Y. Duh, Studies of Microplasmas and High-Field Effects in Silicon, Stanford Electronics Labs, California, 1964.

[10] F.P. Incropera, D.P. DeWitt, Fundamentals of Heat and Mass Transfer, fifth ed., Wiley, Hoboken, 2001.

[11] Y. Ju, M.-T. Hung, T. Usui, Nanoscale heat conduction across metal-dielectric interfaces, J. Heat Transfer 128 (9) (2006) 919–925.

[12] R. Stedman, L. Almqvist, G. Nilsson, Phonon-frequency distributions and heat capacities of aluminum and lead, Phys. Rev. 162 (3) (1967) 549–557.

[13] A. Ward, D. Broido, D. Stewart, G. Deinzer, Ab initio theory of the lattice thermal conductivity in diamond, Phys. Rev. B 80 (12) (2009) 125203.

[14] B.S. Yilbas, A.Y. Al-Dweik, S. Bin Mansoor, Non-equilibrium energy transport in a thin metallic film: analytical solution for radiative transport equation, Phys. B 454 (2014) 15–22.

[15] G.L. Eesley, Generation of nonequilibrium electron and lattice temperatures in copper by picosecond laser pulses, Phys. Rev. B 33 (1986) 2144.

[16] B.S. Yilbas, Improved formulation of electron kinetic theory approach for laser ultra-short-pulse heating, Int. J. Heat Mass Transfer 49 (13–14) (2006) 2227–2238.

[17] E. Hecht, Optics, fourth ed., Pearson Education Ltd, Essex, UK, 2014.

Chapter 7

Thermal Boundary Resistance for Cross-Plane Transport and the Presence of Minute Vacuum Gap at Interface

7.1 INTRODUCTION

The cross-plane phonon transport across thin films gives rise to a thermal boundary resistance at the interface of the films pair, which plays an important role in the transport characteristics while modifying the phonon intensity distribution in the films. The mismatch of the physical properties in between the films pair brings up challenges in modeling the thermal energy transport across the thin films. The thermal energy transport depends on the frequency of phonons in the thin films and the frequency mismatch of phonons across the interface alters the phonon intensity at the interfaces of both films. The correct formulation of the problem is necessary for the sound assessment of energy transport across the film interfaces. The phonon intensity distribution in thin films depends mainly on the physical properties of the films such as geometric size of the films, group velocity, dispersion relations, and the size of the heat source disturbance applied at the film edges. In this case, reducing film size affects the phonon transport across the film, which is associated with the influence of the quasiballistic phonons on the thermal resistance of the film. The similar affects are observed when the heat source size, causing thermal disturbance across the film, is reduced. On the other hand, introducing the small vacuum gap in between the films pair changes the interfacial conditions and alters the thermal resistance for the phonon transport across the interface. Depending on the gap size, some of the long wavelength phonons can jump across the small gap while contributing to the thermal transport across the film.

In the Casimir limit, thermal transport takes place through thermal radiation across the gap in between the film pair. However, the near-field radiation also contributes to the energy transport across the vacuum gap

Heat Transport in Micro- and Nanoscale Thin Films.
DOI: http://dx.doi.org/10.1016/B978-0-323-42979-5.00007-4

in between the films. Therefore, thermal transport across the gap in between the film pair is governed by the ballistic phonons and near-field thermal radiation. In addition, the gap in between the films pair acts like a thermal boundary resistance in between the films while influencing the thermal properties of the films pair. Consequently, correct formulation and predictions of the phonon transport characteristics become essential for the cross-plane thermal energy transfer with presence of the thermal boundary resistance due to mismatch between the physical properties of the film and the minute vacuum gap at the interface. In this chapter, application of phonon transport across the film pairs with different thermophysical properties and transport characteristics across the minute vacuum gap are presented in light of previous study [1–4].

7.2 APPLICATION OF THERMAL BOUNDARY RESISTANCE ACROSS THIN FILMS

A cross-plane phonon transport between silicon and a diamond thin films pair incorporating the thermal boundary resistance at the interface is presented herein. The cut-off mismatch model (CMM) is introduced to formulate the thermal boundary resistance at the interface of the films pair when the film edges are thermally disturbed in line with the previous study [1]. The transmittance and reflection of phonon intensities are incorporated in the analysis while satisfying the energy balance at the interface. The analysis is extended to include the effect of the heat source size, at the film edges, on the phonon transport characteristics in the film. The Boltzmann transport equation (BTE) is used to formulate the frequency-dependent transient phonon transport in a two-dimensional silicon-diamond films pair.

The transient and frequency dependent of the BTE is considered within the two-dimensional silicon-diamond films pair. The boundary conditions used are those of constant temperature (Fig. 7.1). At the interface of the silicon and diamond films, appropriate interface conditions including the reflectance and the transmittance of phonons are imposed, which satisfy the energy balance at the interface.

Fig. 7.1 shows the films pair and the coordinate system used in the analysis. The transient, two-dimensional, frequency-dependent BTE in the relaxation-time approximation and excluding the effects of external forces can be written as

$$\frac{1}{v_\omega}\frac{\partial f_\omega}{\partial t} + \cos\theta\frac{\partial f_\omega}{\partial x} + \sin\theta\sin\phi\frac{\partial f_\omega}{\partial z} = \frac{f_\omega^o - f_\omega}{\Lambda_\omega} \tag{7.1}$$

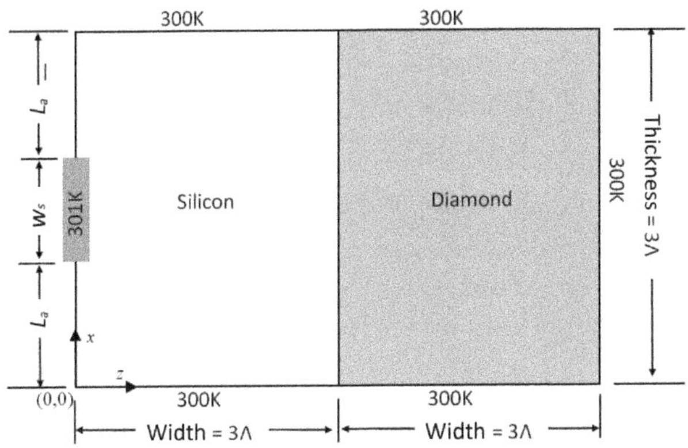

■ **FIGURE 7.1** Schematic view of silicon-diamond films pair, heat source at the high temperature edge of the silicon film, and the coordinate system. Λ (=33.8 nm) is the mean free path of silicon corresponding to the gray medium [1].

In Eq. (7.1), the Cartesian coordinates have been used for the spatial variables, whereas spherical polar coordinates have been used for the velocity components. Here, ω is frequency (in subscript it shows the frequency dependency), f_ω is frequency-dependent probability distribution function in phase space, f_ω^o is frequency-dependent equilibrium probability distribution function in phase space, t is the time, x is the distance in the x direction, z is the distance in the z direction, θ is polar angle, ϕ is azimuthal angle, τ_ω is the frequency-dependent relaxation time, v_ω is the frequency-dependent phonon group velocity, and $\Lambda_\omega = v_\omega \tau_\omega$ is the frequency-dependent mean free path.

The BTE (Eq. 7.1) may be transformed into the equation of phonon radiative transfer (EPRT), which can be given as

$$\frac{1}{v_\omega}\frac{\partial I_\omega}{\partial t} + \cos\theta\frac{\partial I_\omega}{\partial x} + \sin\theta\sin\phi\frac{\partial I_\omega}{\partial z} = \frac{\frac{1}{4\pi}\int_0^{2\pi}\int_0^{\pi} I_\omega \sin\theta\, d\theta\, d\phi - I_\omega}{\Lambda_\omega} \tag{7.2}$$

$$t>0,\quad 0<x<L_x,\quad 0<z<L_z,\quad 0<\theta<\pi,\quad 0<\phi<2\pi$$

In Eq. (7.2), the phonon intensity I_ω is a function of x, z, t, θ, ϕ, and ω; i.e., $I_\omega = I_\omega(x, z, t, \theta, \phi)$.

The details of EPRT are given in Section 4.4.4.

The dispersion relations represent various polarizations and modes of vibrations of the silicon [5] and diamond [6]. Moreover, there are three different polarizations and two different modes of vibrations of the phonons. Waves can propagate in a dielectric crystal in three different polarizations—one longitudinal and two transverse. The dispersion relations presented in this study are for the [001] direction in silicon and the diamond, which in reality is a symmetry direction. Along this direction the two transverse waves coincide and therefore we have only one dispersion curve for the two transverse waves for each mode of vibration. The other curve is for the longitudinal wave. There are two modes of longitudinal and transverse waves, acoustic and optical. Hence we have now six different types of waves, but since the transverse waves are identical, we end up with four different types of waves and hence four different curves in the dispersion relation. The EPRT given in Eq. (7.2) has to be applied to each of these waves individually. The resulting equation is as follows.

$$\frac{1}{v_{ij,\omega}}\frac{\partial I_{ij,\omega}}{\partial t} + \cos\theta\frac{\partial I_{ij,\omega}}{\partial x} + \sin\theta\sin\phi\frac{\partial I_{ij,\omega}}{\partial z} = \frac{\frac{1}{4\pi}\int_0^{2\pi}\int_0^{\pi} I_{ij,\omega}\sin\theta\, d\theta\, d\phi - I_{ij,\omega}}{\Lambda_{ij,\omega}} \tag{7.3}$$

In Eq. (7.3), $i=T$ represents transverse mode and $i=L$ stands for longitudinal mode. In addition, $j=A$ corresponds to acoustic and $j=O$ stands for optical phonons. This description is valid for both silicon and diamond films without any modification. Note that the dispersion relation and hence the thermophysical properties of silicon and diamond are different.

The heat flux vector is defined at all points inside the thin film, which is given as

$$q''_{x,\mathrm{Si/Di}}(x,z,t) = \int_0^{\omega_{\max}}\int_0^{2\pi}\int_0^{\pi}(2I_{\mathrm{Si/Di,TA},\omega} + I_{\mathrm{Si/Di,LA},\omega} + 2I_{\mathrm{Si/Di,TO},\omega} + I_{\mathrm{Si/Di,LO},\omega}) \times \cos\theta\sin\theta\, d\theta\, d\phi\, d\omega \tag{7.4}$$

and

$$q''_{z,\mathrm{Si/Di}}(x,z,t) = \int_0^{\omega_{\max}}\int_0^{2\pi}\int_0^{\pi}(2I_{\mathrm{Si/Di,TA},\omega} + I_{\mathrm{Si/Di,LA},\omega} + 2I_{\mathrm{Si/Di,TO},\omega} + I_{\mathrm{Si/Di,LO},\omega}) \times \sin\phi\sin^2\theta\, d\theta\, d\phi\, d\omega \tag{7.5}$$

The heat flux vector can be written as

$$\mathbf{q}''_{\mathrm{Si/Di}} = q''_{x,\mathrm{Si/Di}}\mathbf{i} + q''_{z,\mathrm{Si/Di}}\mathbf{k} \tag{7.6}$$

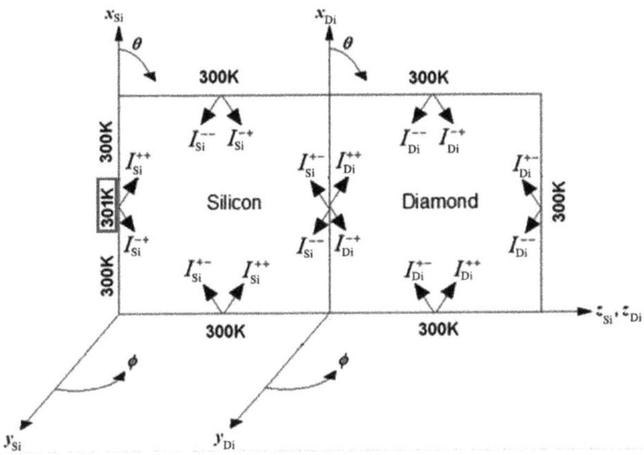

■ **FIGURE 7.2** Schematic view of emitted, transmitted, and reflected phonon intensities at the boundaries [1].

7.2.1 Initial and Boundary Conditions

There are four boundaries introduced for the silicon and diamond film pair: bottom, top, left, and right (Fig. 7.2). In addition, the right boundary for the silicon and the left boundary for diamond make a single interface. At this interface an appropriate condition has to be defined, which is derived from the energy balance condition at the interface. The phonon intensity specification at the boundaries, the interface condition, and the initial condition should be prescribed to complete the problem specification. The geometry of the physical domain and the boundary conditions do not change with time. Initially, the silicon and diamond films are assumed to be at a uniform temperature of 300K and after that for all times the following boundary conditions are imposed. The left boundary of silicon are assumed to be at 301K while bottom and top boundaries of the silicon layer are assumed to be at a constant temperature of 300K. The top, right, and bottom boundaries of the diamond layer are assumed to be at a constant temperature at 300K. At the interface, the temperatures in the silicon and diamond layers are variable and they are determined by means of the interface condition. The initial, boundary, and interface conditions are described next.

7.2.1.1 *Initial Conditions*

The initial condition for all intensities in a Si-Di film pair is given as

$$I_{Si/Di,ij,\omega}(x_{Si/Di}, z_{Si/Di}, t=0, \theta, \phi) = \frac{C_{Si/Di,ij,\omega} v_{Si/Di,ij,\omega} T_{initial}}{4\pi}$$
$$0 < x_{Si/Di} < L_{x,Si/Di},\ 0 < z_{Si/Di} < L_{z,Si/Di},\ 0 < \theta < \pi,\ 0 < \phi < 2\pi \quad (7.7)$$
$$\text{and } T_{initial} = 300\text{K}$$

7.2.1.2 *Boundary Conditions*

In an acoustically thin medium, the photon mean free path (Λ) is comparable or larger than the physical thickness of the medium. Mean free path (Λ) is a statistical quantity, which can be related to $p = \exp(-x/\Lambda)$, where p is the probability of a wave, which would travel a distance x without undergoing scattering [7]. The probability of a phonon, which is emerging from one boundary and being nonscattered until reaching the other boundary, is greater than zero. This situation is well described previously such that for an acoustically thin regime, the boundary thermalizes the phonons, which can be expressed in terms of blackbody phonon radiation [8].

The phonon intensities at the boundaries of Si and Di film are given as

$$I_{\mathrm{Si/Di},ij,\omega}(x_{\mathrm{Si/Di}}, z_{\mathrm{Si/Di}}, t, \theta, \phi) = \frac{C_{\mathrm{Si/Di},ij,\omega} v_{\mathrm{Si/Di},ij,\omega} T_{\mathrm{boundary}}}{4\pi} \tag{7.8}$$

where $t > 0$. Details of $x_{\mathrm{Si/Di}}, z_{\mathrm{Si/Di}}, t, \theta, \phi$, and T_{boundary} are given in Table 7.1.

In Eqs. (7.7) and (7.8), $i = T$ represents transverse mode and $i = L$ stands for longitudinal mode. In addition, $j = A$ corresponds to acoustic and $j = O$ stands for optical phonons. $C_{\mathrm{Si/Di},ij,\omega}$ is the frequency-dependent volumetric heat capacity at the nominal temperature of $T_{\mathrm{o}} = 300\mathrm{K}$, and $v_{\mathrm{Si/Di},ij,\omega}$ is the frequency-dependent phonon group velocity for the ij branch for the [001] direction in silicon/diamond. $C_{\mathrm{Si/Di},ij,\omega}$ can calculated from [9]

$$C_{\mathrm{Si/Di},ij,\omega} = k_{\mathrm{B}} \left(\frac{\hbar\omega}{k_{\mathrm{B}}T}\right)^2 \frac{e^{\hbar\omega/k_{\mathrm{B}}T}}{(e^{\hbar\omega/k_{\mathrm{B}}T} - 1)^2} D_{\mathrm{Si/Di},ij,\omega}(\omega) \tag{7.9}$$

where k_{B} is the Boltzmann's constant.

Table 7.1 Conditions Used to Calculate the Intensities at Boundaries

Boundary	x	z	θ	ϕ	T_{boundary} (K)
Silicon Film					
Top	$L_{x,\mathrm{Si}}$	$0 - L_{z,\mathrm{Si}}$	$\pi/2 - \pi$	$0 - 2\pi$	300
Bottom	0	$0 - L_{z,\mathrm{Si}}$	$0 - \pi/2$	$0 - 2\pi$	300
Left	$L_a - L_a + w_s$	0	$0 - \pi$	$0 - \pi$	301
	$0 - L_a$ and $L_a - L_a + w_s$	0	$0 - \pi$	$0 - \pi$	300
Diamond Film					
Top	$L_{x,\mathrm{Di}}$	$0 - L_{z,\mathrm{Di}}$	$\pi/2 - \pi$	$0 - 2\pi$	300
Bottom	0	$0 - L_{z,\mathrm{Di}}$	$0 - \pi/2$	$0 - 2\pi$	300
Right	$0 - L_{x,\mathrm{Di}}$	0	$0 - \pi$	$\pi - 2\pi$	300

7.2.1.3 *Interface Condition*

In order to obtain detailed information about the phonon intensities at interface, it is necessary to specify the incoming/outgoing phonon intensities distribution for interface. Fig. 7.2 shows that in the silicon film, I_{Si}^{++} and I_{Si}^{-+} are the phonon intensities that are coming toward the interface, and the phonon intensities that are going from the interface are I_{Si}^{--} and I_{Si}^{+-}. In the diamond film, the I_{Di}^{--} and I_{Di}^{+-} are the phonon intensities that are coming toward the interface, and phonon intensities that are going from the interface are I_{Di}^{++} and I_{Di}^{-+}. The heat flux that is coming at the interface due to the incoming phonon intensities from both films is either reflected or transmitted. Boundary condition at the edges of silicon and diamond film give us the values of the incoming intensities (I_{Si}^{++}, I_{Si}^{-+}, I_{Di}^{--}, and I_{Di}^{+-}) toward the interface. In order to determine the unknown phonon intensities distribution at the interface—the outgoing phonon intensities from the interface (I_{Si}^{--}, I_{Si}^{+-}, I_{Di}^{++}, and I_{Di}^{-+})—the following models were adopted to model the thermal boundary resistance across the interface: diffuse mismatch model (DMM) and CMM.

7.2.1.3.1 **Diffuse Mismatch Model**

In the DMM, the intensities are integrated over all frequencies and then the energy balance is applied at the interface [10]. In this case, energy transfer is considered independent of frequencies [10]. Therefore, all phonons at the silicon film interface can possibly transfer their energy to the diamond film interface irrespective of the phonon frequency mismatch.

Heat fluxes at the interface can be written as

$$q''^{+/-}_{z,\mathrm{Si}}(x_{\mathrm{Si}}, z_{\mathrm{Si}} = L_{z,\mathrm{Si}}, t) = \int_0^{\omega_{\max,Si}} \left(\int_0^{\pi} \int_0^{\pi/2} I_{\mathrm{Si},\omega}^{++/+-} \sin\phi \sin^2\theta \, d\theta \, d\phi + \int_0^{\pi} \int_{\pi/2}^{\pi} I_{\mathrm{Si},\omega}^{-+/--} \sin\phi \sin^2\theta \, d\theta \, d\phi \right) d\omega \tag{7.10}$$

and

$$q''^{+/-}_{z,\mathrm{Di}}(x_{\mathrm{Di}}, z_{\mathrm{Di}} = 0, t) = \int_0^{\omega_{\max,Di}} \left(\int_0^{\pi} \int_0^{\pi/2} I_{\mathrm{Di},\omega}^{++/+-} \sin\phi \sin^2\theta \, d\theta \, d\phi + \int_0^{\pi} \int_{\pi/2}^{\pi} I_{\mathrm{Di},\omega}^{-+/--} \sin\phi \sin^2\theta \, d\theta \, d\phi \right) d\omega \tag{7.11}$$

where $I^{++/--/+-/-+}_{\text{Si/Di},\omega}$ is sum of all frequency branches and modes. For example $I^{++}_{\text{Si},\omega}$ is

$$I^{++}_{\text{Si},\omega} = I^{++}_{\text{LA,Si},\omega} + 2I^{++}_{\text{TA,Si},\omega} + I^{++}_{\text{LO,Si},\omega} + 2I^{++}_{\text{TO,Si},\omega} \tag{7.12}$$

Using the energy balance at the interface, we can write the outgoing heat fluxes at the interface in terms of incoming heat fluxes:

$$q''^{-/+}_{z,\text{Si/Di}} = T_{n/m} q''^{-/+}_{z,\text{Di/Si}} - R_{m/n} q''^{+/-}_{z,\text{Si/Di}} \tag{7.13}$$

where T is the transmittance and R is used for the reflectance at the interface. Here, the subscript m indicates from silicon to diamond (Si-Di) and n indicates from diamond to silicon (Di-Si). Considering the diffuse scattering and utilizing Eq. (7.13) with Eqs. (7.10) and (7.11), we can get the following relations:

$$I^{--}_{\text{Si},\omega} = I^{+-}_{\text{Si},\omega} = \frac{1}{\pi\omega_{\text{max,Si}}}\left[-T_{\text{Di-Si}} q''^{-}_{z,\text{Di}} + R_{\text{Si-Di}} q''^{+}_{z,\text{Si}}\right] \tag{7.14}$$

and

$$I^{++}_{\text{Di},\omega} = I^{-+}_{\text{Di},\omega} = \frac{1}{\pi\omega_{\text{max,Di}}}\left[T_{\text{Si-Di}} q''^{+}_{z,\text{Si}} - R_{\text{Di-Si}} q''^{-}_{z,\text{Di}}\right] \tag{7.15}$$

The outgoing intensities $I^{--/+-}_{\text{Si},\omega}$ and $I^{++/-+}_{\text{Di},\omega}$, which can be calculated from Eqs. (7.14) and (7.15), are distributed among different frequency modes and branches using

$$I^{--/+-/++/-+}_{\text{Si/Di},ij,\omega} = \frac{C_{\text{Si/Di},ij,\omega} v_{\text{Si/Di},ij,\omega} \omega_{\text{max,Si/Di}}}{\langle C_{\text{Si/Di}} v_{\text{Si/Di}} \rangle} I^{--/+-/++/-+}_{\text{Si/Di},\omega} \tag{7.16}$$

where

$$\langle C_{\text{Si/Di}} v_{\text{Si/Di}} \rangle = \int_0^{\omega_{\text{max,Si/Di}}} \begin{pmatrix} C_{\text{Si/Di,LA}}(\omega) v_{\text{Si/Di,LA}}(\omega) + 2C_{\text{Si/Di,TA}}(\omega) v_{\text{Si/Di,TA}}(\omega) \\ + C_{\text{Si/Di,LO}}(\omega) v_{\text{Si/Di,LO}}(\omega) \\ + 2C_{\text{Si/Di,LO}}(\omega) v_{\text{Si/Di,LO}}(\omega) \end{pmatrix} d\omega \tag{7.17}$$

where, in Eqs. (7.16) and (7.17), $i = T$ represents transverse mode and $i = L$ stands for longitudinal mode. In addition, $j = A$ corresponds to acoustic and $j = O$ stands for optical phonons.

The transmittance and reflectance are calculated based on the energy conservation at the interface. In this case, energy conservation gives

$$T_{\text{Si-Di}} \langle C_{\text{Si}} v_{\text{Si}} \rangle = T_{\text{Di-Si}} \langle C_{\text{Di}} v_{\text{Di}} \rangle \tag{7.18}$$

The DMM gives

$$T_{\text{Si-Di}} = 1 - T_{\text{Di-Si}} \tag{7.19}$$

Simultaneous solutions of the previous two equations yield the transmissions

$$T_{\text{Si-Di/Di-Si}} = \frac{\langle C_{\text{Di/Si}} v_{\text{Di/Si}} \rangle}{\langle C_{\text{Di}} v_{\text{Di}} \rangle + \langle C_{\text{Si}} v_{\text{Si}} \rangle} \tag{7.20}$$

The reflectance can be written as

$$R_{\text{Di-Si/Si-Di}} = 1 - T_{\text{Di-Si/Si-Di}} \tag{7.21}$$

7.2.1.3.2 **Cut-Off Mismatch Model**

In the CMM, the mismatch due to the frequency and branch is considered. Fig. 7.3 shows the optical branch frequencies of silicon not matching with the optical branch frequencies of diamond. Therefore longitudinal and transverse optical branch phonons of both layers are reflected from the interface; i.e., transmittance becomes zero. Similarly, above the cut-off frequency, the transmittance remains zero for longitudinal and transverse acoustic branches of phonons in the diamond thin film, as shown in Fig. 7.3. We can, therefore, apply the energy balance at each frequency.

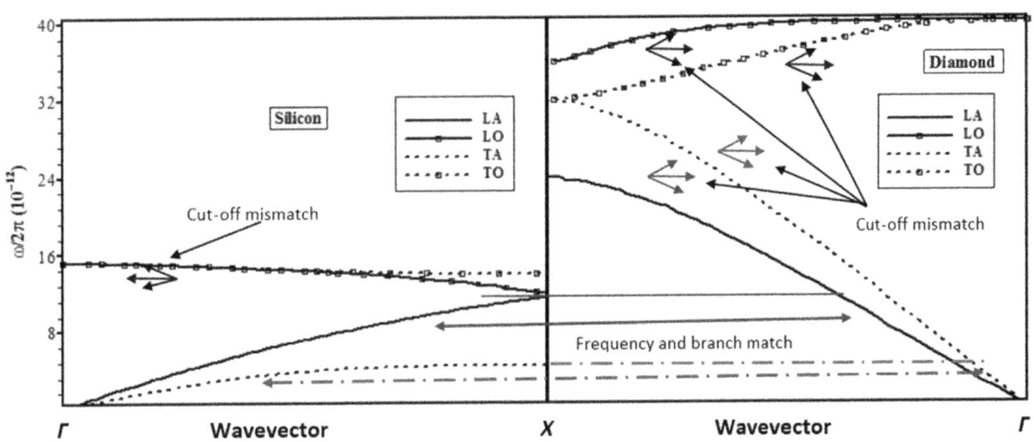

FIGURE 7.3 Dispersion relations for silicon and diamond films and wavevectors for cut-off mismatch. Phonon dispersion of silicon is on the left and diamond is on the right. Note that the scattering is permitted between phonons of the same polarization and frequency [1].

Heat fluxes at the interface can be written as

$$q''^{+/-}_{z,\mathrm{Si},ij,\omega}(x_{\mathrm{Si}}, z_{\mathrm{Si}} = L_{z,\mathrm{Si}}, t) = \int_0^{\pi}\int_0^{\pi/2} I^{++/+-}_{\mathrm{Si},ij,\omega} \sin\phi \sin^2\theta \, d\theta \, d\phi + \int_0^{\pi}\int_{\pi/2}^{\pi} I^{-+/--}_{\mathrm{Si},ij,\omega} \sin\phi \sin^2\theta \, d\theta \, d\phi \tag{7.22}$$

and

$$q''^{+/-}_{z,\mathrm{Di},ij,\omega}(x_{\mathrm{Di}}, z_{\mathrm{Di}} = 0, t) = \int_0^{\pi}\int_0^{\pi/2} I^{++/+-}_{\mathrm{Di},ij,\omega} \sin\phi \sin^2\theta \, d\theta \, d\phi + \int_0^{\pi}\int_{\pi/2}^{\pi} I^{-+/--}_{\mathrm{Di},ij,\omega} \sin\phi \sin^2\theta \, d\theta \, d\phi \tag{7.23}$$

where subscript ω is used to indicate that these equations are written for each frequency of the phonons, $i = T$ represents transverse mode and $i = L$ stands for longitudinal mode. In addition, $j = A$ corresponds to acoustic and $j = O$ stands for optical phonons.

Using the energy balance at the interface, we can write the outgoing heat fluxes at the interface for each frequency in terms of incoming heat fluxes:

$$q''^{-/+}_{z,\mathrm{Si/Di},ij,\omega} = T_{n/m}\, q''^{-/+}_{z,\mathrm{Di/Si},ij,\omega} - R_{m/n}\, q''^{+/-}_{z,\mathrm{Si/Di},ij,\omega} \tag{7.24}$$

where T is the transmittance and R is used for reflectance. The subscript Si-Di indicates from silicon to diamond and Di-Si indicates from diamond to silicon.

Considering the diffuse scattering and utilizing Eq. (7.24) with Eqs. (7.22) and (7.23) we can get the following relations:

$$I^{--}_{\mathrm{Si},ij,\omega} = I^{+-}_{\mathrm{Si},ij,\omega} = \frac{1}{\pi}\left[-T_{\text{Di-Si},ij,\omega}\, q''^{-}_{z,ij,\mathrm{Di},\omega} + R_{\text{Si-Di},ij,\omega}\, q''^{+}_{z,\mathrm{Si},ij,\omega}\right] \tag{7.25}$$

and

$$I^{++}_{\mathrm{Di},ij,\omega} = I^{-+}_{\mathrm{Di},ij,\omega} = \frac{1}{\pi}\left[T_{\text{Si-Di},ij,\omega}\, q''^{+}_{z,\mathrm{Si},ij,\omega} - R_{\text{Di-Si},ij,\omega}\, q''^{-}_{z,\mathrm{Di},ij,\omega}\right] \tag{7.26}$$

The transmittance and reflectance can be calculated from the following:

$$T_{\text{Si-Di/Di-Si},ij,\omega} = \frac{C_{\mathrm{Di/Si},ij,\omega} v_{\mathrm{Di/Si},ij,\omega}}{C_{\mathrm{Di},ij,\omega} v_{\mathrm{Di},ij,\omega} + C_{\mathrm{Si},ij,\omega} v_{\mathrm{Si},ij,\omega}} \tag{7.27}$$

The reflectance can be determined from

$$R_{\text{Di-Si/Si-Di},ij,\omega} = 1 - T_{\text{Di-Si/Si-Di},ij,\omega} \tag{7.28}$$

7.2.1.4 ***Thermal Boundary Resistance***

The thermal boundary resistance at the interface can be defined as

$$\mathrm{TBR} = \frac{\Delta T_{\text{eq,Si-Di}}}{q''_{z,\text{interface}}} \tag{7.29}$$

In Eq. (7.29), $\Delta T_{\text{eq,Si-Di}} = T_{\text{eq,Si}} - T_{\text{eq,Di}}$ is the equivalent equilibrium temperature difference at the interface for silicon and diamond thin films and $q''_{z,\text{interface}}$ is the heat flux at the interface. Equivalent equilibrium temperature is the analog of the usual thermodynamic temperature defined in the diffusive limit in any medium. It represents the average energy of all phonons around a local point and it is equivalent to the equilibrium temperature of phonons when they redistribute adiabatically to an equilibrium state. It is defined as

$$T_{\text{eq,Si/Di}}(x,z,t) = \frac{\displaystyle\int_0^{\omega_{\max}}\int_0^{2\pi}\int_0^{\pi}\begin{pmatrix} 2I_{\text{Si/Di,TA},\omega} + I_{\text{Si/Di,LA},\omega} \\ + 2I_{\text{Si/Di,TO},\omega} + I_{\text{Si/Di,LO},\omega}\end{pmatrix}\sin\theta\, d\theta\, d\phi\, d\omega}{\displaystyle\int_0^{\omega_{\max}}\begin{pmatrix} 2C_{\text{Si/Di,TA},\omega}v_{\text{TA},\omega} + C_{\text{Si/Di,LA},\omega}v_{\text{LA},\omega} \\ + 2C_{\text{Si/Di,TO},\omega}v_{\text{TO},\omega} + C_{\text{Si/Di,LO},\omega}v_{\text{Si/Di,LO},\omega}\end{pmatrix} d\omega} \tag{7.30}$$

and

$$q''_{z,\text{interface}} = q''_{z,\text{Si}}(x_{\text{Si}}, z_{\text{Si}} = L_{z,\text{Si}}, t) = q''_{z,\text{Di}}(x_{\text{Di}}, z_{\text{Di}} = 0, t) \tag{7.31}$$

In order to assess the phonon transport across thin film pairs incorporating the cut-off mismatch and diffusive mismatch models, a ratio of equilibrium intensities by two models are introduced for the longitudinal acoustic and transverse acoustic waves, which are

$$\eta_{\text{LA/TA}} = \frac{\left(I^0_{\text{Si/Di,LA/TA}}\right)_{\text{CMM}}}{\left(I^0_{\text{Si/Di,LA/TA}}\right)_{\text{DMM}}} \tag{7.32}$$

where $I^0_{\text{Si/Di, LA}}$ represents the equilibrium intensities for the longitudinal acoustic branch of phonons for silicon/diamond and $I^0_{\text{Si/Di,TA}}$ represents the equilibrium intensities for the transverse acoustic branch of phonons for silicon/diamond. Equilibrium intensities are calculated from

$$I^0_{\text{Si/Di,LA/TA}} = \int_0^{\omega_{\max}} I^0_{\text{Si/Di,LA/TA},\omega}\, d\omega \tag{7.33}$$

The expression for the effective thermal conductivity of the film in terms of the wavenumber (k) can be written as [6]

$$k(T) = \frac{1}{3}\sum_{q=1}^{2}\sum_{p=1}^{3}\int_0^{k_{\max}} C_k v_k^2 \tau_k\, dk \tag{7.34}$$

where p is the polarization, q is the modes of phonons, the τ_k is the wavenumber (frequency)-dependent relaxation time, v_k is the wavenumber (frequency)-dependent phonon group velocity, and ($\Lambda_k = v_k \tau_k$) is the wavenumber (frequency)-dependent mean free path.

7.2.2 Numerical Solution

The EPRT in the silicon and diamond films with the associated initial and boundary conditions are solved numerically using the finite-difference method. These are integrodifferential equations and can be solved using an iterative technique, given in detail in Section 6.2.3.

The solution of EPRT starts by first initializing the intensities at each of the grid points to the initial condition, as well as by initializing the unknown $(\text{integral})_{i,j}$ at each of the grid points (x_i, z_j) to the corresponding initial value as described in the section on initial condition. Then at each time-step, starting from the appropriate boundary equilibrium intensities, and after assuming all unknown variables at the current time-step to be equal to their corresponding values in the previous time-step, each of the previous four equations are solved to obtain better estimates of the intensities inside the domain. Subsequently, a better value of the integral is evaluated from these new estimates of the intensities at each spatial grid point. This whole procedure is to be repeated for each discrete frequency and for each of the four phonon branches. This procedure is incorporated to solve both thin films, silicon and diamond. Once this is done, the unknown intensities at the silicon and diamond interface are calculated from the interface conditions. This whole procedure now constitutes a single iteration. Several iterations are performed until a converged solution is obtained for the current time-step. The criterion for convergence is that the maximum absolute difference everywhere inside the domain, between the equilibrium intensities, $(\text{integral})_{i,j}$, calculated at the current iteration and that at the previous iteration should become smaller than a preassigned small positive number. Mathematically, $\max|((\text{integral})_{i,j})^h - ((\text{integral})_{i,j})^{h-1}| < \varepsilon$, where h is the iteration index and $\varepsilon = 10^{-6}$. Once a converged solution is obtained, the algorithm moves on to the next time-step. Table 7.2 gives the properties of silicon and diamond.

Table 7.2 Properties of Silicon and Diamond Films Used in the Simulations

	Phonon Speed, (ms^{-1})	Average Mean Free Path, (nm)
Silicon	8430	33.8
Diamond	12,288	447

7.2.3 Findings of Thermal Boundary Resistance Application

Phonon transport across a silicon-diamond films pair is considered for the case of temperature disturbance at the film edges. Thermal boundary resistance at the films pair interface is incorporated through the diffusive mismatch model and the CMM, which is based on the phonon frequencies mismatch between the films for different polarizations. The influence of heat source size located at the silicon film edge on the transport characteristics are analyzed for each branch of the phonon polarizations.

Fig. 7.4 shows temporal behavior of thermal boundary resistance obtained from the diffusive mismatch and CMMs for various heat source sizes. The location is the mid-height of the film thickness (3Λ) at the films pair interface. Note that the heat source size is normalized with the mean free path of silicon based on the gray medium ($\Lambda = 33.8$ nm). Thermal boundary resistance attains high values in the early heating period because of low intensity of the emitted phonons from the high temperature (301K) edge of the silicon film (Fig. 7.1); phonons emitted from the high temperature edge do not contribute to film resistance in the early heating period and phonons with large wavelengths can reach the interface during this period. Therefore, phonon intensity remains low at the interface and the rate of transmitted phonons to the diamond film across the interface becomes low. This results in high thermal boundary resistance for both diffusive mismatch and cut-off models. As time progresses, intensity of the emitted phonons reaching the interface increases and the amount of phonons transmitted to the diamond film increases, which in turn results in quasisteady behavior of the thermal boundary resistance with progressing time. Since frequency-dependent phonon transport across the interface is considered in the CMM, phonons having the same frequency range in the silicon and the diamond films can possibly transmit to the diamond film across the interface. However, some of the phonons reflected from the interface, due to frequency mismatch, contribute to thermal boundary resistance enhancement at the interface. Note that energy balance is satisfied among the emitted, transmitted, and reflected phonons at the interface. Since the diffusive mismatch model relies on the gray body temperature based on the equilibrium phonon intensity [11], phonons with all frequencies can possibly transmit to the diamond film across the interface, which suppresses the thermal boundary resistance enhancement at the interface. This causes thermal boundary resistance reduction at the interface for the diffusive mismatch model. Consequently, when comparing thermal boundary resistance due to the diffusive mismatch model and the CMM, the diffusive mismatch model

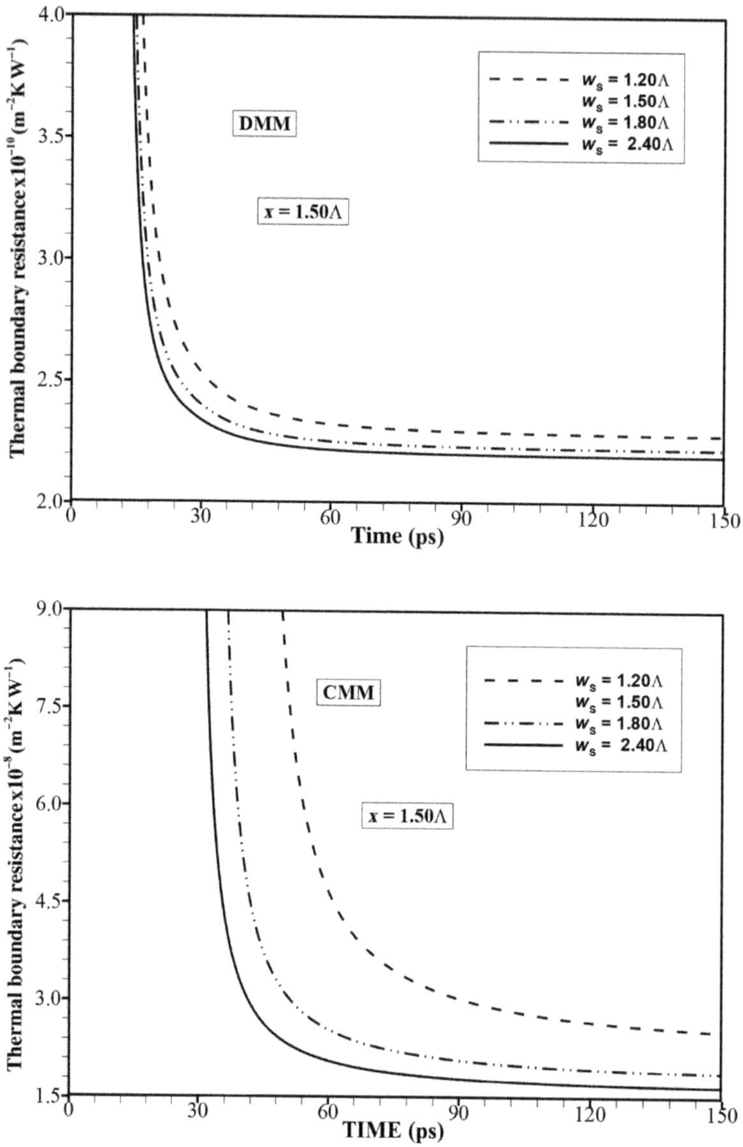

FIGURE 7.4 Thermal boundary resistance predicted from the diffusive mismatch (DMM) and cut-off (CMM) mismatch models for various heat source sizes [1].

results in relatively lower thermal boundary resistance than that corresponding to the CMM at the interface, which is particularly true for the late heating periods $t \geq 80$ ps. The effect of heat source size on the thermal boundary resistance is significant, which is more pronounced for the CMM.

This is attributed to the influence of ballistic and quasiballistic phonons on the phonon transport in the silicon film. Since reducing the heat source size generates an effect similar to the film size effect on the phonon transport [12], phonon equilibrium intensity at the film interface reduces with reducing heat source size. This in turn increases the thermal boundary resistance at the interface. In addition, all the ballistic and quasiballistic phonons reaching the interface do not transmit across the interface because of the frequency mismatch, which is considered in the CMM. Therefore, some of the quasiballistic and ballistic phonons contribute to the thermal boundary resistance enhancement at the interface. Since the intensity of phonons emitted from the high temperature edge of the silicon film and reaching the silicon interface becomes low for a small heat source size, the equilibrium phonon intensity incorporating all the phonon frequencies remains low at the silicon interface. This causes a thermal boundary resistance increase at the interface for a diffusive mismatch model with decreasing heat source size.

Fig. 7.5 shows the phonon intensity ratio $\eta_{LA} = \frac{(I^0_{LA})_{CMM}}{(I^0_{LA})_{DMM}}$, where $\left(I^0_{LA}\right)_{CMM}$ is the longitudinal acoustic phonon equilibrium intensity obtained from the CMM, and $(I^0_{LA})_{DMM}$ is the longitudinal acoustic phonon equilibrium intensity obtained from the diffusive mismatch model of equilibrium phonon intensities due to the CMM over the diffusive mismatch model along the width of the silicon for various heating periods. Note that I^o is the equilibrium intensity corresponding to a blackbody intensity [8], which is frequency dependent. The edges of the silicon film are considered to be at 300K except at the high temperature edge, which is 301K, to resemble the temperature disturbance initiating the phonon transport in the films pair; intensity ratio along the mid-thickness is shown in Fig. 7.1. In general, the intensity ratio increases with increasing film width and the heat source size located at the high temperature (301K) edge of the silicon film. Since the ratio is greater than 1, phonon equilibrium intensity corresponding to the longitudinal acoustic branch attains higher values at the interface for the cut-off mismatch consideration for phonon transmittance across the interface as compared to its counterpart corresponding to the diffusive mismatch model. This is attributed to the formulation of the thermal boundary resistance at the interface; the large wavelengths of the longitudinal phonons do not contribute to the film resistance and those phonons reaching the silicon film edge at the interface is less likely transmit to the diamond film due to the mismatch between silicon and diamond dispersion relations at high frequencies.

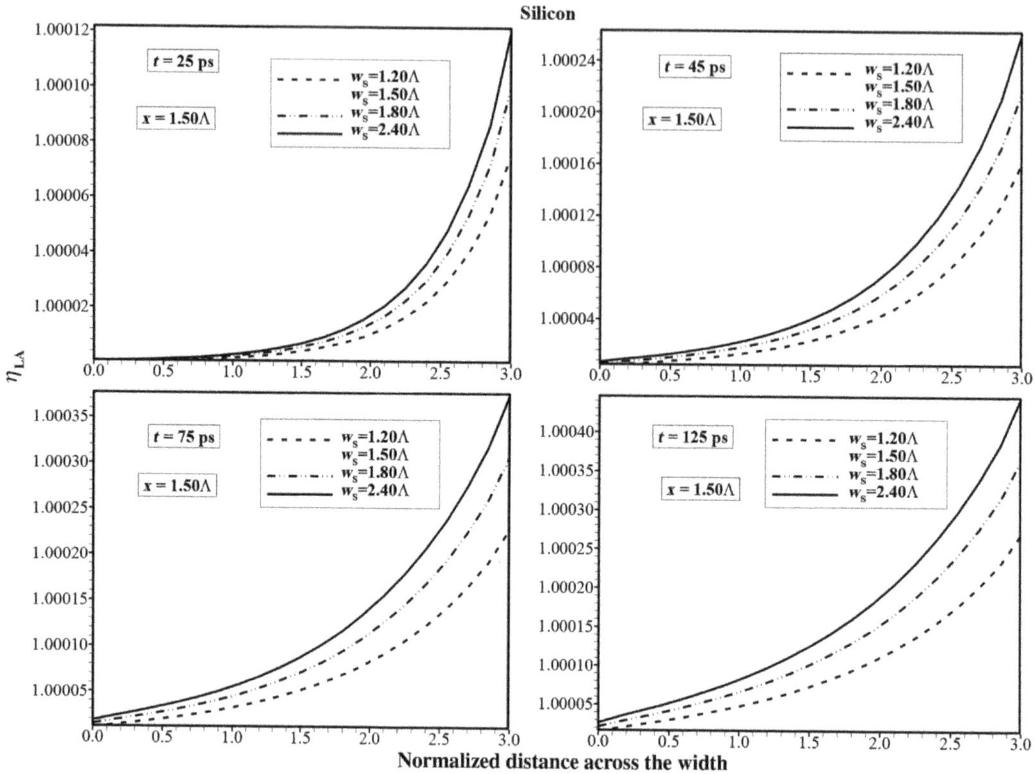

■ **FIGURE 7.5** Phonon intensity ratio $\left(\eta_{LA}=\frac{(I^0_{LA})_{CMM}}{(I^0_{LA})_{DMM}}\right)$ along the silicon film width for different heat source sizes and heating durations. $(I^0_{LA})_{CMM}$ is the longitudinal acoustic phonon equilibrium intensity obtained from the CMM, and $(I^0_{LA})_{DMM}$ is the longitudinal acoustic phonon equilibrium intensity obtained from the diffusive mismatch model [1].

The phonon thermalization in the diffusive mismatch model results in a fixed uniform temperature, and the film edges at the interface are in thermodynamic equilibrium at their respective temperatures [13]. Consequently, phonon equilibrium intensity attains low value across the interface due to a fixed value of uniform temperature at the interface. However, phonons having the wavelength matching across the silicon and diamond films can possibly transfer across the interface while reducing the longitudinal acoustic phonon intensities at the interface of the silicon film. This reduction is not substantial to reduce equilibrium phonon intensity at the silicon interface. Moreover, increasing heat source size at the high temperature (301K) silicon edge enhances the intensity ratio, indicating the effect of cut-off mismatch consideration on the phonon intensities at the silicon film interface, which may be more realistic as compared to that of the diffusive mismatch consideration. This is because of the fact that the thermal boundary resistance measured in the previous study [14] agrees well with the predictions of the present study for the CMM (thermal boundary resistance reported in the early study [14] is $(1.7 \pm 0.5) \times 10^{-8}$ $m^2K\ W^{-1}$ while the predicted quasisteady state values after incorporating the CMM is 1.62×10^{-8} $m^2K\ W^{-1}$ and diffusive mismatch model is 2.18×10^{-10} $m^2K\ W^{-1}$). Note that increasing heat source size enhances the amount of longitudinal ballistic phonons emitted from the high temperature edge, which does not contribute to temperature increase in the film including the film edges. These phonons having high frequencies cannot transfer to the diamond film due to the high frequency mismatching of phonons between silicon and diamond films. Therefore, longitudinal phonon intensity attains high values at the silicon interface while increasing the equilibrium intensity ratio. Increasing heating duration increases the intensity ratio slightly.

Fig. 7.6 shows the longitudinal equilibrium phonon intensity ratio $\left(\eta_{LA} = \frac{(I^0_{LA})_{CMM}}{(I^0_{LA})_{DMM}}\right)$ in the diamond film for different values of the heat source size located at the high temperature (301K) edge of the silicon film for various heating periods. Since the transmitted longitudinal phonons contribute to the phonon intensity in the diamond film, the CMM results in a reduced amount of phonon transmittance to the diamond film, due to phonon frequency mismatch between the silicon and diamond films, than that of the diffusive mismatch model. Therefore, the equilibrium intensity ratio remains less than 1 in the diamond film. As the film width increases, longitudinal phonon intensity ratio increases. This behavior is attributed to the phonon scattering in the film due to film resistance while increasing the equilibrium intensity ratio toward 1 along the film width.

Fig. 7.7 shows the phonon intensity ratio $\eta_{TA} = \frac{(I^0_{TA})_{CMM}}{(I^0_{TA})_{DMM}}$, where $(I^0_{TA})_{CMM}$ is transverse acoustic phonon equilibrium intensity obtained from the cutoff mismatch, and $(I^0_{TA})_{DMM}$ is transverse acoustic phonon equilibrium intensity obtained from the diffusive mismatch model) due to the transverse acoustic branch for different heat source sizes located at the high temperature (301K) edge of the silicon film and various heating durations. As similar to those shown in Fig. 7.5, the intensity ratio remains higher than 1 in the silicon film, which is particularly true in the region close to the silicon film edge at the interface. In this case, quasiballistic phonons do not contribute to the phonon intensity increase in the film, except toward the film edge at the interface. In addition, phonons having high frequencies do not transmit to the diamond film because of the frequency mismatch between the silicon and the diamond films according

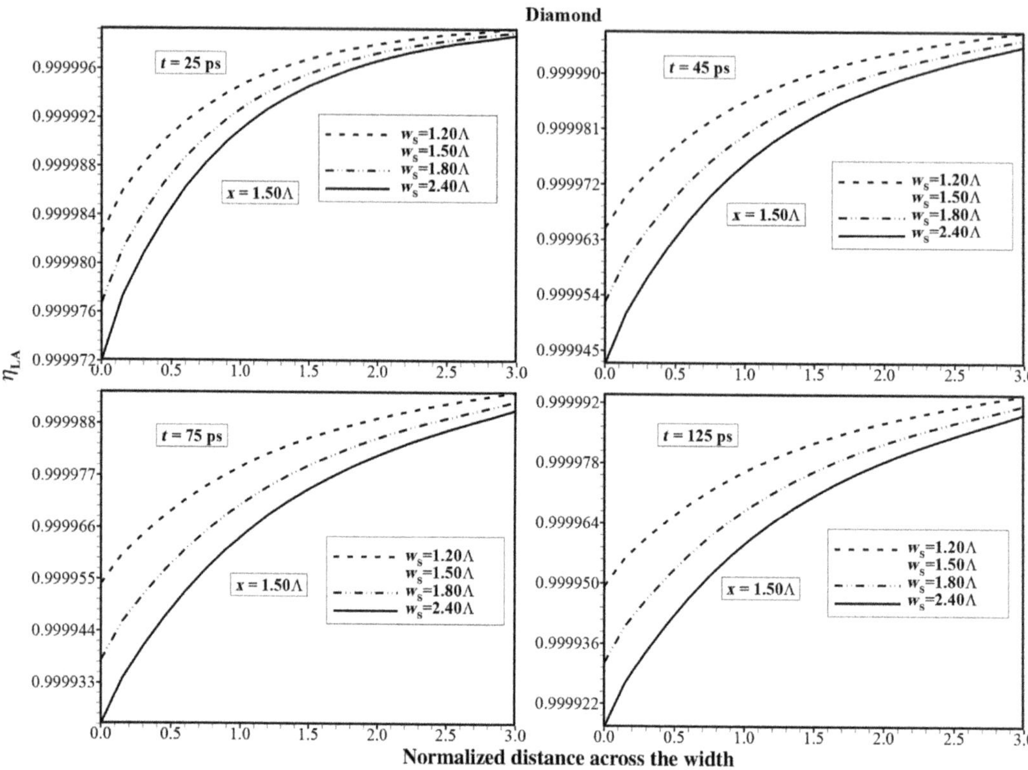

■ **FIGURE 7.6** Phonon intensity ratio $\left(\eta_{LA} = \frac{(I^0_{LA})_{CMM}}{(I^0_{LA})_{DMM}}\right)$ along the diamond film width for different heat source sizes and heating durations. $(I^0_{LA})_{CMM}$ is the longitudinal acoustic phonon equilibrium intensity obtained from the CMM, and $(I^0_{LA})_{DMM}$ is the longitudinal acoustic phonon equilibrium intensity obtained from the diffusive mismatch model [1].

to the dispersion relations of both films. Consequently, the transverse equilibrium phonon intensity increases at the silicon film edge for the CMM. Increasing heat source size causes slight changes in the intensity ratio; decreasing heat source size lowers the intensity ratio. This behavior is attributed to the emitted transverse phonons from the high-temperature (301K) silicon film edge, which increases with increasing heat source size while contributing to the phonon intensity at the silicon interface. As the heating period progresses, the intensity ratio due to different heat source size slightly changes the trend, which is more pronounced for the heating period of 125 ps. This is associated with the phonon intensity increase at the silicon interface with processing time.

Fig. 7.8 shows the phonon intensity ratio $\left(\eta_{TA} = \frac{(I^0_{TA})_{CMM}}{(I^0_{TA})_{DMM}}\right)$ due to transverse acoustic phonon intensities in the diamond film. The intensity ratio attains less than 1 for all the heat sources sizes and heating durations.

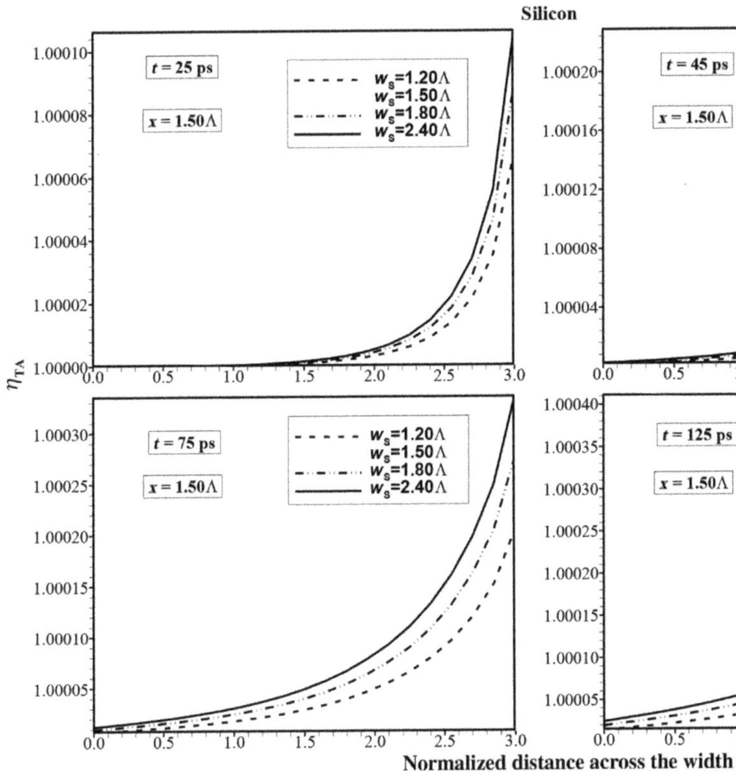

■ **FIGURE 7.7** Transverse equilibrium phonon intensity ratio $\left(\eta_{TA} = \frac{(I^{o}_{TA})_{CMM}}{(I^{o}_{TA})_{DMM}}\right)$ in the silicon film for different values of the heat source size located at the high-temperature (301K) edge of the silicon film for various heating periods [1].

The phonon intensity ratio increases toward the low temperature edge (300K) of the diamond film. In this case, phonon intensity distributions at the film edge due to cut-off mismatch and diffusive mismatch models become the same. This behavior is related to low temperature at the film edge, which acts like a phonon sink reducing the phonon intensity significantly in this region. Therefore, the phonon intensities due to cut-off and diffusive mismatch models become almost similar.

7.3 APPLICATIONS IN CROSS-PLANE TRANSPORT ACROSS DIELECTRIC THIN FILMS WITH MINUTE GAP

The phonon transport across the silicon thin films pair incorporating the minute gap in between the film pair is examined. A numerical solution of the frequency-dependent BTE is introduced to obtain phonon intensity

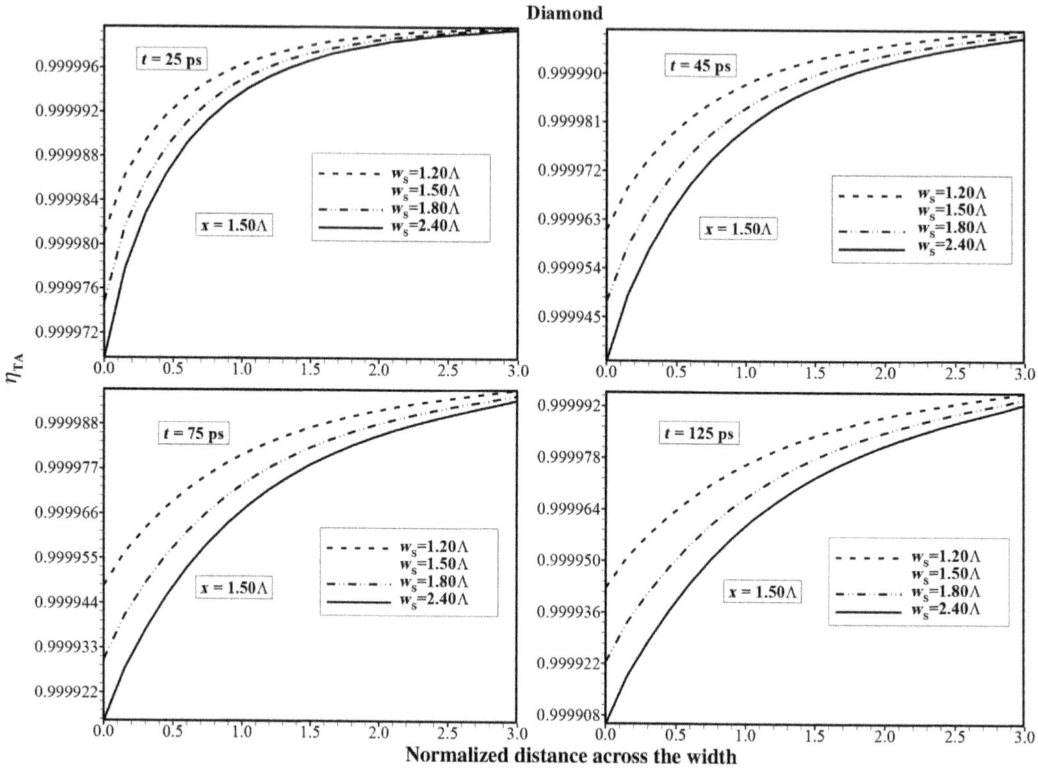

■ **FIGURE 7.8** Transverse equilibrium phonon intensity ratio $\left(\eta_{TA} = \frac{(I^0_{TA})_{CMM}}{(I^0_{TA})_{DMM}}\right)$ in the diamond film for different values of the heat source size located at the high temperature (301K) edge of the silicon film for various heating periods [1].

distribution in the silicon films pair. Since the gap allocated in between the films pair is considered to be smaller than the mean free path of silicon (Casimir limit), the radiation boundary condition together with transmittance and reflectance of phonons satisfying the energy equation are incorporated across the gap in line with the previous study [2]. The effect of the gap size on the phonon transport characteristics is include here.

The physical domain considered consists of the two silicon thin films (pair) separated by a considerably small vacuum gap (Casimir limit) as shown in Fig. 7.9. In the analysis, the transient, wavenumber (frequency)-dependent BTE is incorporated to obtain the phonon intensity distribution in the silicon thin films pair. The boundary conditions used are those of constant temperature (Fig. 7.9) resembling temperature disturbance across the films pair at the exterior edges. At the interface of the thin silicon films pair appropriate interface conditions are imposed, satisfying the energy balance.

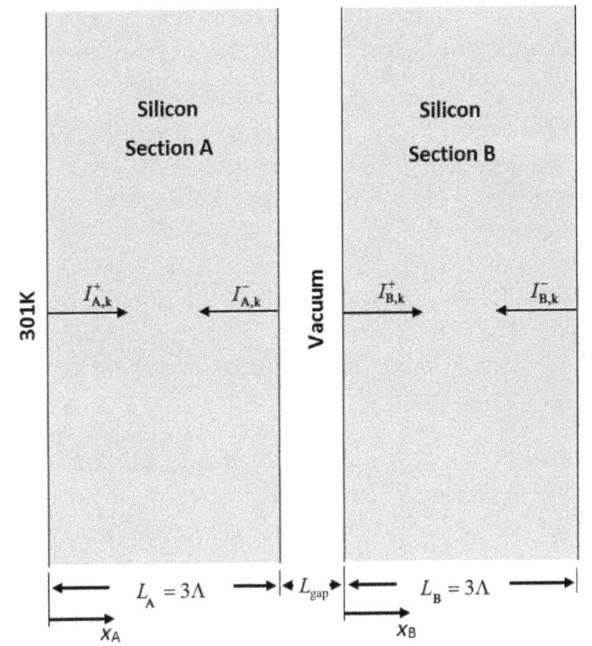

■ **FIGURE 7.9** Schematic view of thin silicon films pair consisting of two silicon films separated by a small gap. Λ is the mean free path of silicon corresponding to a gray medium [2].

It is assumed that the film thickness and the gap size are significantly smaller than the width and the thickness of the film; one-dimensional consideration of phonon transport is assumed. The relaxation-time approximation to the EPRT reduces the complications faced during the numerical solution of the BTE with the appropriate boundary and interface conditions. Therefore, a transient form of the EPRT is used to obtained phonon transport characteristics.

The one-dimensional EPRT is given as [8]

$$\frac{1}{v_k}\frac{\partial I_k}{\partial t} + \mu \frac{\partial I_k}{\partial x} = \frac{\frac{1}{2}\int_{-1}^{1} I_k d\mu - I_k}{\Lambda_k} \quad t > 0, \quad 0 < x < L, \quad -1 < \mu < 1 \quad (7.35)$$

In Eq. (7.35), the phonon intensity I_k is a function of x, μ, t and k:

$$I_k = I_k(x, \mu, t) \quad (7.36)$$

where $\mu = \cos\theta$, θ is the polar angle, v_k is the wavenumber (frequency)-dependent phonon group velocity, x is distance in spatial coordinate, t is time, Λ_k is the wavenumber (frequency)-dependent phonon mean free path, L is the length of the physical domain, and k is the wavenumber.

In order to account for the frequency-dependent formulation of the phonon transport, the isotropic assumptions of dispersion relations are considered. In this case, dispersion relations of silicon in the [001] direction (as shown in Fig. 7.10) are introduced. This provides three polarizations and two different modes of vibrations of phonons in the silicon film. Since two transverse waves are degenerate, only one dispersion curve is sufficient for two transverse waves; the other is for the longitudinal wave for silicon. In the thin film, two sets of longitudinal and transverse waves are present. The first set is called the acoustic waves and the second set is called the optical waves. Therefore, six different types of waves take place in the film. The transverse waves are degenerate; four different types of waves corresponding to four different curves are present in the dispersion relation. The EPRT, given in Eq. (7.36), needs to be applied for each of these waves individually:

$$\frac{1}{v_{ij,k}}\frac{\partial I_{ij,k}}{\partial t} + \mu\frac{\partial I_{ij,k}}{\partial x} = \frac{\frac{1}{2}\int_{-1}^{1} I_{ij,k}d\mu - I_{ij,k}}{\Lambda_{ij,k}} \tag{7.37}$$

In Eq. (7.37), $i = T$ represents transverse mode and $i = L$ stands for longitudinal mode. In addition, $j = A$ corresponds to acoustic and $j = O$ stands for optical phonons. Eq. (7.37) can be equally applicable for both silicon thin films; i.e., Section A and Section B in Fig. 7.9.

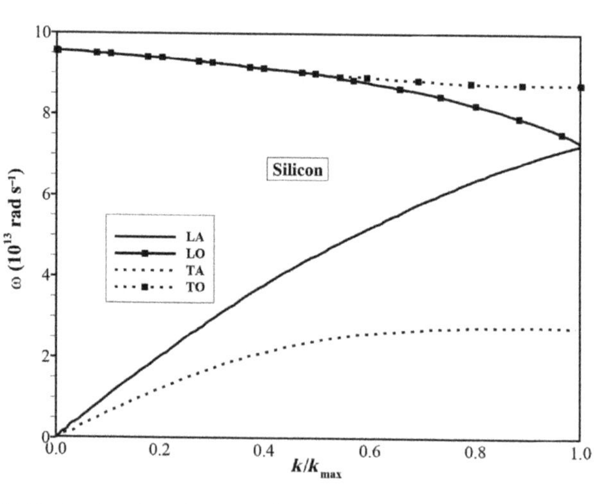

FIGURE 7.10 Dispersion relation for silicon [15].

The group velocity is calculated from the following:

$$v_{ij,\mathrm{k}} = \frac{\partial \omega_{ij,\mathrm{k}}}{\partial k_{ij,\mathrm{k}}} \tag{7.38}$$

and the volumetric heat capacity is calculated from [9]

$$C_{ij,\mathrm{k}} = k_{\mathrm{B}} \left(\frac{\hbar\omega}{k_{\mathrm{B}}T} \right)^2 \frac{e^{\hbar\omega/k_{\mathrm{B}}T}}{\left(e^{\hbar\omega/k_{\mathrm{B}}T} - 1\right)^2} D_{ij,\mathrm{k}}(\omega) \tag{7.39}$$

where k_{B} is the Boltzmann's constant, $\hbar(=h/2\pi)$ is the reduced Plank's constant, T is the temperature, $D_{ij}(\omega)$ is the density of states, which is number of phonon per unit volume per unit energy interval, which can be written as [15]

$$D_{ij,\mathrm{k}}(\omega) = \frac{k_{ij,\mathrm{k}}^2}{2\pi^2} \frac{dk_{ij,\mathrm{k}}}{d\omega_{ij,\mathrm{k}}} \tag{7.40}$$

The mean free path ($\Lambda_{ij,\mathrm{k}}$) is the product of the phonon group velocity ($v_{ij,\mathrm{k}}$) and relaxation time ($\tau_{ij,\mathrm{k}}$). Details for the estimation of relaxation time can be found in Section 6.2.1.

7.3.1 Initial, Boundary, and Interface Conditions

Since the radiative phonon transport equation is transient, the initial condition should be introduced. In addition, the solution domain consists of two silicon thin films, and the boundary conditions for each film need to be incorporated accordingly. In order to satisfy the energy balance at the interface of thin films the interface conditions are also implemented.

7.3.1.1 *Section A (Fig. 7.9)*

7.3.1.1.1 Initial Condition

The initial films are considered to be isotropic and the diffuse medium is assumed initially. In this case, the radiative equilibrium over all frequencies takes place in the film initially. This yields the condition,

$$I_{\mathrm{A},ij,\mathrm{k}}(x_{\mathrm{A}}, \mu, t=0) = \frac{C_{ij,\mathrm{k}} v_{ij,\mathrm{k}} T_{\mathrm{initial}}}{4\pi} \tag{7.41}$$

7.3.1.1.2 Left Boundary Condition

The film surface at the boundary is considered to be diffusive. Therefore, the left boundary condition is

$$I_{\mathrm{A},ij,\mathrm{k}}(x_{\mathrm{A}} = 0, \mu, t) = \frac{C_{ij,\mathrm{k}} v_{ij,\mathrm{k}} T_{\mathrm{left}}}{4\pi}, \tag{7.42}$$

7.3.1.2 *Section B (Fig. 7.9)*

7.3.1.2.1 Initial Condition

After considering the diffusive medium of the film initially, the initial condition becomes

$$I_{B,ij,k}(x_B, \mu, t=0) = \frac{C_{ij,k} v_{ij,k} T_{initial}}{4\pi} \tag{7.43}$$

7.3.1.2.2 Right Boundary Condition

$$I_{B,ij,k}(x_B = 3\Lambda, \mu, t) = \frac{C_{ij,k} v_{ij,k} T_{right}}{4\pi}, \tag{7.44}$$

In Eqs. (7.41)–(7.44), $i = T$ represents transverse mode and $i = L$ stands for longitudinal mode. In addition, $j = A$ corresponds to acoustic and $j = O$ stands for optical phonons, and $C_{ij,k}$ is the wavenumber-dependent volumetric heat capacity at the nominal temperature of $T_o = 300$ K, and $v_{ij,k}$ is the wavenumber-dependent phonon group velocity for the ijth branch for the [001] direction in the silicon. Note that in order to generate a temperature disturbance across the film pairs, temperature gradient is generated. In this case, temperature is fixed at 300K at the right side of Section B, which is the right edge of the second silicon film. This in turn acts as a driving force for the phonon transport. On the other hand, setting temperature at the right boundary of the silicon film at 301K could also cause a transient temperature disturbance across the film, since initial temperature in the films pair is considered to be 300K. However, this setting can result in shorter duration of thermal communication within the films, due to the transient nature of the phonon transport, rather than the cross-plane transport across the gap. This is because both films are considered to be silicon in the films pair and there is no mismatch of dispersion relations of the films at the gap interface.

7.3.1.3 *Interface Conditions*

In order to complete the solution of the domain (Fig. 7.9), interface conditions should be incorporated. Interface conditions are obtained by applying the energy balance at the interface with consideration of radiation heat transfer due to the Casimir limit. The radiation heat transfer is incorporated between Sections A and B, which is given by

$$q_{rad,total}(t) = \sigma(T_A^4(x_{A\text{-Interface}}, t) - T_B^4(x_{B\text{-Interface}}, t)) \tag{7.45}$$

where $x_{A\text{-Interface}} = 3\Lambda$ and $x_{B\text{-Interface}} = 0$.

Energy transfer from Section A to Section B by radiation is disturbed into all the branches of each mode of phonons. This can be expressed as follows:

$$q_{\mathrm{rad},ij,\mathrm{k}}(t) = \frac{\int_0^1 I^+_{\mathrm{A},ij,\mathrm{k}}(x_{A\text{-Interface}}, \mu, t)d\mu}{I^+_{\mathrm{A,total}}(t)} q_{\mathrm{rad,total}}(t) \tag{7.46}$$

where

$$I^+_{\mathrm{A,total}}(t) = \int_0^{k_{\max}} \int_0^1 \begin{pmatrix} I^+_{\mathrm{A,LA,k}}(x_{A\text{-Interface}}, \mu, t) + I^+_{\mathrm{A,TA,k}}(x_{A\text{-Interface}}, \mu, t) \\ + I^+_{\mathrm{A,LO,k}}(x_{A\text{-Interface}}, \mu, t) + I^+_{\mathrm{A,TO,k}}(x_{A\text{-Interface}}, \mu, t) \end{pmatrix} d\mu dk$$

The heat fluxes $q^+_{\mathrm{A},ij,\mathrm{k}}, q^-_{\mathrm{A},ij,\mathrm{k}}, q^+_{\mathrm{B},ij,\mathrm{k}},$ and $q^-_{\mathrm{B},ij,\mathrm{k}}$ (as shown in Fig. 7.11) are given as

$$q^+_{\mathrm{A},ij,\mathrm{k}}(t) = 2\pi \int_0^1 I^+_{\mathrm{A},ij,\mathrm{k}}(x_{A\text{-Interface}}, \mu, t)\mu d\mu \tag{7.47}$$

$$q^-_{\mathrm{A},ij,\mathrm{k}}(t) = 2\pi \int_{-1}^0 I^-_{\mathrm{A},ij,\mathrm{k}}(x_{A\text{-Interface}}, \mu, t)\mu d\mu \tag{7.48}$$

$$q^+_{\mathrm{B},ij,\mathrm{k}}(t) = 2\pi \int_0^1 I^+_{\mathrm{B},ij,\mathrm{k}}(x_{B\text{-Interface}}, \mu, t)\mu d\mu \tag{7.49}$$

$$q^-_{\mathrm{B},ij,\mathrm{k}}(t) = 2\pi \int_{-1}^0 I^-_{\mathrm{B},ij,\mathrm{k}}(x_{B\text{-Interface}}, \mu, t)\mu d\mu \tag{7.50}$$

The superscripts + and − represent the energy in positive and negative x directions, respectively. When the phonon wavelength is greater than the gap length $\lambda \geq L_{\mathrm{gap}}$, the phonon wave jumps over the gap from one section to another ($\mathrm{A} \rightleftarrows \mathrm{B}$). Here, λ is the phonon wavelength and L_{gap} is the gap size (length) between the sections A and B. The wavenumber

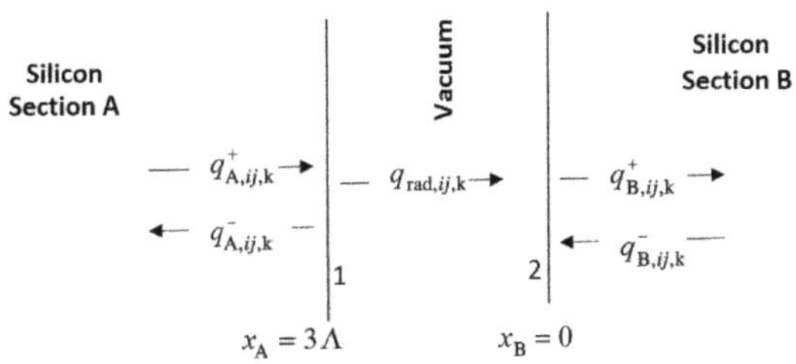

FIGURE 7.11 Schematic view of gap located in between two-silicon films pair and heat flux vectors (k is wavenumber; and $i = T$ represents transverse mode of phonons, $i = L$ stands for longitudinal mode of phonons, $j = A$ corresponds to acoustic and $j = O$ stands for optical phonons). The space between $x_A = 3\Lambda$ and $x_B = 0$ corresponds to the gap size. Λ is the mean free path of silicon corresponding to a gray medium [2].

corresponding to the L_{gap} is $k_{cut\text{-}off} = 2\pi/L_{gap}$, which works as the cut-off wavenumber. All the phonons having wavenumbers less than the cut-off wavenumber ($k < k_{cut\text{-}off}$) can possibly jump from one section to another ($A \rightleftarrows B$). Note that we have applied energy balance at each wavelength of transverse acoustic, longitudinal acoustic, transverse optical, and longitudinal optical branches of phonon waves.

7.3.1.3.1 Energy Balance at Point 1 (Fig. 7.11)

Although energy balance incorporating the CMM was used previously to formulate the interface conditions across the vacuum gap, the main concern was the mismatch due to wavenumber-dependent dispersion relations of two different thin films, namely, silicon and diamond. In the case of silicon-silicon thin films pair with a presence of the vacuum gap, the cut-off model is mainly associated with the ballistic limit because both films have the same dispersion relations. In this case, energy balance has to be satisfied at the interface, which requires that the following conditions be fulfilled.

When $k < k_{cut\text{-}off}$

$$q^{-}_{A,ij,k}(t) = q^{-}_{B,ij,k}(t) + q_{rad,ij,k}(t) \tag{7.51}$$

Considering the diffusive condition at the inference; incorporating Eq. (7.48) into the preceding equation yields the following.

$$I^{-}_{A,ij,k}(x_{A\text{-Interface}}, \mu, t) = (-q^{-}_{B,ij,k}(t) - q_{rad,ij,k}(t))/\pi \tag{7.52}$$

When $k > k_{cut\text{-}off}$

$$q^{-}_{A,ij,k}(t) = -q^{+}_{A,ij,k}(t) + q_{rad,ij,k}(t) \tag{7.53}$$

Using the diffusive boundary condition at the interface along with Eq. (7.48), Eq. (7.53) reduces to the following:

$$I^{-}_{A,ij,k}(x_{A\text{-Interface}}, \mu, t) = (q^{+}_{A,ij,k}(t) - q_{rad,ij,k}(t))/\pi \tag{7.54}$$

7.3.1.3.2 Energy Balance at Point 2 (Fig. 7.11)

When $k < k_{cut\text{-}off}$

$$q^{+}_{B,ij,k}(t) = q^{+}_{A,ij,k}(t) + q_{rad,ij,k}(t) \tag{7.55}$$

Considering the diffuse condition at the boundary that is $I^{+}_{B,ij,k}$ is independent of μ; incorporating Eq. (7.49) into the preceding equation yields the following.

$$I^{+}_{B,ij,k}(x_{B\text{-Interface}}, \mu, t) = (q^{+}_{A,ij,k}(t) + q_{rad,ij,k}(t))/\pi \tag{7.56}$$

When $k > k_{\text{cut-off}}$

$$q^{+}_{B,ij,k}(t) = -q^{-}_{B,ij,k}(t) + q_{\text{rad},ij,k}(t) \tag{7.57}$$

Using Eq. (7.49), the preceding equation can be reduced to

$$I^{+}_{B,ij,k}(x_{B\text{-Interface}}, \mu, t) = \left(-q^{-}_{B,ij,k}(t) + q_{\text{rad},ij,k}(t)\right)/\pi \tag{7.58}$$

In Eqs. (7.45)–(7.58), $i = T$ represents transverse mode and $i = L$ stands for longitudinal mode. In addition, $j = A$ corresponds to acoustic and $j = O$ stands for optical phonons.

In order to calculate the intensities at the interface of the film pair ($I^{-}_{A,ij,k}$ and $I^{+}_{B,ij,k}$), we need to solve Eqs. (7.52), (7.54), (7.56), and (7.58) along with Eqs. (7.46)–(7.50).

The equivalent equilibrium phonon temperature is used to quantify the phonon intensity distribution in the film.

The heat flux due to the contribution of the sum of all branches of the acoustic and optical phonons is given as

$$q_{A/B}(x,t) = 2\pi \int_0^{k_{\max}} \int_{-1}^{1} I_{A/B,k}\,\mu\, d\mu\, dk \tag{7.59}$$

where $k_{\max}$ represents the maximum wavenumber in the dispersion relation and the phonon intensity over the all branches of the acoustic and optical phonons, which is

$$I_{A/B,k} = (2I_{A/B,TA,k} + I_{A/B,LA,k} + 2I_{A/B,TO,k} + I_{A/B,LO,k}) \tag{7.60}$$

In order to examine the effective thermal conductivity (k_{eff}) of the films pair with the gap size, the Fourier analogy is considered for the temperature data obtained at the steady state situation. The heat flux data, which are calculated from the steady state BTE across the films pair, are used for this purpose.

The expression for the effective thermal conductivity of the film in terms of the wavenumber (k) can be written as [15]

$$k_{\text{eff}}(T) = \frac{1}{3}\begin{bmatrix} \int_0^{k_{\max,\,LA}} C_{k,LA} v_{k,LA}^2 \tau_{k,LA}\, dk + 2\int_0^{k_{\max,\,TA}} C_{k,TA} v_{k,TA}^2 \tau_{k,TA}\, dk \\ \int_0^{k_{\max,\,LO}} C_{k,LO} v_{k,LO}^2 \tau_{k,LO}\, dk + 2\int_0^{k_{\max,\,TO}} C_{k,TO} v_{k,TO}^2 \tau_{k,TO}\, dk \end{bmatrix} \tag{7.61}$$

In addition, to assess the phonon transport characteristics in the film, the dimensionless temperature is introduced, which is given by

$$\text{Dimensionless Temperature} = \frac{T_{A,B}(x,t) - T_{Low}}{T_{High} - T_{Low}} \tag{7.62}$$

where $T_{High} = 301\text{K}$ and $T_{Low} = 300\text{K}$, which correspond to high and low temperature at the external edges of the films pair. Note that the dimensionless temperature varies between 0 and 1.

7.3.2 Numerical Solution

The equations of phonon radiative transfer in both thin silicon films with the associated initial, boundary, and interface conditions are incorporated in the numerical analysis using the finite-difference method. The EPRT is an integrodifferential equation and can be solved numerically using an iterative technique. Since the equation is time-dependent, the implicit scheme can be adopted to solve the transient equation. Consider the EPRT for a particular phonon branch, in the silicon film:

$$\frac{1}{v_k}\frac{\partial I_k}{\partial t} + \mu\frac{\partial I_k}{\partial x} = \frac{\int_{-1}^{1} I_k d\mu - I_k}{\Lambda_k} \tag{7.63}$$

The variable μ varies between $-1 \le \mu \le 1$. To facilitate the solution of the EPRT, it is customary to break the intensity into two parts. When $-1 \le \mu \le 0$, $I_k = I_k^-(x, \mu, t)$, and when $0 \le \mu \le 1$, $I_k = I_k^+(x, \mu, t)$. Therefore, we may write separate equations for the two intensities:

$$\frac{1}{v_k}\frac{\partial I_k^-}{\partial t} + \mu\frac{\partial I_k^-}{\partial x} = \frac{\int_{-1}^{0} I_k^- d\mu + \int_{0}^{1} I_k^+ d\mu - I_k^-}{\Lambda_k} \qquad -1 \le \mu \le 0 \tag{7.64}$$

and

$$\frac{1}{v_k}\frac{\partial I_k^+}{\partial t} + \mu\frac{\partial I_k^+}{\partial x} = \frac{\int_{-1}^{0} I_k^- d\mu + \int_{0}^{1} I_k^+ d\mu - I_k^+}{\Lambda_k} \qquad 0 \le \mu \le 1 \tag{7.65}$$

In this case, the implicit method is used for the time discretization and the discrete ordinates method is used for the discretization of the angular term. The time-derivative in each equation is approximated by means of the first-order backward-difference. The partial derivative with respect to the coordinate variable x is discretized using a forward-difference for the first equation in which $-1 \le \mu \le 0$, and is discretized using a backward-difference for the second equation in which $0 \le \mu \le 1$. This procedure is important to obtain stable discretization of the EPRT. The integrals are calculated by first discretizing the cosine of the azimuthal angle μ and then using Simpson's rule to evaluate the integrals numerically for each time-step and each coordinate location. It is important to keep in mind that the intensities I_k^- and I_k^+ are two separate variables and each needs an initial condition and a single boundary or interface condition. For both the intensities I_k^- and I_k^+, the initial condition, boundary, and interface conditions are used as described in the previous section. This procedure is to be performed for the four different polarizations as well as for each wavenumber. The wavenumber ranges from 0 to $k_{max} = 2\pi/a$, where a is the lattice constant for the silicon crystal.

It starts by first initializing the unknown intensities at first time-step on each of the grid points (x, μ) to some value in both thin silicon films. Then, Eqs. (7.64) and (7.65) are solved for each discrete wavenumber and for each of the four phonon branches for both films. Once this is completed, the equivalent equilibrium temperature is calculated at each grid point within the complete domain. Then, the radiation heat transfer is calculated from Eq. (7.45), and all the interface conditions are updated using Eqs. (7.52), (7.54), and (7.58). The whole procedure now constitutes a single iteration. Now again solve Eqs. (7.64) and (7.65) for both silicon films incorporating the updated interface conditions. Several iterations are performed until a converged solution is obtained. This completes the first time-step. Similarly, a solution is obtained at the time-steps. The criterion for convergence is that the maximum absolute difference inside the domain, between the equivalent equilibrium temperature calculated at the current iteration and that at the previous iteration, should become smaller than a preassigned small positive number. Mathematically, $\max\left|T^h_{A/B,x_A/x_A} - T^{h-1}_{A/B,x_A/x_A}\right| < \varepsilon$, where h is the iteration index and $\varepsilon = 1 \times 10^{-6}$. Once the simulation is complete, the equivalent equilibrium phonon temperature may be evaluated at each time-step and at each location inside the two films. Table 7.3 shows the summary of parameters used in the simulation.

Table 7.3 Parameters Used in Simulation ($\Lambda = 33.8$ nm)

		Temperature (K)		
Section	**Thickness**	**Left Edge**	**Right Edge**	**Gap Size**
A	3Λ	301	Interface condition	$0-0.03\Lambda$
B	3Λ	Interface condition	300	

7.3.3 Findings of Application of Minute Vacuum Gap

Phonon transport across the silicon thin films pair with presence of the minute vacuum gap in between the films is considered. The frequency-dependent EPRT is solved numerically to determine the transient phonon intensity distribution in the films pair. Equivalent equilibrium temperature is introduced to assess the phonon transport characteristics.

Fig. 7.12 shows dimensionless temperature along the film pairs and the gap for 100 ps heating duration. In addition, equivalent equilibrium temperature with no gap in between the films pair is shown for comparison purposes. Temperature jump is evident at the edges of the films in the films pair because of the boundary scattering taking place at the edges. Since the boundary scattering remains the same at the high-temperature edge of the film, temperature jump remains almost the same for all gap sizes in between the silicon films pair. Note that the reflected phonons from the first film edge due to their small wavelength (smaller than the gap size) contributes to the temperature jump at the high-temperature film edge; however, their contribution is not significantly high to resemble a large temperature jump at the high-temperature edge

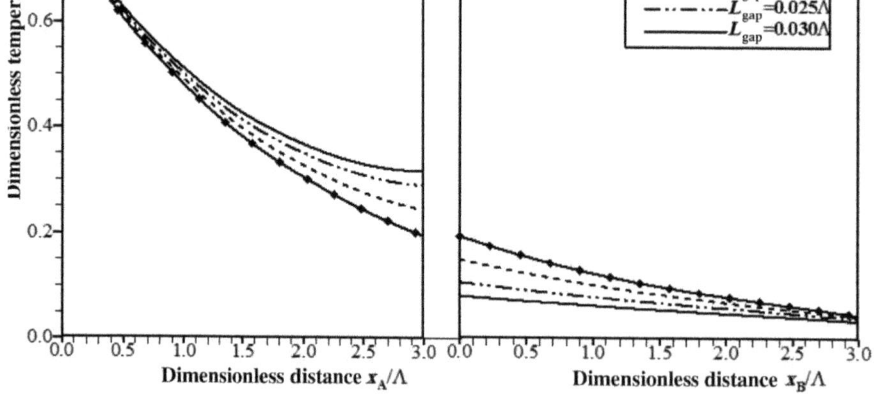

FIGURE 7.12 Equivalent equilibrium temperature along the thickness of films pair for heating duration of 100 ps. Section A represents the first film and Section B is the second film. The gap is located in between two films. Λ is the mean free path of silicon corresponding to a gray medium [2].

due to differences in the gap size. Phonons emitted from the high temperature edge undergo scattering in the film and cause equivalent equilibrium temperature increase prior to reaching the film interface. Phonons with large wavelengths (larger than the gap size) can possibly be transmitted to the second film. However, some of those phonons contribute to the boundary scattering at the gap edge, which adds to temperature jump at the film pair interface. Consequently, reflected phonons and those phonons having smaller wavelengths than the gap size contribute to the boundary scattering at the interface. This results in varying temperature jump with the gap size. In this case, increasing gap size causes a large temperature jump at the first film interface due to the attainment of high phonon intensity at the edge. High phonon intensity results in increased equivalent equilibrium temperature at the first film edge. Radiation heat transfer (Casimir limit) from the first film edge to the second film edge across the gap contributes to temperature increase at the second film edge. However, phonons transmitted have a significant effect on the equivalent equilibrium temperature increase at the second film edge because of the boundary scattering in this region.

Although increasing gap size causes the attainment of high equivalent equilibrium temperature at the first film edge, the rate of phonons transmitted to the second film becomes small. This, in turn, results in a small temperature rise at the second film edge despite the fact that the amount of heat transferred via radiation is high at the large gap size. This behavior reveals that equivalent equilibrium temperature increase is mainly because of those phonons transmitted to the second film rather than the net radiation heat transfer taking place between the film edges across the gap. This situation can be seen in Fig. 7.13, in which the ratio of heat fluxes is due to radiation transport over the phonon transport. Temperature jump at the film pair disappears for the case of the continuous film consideration (no gap case). In the case of the late heating period (500 ps), temperature jump at the high temperature edge of the film varies for different gap sizes. This situation is seen in Fig. 7.14, in which dimensionless equivalent equilibrium temperature is shown along the film thickness for the heating period of 500 ps. The effect of the gap size on temperature jump with progressing heating duration is associated with the amount of phonon scattered at the high temperature edge, which increases with progressing time. In addition, the contribution of phonons reflected from the interface edge of the first film to the amount of phonons scattered at the high-temperature edge increases with progressing heating time.

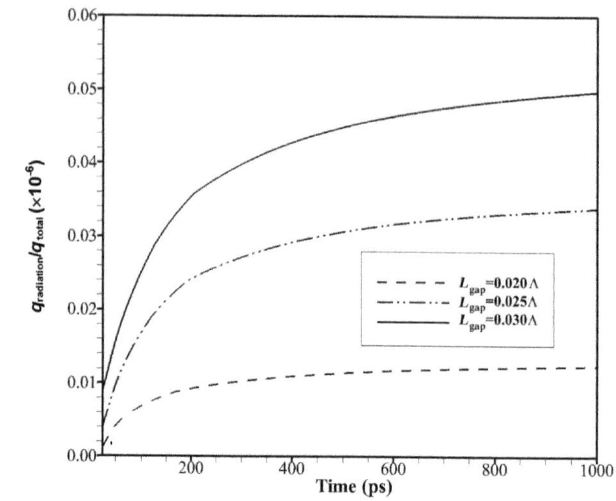

■ **FIGURE 7.13** Heat flux ratio due the radiation heat transfer to the total phonon transfer across the vacuum gap in between the films pair (Λ is the mean free path of silicon corresponding to a gray medium) [2].

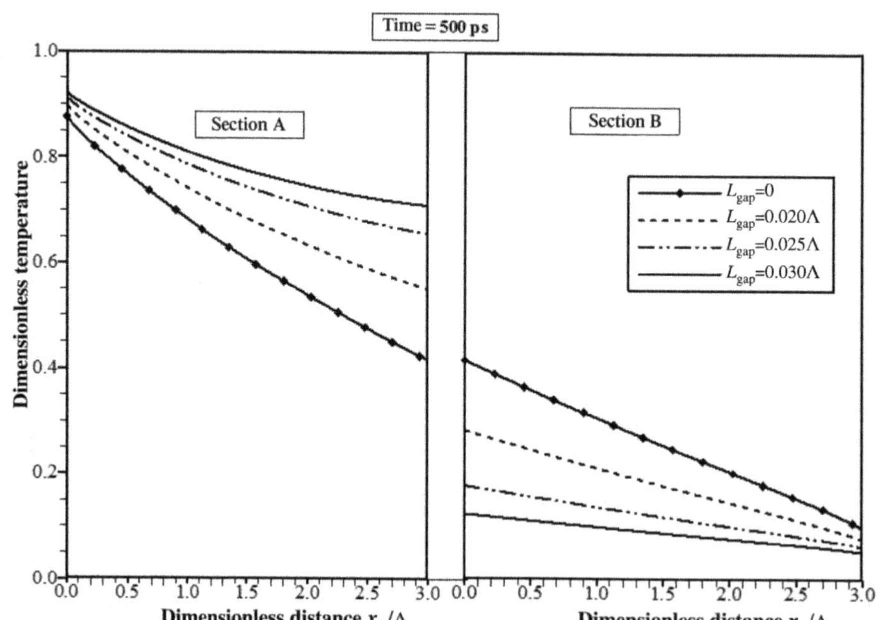

■ **FIGURE 7.14** Equivalent equilibrium temperature along the thickness of films pair for heating duration of 500 ps. Section A represents the first film and Section B is the second film. The gap is located in between two films. Λ is the mean free path of silicon corresponding to a gray medium [2].

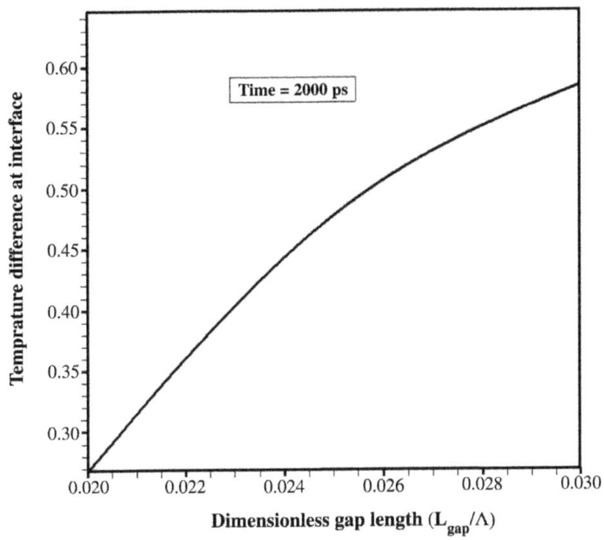

■ **FIGURE 7.15** Equivalent equilibrium temperature difference between the first film edge ($x_A = 3\Lambda$, where Λ is the mean free path of silicon corresponding to a gray medium) and the second film edge ($x_B = 0$) with dimensionless gap length (L_{gap}/Λ, where L_{gap} is the gap size in between the films) [2].

Fig. 7.15 shows equivalent equilibrium temperature difference at the edges of the first film interface ($x_A = 3\Lambda$, Fig. 7.9) and second film interface ($x_B = 0$, Fig. 7.9) at the heating duration of 2000 ps. Note that equivalent equilibrium temperature attains almost quasisteady state at the interfaces of the film pair after 2000 ps of heating duration. Increasing gap size at the film interface increases the temperature difference, which is more pronounced for the large gap sizes. In the case of small gap sizes ($L_{gap}/\Lambda \leq 0.025$), variation of temperature difference is almost linear with increasing gap size. However as the gap size increases further, this variation becomes almost nonlinear. Consequently, phonon transmittance and net radiative heating at the edges of the films pair become complex, in terms of the phonon scattering at the interface boundaries, for the large gap sizes. Although the effect of net radiative heating on equivalent equilibrium temperature variation at the interfaces of the film edges is not significant, this effect may become important when the gap size becomes large. This is due to the fact that equivalent equilibrium temperature increases at the first film edge with increasing film gap size while altering the net radiation heat transfer across the gap in the interface of the films pair. In addition, attainment of high values of equivalent equilibrium temperature at the first film edge and low values of temperature

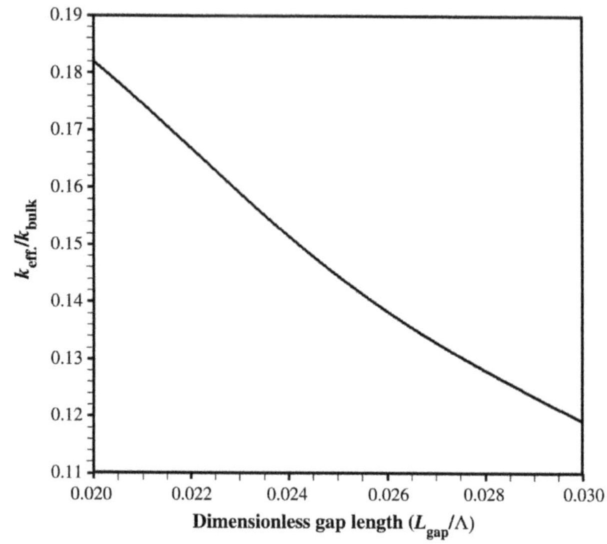

■ **FIGURE 7.16** Thermal conductivity ratio (k_{eff}/k_{Bulk}) with dimensionless gap length (L_{gap}/Λ, where L_{gap} is the gap size in between the films and Λ is the mean free path of silicon corresponding to a gray medium) [2].

at the second film edge due to the large gap sizes enhance the temperature differences for the large gap sizes. Consequently, a high rate of reflected phonons from the first film interface and low rate of phonon transmittance at the second film interface modify equivalent equilibrium temperature rise at both interfaces, which is more pronounced for the large gap sizes. The ballistic phonons transmitted across the interface suppresses temperature rise at the first film interface, which in turn lowers the equivalent equilibrium temperature difference across the gap in the films pair.

Fig. 7.16 shows the thermal conductivity ratio (k_{eff}/k_{Bulk}, where k_{eff} is the thermal conductivity of the film pair and k_{Bulk} is the thermal conductivity of the bulk silicon) with the dimensionless gap size. Note that k_{eff} is obtained from the steady state solution of the BTE across the films pair. The thermal conductivity ratio reduces with increasing gap size, since the k_{Bulk} remains the same, and k_{eff} reduces with increasing gap size. This is attributed to the temperature variation at the film edges across gap size; temperature difference between the first film edge and the second film edge increases with increasing gap size. This, in turn, suppresses the heat transfer across the film pair, particularly toward the second film. In addition, those phonons with large wavelengths can possibly transfer to the second film across the gap. Since these phonons resemble only a fraction of the total phonons emitted from the high-temperature edge of the first film, phonon energy transport into the second film becomes small. However, thermal energy transfer from the first film edge to the second film edge via radiation (Casimir limit) does alter temperature at the second film edge. Consequently, the presence of a gap resembles the thermal boundary resistance in between the thin films. However, changing the gap size within the film pair, thermal conductivity can be reduced significantly, which can be of interest for the thermal insulation applications.

7.4 APPLICATIONS IN CROSS-PLANE TRANSPORT ACROSS DIELECTRIC THIN FILMS WITH MINUTE GAP: CONSIDERATION OF NEAR-FIELD RADIATION

In dielectric thin films, the absence of carriers in the conduction band uplifts the role of phonons in the thermal transport within the film. Since phonons undergo scattering due to boundary resistance and lattice imperfections, transport characteristics become highly dependent on the dispersion relations of the film material under consideration. In the case of composite films with presence of nanoscale voids, near-field radiation becomes part of the process due to the surface waves. Phonons, having larger wavelengths than the size of voids, can contribute to the thermal transport via creating ballistic affects across the voids. Although thermal transport characteristics can be analyzed through the molecular dynamics, the assumption of periodicity at boundaries may not capture exactly thermal characteristics across the voids. On the other hand, the first principle solution, incorporating the phonon scattering and dispersions for the thermal transport, provides physical insight into the process and lets us predict the film thermal characteristics correctly. In this section, the cross-plane phonon transport with presence of a minute gap in between two thin films is investigated for different film thicknesses in line with the previous study [3]. Temperature disturbance across the films pair is introduced to activate the phonon cross-plane transport. Since the gap size is small, the effect of surface waves becomes important; therefore, near-field radiation heat transfer is incorporated across the film edges. The BTE is incorporated to formulate the transient frequency-dependent phonon transport in the films pair while the CMM based on the phonon wavelengths are considered across the gap in between the films [16]. The BTE is solved numerically, incorporating the discrete ordinate method to predict the phonon intensity distributions within the films pair.

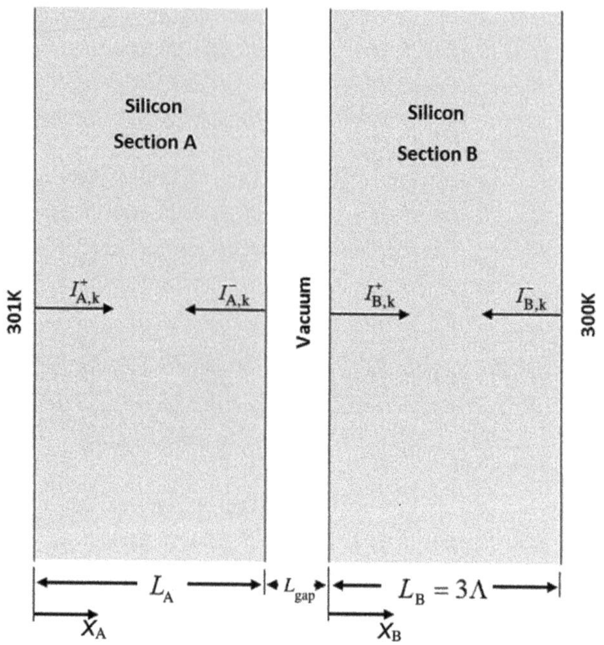

■ **FIGURE 7.17** Schematic view of thin silicon films pair consisting of two silicon films separated by a small gap. Λ is the mean free path of silicon corresponding to a gray medium. *k* is wavenumber, where $\Lambda = 33.8$ nm [3].

In the analysis, transient, wavenumber (frequency)-dependent BTE is incorporated to obtain the phonon intensity distribution in the silicon thin films pair. The boundary conditions used are those of constant temperature (Fig. 7.17) resembling temperature disturbance across the films pair at the exterior edges. At the interface of the thin silicon films pair appropriate interface conditions are imposed, satisfying the energy balance. The relaxation-time approximation to the EPRT reduces the complications faced during the numerical solution of the BTE with the appropriate boundary and interface conditions.

Although the formulation of the BTE is similar to that given in the previous section, it is also given herein briefly for the completeness of the arguments. The transient and frequency (wavenumber)-dependent BTE, without considering the external forces and introducing the relaxation-time approximation, can be written as

$$\frac{1}{v_k}\frac{\partial I_k}{\partial t} + \mu\frac{\partial I_k}{\partial x} = \frac{\frac{1}{2}\int_{-1}^{1} I_k d\mu - I_k}{v_k \tau_k} \tag{7.66}$$

Eq. (7.66) is analogous to the equation of radiative heat transfer known as the EPRT. The dispersion relations of silicon [15] are incorporated into the analysis. There are six phonon branches present: two acoustic-transverse, two acoustic-optical, one acoustic-longitudinal, and one optical-longitudinal. The equation for phonon branches can be represented as

$$\frac{1}{v_{ij,k}}\frac{\partial I_{ij,k}}{\partial t}+\mu\frac{\partial I_{ij,k}}{\partial x}=\frac{\frac{1}{2}\int_{-1}^{1} I_{ij,k}d\mu - I_{ij,k}}{v_{ij,k}\tau_{ij,k}} \qquad \begin{matrix} 0<x<L \\ -1<\mu<1 \end{matrix} \tag{7.67}$$

In this equation, phonon intensity I is the function of t, x, μ and also is dependent on wavenumber (k) and phonon branch. In Eq. (7.67), $i=T$ represents transverse mode and $i=L$ stands for longitudinal mode. In addition, $j=A$ corresponds to acoustic and $j=O$ stands for optical phonons.

Equivalent equilibrium phonon temperature is defined anywhere in the thin silicon films to quantify the phonon intensity distribution inside the film. The heat flux due to the contribution of the sum of all branches of the acoustic and optical phonons is given as

$$q_{A/B}(x,t)=2\pi\int_{0}^{k_{max}}\int_{-1}^{1} I_{A/B,k}\mu d\mu dk \tag{7.68}$$

where k_{max} represents the maximum wavenumber in the dispersion relation and the phonon intensity over the all branches of the acoustic and optical phonons, which is

$$I_{A/B,k}=(2I_{A/B,TA,k}+I_{A/B,LA,k}+2I_{A/B,TO,k}+I_{A/B,LO,k}) \tag{7.69}$$

7.4.1 Initial, Boundary, and Interface Conditions

Since the radiative phonon transport equation is transient, an initial condition should be introduced. In addition, the solution domain consists of two silicon thin films, so the boundary conditions for each film need to be incorporated accordingly. In order to satisfy the energy balance at the interface of thin films the interface conditions are also implemented.

7.4.1.1 *Section A (Fig. 7.17)*

7.4.1.1.1 Initial Condition

$$I_{A,ij,k}(x_A,\mu,t=0)=\frac{C_{ij,k}v_{ij,k}T_{initial}}{4\pi} \tag{7.70}$$

7.4.1.1.2 Left Boundary Condition

$$I_{A,ij,k}(x_A = 0, \mu, t) = \frac{C_{ij,k} v_{ij,k} T_{left}}{4\pi}, \tag{7.71}$$

where $T_{left} = 301$ K

7.4.1.2 *Section B (Fig. 7.17)*

7.4.1.2.1 Initial Condition

$$I_{B,ij,k}(x_B, \mu, t = 0) = \frac{C_{ij,k} v_{ij,k} T_{initial}}{4\pi} \tag{7.72}$$

7.4.1.2.2 Right Boundary Condition

$$I_{B,ij,k}(x_B = 3\Lambda, \mu, t) = \frac{C_{ij,k} v_{ij,k} T_{right}}{4\pi}, \tag{7.73}$$

where $T_{right} = 300$ K.

In Eqs. (7.70)–(7.73), $i = T$ represents transverse mode and $i = L$ stands for longitudinal mode. In addition, $j = A$ corresponds to acoustic and $j = O$ stands for optical phonons, $C_{ij,k}$ is the wavenumber-dependent volumetric heat capacity at the nominal temperature of $T_o = 300$ K, and $v_{ij,k}$ is the wavenumber-dependent phonon group velocity for the ijth branch in the silicon.

7.4.1.3 *Interface Conditions*

Interface conditions are required to satisfy the energy balance across the minute vacuum gap and the phonon intensity distributions at the edges of both films. The ballistic phonon transport is a parallel energy transferring process to near-field radiation because of the minute vacuum gap in between the films [17]. Note that when the characteristic length of separation (vacuum gap size – length) becomes comparable to the thermal wavelength, the influence of surface waves on the energy transport dominates over the thermal radiation [18]. Moreover, it was demonstrated that energy transport by the near field thermal radiation becomes larger than that of Planck's blackbody radiation by several orders of magnitude [19]. Energy balance along with the consideration of near-field radiation heat transfer is included to obtain the phonon intensities (I_A^- and I_B^+) at the interface of the silicon films.

The radiative transfer can take place via two types of waves: propagating electromagnetic waves and evanescent waves. Evanescent waves are the surface waves generated at the interface of the media and exponentially

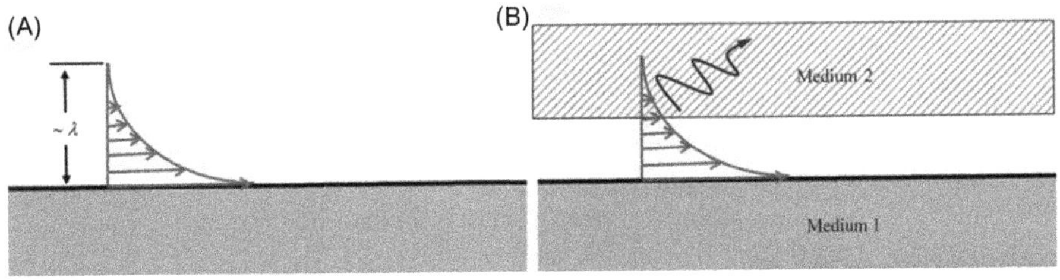

■ **FIGURE 7.18** (A) Evanescent waves at an interface. (B) Tunneling of an evanescent wave.

decaying over a distance of about a wavelength λ normal to that interface, as shown in Fig. 7.18(A).

There is no net energy radiated from the surface in the absent structure interacting with the evanescent wavefield [20], as shown in Fig. 7.18(A). When the other surface is present, which is intersecting the evanescent wavefield, the energy transfer takes place due to the tunneling of the evanescent wave, as shown in Fig. 7.18(B). When the distance between the two bodies is greater than the dominant wavelength of the thermal radiation, thermal energy is transferred via the propagating electromagnetic waves only, known as far-field radiative transport. When the distance is smaller than the dominant wavelength of the thermal radiation, the energy transferred by the evanescent waves is known as near-field radiative transport. The classic theory of thermal radiation is based on the basic assumption that distance between the bodies exchanging thermal radiation is greater than the dominant wavelength of thermal emission [21]. Therefore, classic theory of thermal radiation is only applicable for the far-field radiative transport. For the modeling of near-field radiative transport we can incorporate the fluctuational electrodynamics [18]. When a minute-sized vacuum gap (comparable to the dominant thermal wavelength) is present between the two films, the influence of evanescent waves on the energy transport dominates over the thermal radiation [19]. Moreover, it was demonstrated that energy transport by the near-field thermal radiation becomes larger than that of Planck's blackbody radiation by several orders of magnitude [22].

Silicon films A and B are separated by a minute vacuum gap. At the length scale adopted for the minute vacuum gap, the effect of evanescent wave also becomes significant. Therefore, we need to include the effect of the contribution of evanescent waves along with the propagating waves in near-field thermal radiation. The near-field radiation across the vacuum gap in between the edges of the aluminum and silicon films is modeled using fluctuational electrodynamics [17]. In this case, radiative heat flux can be written as [22]

$$q_{\text{rad},A\text{-}B}(t) = \int_0^{k_{\max,Al}} (q_{evan,\text{A,k}}(t) + q_{\text{prop,A,k}}(t))dk \tag{7.74}$$

and

$$q_{\text{rad},B\text{-}A}(t) = \int_0^{k_{\max,Si}} (q_{evan,\text{B,k}}(t) + q_{\text{prop,B,k}}(t))dk \tag{7.75}$$

Eqs. (7.74) and (7.75) represent the heat flux due to near-field radiation transfer between the edges of film A to film B, and film B to film A, respectively. $q_{evan,\text{A/B,k}}$ is the energy transfer due to the evanescent waves, $q_{\text{prop,A}/B,k}$ is the energy transfer due to the propagating waves, and k is the wavenumber. The radiation heat transfer due to propagative ($q_{\text{prop,A}/B,k}$) and evanescent modes ($q_{\text{prop,A}/B,k}$) can be written as [6]

$$q_{\text{prop},A/\text{B,k}}(t) = \frac{\Theta(\omega(k), T_{\text{A/B}}(x_{\text{A}/B\text{-Interface}}, t))}{4\pi^2} \times \int_0^{k_v} k_p dk_p x \sum_{\gamma=\text{TE,TM}} \frac{\left(1 - |R_{\text{A}}^{\gamma}|^2 - |T_{\text{A}}^{\gamma}|^2\right)\left(1 - |R_{\text{B}}^{\gamma}|^2 - |T_{\text{B}}^{\gamma}|^2\right)}{\left|1 - R_{\text{A}}^{\gamma} R_{\text{B}}^{\gamma} e^{2ik_z L_{\text{gap}}}\right|^2} \tag{7.76}$$

$$q_{evan,\text{A/B,k}}(t) = \frac{\Theta(k, T_{\text{A/B}}(x_{\text{A}/B\text{-Interface}}, t))}{\pi^2} \times \int_{k_v}^{\infty} k_p dk_p e^{-2k_z'' L_{\text{gap}}} x \sum_{\gamma=\text{TE,TM}} \frac{\text{Im}(R_{\text{A}}^{\gamma})\text{Im}(R_{\text{B}}^{\gamma})}{\left|1 - R_{\text{A}}^{\gamma} R_{\text{B}}^{\gamma} e^{2ik_z L_{\text{gap}}}\right|^2} \tag{7.77}$$

where $x_{A\text{-Interface}} = L_{\text{A}}$ and $x_{B\text{-Interface}} = 0$. k_{p} is the wavevector parallel to the surfaces, $R_{\text{A/B}}^{\gamma}$ and $T_{\text{A/B}}^{\gamma}$ represent the reflectance and transmittance, respectively. Θ is the mean energy of an electromagnetic state, k_z is the z component of wavevector in a vacuum, and L_{gap} is the size of the vacuum gap. TE and TM represent the transverse electric and transverse magnetic, respectively. Mean energy of an electromagnetic state Θ can be given as [7]

$$\Theta(k, T_{\text{A/B}}(x_{\text{A}/B\text{-Interface}}, t)) = \frac{\hbar\omega(k)}{\exp(\hbar\omega(k)/k_{\text{B}} T_{\text{A/B}}(x_{\text{A}/B\text{-Interface}}, t)) - 1} \tag{7.78}$$

The reflectance $R_{\text{A/B}}^{\gamma}$ and transmittance $T_{\text{A/B}}^{\gamma}$ can be given by [4]

$$R_{\text{A/B}}^{\gamma} = \frac{r_{\text{vac.gap,A/B}}^{\gamma} + r_{\text{A/B,vac.gap}}^{\gamma} e^{2ik_{z\text{A/B}} t_{\text{A/B}}}}{1 + r_{\text{vac.gap,A/B}}^{\gamma} r_{\text{A/B,vac.gap}}^{\gamma} e^{2ik_{z\text{A/B}} t_{\text{A/B}}}} \tag{7.79}$$

and

$$T^{\gamma}_{A/B} = \frac{t^{\gamma}_{vac.gap,A/B} + t^{\gamma}_{A/B,vac.gap} e^{2ik_{zA/B}t_{A/B}}}{1 + r^{\gamma}_{vac.gap,A/B} r^{\gamma}_{A/B,vac.gap} e^{2ik_{zA/B}t_{A/B}}} \tag{7.80}$$

where r^{γ} and t^{γ} are the Fresnel reflection and transmission coefficients in the γ polarization, respectively.

The total net thermal radiative energy transfer becomes

$$q_{rad,total}(t) = q_{rad,A\text{-}B}(t) - q_{rad,B\text{-}A}(t) \tag{7.81}$$

where, $q_{rad,total}$ is the total net thermal radiation between Sections A and B.

Energy transfer from Section A to Section B by radiation is disturbed into all the branches of each mode of phonons. This can be expressed as follows:

$$q_{rad,ij,k}(t) = \frac{\int_0^1 I^{+}_{A,ij,k}(x_{A\text{-Interface}}, \mu, t) d\mu}{I^{+}_{A,total}(t)} q_{rad,total}(t) \tag{7.82}$$

where

$$I^{+}_{A,total}(t) = \int_0^{k_{max}} \int_0^1 \begin{pmatrix} I^{+}_{A,LA,k}(x_{A\text{-Interface}}, \mu, t) + I^{+}_{A,TA,k}(x_{A\text{-Interface}}, \mu, t) \\ + I^{+}_{A,LO,k}(x_{A\text{-Interface}}, \mu, t) + I^{+}_{A,TO,k}(x_{A\text{-Interface}}, \mu, t) \end{pmatrix} d\mu dk \tag{7.83}$$

The heat fluxes $q^{+}_{A,ij,k}$, $q^{-}_{A,ij,k}$, $q^{+}_{B,ij,k}$, and $q^{-}_{B,ij,k}$ (as shown in Fig. 7.17) are given as

$$q^{+}_{A,ij,k}(t) = 2\pi \int_0^1 I^{+}_{A,ij,k}(x_{A\text{-Interface}}, \mu, t) \mu d\mu \tag{7.84}$$

and

$$q^{-}_{A,ij,k}(t) = 2\pi \int_0^1 I^{-}_{A,ij,k}(x_{A\text{-Interface}}, \mu, t) \mu d\mu \tag{7.85}$$

and

$$q^{+}_{B,ij,k}(t) = 2\pi \int_0^1 I^{+}_{B,ij,k}(x_{B\text{-Interface}}, \mu, t) \mu d\mu \tag{7.86}$$

and

$$q^{-}_{B,ij,k}(t) = 2\pi \int_0^1 I^{-}_{B,ij,k}(x_{B\text{-Interface}}, \mu, t) \mu d\mu \tag{7.87}$$

When the phonon wavelength is greater than the gap length $\lambda \geq L_{gap}$, the phonon wave jumps over the gap from one section to another (A⇄B). Here, λ is the phonon wavelength and L_{gap} is the gap size (length) between Sections A and B. The wavenumber corresponding to the L_{gap} is $k_{cut\text{-}off} = 2\pi/L_{gap}$, which works as the cut-off wavenumber. All the phonons having wavenumbers less than the cut-off wavenumber ($k < k_{cut\text{-}off}$) can possibly jump from one section to another (A⇄B). Note that we have applied energy balance at each wavelength of transverse acoustic, longitudinal acoustic, transverse optical, and longitudinal optical branches of phonon waves.

7.4.1.3.1 **Energy Balance at Point 1 (Fig. 7.19)**

Energy balance has to be satisfied at the interface, which requires that the following conditions be fulfilled.

When $k < k_{cut\text{-}off}$

$$q^{-}_{A,ij,k}(t) = q^{-}_{B,ij,k}(t) + q_{rad,ij,k}(t) \tag{7.88}$$

Considering the diffusive condition at the inference and incorporating Eq. (7.85) into the previous equation yields the following:

$$I^{-}_{A,ij,k}(x_{A\text{-Interface}}, \mu, t) = (-q^{-}_{B,ij,k}(t) - q_{rad,ij,k}(t))/\pi \tag{7.89}$$

When $k > k_{cut\text{-}off}$

$$q^{-}_{A,ij,k}(t) = -q^{+}_{A,ij,k}(t) + q_{rad,ij,k}(t) \tag{7.90}$$

Diffuse boundary condition at the interface results in the following equation:

$$I^{-}_{A,ij,k}(x_{A\text{-Interface}}, \mu, t) = (q^{+}_{A,ij,k}(t) - q_{rad,ij,k}(t))/\pi \tag{7.91}$$

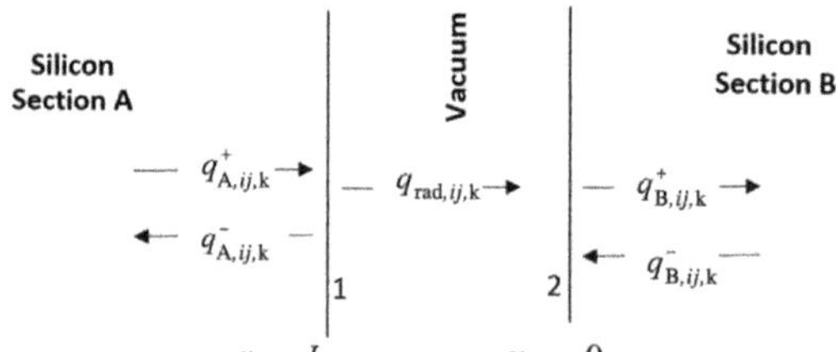

FIGURE 7.19 Schematic view of gap located in between two-silicon films pair and heat flux vectors (k is wavenumber, and $i = T$ represents transverse mode of phonons, $i = L$ stands for longitudinal mode of phonons, $j = A$ corresponds to acoustic, and $j = O$ stands for optical phonons). The space between $x_A = L_A$ and $x_B = 0$ corresponds to the gap size [3].

7.4.1.3.2 Energy Balance at Point 2 (Fig. 7.19)

When $k < k_{\text{cut-off}}$

$$q^{+}_{B,ij,k}(t) = q^{+}_{A,ij,k}(t) + q_{rad,ij,k}(t) \tag{7.92}$$

Considering the diffuse condition at the boundary, that is $I^{+}_{B,ij,k}$, independent of μ and incorporating Eq. (7.86) into the previous equation yields the following:

$$I^{+}_{B,ij,k}(x_{B\text{-Interface}}, \mu, t) = (q^{+}_{A,ij,k}(t) + q_{rad,ij,k}(t))/\pi \tag{7.93}$$

When $k > k_{\text{cut-off}}$

$$q^{+}_{B,ij,k}(t) = -q^{-}_{B,ij,k}(t) + q_{rad,ij,k}(t) \tag{7.94}$$

This can be reduced into the following:

$$I^{+}_{B,ij,k}(x_{B\text{-Interface}}, \mu, t) = (-q^{-}_{B,ij,k}(t) + q_{rad,ij,k}(t))/\pi \tag{7.95}$$

In Eqs. (7.82)–(7.95), $i = T$ represents transverse mode and $i = L$ stands for longitudinal mode. In addition, $j = A$ corresponds to acoustic and $j = O$ stands for optical phonons. Numerical treatment of the EPRT is similar to that given in Section 7.3. Therefore, we can refer to Section 7.3 for a numerical solution. The value of bulk properties of silicon, which are used in the simulation, are given in Table 7.4.

7.4.2 Findings of Dielectric-Dielectric Films Pair and Presence of Vacuum Gap at Interface

Thermal energy transport across the silicon films pair with the presence of a small vacuum gap is investigated. Near-field radiation heat transfer together with the CMM is incorporated at the edges of the film across the gap to account for the thermal resistance. Phonon transport characteristics are assessed through equivalent equilibrium temperature.

Fig. 7.20 shows the dimensionless equivalent equilibrium temperature $\left(\frac{T - T_{right}}{T_{left} - T_{right}}, \text{ where } T_{right} = 300\text{ K and } T_{left} = 301\text{ K}\right)$ variation along the

Table 7.4 Bulk Properties of Silicon Used in the Simulation

	Silicon Bulk Properties
Thermal conductivity, ($Wm^{-1}K^{-1}$)	157
Volumetric specific heat, ($Jm^{-3}K^{-1}$)	1,653,000
Speed (ms^{-1})	8430
Mean free path, $\Lambda = 3k/Cv$, (m)	33.8×10^{-9}
Relaxation time $\tau = \Lambda/v$, (s)	4.01×10^{-12}

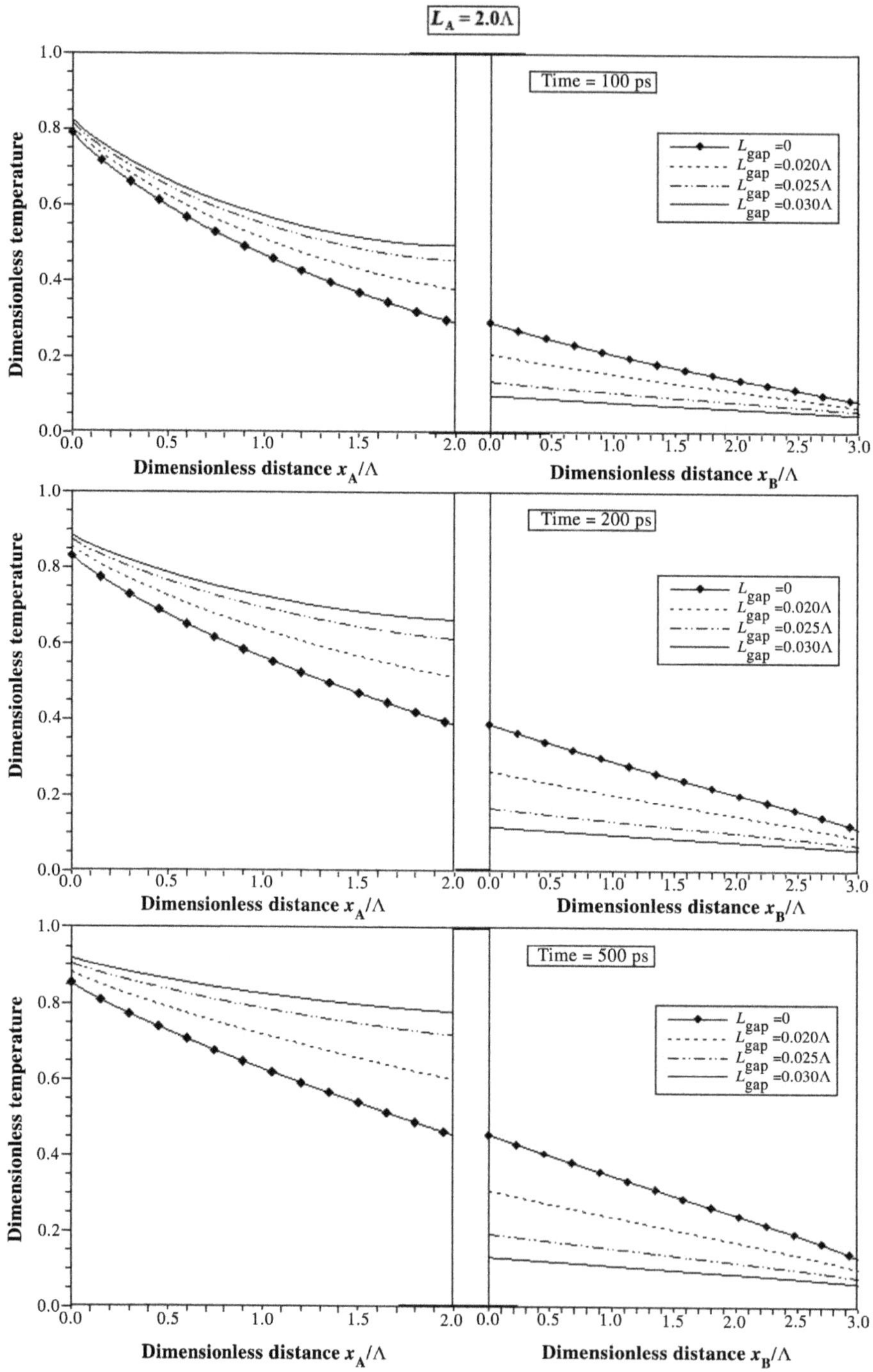

■ **FIGURE 7.20** Dimensionless temperature along the films pair for different gap sizes and heating periods. First film thickness is 2Λ, where Λ is the mean free path of silicon corresponding to a gray medium [3].

film thickness for different gap sizes and heating periods. Note that the first film thickness is twice that of Λ, which is the mean free path of silicon corresponding to a gray medium. Note that the mean is based on the gray body assumption of the bulk properties of silicon. Moreover, temperature data for the zero gap spacing between the films are also plotted for comparison. Temperature jump at the film edges across the gap increases as the gap size increases. In this case, phonons emitted from the high-temperature edge and having wavelengths larger than the gap size can possibly transmit to the second film. Phonons with smaller wavelengths than the gap size reflect from the first silicon film edge, located at the gap interface, while contributing to the film resistance in the close region of the edge. Therefore, phonon intensity increases in the region close to the first film edge while increasing the equivalent equilibrium temperature. Note that equivalent equilibrium temperature represents the average energy of all phonons around a local point when they redistribute adiabatically to an equilibrium state. Consequently, equivalent equilibrium temperature decay becomes gradual with increasing gap size along the first film thickness.

Although near-field radiation heat transfer from the first film edge toward the second film edge is incorporated in the analysis across the gap, equivalent equilibrium temperature remains low with increasing gap size despite the fact that temperature increases at the first film edge with increasing gap size. This indicates that energy transport by near-field radiation from the high-temperature film edge across the gap is less than the energy transport to the second film due to transmission of the long wavelength phonons. Therefore, the gap behaves like thermal boundary resistance and only phonons with larger wavelengths than the gap size can transmit to the edge of the second film. Moreover, reducing gap size allows more phonons to transmit from the first film edge to the second film edge across the gap. This in turn increases the phonon intensity decay sharply in the first film across its thickness. This behavior is observed from the decay of equivalent equilibrium temperature toward the first film edge located at the gap interface. As the heating period progresses, temperature jump across the gap increases because of the emitted phonon intensity, which increases with progressing time in the first film. As time progresses further, phonon intensity distribution in the films pair attains quasisteady and equivalent equilibrium temperature variation becomes almost linear across the film thickness; in this case, temperature jump across the gap reaches the highest.

Fig. 7.21 shows dimensionless equivalent equilibrium temperature along the film thickness for different gap sizes and heating periods, as similar

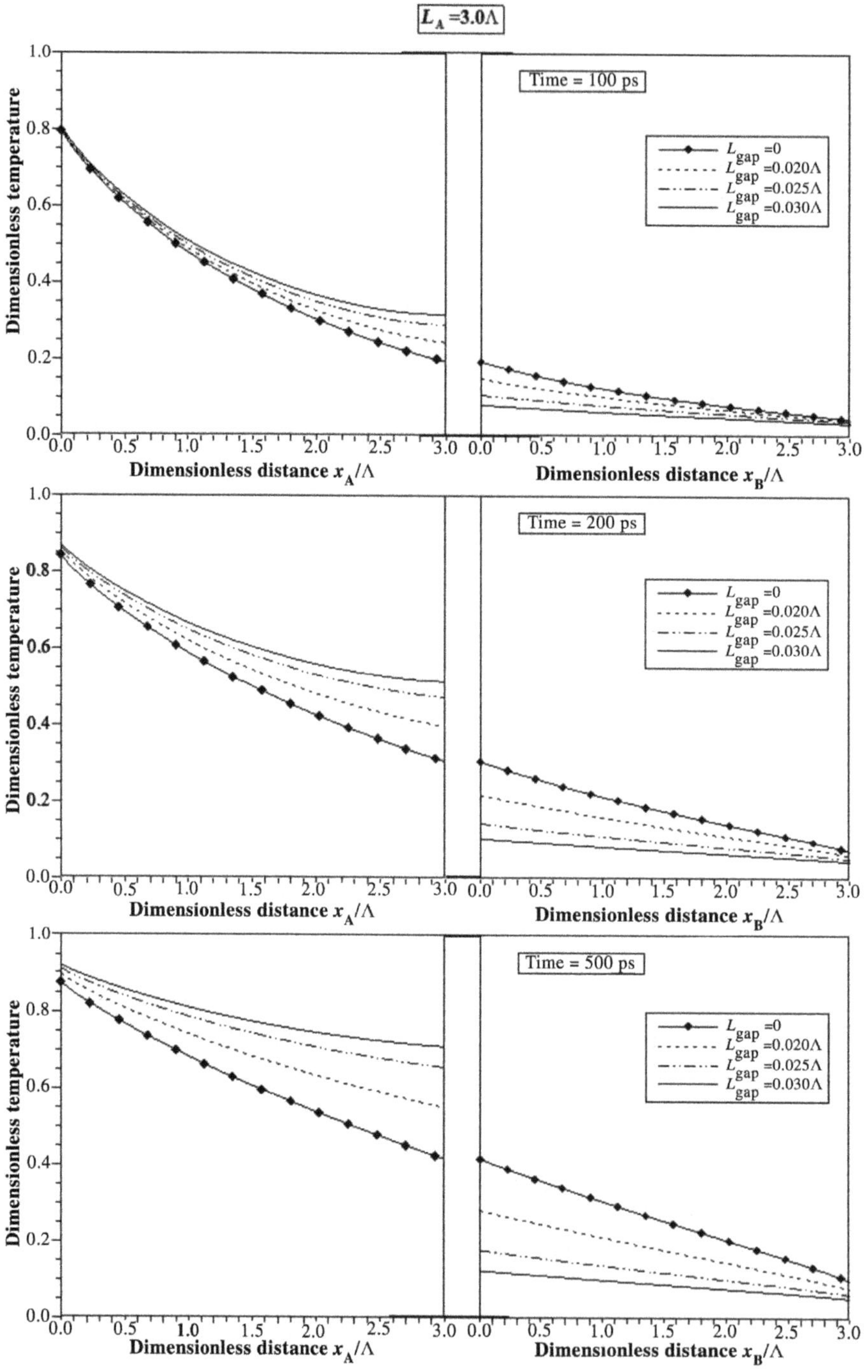

■ **FIGURE 7.21** Dimensionless temperature along the films pair for different gap sizes and heating periods. First film thickness is 3Λ, where Λ is the mean free path of silicon corresponding to a gray medium [3].

to Fig. 7.20; however, the first film thickness is 3Λ in Fig. 7.21. In general, equivalent equilibrium temperature behaves similarly to that shown in Fig. 7.20, provided that the decay of temperature in the films pair differs significantly. In addition, temperature jumps due to the first and the second films are different. In this case, temperature decay in the film becomes sharp and temperature jump across the gap becomes smaller with increasing the first film thickness. This behavior is associated with the phonon intensity distribution, which changes in the film with increasing film size due to increase of film resistance to the phonon transport. Phonons emitted from the high-temperature (301K) edge of the first silicon film undergoes scattering in the film and the ballistic contribution of the large wave phonons to the phonon transport reduces because of the increasing film thickness. Consequently, phonon intensity in the film reduces sharply with increasing film thickness. Increasing gap size increases temperature jump at the gap interface as similar those shown in Fig. 7.20.

Fig. 7.22 shows equivalent equilibrium temperature difference at the edges of the films, located at the gap interface, with the gap size for different film thicknesses for the heating period of 500 ps when the quasi steady heating is established. Temperature difference increases with increasing gap size, which is associated with the phonon wavelength. In this case, only phonons with larger wavelengths than the gap size can

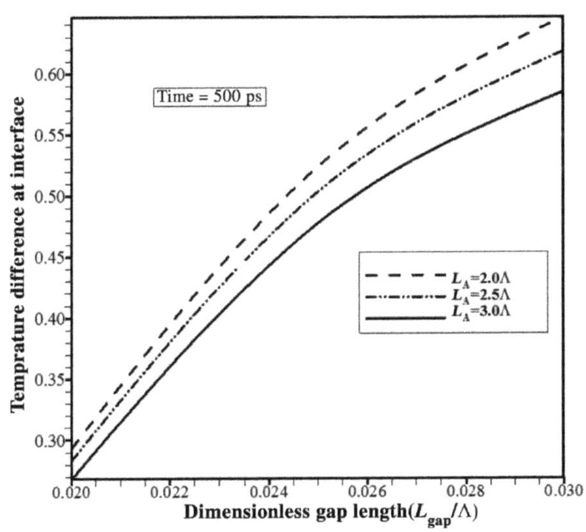

■ **FIGURE 7.22** Dimensionless temperature with different gap sizes for different first film thicknesses [3]. The heating period is 500 ps.

transmit from the first film edge to the second film edge. Although near-field radiation heating due to temperature difference at both edges across the gap enhances the thermal energy transport and reduces the temperature difference, contribution of the radiative thermal energy transport to temperature difference is considerably small as observed from Figs. 7.20 and 7.21. As the gap size increases, increase in temperature difference becomes gradual as compared to the small-sized gaps. This behavior is attributed to the boundary scattering of phonons at the film edges, which contributes to temperature jump taking place at each edge of the film. In this case, increasing intensity at the first film edge, due to a quasisteady heating situation, results in an increased temperature jump. Since phonon intensity in the second film edge is small for a large gap size due to reduced phonon transmission, temperature jump also becomes less. Therefore, rise of temperature difference due to the edges of the films, located at the gap interface, becomes less with increasing gap size. On the other hand, reducing the first film thickness increases temperature difference at the film edges, and is more pronounced with increasing gap size. Increasing film size modifies phonon intensity distribution in the first film and alters equivalent equilibrium temperature.

Fig. 7.23 shows the thermal conductivity ratio (k_{eff}/k_{Bulk}, where k_{eff} is the thermal conductivity of the film pair and k_{Bulk} is the thermal conductivity

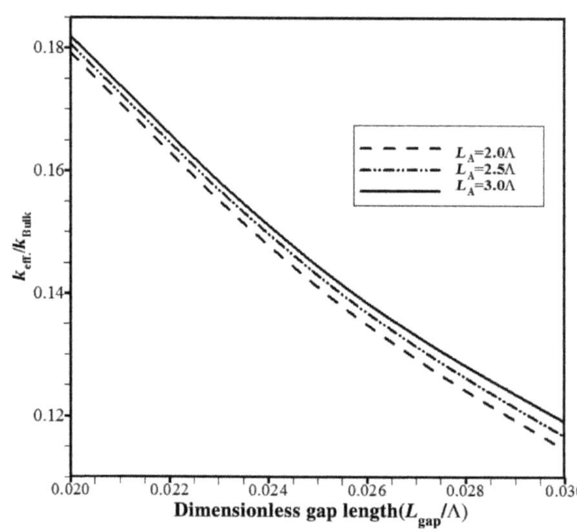

■ **FIGURE 7.23** Thermal conductivity ratio (k_{eff}/k_{Bulk}) with dimensionless gap length for different first film thicknesses. L_{gap}/Λ, where L_{gap} is the gap size in between the films and Λ is the mean free path of silicon corresponding to a gray medium [3].

of the bulk silicon ($k_{Bulk} = 157\ W\ m^{-1}K^{-1}$)) with the gap size. Note that k_{eff} is obtained from the steady state solution of the BTE across the films pair including the gap. k_{eff}/k_{Bulk} reduces with increasing gap size. In this case, k_{Bulk} remains the same; however, k_{eff} reduces with increasing gap size because of the phonon intensity distribution in the second film, which reduces with increasing gap size. Therefore, temperature variation at the film edges across gap increases with increasing gap size (Fig. 7.22), and in turn, has an adverse effect on the amount of heat transfer across the films pair, particularly in the second film. Moreover, those phonons having large wavelengths can possibly transmit to the second film across the gap. However, these phonons correspond to only a fraction of the total phonons emitted from the high-temperature edge of the first film; phonon energy transport toward the second film becomes small. Since energy transfer due to the near-field thermal radiation from the first film edge to the second film edge does change temperature considerably at the second film edge, the gap resembles the thermal boundary resistance in between the thin films while suppressing the amount of heat flow into the second film. Since altering the gap size within the films pair modifies the thermal conductivity, the findings become of interest to the thermal insulation applications. In addition, increasing the first film thickness results in the enhancement of effective thermal conductivity.

7.5 APPLICATIONS IN CROSS-PLANE TRANSPORT ACROSS METALLIC-DIELECTRIC THIN FILMS COMPOSITE WITH MINUTE GAP UNDER LASER IRRADIATION

The cross-plane energy transport across the metallic-dielectric thin films pair with the existence of a minute vacuum gap in between them introduces complicated transport properties and transport characteristics. When the size of the films in the films pair becomes comparable or slightly larger than the mean free path of constituting films, energy transport is governed by phonons in the dielectric film and by the collisional process defined through the electron–phonon coupling in the metallic film. However, the presence of a vacuum gap at the interface of the films pair modifies energy transfer from one film to the other. When the gap size reduces to less than the average mean free path of constituting film (Casimir limit), the thermal radiative transport across the gap takes place. In addition, some of the ballistic phonons from the metallic film can jump across the vacuum gap while contributing to energy transport across the interface of the films pair. Energy transport in the metallic film depends on the absorption characteristics of the incident laser irradiation and electron–phonon coupling; however, ballistic phonon and thermal radiation transport across the vacuum gap determines the energy state in the dielectric film. In this section, the phonon transport across aluminum-silicon thin films with the presence of a vacuum gap is considered in line with the previous study [4].

Heat generation in the aluminum film due to a laser short pulse irradiation is incorporated in the analysis. The frequency-dependent BTE is used to formulate the energy transport in electron and lattice subsystems in the aluminum film, which is subjected to a laser short-pulse irradiation. Energy transport across the vacuum gap is formulated incorporating the contributions of the ballistic phonons together with thermal radiation, due to the Casimir limit, while satisfying the energy balance across the interface. Equilibrium equivalent temperature is introduced to quantify the phonon intensities in the films pair. Electron–phonon coupling is introduced to model the thermal transport characteristics in the aluminum film.

Fig. 7.24 shows the arrangement of thin films; a minute vacuum is separating the aluminum and silicon films. The laser heating is

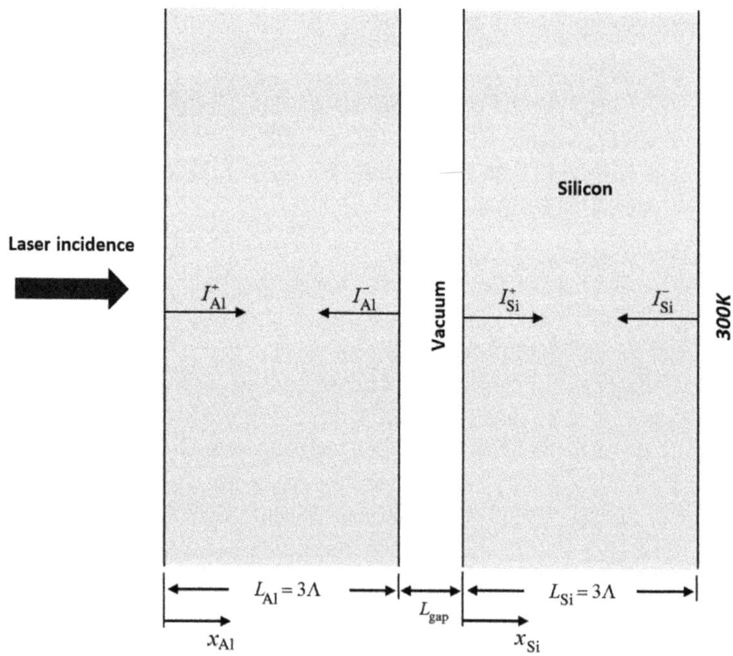

■ **FIGURE 7.24** Schematic view of aluminum-silicon film pair and laser irradiation. Λ is the mean free path of silicon corresponding to the gray medium ($\Lambda = 33.8$ nm) [4].

introduced at the left edge of the aluminum film as shown in Fig. 7.24. In order to get the solution of physical domain, the proper interface condition at the interface of aluminum and silicon are incorporated, which satisfies the energy balance.

Although formulation of the phonon radiative transport equation is given in the previous sections, it is also presented briefly here due to completeness of the mathematical arguments. The transient, one-dimensional, frequency-dependent EPRT in the relaxation-time approximation and excluding the effects of external forces can be written as [8]

$$\frac{1}{v_k}\frac{\partial I_k}{\partial t} + \mu\frac{\partial I_k}{\partial x} = \frac{\frac{1}{2}\int_{-1}^{1} I_k d\mu - I_k}{v_k \tau_k} \tag{7.96}$$

In Eq. (7.96), $\mu = \cos\theta$, θ is polar angle, t is time, x is distance in spatial coordinate, τ_k is wavenumber (frequency)-dependent relaxation time, v_k is wavenumber (frequency)-dependent group velocity, I_k^0 wavenumber (frequency)-dependent equilibrium phonon intensity, I_k wavenumber (frequency)-dependent phonon intensity, and k is the wavenumber.

7.5.1 Aluminum Film

In the metal thin films, thermal separation occurs between the electron and phonon subsystems, which requires incorporating the restructuring of the energy equations in line with the BTE. In both subsystems the energy transport is formulated using the BTE. In this case, separate EPRTs for both subsystems need to be developed. During the energy transport the electron and lattice (phonon) subsystems thermally communicate with each other. Therefore, electron–phonon coupling is included in the EPRTs of both subsystems. Moreover, the laser heating of the aluminum thin film should be modeled by means of a volumetric heat source term in the EPRT for the electron subsystem. The electron–phonon coupling term as well as the volumetric heat source term are incorporated in such a way that in the diffusive limit, the EPRT becomes the standard two-temperature model with a volumetric heat generation [23].

7.5.1.1 *Electron Subsystem*

The relaxation time of electrons is very short as compare to the relaxation time of phonons; therefore a gray body assumption is incorporated for radiative energy transfer in electron subsystems. Since the laser irradiated energy is absorbed by electrons within the absorption depth of aluminum, the Beer–Lambert law can be adopted to formulate the volumetric source resembling the absorbed laser power inside the substrate material. The proposed energy transport equation for electron subsystem, which includes the electron–phonon coupling and volumetric source, should satisfy the conservation of energy. For the electron subsystem the governing equation can be written as

$$\frac{1}{v_e}\frac{\partial I_e}{\partial t}+\mu\frac{\partial I_e}{\partial x_{Al}}=\frac{\frac{1}{2}\int_{-1}^{1}I_e d\mu-I_e}{v_e\tau_e}-\frac{G}{2}\left(\frac{1}{C_e v_e}\int_{-1}^{1}I_e d\mu-\frac{1}{C_{Al}v_{Al}}\int_{-1}^{1}I_{Al}d\mu\right)+\frac{1}{4\pi}S(x,t) \tag{7.97}$$

where

$$S(x,t)=(1-R)I_0\alpha\exp\left[-\alpha x-2.77(t_h/t_p)^2\right] \tag{7.98}$$

In Eq. (7.97), I_e is the electron intensity, v_e is the electron speed, τ_e is the electron relaxation time, C_e is electron volumetric specific heat, I_{Al} is the phonon intensity, v_{Al} is the phonon group velocity, and C_{Al} is the phonon volumetric specific heat. S is the laser volumetric heat source term, R is the reflectance, α is the absorption coefficient, I_0 is the laser beam intensity, t_p is the laser pulse duration, and G is the electron–phonon coupling. The rigorous form of temperature-dependent electron–phonon coupling *(G)* can be expressed as [24]

$$G(T_e)=\frac{\pi\hbar k_B\lambda\langle\omega^2\rangle}{D_e(\varepsilon_f)}\int_{-\infty}^{\infty}D_e^2(\varepsilon)\left(-\frac{\partial fe}{\partial\varepsilon}\right)d\varepsilon \tag{7.99}$$

$\langle\omega^2\rangle$ is the second moment of the phonon spectrum [25], λ is the electron–phonon mass enhancement parameter [24], D_e is the density of the state of electron, ε is the energy level, ε_f is the Fermi level energy, T_e is the electron temperature, fe is the Fermi distribution function, which is given as [24]

$$f_e=\exp[(\varepsilon-\mu_e)/k_B T_e] \tag{7.100}$$

In this equation, μ_e represents the chemical potential of electron. Electron–phonon coupling (Eq. 7.99) varies at high temperature [26]. At low temperatures, the variation of electron–phonon coupling with temperature is shown to be considerably small [27].

Using the definitions of the phonon and electron temperatures T_{Al} and T_e, which are, $C_{Al}v_{Al}T_{Al}/4\pi=\frac{1}{2}\int_{-1}^{1}I_{Al}d\mu$ and $C_e v_e T_e/4\pi=\frac{1}{2}\int_{-1}^{1}I_e d\mu$, Eq. (7.97) becomes

$$\frac{1}{v_e}\frac{\partial I_e}{\partial t}+\mu\frac{\partial I_e}{\partial x_{Al}}=\frac{\frac{1}{2}\int_{-1}^{1}I_e d\mu-I_e}{v_e\tau_e}-\frac{G}{4\pi}(T_e-T_{Al})+\frac{1}{4\pi}S(x,t)\qquad \begin{matrix}0<x_{Al}<3\Lambda\\ -1<\mu<1\end{matrix} \tag{7.101}$$

Table 7.5 Properties Used in the Simulations

	Aluminum Electron subsystem
Thermal conductivity, ($Wm^{-1}K^{-1}$)	225.7
Volumetric specific heat, ($Jm^{-3}K^{-1}$)	23,659
Speed (ms^{-1})	2.03×10^{6}
Mean free path, $\Lambda = 3k/Cv$, (m)	14.1×10^{-9}
Relaxation time $\tau = \Lambda/v$, (s)	6.945×10^{-15}
Electron–phonon coupling constant in aluminum, (Wm^{-3})	2.4642×10^{17}
Laser source parameters	
$R = 0.5$	
$\alpha = 1.3 \times 10^{8}$ 1 m^{-1}	
$t_p = 92$ fs	
$I_{laser} = 0.1 \times 10^{12}$ W m^{-2}	

In the conduction limit, Eq. (7.101) can be reduced to a modified form of the two-equation model; details are in Section 4.4.6.4. The properties of electron subsystem (v_e, C_e, and τ_e) are given in Table 7.5.

7.5.1.2 *Lattice Subsystem*

The radiative transport equation for the lattice subsystem can be written as [28]

$$\frac{1}{v_{Al}}\frac{\partial I_{Al}}{\partial t} + \mu\frac{\partial I_{Al}}{\partial x_{Al}} = \frac{\frac{1}{2}\int_{-1}^{1} I_{Al}d\mu - I_{Al}}{v_{Al}\tau_{Al}} - \frac{G}{2}\left(\frac{1}{C_{Al}v_{Al}}\int_{-1}^{1} I_{Al}d\mu - \frac{1}{C_e v_e}\int_{-1}^{1} I_e d\mu\right) \tag{7.102}$$

In Eq. (7.102), I_{Al} is the phonon intensity, v_{Al} is the phonon group velocity, τ_{Al} is the phonon relaxation time, and C_{Al} is the phonon volumetric specific heat.

Using the definitions of the phonon and electron temperatures, Eq. (7.102) becomes

$$\frac{1}{v_{Al}}\frac{\partial I_{Al}}{\partial t} + \mu\frac{\partial I_{Al}}{\partial x_{Al}} = \frac{\frac{1}{2}\int_{-1}^{1} I_{Al}d\mu - I_{Al}}{v_{Al}\tau_{Al}} - \frac{G}{4\pi}(T_{Al} - T_e) \qquad \begin{matrix} 0 < x_{Al} < 3\Lambda \\ -1 < \mu < 1 \end{matrix} \tag{7.103}$$

To incorporate the frequency dependency in the lattice (phonon) subsystem of aluminum, the dispersion relations are used [29]. The dispersion relations provide information about the acoustic mode of vibrations of phonons, which contains three different polarizations—two transverse and one longitudinal. Because of the degeneracy, the two transverse waves are represented by one curve. The second curve represents the longitudinal wave in dispersion relation of aluminum [29]. As the electron relaxation time is much shorter than the phonon relaxation time, we can incorporate the gray body assumption in the electron subsystem; the radiative energy transport in the electron subsystem is independent of frequency. Therefore, the overall contribution of the electron–phonon coupling term (the second term on right-hand side of Eq. (7.103)) is first calculated and then redistributed equally among all the frequency-dependent branches of the lattice subsystem. The electron–phonon coupling term for each frequency-dependent branch can be written as

$$G_{\mathrm{k}} = \frac{G}{4\pi}\frac{(T_{\mathrm{Al}} - T_{\mathrm{e}})}{3k_{\max}} \tag{7.104}$$

The governing equations for frequency-dependent transient phonon radiative transport in a lattice (phonon) subsystem can be written as

$$\frac{1}{v_{\mathrm{Al},i\mathrm{A,k}}}\frac{\partial I_{\mathrm{Al},i\mathrm{A,k}}}{\partial t} + \mu\frac{\partial I_{\mathrm{Al},i\mathrm{A,k}}}{\partial x_{\mathrm{Al}}} = \frac{\frac{1}{2}\int_{-1}^{1} I_{\mathrm{Al},i\mathrm{A,k}}d\mu - I_{\mathrm{Al},i\mathrm{A,k}}}{v_{\mathrm{Al},i\mathrm{A,k}}\tau_{\mathrm{Al},i\mathrm{A,k}}} - G_{\mathrm{k}} \qquad \begin{matrix} 0 < x_{\mathrm{Al}} < 3\Lambda \\ -1 < \mu < 1 \end{matrix} \tag{7.105}$$

where $i = T$ represents transverse mode, $i = L$ stands for longitudinal mode, and A represents the acoustic phonons.

The frequency-dependent phonon group velocity can be calculated from the following:

$$v_{\mathrm{p}} = \frac{\partial \omega}{\partial k} \tag{7.106}$$

Here, the $\omega(k)$ is the wavenumber-dependent frequency, which can be obtained from dispersion relation. The constant relaxation time is assumed for all frequency-dependent branches of the lattice(phonon) subsystem. The value $\tau = 2.73$ ps is selected so that the phonon thermal conductivity of aluminum $\left(k_{\mathrm{p}} = (1/3)\sum_{p=1}^{2}\int_{0}^{k_{\max,\mathrm{Al}}} C_{\mathrm{p,TA/LA}}v_{\mathrm{p,TA/LA}}^{2}\tau dk\right)$ is $k_{\mathrm{p}} \approx 11.3\ \mathrm{W\,m^{-1}K^{-1}}$ [30].

7.5.2 Silicon Film

Eq. (7.96) is applicable for silicon film. The dispersion relation of silicon [15] is used to solve the frequency-dependent phonon transport equation.

In the case of silicon, six phonon branches are present: two acoustic-transverse, two acoustic-optical, one acoustic-longitudinal, and one optical-longitudinal. In order to determine the transport characteristics of the phonon we need to solve a separate EPRT for each phonon branch.

In the silicon film the acoustic-transverse and optical-transverse branches are degenerate. Therefore, four phonon branches represent all six branches. The equations for phonon branches can be given as

$$\frac{1}{v_{\mathrm{Si},ij,\mathrm{k}}}\frac{\partial I_{\mathrm{Si},ij,\mathrm{k}}}{\partial t}+\mu\frac{\partial I_{\mathrm{Si},ij,\mathrm{k}}}{\partial x_{\mathrm{Si}}}=\frac{\frac{1}{2}\int_{-1}^{1}I_{\mathrm{Si},ij,\mathrm{k}}d\mu-I_{\mathrm{Si},ij,\mathrm{k}}}{v_{\mathrm{Si},ij,\mathrm{k}}\tau_{\mathrm{Si},ij,\mathrm{k}}}\quad\begin{array}{c}0<x_{\mathrm{Si}}<3\Lambda\\-1<\mu<1\end{array}\tag{7.107}$$

where $i=T$ represents transverse mode and $i=L$ stands for longitudinal mode. In addition, $j=A$ corresponds to acoustic and $j=O$ stands for optical phonons.

In the previous equation phonon intensity I is the function of t, x, μ and also is dependent on is wavenumber (k) and phonon branch.

7.5.3 Definition of Temperature and Heat Flux

Equivalent equilibrium phonon temperature is used anywhere in the thin aluminum and silicon film. In this case:

7.5.3.1 Phonon Temperature in Aluminum Film

$$T_{\mathrm{Al}}(x,t)=\frac{2\pi\int_{0}^{k_{\max,Al}}\int_{-1}^{1}(2I_{\mathrm{Al,TA,k}}+I_{\mathrm{Al,LA,k}})d\mu dk}{\int_{0}^{k_{\max,Al}}(2C_{\mathrm{Al,TA,k}}v_{\mathrm{Al,TA,k}}+C_{\mathrm{Al,LA,k}}v_{\mathrm{Al,LA,k}})dk}\tag{7.108}$$

7.5.3.2 Phonon Temperature in Silicon Film

$$T_{\mathrm{Si}}(x,t)=\frac{2\pi\int_{0}^{k_{\max,\mathrm{Si}}}\int_{-1}^{1}(2I_{\mathrm{Si,TA,k}}+I_{\mathrm{Si,LA,k}}+2I_{\mathrm{Si,TO,k}}+I_{\mathrm{Si,LO,k}})d\mu dk}{\int_{0}^{k_{\max,\mathrm{Si}}}\left(\begin{array}{l}2C_{\mathrm{Si,TA,k}}v_{\mathrm{Si,TA,k}}+C_{\mathrm{Si,LA,k}}v_{\mathrm{Si,LA,k}}\\+2C_{\mathrm{Si,TO,k}}v_{\mathrm{Si,TO,k}}+C_{\mathrm{Si,LO,k}}v_{\mathrm{Si,LO,k}}\end{array}\right)dk}\tag{7.109}$$

7.5.3.3 Electron Temperature in Aluminum Film

$$T_{\mathrm{e}}(x,t)=\frac{2\pi\int_{-1}^{1}I_{\mathrm{e}}d\mu}{C_{\mathrm{e}}v_{\mathrm{e}}}\tag{7.110}$$

7.5.3.4 *Phonon Heat Flux in Aluminum Film*

The heat flux is the sum of contribution from all branches of the phonons, and can be written as

$$q_{\mathrm{Al}}(x,t) = 2\pi \int_0^{k_{\mathrm{max,Al}}} \int_{-1}^{1} (2I_{\mathrm{Al,TA,k}} + I_{\mathrm{Al,LA,k}})\mu d\mu dk \quad (7.111)$$

where k_{max} is the maximum wavenumber, which can be obtained from the dispersion relation, and the subscript Al represents aluminum.

7.5.3.5 *Phonon Heat Flux Silicon Film*

The heat flux is the sum of contribution from all branches of the phonons, and can be written as

$$q_{\mathrm{Si}}(x,t) = 2\pi \int_0^{k_{\mathrm{max,Si}}} \int_{-1}^{1} (2I_{\mathrm{Si,TA,k}} + I_{\mathrm{Si,LA,k}} + 2I_{\mathrm{Si,TO,k}} + I_{\mathrm{Si,LO,k}})\mu d\mu dk \quad (7.112)$$

where k_{max} is maximum wavenumber, which can be obtained from the dispersion relation, and the subscript Si represents silicon.

7.5.3.6 *Electron Heat Flux in Aluminum Film*

The heat flux of the electron subsystem can be written as

$$q_{\mathrm{e}}(x,t) = 2\pi \int_{-1}^{1} I_{\mathrm{e}}\mu d\mu \quad (7.113)$$

7.5.4 **Initial, Boundary, and Interface Conditions**

Since we are solving the transient radiative phonon transport equation, the initial condition is incorporated. The solution domain is initialized at 300K. Boundary conditions are also introduced at the boundaries of the aluminum and silicon thin films. In addition to the initial and boundary conditions, the interface conditions are also implemented such that energy balance is satisfied at the interface of thin films.

7.5.4.1 *Aluminum Film (Fig. 7.24)*

7.5.4.1.1 **Phonon Subsystem**

The initial condition in the phonon subsystem of aluminum is given as

$$I_{\mathrm{Al,TA/LA,k}}(x_{\mathrm{Al}}, \mu, t=0) = \frac{C_{\mathrm{Al,TA/LA,k}} v_{\mathrm{Al,TA/LA,k}} T_{\mathrm{initial}}}{4\pi} \quad (7.114)$$

where $T_{\mathrm{initial}} = 300\mathrm{K}$

The aluminum film left boundary is considered to be insulated.

$$q_{\mathrm{Al,TA/LA}}(x=0,t) = 2\pi \int_0^{k_{\mathrm{max,Al}}} \int_{-1}^{1} I_{\mathrm{Al,TA/LA,k}}(x=0,\mu,t)\mu d\mu dk = 0 \quad (7.115)$$

7.5.4.1.2 **Electron Subsystem**

The initial condition in the electron subsystem of aluminum is given as

$$I_{\mathrm{e}}(x_{\mathrm{Al}},\mu,t=0) = \frac{C_{\mathrm{e}} v_{\mathrm{e}} T_{\mathrm{initial}}}{4\pi} \quad (7.116)$$

where $T_{\mathrm{initial}} = 300\mathrm{K}$

The electrons do not exchange the thermal energy from the phonon of different materials. Hence, at the interface of metal and dielectric thin film, the electron subsystem of metallic film does not exchange thermal energy with the phonons of the dielectric thin film. Therefore, for the electron subsystem, the interface boundary is considered to be insulated.

$$q_{\mathrm{e}}(x_{\mathrm{Al}}=0,t) = 2\pi \int_{-1}^{1} I_{\mathrm{e}}(x_{\mathrm{Al}}=0,\mu,t)\mu d\mu = 0 \quad (7.117)$$

$$q_{\mathrm{e}}(x_{\mathrm{Al}}=3\Lambda,t) = 2\pi \int_{-1}^{1} I_{\mathrm{e}}(x_{\mathrm{Al}}=3\Lambda,\mu,t)\mu d\mu = 0 \quad (7.118)$$

7.5.4.2 ***Silicon Film (Fig. 7.24)***

The initial condition in silicon is given as

$$I_{\mathrm{Si},ij,\mathrm{k}}(x_{\mathrm{Si}},\mu,t=0) = \frac{C_{\mathrm{Si},ij,\mathrm{k}} v_{\mathrm{Si},ij,\mathrm{k}} T_{\mathrm{initial}}}{4\pi} \quad (7.119)$$

where $T_{\mathrm{initial}} = 300\mathrm{K}$

The boundary condition on the right side of silicon can be given as

$$I_{\mathrm{Si},ij,\mathrm{k}}(x_{\mathrm{Si}}=3\Lambda,\mu,t) = \frac{C_{\mathrm{Si},ij,\mathrm{k}} v_{\mathrm{Si},ij,\mathrm{k}} T_{\mathrm{right}}}{4\pi} \quad (7.120)$$

where $T_{\mathrm{right}} = 300\mathrm{K}$

In Eqs. (7.119) and (7.120), $i = T$ represents transverse mode and $i = L$ stands for longitudinal mode. In addition, $j = A$ corresponds to acoustic and $j = O$ stands for optical phonons.

7.5.4.3 ***Interface Conditions***

Interface conditions are required to get the phonon intensities distribution in both thin films. The intensities that are going in a positive x direction are represented by (I_{Al}^{+} and I_{Si}^{+}) and the intensities going in a negative

x direction are represented by (I_{Al}^{-} and I_{Si}^{-}), as shown in the Fig. 7.24. Energy balance along with the consideration of radiation heat transfer is applied in order to get the phonon intensities at the aluminum and silicon film interface.

The radiation heat transfer between the aluminum and silicon thin films is given by

$$q_{\text{rad,total}}(t) = \sigma(T_{\text{Al}}^4(x_{\text{Al-Interface}}, t) - T_{\text{Si}}^4(x_{\text{Si-Interface}}, t)) \tag{7.121}$$

where σ is the Stefan–Boltzmann constant $x_{\text{Al-Interface}} = 3\Lambda$ and $x_{\text{Si-Interface}} = 0$

Those phonon waves, where the wavelength ($\lambda_{\text{Al,Si}}$) is larger than the gap length (L_{gap}) (i.e., $\lambda_{\text{Al,Si}} \geq L_{\text{gap}}$), can jump from one film edge to another film edge (Aluminum$\rightleftarrows$Silicon). For the phonon transmittance, the cut-off wavenumber is defined, which is based on the gap length (L_{gap}) (i.e., $k_{\text{cut-off}} = 2\pi/L_{\text{gap}}$). Phonon waves can jump over the vacuum gap, if the wavenumber of phonon waves is less than the cut-off wavenumber ($k < k_{\text{cut-off}}$). Depending upon the wavenumber of phonons, the energy of phonons is subdivided into two parts. Part 1 represents the phonon waves with a wavenumber less than the cut-off wavenumber ($k < k_{\text{cut-off}}$); i.e., phonon waves can jump over the gap. Part 2 is representative of phonon waves that cannot jump over the gap; i.e., ($k < k_{\text{cut-off}}$).

The radiation heat transfer between two thin films is subdivided into two parts, which can be given by

$$q_{\text{rad,1}}(t) = \frac{\langle C_{\text{Al}} v_{\text{Al}} \rangle \big|_0^{k_{\text{cut-off}}}}{\langle C_{\text{Al}} v_{\text{Al}} \rangle \big|_0^{k_{\text{max,Al}}}} q_{\text{rad,total}}(t) \tag{7.122}$$

and

$$q_{\text{rad,2}}(t) = \frac{\langle C_{\text{Al}} v_{\text{Al}} \rangle \big|_{k_{\text{cut-off}}}^{k_{\text{max,Al}}}}{\langle C_{\text{Al}} v_{\text{Al}} \rangle \big|_0^{k_{\text{max,Al}}}} q_{\text{rad,total}}(t) \tag{7.123}$$

where

$$\langle C_{\text{Al}} v_{\text{Al}} \rangle \big|_0^{k_{\text{cut-off}}} = \int_0^{k_{\text{cut-off}}} (2C_{\text{Al,TA,k}} v_{\text{Si,TA,k}} + C_{\text{Al,LA,k}} v_{\text{Si,LA,k}}) dk \tag{7.124}$$

$$\langle C_{\text{Al}} v_{\text{Al}} \rangle \big|_0^{k_{\text{max,Si}}} = \int_0^{k_{\text{max,Si}}} (2C_{\text{Al,TA,}k} v_{\text{Si,TA,}k} + C_{\text{Al,LA,}k} v_{\text{Si,LA,}k}) dk \tag{7.125}$$

$$\langle C_{\text{Al}} v_{\text{Al}} \rangle \big|_{k_{\text{cut-off}}}^{k_{\text{max,Si}}} = \int_{k_{\text{cut-off}}}^{k_{\text{max,Si}}} (2C_{\text{Al,TA,}k} v_{\text{Si,TA,}k} + C_{\text{Al,LA,}k} v_{\text{Si,LA,}k}) dk \tag{7.126}$$

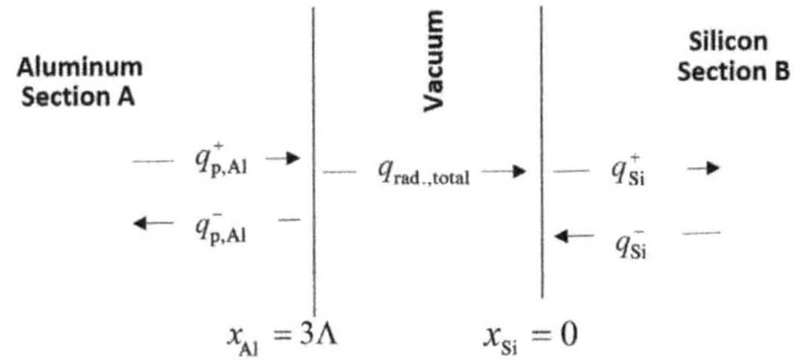

■ **FIGURE 7.25** Schematic view of energy flow across the vacuum gap in between the films pair [4].

7.5.4.3.1 Energy Balance for Part 1 ($k < k_{cut\text{-}off}$)

The heat fluxes at interface $q^+_{Al,1}(t)$ and $q^-_{Si,1}(t)$ (as shown in Fig. 7.25) can be written as

$$q^+_{Al,1}(t) = 2\pi \int_0^{k_{cut\text{-}off}} \int_0^1 \left(I^+_{Al,LA,k}(x_{Al\text{-}Interface}, \mu, t) + 2I^+_{Si,Al,k}(x_{Al\text{-}Interface}, \mu, t)\right) \mu d\mu dk \tag{7.127}$$

$$q^-_{Si,1}(t) = 2\pi \int_0^{k_{cut\text{-}off}} \int_{-1}^0 \begin{pmatrix} I^-_{Si,LA,k}(x_{Si\text{-}Interface}, \mu, t) + 2I^-_{Si,TA,k}(x_{Si\text{-}Interface}, \mu, t) \\ + I^-_{Si,LO,k}(x_{Si\text{-}Interface}, \mu, t) \\ + 2I^-_{Si,TO,k}(x_{Si\text{-}Interface}, \mu, t) \end{pmatrix} \mu d\mu dk \tag{7.128}$$

At the interface, the energy balance provides the following:

$$q^-_{Al,1}(t) = T_{Si\text{-}Al} q^-_{Si,1}(t) - R_{Al\text{-}Si} q^+_{Al,1}(t) + q_{rad,1}(t) \tag{7.129}$$

$$q^+_{Si,1}(t) = T_{Al\text{-}Si} q^+_{Al,1}(t) - R_{Si\text{-}Al} q^-_{Si,1}(t) + q_{rad,1}(t) \tag{7.130}$$

where T represents the transmittance and R represent the reflectance. The diffusive mismatch model is used to calculate the transmittance and the reflectance, which can be written as

$$T_{Al\text{-}Si/Si\text{-}Al} = \frac{\left\langle C_{Si/Al} v_{Si/Al} \right\rangle \Big|_0^{k_{cut\text{-}off}}}{\left\langle C_{Al} v_{Al} \right\rangle \Big|_0^{k_{cut\text{-}off}} + \left\langle C_{Si} v_{Si} \right\rangle \Big|_0^{k_{cut\text{-}off}}} \tag{7.131}$$

where

$$\left\langle C_{Si} v_{Si} \right\rangle \Big|_0^{k_{cut\text{-}off}} = \int_0^{k_{cut\text{-}off}} \begin{pmatrix} 2C_{Si,TA,k} v_{Si,TA,k} + C_{Si,LA,k} v_{Si,LA,k} \\ + 2C_{Si,TO,k} v_{Si,TO,k} + C_{Si,LO,k} v_{Si,LO,k} \end{pmatrix} dk \tag{7.132}$$

The reflectance can be written as

$$R_{Si\text{-}Al/Al\text{-}Si} = 1 - T_{Si\text{-}Al/Al\text{-}Si} \tag{7.133}$$

7.5.4.3.2 **Energy Balance for Part 2 ($k > k_{cut\text{-}off}$)**

The heat fluxes at interface $q^{+}_{Al,2}(t)$ and $q^{-}_{Si,2}(t)$ (as shown in Fig. 7.25) can be written as

$$q^{+}_{Al,2}(t) = 2\pi \int_{k_{cut-off}}^{k_{max,Al}} \int_{0}^{1} (I^{+}_{Al,LA,k}(x_{Al\text{-}Interface}, \mu, t) + 2I^{+}_{Al,TA,k}(x_{Al\text{-}Interface}, \mu, t))\mu d\mu dk \tag{7.134}$$

$$q^{-}_{Si,2}(t) = 2\pi \int_{k_{cut-off}}^{k_{max,Si}} \int_{-1}^{0} \begin{pmatrix} I^{-}_{Si,LA,k}(x_{Si\text{-}Interface}, \mu, t) + 2I^{-}_{Si,TA,k}(x_{Si\text{-}Interface}, \mu, t) \\ + I^{-}_{Si,LO,k}(x_{Si\text{-}Interface}, \mu, t) \\ + 2I^{-}_{Si,TO,k}(x_{Si\text{-}Interface}, \mu, t) \end{pmatrix} \mu d\mu dk \tag{7.135}$$

In this part the phonon wavelength $\lambda_{Al,Si}$ is smaller than gap length L_{gap}, therefore all the phonon energy will reflect at the interface. The energy balance at the interface can be written as

$$q^{-}_{Al,2}(t) = -q^{+}_{Al,2}(t) + q_{rad,2}(t) \tag{7.136}$$

$$q^{+}_{Si,2}(t) = -q^{-}_{Si,2}(t) + q_{rad,2}(t) \tag{7.137}$$

7.5.4.3.3 **Aluminum Interface Condition**

The outgoing phonon intensities (I^{-}_{Al}) from the aluminum interface can be determined from the total amount of energy, which is leaving the film interface and coming toward the aluminum, and can be written as follows:

$$q^{-}_{Al,total}(t) = q^{-}_{Al,1}(t) + q^{-}_{Al,2}(t) \tag{7.138}$$

where

$$q^{-}_{Al,total}(t) = 2\pi \int_{0}^{k_{max,Al}} \int_{-1}^{0} (I^{-}_{Al,total}(t))\mu d\mu dk \tag{7.139}$$

$I^{-}_{Al,total}(t)$ is the total of phonon intensities at the interface, which is leaving the film interface and coming toward the aluminum, given as

$$I^{-}_{Al,total}(t) = I^{-}_{Al,LA,k}(x_{Al\text{-}Interface}, \mu, t) + 2I^{-}_{Al,TA,k}(x_{Al\text{-}Interface}, \mu, t) \tag{7.140}$$

In Eq. (7.139), by incorporating the diffusive condition and using Eqs. (7.129), (7.136), and (7.138), we get

$$I^{-}_{Al,total}(t) = (q^{+}_{Al,1}(t) + q^{+}_{Al,2}(t) - q_{rad,1}(t) - q_{rad,2}(t))/\pi k_{max,Al} \tag{7.141}$$

$I^{-}_{\mathrm{Al,total}}(t)$ is redistributed among the all branches of phonons at interface, which can be written as

$$I^{-}_{\mathrm{Al,TA/LA},k}(x_{\mathrm{Al\text{-}Interface}},\mu,t)=\frac{C_{\mathrm{Al,TA/LA,k}}v_{\mathrm{Al,TA/LA,k}}k_{\mathrm{max,Al}}}{\langle C_{\mathrm{Al}}v_{\mathrm{Al}}\rangle\big|_{0}^{k_{\mathrm{max,Si}}}}I^{-}_{\mathrm{Al,total}}(t) \quad (7.142)$$

7.5.4.3.4 **Silicon Interface Condition**

The outgoing phonon intensities (I^{+}_{Si}) from the silicon interface can be determined from the total amount of outgoing energy from the film interface toward the aluminum, which can be given by

$$q^{+}_{\mathrm{Si,total}}(t)=q^{+}_{\mathrm{Si},1}(t)+q^{+}_{\mathrm{Si},2}(t) \quad (7.143)$$

where

$$q^{-}_{\mathrm{Si,total}}(t)=2\pi\int_{0}^{k_{\mathrm{max,Si}}}\int_{0}^{1}(I^{+}_{\mathrm{Si,total}}(t))\mu d\mu dk \quad (7.144)$$

$I^{+}_{\mathrm{Si,total}}(t)$ is the total of phonon intensities at the interface, leaving the film interface and coming toward the silicon, and is given as

$$I^{+}_{\mathrm{Si,total}}(t)=\begin{pmatrix} I^{+}_{\mathrm{Si,LA},k}(x_{\mathrm{Si\text{-}Interface}},\mu,t)+2I^{+}_{\mathrm{Si,TA},k}(x_{\mathrm{Si\text{-}Interface}},\mu,t)\\ +I^{+}_{\mathrm{Si,LO},k}(x_{\mathrm{Si\text{-}Interface}},\mu,t)+2I^{+}_{\mathrm{Si,TO},k}(x_{\mathrm{Si\text{-}Interface}},\mu,t)\end{pmatrix} \quad (7.145)$$

At the interface, incorporating the diffusive condition in Eq. (7.144) and using Eqs. (7.130) and (7.137) into Eq. (7.143) results in

$$I^{+}_{\mathrm{Si,total}}(t)=(-q^{-}_{\mathrm{Si},1}(t)-q^{-}_{\mathrm{Si},2}(t)+q_{\mathrm{rad},1}(t)+q_{\mathrm{rad},2}(t))/\pi k_{\mathrm{max,Si}} \quad (7.146)$$

$I^{+}_{\mathrm{Si,total}}(t)$ is redistributed among all the branches of phonons at the interface, which can be written as

$$I^{+}_{\mathrm{Si},ij,\mathrm{k}}(x_{\mathrm{Si\text{-}Interface}},\mu,t)=\frac{C_{\mathrm{Si},ij,\mathrm{k}}v_{\mathrm{Si},ij,\mathrm{k}}k_{\mathrm{max,Si}}}{\langle C_{\mathrm{Si}}v_{\mathrm{Si}}\rangle\big|_{0}^{k_{max,\mathrm{Si}}}}I^{+}_{\mathrm{Si,total}}(t) \quad (7.147)$$

In Eq. (7.147), $i=T$ represents transverse mode and $i=L$ stands for longitudinal mode. In addition, $j=A$ corresponds to acoustic and $j=O$ stands for optical phonons. Numerical treatment of the EPRT is similar to that given in Section 7.3; refer to Section 7.3 for the numerical solution. The values of bulk properties of silicon, used in the simulation, are given in Table 7.4.

7.5.5 Findings of Laser Heating of Films Pair With Presence of Minute Vacuum Gap at the Interface

Fig. 7.26 shows the equivalent equilibrium temperature distribution in the lattice subsystem of aluminum and silicon films for various gap sizes and the heating period of 100 ps. Equivalent equilibrium temperature for the zero thickness vacuum gap is also shown in Fig. 7.26 for comparison. Equivalent equilibrium temperature reduces with the increasing normalized distance (x/Λ, where Λ is the mean free path of silicon) in the aluminum film, which is more pronounced for reduced film thicknesses. This behavior is associated with the following: (1) energy gain of phonons from electrons due to the electron–phonon coupling, and (2) energy transport across the silicon film by the ballistic phonons. Electrons gain energy from the irradiated field according to the Beer–Lambert law while losing some of their excess energy to the lattice site through the coupling between electrons and phonons. Phonon intensity attains high values in the surface region of aluminum where the energy gained

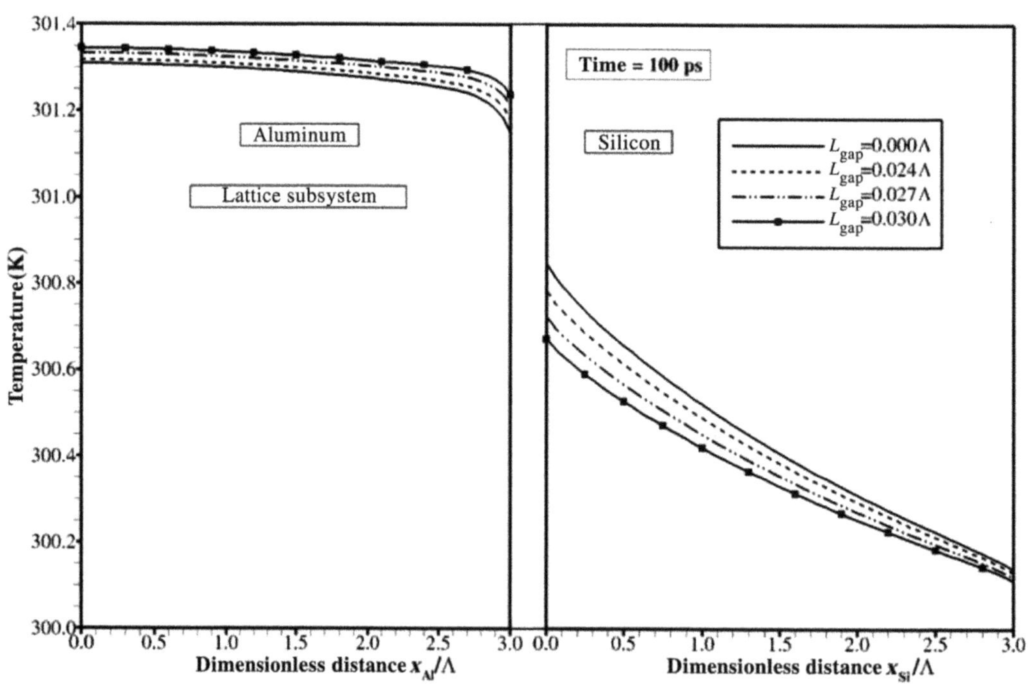

■ **FIGURE 7.26** Equivalent equilibrium temperature across the films with the presence of a vacuum gap at the interface in the lattice subsystem for various gap sizes and heating duration of 100 ps [4].

by electrons is high due to the absorbed irradiated energy in this region. Note that the absorption of the irradiated energy is assumed to be governed by the Beer–Lambert law in the aluminum film, which demonstrates the exponential decay of irradiated energy within the absorption depth of aluminum. The thickness of the aluminum film ($3\Lambda = 101.4 \times 10^{-9}$ m, where Λ is the mean free path of silicon) is considered to be larger than that of the absorption depth in the aluminum (7.62×10^{-9} m) to ensure the irradiated energy is absorbed within the aluminum film.

Electron temperature in the aluminum film is also shown in Fig. 7.27 for different vacuum gap sizes and heating duration of 100 ps. Note that electrons do not jump across the vacuum gap because of the silicon, which is considered to be demonstrating dielectric characteristics. Therefore, electrons reflect from the aluminum edge at the interface, which in turn results in insulated boundary for the electron subsystem at the aluminum–silicon interface. However, the absorption of laser-irradiated energy in the electron subsystem of the aluminum film gives rise to almost exponential decay of electron temperature in the film, particularly for a zero-thickness vacuum gap. As the vacuum gap

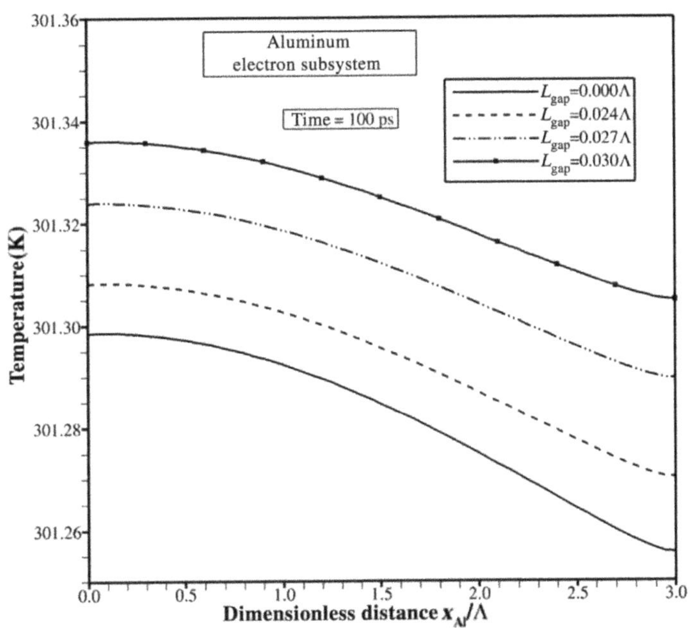

■ **FIGURE 7.27** Equivalent equilibrium temperature in the electron subsystem of aluminum film for various gap sizes and heating duration of 100 ps [4].

thickness increases, electron temperature behavior changes along the film thickness. In this case, energy transfer to the lattice site through electron–phonon coupling modifies the electron temperature distribution such that electron temperature increases while lattice site temperature reduces with increasing vacuum gap thickness (Fig. 7.27). In this case, as the gap thickness increases, the possibility of ballistic phonons, which can jump to the silicon film from the aluminum film edge, reduces; the gap size becomes larger than the wavelengths of some the phonons that could not jump across the vacuum gap. Therefore, as the vacuum size increases, phonon temperature in the aluminum film increases. Note that phonons at the aluminum film interface that cannot jump to the silicon film reflect from the interface while contributing to the phonon intensity increase in the near region of the aluminum film interface. On the other hand, the amount of energy transfer from the electron subsystem toward the lattice subsystem is governed by the electron and phonon energy differences. Consequently, increasing phonon intensity in the near region of the aluminum interface suppresses the energy transfer from electron subsystem to phonon subsystem, which in turn gives rise to the attainment of high electron temperature in this region.

In the case of the silicon film, energy transfer from the edge of the aluminum film toward the silicon film edge takes place due to (1) thermal radiation, due to the Casimir limit because of the small thickness of the vacuum gap, which is less than the mean free path; and (2) ballistic phonon transfer from the aluminum film edge toward the silicon film. Moreover, it is observed that the main driving force for the energy transfer across the vacuum gap is the ballistic phonons contribution, which remains higher than that of energy transfer by thermal radiation across the gap. This situation can be seen in Fig. 7.28, in which the ratio of the energy flux due to thermal radiation over the ballistic phonon transport across the gap is shown. The ratio remains less than unity for all the vacuum gap sizes. However, reduced vacuum gap size results in decreased intensity ratio. Note that reflected phonons from the edge of the aluminum film, due to the large gap size, contributes to equivalent equilibrium temperature increase at the aluminum film edge. This enhances the thermal radiation emitted from the aluminum edge toward the silicon edge. Therefore, despite the increase in the thermal radiation emitted from the aluminum edge, temperature remains low at the silicon edge as the vacuum gap size increases. This indicates that the ballistic phonon contribution dominates over the thermal radiation contribution in terms of phonon intensity increase at the silicon edge. However, increasing the vacuum gap size lowers the ballistic phonon contribution

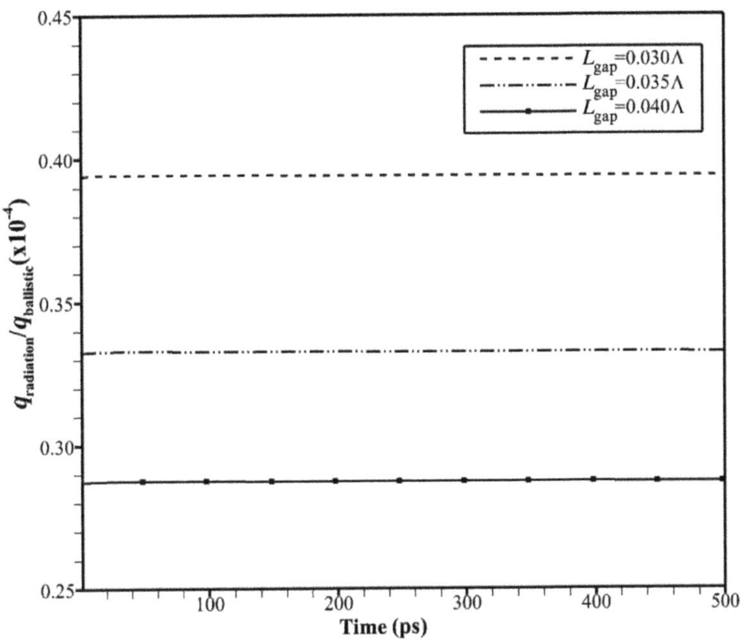

■ **FIGURE 7.28** Temporal variation of the flux ratio of radiation heat transfer to ballistic phonon transfer across the vacuum gap for various gap sizes [4].

to temperature increase at the silicon film edge. Moreover, equivalent equilibrium temperature decays inside the silicon film because of the constant temperature (300K) boundary condition incorporated at the far edge of the silicon film.

Fig. 7.29 shows temporal variation of equivalent equilibrium temperature difference. Temperature difference corresponds to equivalent equilibrium temperature change in between the aluminum and silicon edges across the vacuum gap. Temperature difference remains high in the early heating period, indicating that the energy transferred from the aluminum film edge toward the silicon edge remains low across the vacuum gap in the early heating period. This situation is more pronounced in the case of increased gap sizes. Although equivalent equilibrium temperature for phonons remains high at the edge of the aluminum film, equivalent equilibrium temperature at the silicon edge remains low. This is attributed to the contribution of the ballistic phonons to the energy transfer across the vacuum gap, which is higher than that transferred by the thermal radiation due to the Casimir limit (Fig. 7.28). As the heating period progresses, temperature difference reduces slightly for the large gap sizes;

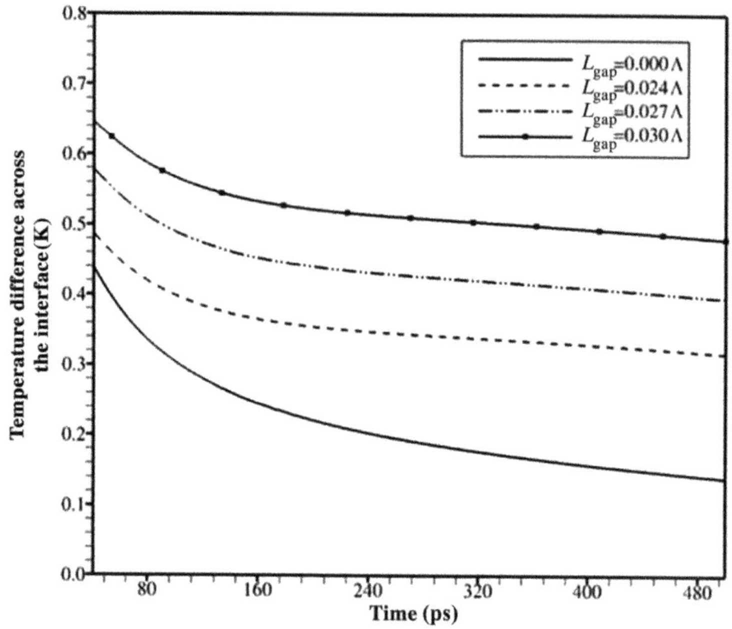

■ **FIGURE 7.29** Temporal variation of equivalent equilibrium temperature difference across the vacuum gap in the lattice subsystems of aluminum and silicon films pair for various gap sizes [4].

however, considerable reduction in temperature difference occurs for the case of nonpresence of the gap at the interface. Note that energy balance is satisfied for each phonon frequency at the interface when modeling the interface conditions across the vacuum gap. In the case of nonpresence of a vacuum gap, thermal boundary resistance is the only resistance causing temperature jump across the interface in the lattice subsystem. However, in the case of presence of the vacuum gap, it is the ballistic phonons and thermal radiation that contribute to the energy transfer across the vacuum gap. This results in a large temperature difference in the lattice subsystem across the vacuum gap.

7.6 CLOSING REMARKS

Thermal energy transport suffers from resistance when micro/nanoscale structures with different physical properties are in contact. This is particularly true for thin film pairs. Since the transport characteristics are governed by phonons, boundary scattering takes place at the films pair interface while lowering temperatures at the film edges across the interface. This situation can be simplified through introducing thermal boundary resistance across the films pair interface. Mainly three approaches are considered: diffusive mismatch, acoustic mismatch, and cut-off mismatch. In the case of diffusive mismatch, the gray body assumption is adopted at the interface of the films pairs. In this case, phonon frequency and the effect of the ballistic phonon transport are overlooked across the interface. The acoustic mismatch considers the phonon frequencies, and those phonons having the acoustic frequency matching for both films can transfer across the interface. However, experimental findings revealed that the predictions do not agree well with the experimental data and the model needs further refinements. The cut-off frequency match model relies on the ballistic phonon transport; the ballistic phonons with higher wavelengths can jump across the interface and others undergo reflection and boundary scattering at the interface. However, the energy is balanced among the ballistic phonons, reflected phonons, scattered phonons, and transmitted phonons. The findings reveal that the CMM predicts realistic phonon transport characteristics across the interface and the predictions agree well with the experimental results.

On the other hand, the presence of the minute-sized vacuum gap at the films pair interface provides additional resistance across the interface. This situation is formulated and a case study is introduced to demonstrate the transport characteristics when the vacuum gap is incorporated at the interface. Near-field radiation across the vacuum is also considered as well as its contribution to the thermal energy transport across the interface. The case study that introduced the energy transport across the vacuum gap due to the ballistic phonons and the near-field radiation is discussed. It is demonstrated that the ballistic phonons are the major contributors to the energy transport across the vacuum gap.

REFERENCES

[1] H. Ali, B.S. Yilbas, Phonon cross-plane transport and thermal boundary resistance: effect of heat source size and thermal boundary resistance on phonon characteristics, Continuum Mech. Thermodyn. 28 (5) (2016) 1373–1393.

[2] H. Ali, B.S. Yilbas, Phonon transport characteristics across silicon thin film pair: presence of a gap between the films, J. Non-Equilib. Thermodyn. 40 (3) (2015) 153–170.

[3] H. Ali, B.S. Yilbas, Thermal transport across a pair of thin silicon films with the presence of minute vacuum gap: effect of film thickness on thermal characteristics, Can. J. Phys. 94 (9) (2016) 933–944.

[4] H. Ali, B.S. Yilbas, Energy transport across the thin films pair with presence of minute vacuum gap at interface, J. Non-Equilib. Thermodyn. (2016). Available from: http://dx.doi.org/10.1515/jnet-2016.

[5] A. Ward, D.A. Broido, Intrinsic phonon relaxation times from first-principles studies of the thermal conductivities of Si and Ge, Phys. Rev. B 81 (8) (2010) 85205.

[6] A. Ward, D. Broido, D. Stewart, G. Deinzer, Ab initio theory of the lattice thermal conductivity in diamond, Phys. Rev. B 80 (12) (2009) 125203.

[7] J.H. Jeans, The Dynamical Theory of Gases, fourth ed., Dover, New York, 1954.

[8] A. Majumdar, Microscale heat conduction in dielectric thin films, J. Heat Transfer 115 (1) (1991) 7–16.

[9] N.W. Arhcroft, N.D. Mermin, Solid State Physics, Harcourt College Publihers, 1976.

[10] E.T. Swartz, R.O. Pohl, Thermal boundary resistance, Rev. Mod. Phys. 61 (3) (1989) 605–668.

[11] L. Pilon, K.M. Katika, Modified method of characteristics for simulating microscale energy transport, J. Heat Transfer 126 (5) (2004) 735–743.

[12] B.S. Yilbas, S.B. Mansoor, Influence of heat source size on phonon transport in thin silicon film, Transp. Theory. Stat. Phys. 42 (2–3) (2014) 65–84.

[13] K. Sääskilahti, J. Oksanen, J. Tulkki, S. Volz, Role of anharmonic phonon scattering in the spectrally decomposed thermal conductance at planar interfaces, Phys. Rev. B: Condens. Matter Mater. Phys. 90 (13) (2014) 134312.

[14] B.S. Yilbas, S.B. Mansoor, Frequency dependent phonon transport in two-dimensional silicon and diamond films, Mod. Phys. Lett. B 26 (17) (2012) 1250104.

[15] B.N. Brockhouse, Lattice vibrations in silicon and Germanium, Phys. Rev. Lett. 2 (6) (1959) 256–258.

[16] A.J. Minnich, G. Chen, S. Mansoor, B.S. Yilbas, Quasiballistic heat transfer studied using the frequency-dependent Boltzmann transport equation, Phys. Rev. B 84 (23) (2011) 235207.

[17] D.P. Sellan, E.S. Landry, K. Sasihithlu, A. Narayanaswamy, A.J.H. McGaughey, C.H. Amon, Phonon transport across a vacuum gap, Phys. Rev. B: Condens. Matter Mater. Phys. 85 (2) (2012) 24118.

[18] Y. Zheng, PhD Thesis, Near-Field Radiative Momentum, Energy and Entropy Transfer in Fluctuational Electrodynamics, Columbia University, 2014.

[19] L. Hu, A. Narayanaswamy, X. Chen, G. Chen, Near-field thermal radiation between two closely spaced glass plates exceeding Planck's blackbody radiation law, Appl. Phys. Lett. 92 (13) (2008) 133106.

[20] G. Chen, Nanoscale Energy Transport and Conversion, Oxford University Press, New York, 2005.

[21] M. Planck, The Theory of Heat Radiation, P. Blakiston's Son & Co., 1914.

[22] M. Francoeur, M. Pinar Mengüç, R. Vaillon, Spectral tuning of near-field radiative heat flux between two thin silicon carbide films, J. Phys. D. Appl. Phys. 43 (7) (2010) 75501.

[23] B.S. Yilbas, A.Y. Al-Dweik, S.B. Mansoor, Non-equilibrium energy transport in a thin metallic film: analytical solution for radiative transport equation, Phys. B Condens. Matter 454 (2014) 15–22.

[24] Z. Lin, L.V. Zhigilei, V. Celli, Electron-phonon coupling and electron heat capacity of metals under conditions of strong electron-phonon nonequilibrium, Phys. Rev. B 77 (7) (2008) 75133.

[25] W.L. McMillan, Transition temperature of strong-coupled superconductors, Phys. Rev. 167 (1968) 331–345.

[26] J. Bin Lee, S.H. Lee, Thermal boundary resistance effect on non-equilibrium energy transport in metal-dielectric thin films heated by femtosecond pulse lasers, Mater. Trans. 52 (7) (2011) 1492–1499.

[27] T.Q. Qiu, C.L. Tien, Heat transfer mechanisms during short-pulse laser heating of metals, J. Heat Transfer 115 (4) (1993) 835–841.

[28] S.B. Mansoor, B.S.S. Yilbas, Radiative phonon transport in silicon and collisional energy transfer in aluminum films due to laser short-pulse heating: Influence of laser pulse intensity on temperature distribution, Opt. Laser Technol. 44 (1) (2012) 43–50.

[29] R. Stedman, G. Nilsson, Dispersion relations for phonons in aluminum at 80 and 300K, Phys. Rev. 145 (2) (1966) 492–500.

[30] A.C. Bouley, N.S. Mohan, D.H. Damon, The lattice thermal conductivity of copper and aluminum alloys at low temperatures, Therm. Conduct. 14 (1976) 81–88.

Chapter 8

Phonon Radiative Transfer in Curvilinear Coordinate Systems

8.1 INTRODUCTION

The problem of microscale heat transfer in rectangular domains can be dealt with by using the equation of phonon radiative transfer (EPRT) as described in previous chapters. Microscale heat transfer in domains that are not of rectangular shape is also of practical importance, e.g., in problems involving spherical nanoparticles or curved thin films. One of the approaches to deal with complex geometries is to use body-fitted coordinates to cover the domain of interest. The term "body-fitted coordinates" actually implies general curvilinear coordinates that may be orthogonal or nonorthogonal. Cylindrical and spherical coordinates are familiar examples of such coordinate systems. It is also possible to cover more general two- and three-dimensional domains with orthogonal or nonorthogonal grids. To solve the microscale heat transfer problems in such coordinate systems, we need to have a form of EPRT, which is defined in terms of the coordinates of the curvilinear system. For example in a cylindrical coordinate system the EPRT should be defined in terms of the coordinates (ρ, θ, z).

The EPRT was derived by Majumdar [1] based on the Boltzmann transport equation. An analogy exits between the EPRT and the equation of radiative transfer (ERT). In these equations the dependent variable is the photon/phonon intensity I, which is a function of the position (x, y, z) as well as that of the direction (φ,ψ). φ is the polar angle whereas ψ is the azimuthal angle. In addition, I may also be a function of the photon/phonon energy (frequency, wavenumber) and time. The ERT, until recently, was practically available for the Cartesian coordinate system as well as for some specialized coordinate systems only. The dependence of the intensity I on the direction (φ,ψ) makes it difficult to write the ERT in any other coordinate system. Previously, Vaillon et al. [2] derived the ERT in an orthogonal coordinate system but the derivation

Heat Transport in Micro- and Nanoscale Thin Films.
DOI: http://dx.doi.org/10.1016/B978-0-323-42979-5.00008-6

was not complete. Recently J. Freimanis [3,4] used the methods of tensor analysis to derive a form of ERT that is applicable in any orthogonal or nonorthogonal, curvilinear coordinate system. Using the analogy between the ERT and the EPRT, a simplified derivation of the EPRT for curvilinear coordinate systems is described in this chapter. Furthermore, the general form of the EPRT is specialized for two common systems, the cylindrical and spherical coordinate systems. A procedure is also outlined for the numerical solution of the frequency-independent, steady-state EPRT in an irregular, two-dimensional domain that is covered by a curvilinear grid. It should also be mentioned here that the development of the EPRT in this chapter pertains only to Euclidean spaces and not to Riemannian spaces, in which the curvature tensor is nonvanishing.

8.2 CURVILINEAR COORDINATES

All coordinate systems other than the Cartesian and oblique coordinate systems are termed curvilinear coordinate systems. Oblique coordinate systems, for our discussion, can be regarded as a special case of nonorthogonal curvilinear coordinate systems. Mathematically, they are to be treated by the same formalism as general nonorthogonal systems. Hence, in this chapter only curvilinear coordinate systems are discussed. A very important characteristic of curvilinear coordinate systems is that the basis vectors, in terms of which other vectors can be represented, are not constant but change from point to point. The basis vectors in the cylindrical coordinate system are a familiar example. The basis vectors $\mathbf{e}_\rho$ and $\mathbf{e}_\theta$ are position dependent. Additionally, the basis vectors may not initially be of unit magnitude and may have to be normalized. This is particularly true in the case of nonorthogonal systems. Hence, it is a complicated matter to describe a particular, fixed direction in such coordinate systems. Furthermore, the spatial derivative of a vector has to account for its rate-of-change not only because of the change of position but also because of the change of the basis vectors with position. To deal with all such complications, one of the most suitable mathematical tools is tensor analysis.

Tensor analysis is deeply connected to the study of general curvilinear coordinate systems and has methods that can be used to write the conservation equations in such systems. In this chapter our focus is on the application of such methods to the formulation of the EPRT in general coordinate systems. It is presumed that the reader is familiar with the theory of tensor analysis. However, to ensure that the authors and the reader agree on the basic terminologies, symbols, and conventions, some of these are described in the succeeding pages.

Curvilinear coordinates are represented as (x^1, x^2, x^3), or as x^i, $i = 1, 2, 3$. The *range convention* stipulates that whenever the symbol x^i is written it will be understood that the index i will take on the values 1, 2, 3. So there is no need to write the full expression x^i, $i = 1, 2, 3$, but instead the symbol x^i is enough. The *summation convention* requires that whenever an index appears twice in the same expression, once as a subscript and once as a superscript, then there has to be a summation over its range. This means that the expression $a^i b_i$ actually means $\sum_{i=1}^{3} a^i b_i$. The curvilinear coordinates will usually be defined in terms of the Cartesian coordinates by means of transformation equations,

$$y^i = y^i(x^j) \tag{8.1}$$

In this equation, y^i are the Cartesian coordinates and x^i are the curvilinear coordinates. The symbols $\hat{\mathbf{e}}_i$ will denote the Cartesian basis vectors as well as the unit basis vectors in curvilinear coordinates with respect to which vector $\mathbf{A}$ will be written in terms of its *physical components.* They will constitute a set of three orthonormal vectors; i.e., a set of three unit vectors that are mutually orthogonal. Further terms and symbols are defined as they arise, in later sections.

8.3 EPRT IN ORTHOGONAL COORDINATE SYSTEMS

8.3.1 Formulation of the Path-Length Derivative

Selecting the x axis as the polar axis (Fig. 8.1) and choosing a right-handed coordinate system, a unit vector in the direction (φ, ψ) in Cartesian coordinates can be written as

$$\begin{aligned}
\boldsymbol{\Omega} &= \cos\varphi\, \hat{\mathbf{e}}_1 + \sin\varphi \cos\psi\, \hat{\mathbf{e}}_2 + \sin\varphi \sin\psi\, \hat{\mathbf{e}}_3 \\
\boldsymbol{\Omega} &= \mu \hat{\mathbf{e}}_1 + \eta \hat{\mathbf{e}}_2 + \xi \hat{\mathbf{e}}_3 \\
\boldsymbol{\Omega} &= \Omega^1 \hat{\mathbf{e}}_1 + \Omega^2 \hat{\mathbf{e}}_2 + \Omega^3 \hat{\mathbf{e}}_3 \\
\boldsymbol{\Omega} &= \Omega^i \hat{\mathbf{e}}_i
\end{aligned} \tag{8.2}$$

where $(\hat{\mathbf{e}}_1, \hat{\mathbf{e}}_2, \hat{\mathbf{e}}_3)$ are the basis vectors in the Cartesian system usually written as $(\mathbf{i}, \mathbf{j}, \mathbf{k})$. Range and summation convention are employed in the last line of Eq. (8.2).

A particular, fixed direction $\boldsymbol{\Omega}_o$ is defined by the angles (φ_o, ψ_o) and these angles are independent of the coordinates (x, y, z) as we move from

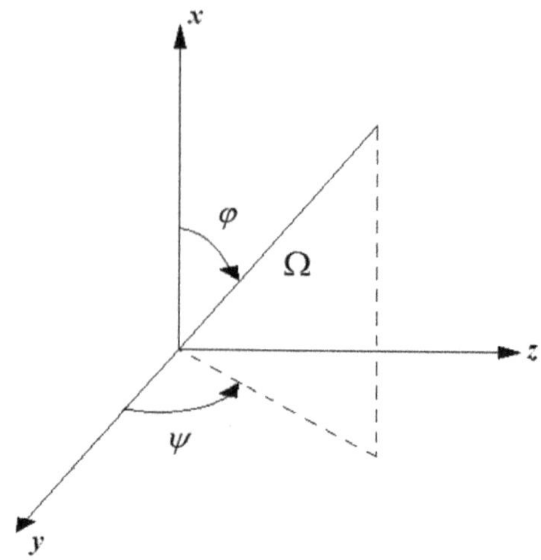

■ **FIGURE 8.1** A unit vector **Ω** at a point in the Cartesian coordinate system defined in terms of the basis vectors that are the same at all points.

point to point in the Cartesian coordinate system. The EPRT in the Cartesian coordinates is written as [1]

$$\frac{1}{v_\omega}\frac{\partial I_\omega}{\partial t} + \nabla \cdot (\boldsymbol{\Omega} I_\omega) = \frac{\frac{1}{4\pi}\iint_{4\pi} I_\omega \sin\varphi d\varphi d\psi - I_\omega}{\Lambda_\omega} \tag{8.3}$$

$$\frac{1}{v_\omega}\frac{\partial I_\omega}{\partial t} + \cos\varphi\frac{\partial I_\omega}{\partial x} + \sin\varphi\cos\psi\frac{\partial I_\omega}{\partial y} + \sin\varphi\sin\psi\frac{\partial I_\omega}{\partial z} = \frac{\frac{1}{4\pi}\iint_{4\pi} I_\omega \sin\varphi d\varphi d\psi - I_\omega}{\Lambda_\omega} \tag{8.4}$$

In the following discussion, as discussed previously, the Cartesian coordinates will be represented as $(x, y, z) \to (y^1, y^2, y^3)$ and any orthogonal curvilinear coordinate as (x^1, x^2, x^3). Using the range convention we will write y^i and x^i for the Cartesian and orthogonal curvilinear coordinates, respectively. The index i takes on the values 1, 2, and 3. Consider now an orthogonal curvilinear coordinate system as defined by the following equations:

$$y^1 = y^1(x^1, x^2, x^3), \quad y^2 = y^2(x^1, x^2, x^3), \quad y^3 = y^3(x^1, x^2, x^3) \tag{8.5}$$

The position vector can be written as

$$\mathbf{r} = y^i \hat{\mathbf{e}}_i \tag{8.6}$$

The covariant basis vectors $\mathbf{E}_i$ in the orthogonal curvilinear coordinate system are defined as [5]

$$\mathbf{E}_i = \frac{\partial \mathbf{r}}{\partial x^i} \tag{8.7}$$

The contravariant basis vectors $\mathbf{E}^j$ in the orthogonal curvilinear coordinate system are defined as [2]

$$\mathbf{E}_i \cdot \mathbf{E}^j = \delta_i^j \tag{8.8}$$

The basis vectors $\mathbf{E}_i$ and $\mathbf{E}^j$, in general, do not have a unit magnitude. In addition, the magnitude and direction of these basis vectors change from point to point as we move in a curvilinear coordinate system. This means that at each point, there is defined a set of unique basis vectors. Selecting the $\mathbf{E}_1$ direction as the polar axis, a unit vector $\mathbf{\Omega}$, at any generic point, in the orthogonal curvilinear coordinate system x^i can be defined in a manner similar to that used for the Cartesian system (Fig. 8.2):

$$\begin{aligned}
\mathbf{\Omega} &= \cos\varphi \frac{\mathbf{E}_1}{\sqrt{\mathbf{E}_1 \cdot \mathbf{E}_1}} + \sin\varphi\cos\psi \frac{\mathbf{E}_2}{\sqrt{\mathbf{E}_2 \cdot \mathbf{E}_2}} + \sin\varphi\sin\psi \frac{\mathbf{E}_3}{\sqrt{\mathbf{E}_3 \cdot \mathbf{E}_3}} \\
\mathbf{\Omega} &= \frac{\mu}{h_1}\mathbf{E}_1 + \frac{\eta}{h_2}\mathbf{E}_2 + \frac{\xi}{h_3}\mathbf{E}_3 \\
\mathbf{\Omega} &= \Omega^i \mathbf{E}_i
\end{aligned} \tag{8.9}$$

In Eq. (8.9), $h_i \rightarrow h_1, h_2, h_3$ are termed as the *scale factors*. Now, a specific value of the angles $(\varphi, \psi) = (\varphi_o, \psi_o)$ in general, will define a different direction at each point x^i. A specific, fixed global direction $\mathbf{\Omega}_o$ will constrain the angles (φ, ψ) according to Eq. (8.9). The path-length derivative $\nabla \cdot (\mathbf{\Omega} I)$ in Eq. (8.3), in an orthogonal curvilinear

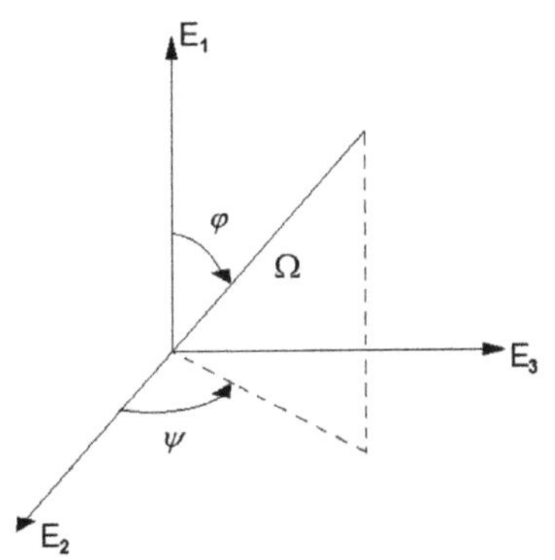

■ **FIGURE 8.2** A unit vector $\mathbf{\Omega}$ at a point in an orthogonal coordinate system defined in terms of the basis vectors at that same point.

coordinate system, can be defined by means of the covariant derivative in tensor analysis [5]:

$$\nabla\cdot(\mathbf{\Omega}I_\omega)=\frac{\partial}{\partial x^j}(\Omega^i\mathbf{E}_i I_\omega)\mathbf{E}^j=(\Omega^i I_\omega)_{|i} \tag{8.10}$$

In Eq. (8.10) the subscript $|i$ denotes covariant differentiation with respect to the coordinate x^i. Using the definition of the covariant derivative, we can write [6]

$$\nabla\cdot(\mathbf{\Omega}I_\omega)=\frac{\partial}{\partial x^i}(\Omega^i I_\omega)+\Gamma^i_{ki}(\Omega^k I_\omega) \tag{8.11}$$

where Γ^k_{ij} are the connection coefficients and are defined in terms of the components of the metric tensor $\mathbf{g}=g_{ij}\mathbf{E}^i\mathbf{E}^j=g^{ij}\mathbf{E}_i\mathbf{E}_j$. The definitions of all these entities for a general curvilinear coordinate system that is not necessarily orthogonal are [6]

$$g_{ij}=\mathbf{E}_i\cdot\mathbf{E}_j;\quad g^{ij}=\mathbf{E}^i\cdot\mathbf{E}^j;\quad [g^{ij}]=[g_{ij}]^{-1} \tag{8.12}$$

$$\Gamma^k_{ij}=\frac{g^{km}}{2}\left\{\frac{\partial g_{im}}{\partial x^j}+\frac{\partial g_{mj}}{\partial x^i}-\frac{\partial g_{ij}}{\partial x^m}\right\} \tag{8.13}$$

Eqs. (8.11)–(8.13) are applicable to an orthogonal as well as nonorthogonal coordinate system. In an orthogonal, curvilinear coordinate system we have the simplified expressions [7]

$$[g_{ij}]=\begin{bmatrix}h_1^2&0&0\\0&h_2^2&0\\0&0&h_3^2\end{bmatrix};\quad [g^{ij}]=\begin{bmatrix}1/h_1^2&0&0\\0&1/h_2^2&0\\0&0&1/h_3^2\end{bmatrix};\quad \Gamma^i_{ki}=\frac{1}{2g}\frac{\partial g}{\partial x^k} \tag{8.14}$$

where g is the determinant of the metric tensor $[g_{ij}]$. Hence,

$$g=(h_1h_2h_3)^2 \tag{8.15}$$

and

$$\Gamma^i_{ki}=\frac{1}{2g}\frac{\partial g}{\partial x^k}=\frac{1}{2(h_1h_2h_3)^2}\frac{\partial}{\partial x^k}(h_1h_2h_3)^2=\frac{1}{h_1}\frac{\partial h_1}{\partial x^k}+\frac{1}{h_2}\frac{\partial h_2}{\partial x^k}+\frac{1}{h_3}\frac{\partial h_3}{\partial x^k} \tag{8.16}$$

Substituting Eq. (8.16) into Eq. (8.11) and rearranging, we obtain

$$\begin{aligned}\nabla\cdot(\mathbf{\Omega}I_\omega)=&\frac{1}{h_1h_2h_3}\left[\mu\frac{\partial h_2h_3I_\omega}{\partial x^1}+\eta\frac{\partial h_1h_3I_\omega}{\partial x^2}+\xi\frac{\partial h_1h_2I_\omega}{\partial x^3}\right]\\&+\left[\frac{1}{h_1}\frac{\partial\varphi}{\partial x^1}\frac{\partial\mu I_\omega}{\partial\varphi}+\frac{1}{h_2}\frac{\partial\varphi}{\partial x^2}\frac{\partial\eta I_\omega}{\partial\varphi}+\frac{1}{h_3}\frac{\partial\varphi}{\partial x^3}\frac{\partial\xi I_\omega}{\partial\varphi}\right]\\&+\left[\frac{1}{h_1}\frac{\partial\psi}{\partial x^1}\frac{\partial\mu I_\omega}{\partial\psi}+\frac{1}{h_2}\frac{\partial\psi}{\partial x^2}\frac{\partial\eta I_\omega}{\partial\psi}+\frac{1}{h_3}\frac{\partial\psi}{\partial x^3}\frac{\partial\xi I_\omega}{\partial\psi}\right]\end{aligned} \tag{8.17}$$

In Eq. (8.17), $\mu=\cos\varphi$, $\eta=\sin\varphi\cos\psi$, $\xi=\sin\varphi\sin\psi$.

8.3.2 Calculation of the Derivatives $\partial\varphi/\partial x^i$ and $\partial\psi/\partial x^i$

In writing the ERT and the EPRT in an orthogonal curvilinear coordinate system, the predominant problem is to be able to derive expressions for the derivatives $\partial\varphi/\partial x^i$ and $\partial\psi/\partial x^i$. These expressions are derived in the following. Consider a specific, global direction $\boldsymbol{\Omega}_o$.

$$\boldsymbol{\Omega}_o = \frac{\cos\varphi}{h_1}\mathbf{E}_1 + \frac{\sin\varphi\cos\psi}{h_2}\mathbf{E}_2 + \frac{\sin\varphi\sin\psi}{h_3}\mathbf{E}_3 \tag{8.18}$$

As we move from point to point in the orthogonal curvilinear coordinate system, the covariant basis vectors $\mathbf{E}_i$ also change, hence to fix a particular direction globally, the angles (φ, ψ) will also change from one point to another. Since $\boldsymbol{\Omega}_o$ is a constant,

$$\frac{\partial\boldsymbol{\Omega}_o}{\partial x^j} = 0 \Rightarrow \frac{\partial\boldsymbol{\Omega}_o}{\partial x^1} = 0, \quad \frac{\partial\boldsymbol{\Omega}_o}{\partial x^2} = 0, \quad \frac{\partial\boldsymbol{\Omega}_o}{\partial x^3} = 0 \tag{8.19}$$

The first task now is to be able to calculate the derivative $\partial\boldsymbol{\Omega}_o/\partial x^j$. To do this, we use the definition from tensor analysis [5],

$$\frac{\partial\mathbf{E}_i}{\partial x^j} = \Gamma_{ij}^k\mathbf{E}_k \tag{8.20}$$

The derivatives $\partial\boldsymbol{\Omega}_o/\partial x^j$ can then be written as

$$\frac{\partial\boldsymbol{\Omega}_o}{\partial x^i} = \frac{\partial(\Omega^i\mathbf{E}_i)}{\partial x^j} = \frac{\partial\Omega^i}{\partial x^j}\mathbf{E}_i + \Omega^i\frac{\partial\mathbf{E}_i}{\partial x^j} \tag{8.21}$$

Using Eq. (8.20) for the last term and rearranging the indices we obtain

$$\frac{\partial\boldsymbol{\Omega}_o}{\partial x^j} = \left(\frac{\partial\Omega^i}{\partial x^j} + \Omega^k\Gamma_{kj}^i\right)\mathbf{E}_i \tag{8.22}$$

Expanding this equation by carrying out the implied summations we finally get

$$\begin{aligned}\frac{\partial\boldsymbol{\Omega}_o}{\partial x^j} = &\left[\frac{-\sin\varphi}{h_1}\frac{\partial\varphi}{\partial x^j} - \frac{\cos\varphi}{h_1^2}\frac{\partial h_1}{\partial x^j} + \frac{\cos\varphi}{h_1}\Gamma_{1j}^1 + \frac{\sin\varphi\cos\psi}{h_2}\Gamma_{2j}^1 + \frac{\sin\varphi\sin\psi}{h_3}\Gamma_{3j}^1\right]\mathbf{E}_1\\
&+\left[\begin{array}{l}\dfrac{1}{h_2}\left[\cos\varphi\cos\psi\dfrac{\partial\varphi}{\partial x^j} - \sin\varphi\sin\psi\dfrac{\partial\psi}{\partial x^j}\right] - \dfrac{\sin\varphi\cos\psi}{h_2^2}\dfrac{\partial h_2}{\partial x^j}\\ + \dfrac{\cos\varphi}{h_1}\Gamma_{1j}^2 + \dfrac{\sin\varphi\cos\psi}{h_2}\Gamma_{2j}^2 + \dfrac{\sin\varphi\sin\psi}{h_3}\Gamma_{3j}^2\end{array}\right]\mathbf{E}_2\\
&+\left[\begin{array}{l}\dfrac{1}{h_3}\left[\cos\varphi\sin\psi\dfrac{\partial\varphi}{\partial x^j} + \sin\varphi\cos\psi\dfrac{\partial\psi}{\partial x^j}\right] - \dfrac{\sin\varphi\sin\psi}{h_3^2}\dfrac{\partial h_2}{\partial x^j}\\ + \dfrac{\cos\varphi}{h_1}\Gamma_{1j}^3 + \dfrac{\sin\varphi\cos\psi}{h_2}\Gamma_{2j}^3 + \dfrac{\sin\varphi\sin\psi}{h_3}\Gamma_{3j}^3\end{array}\right]\mathbf{E}_3\end{aligned} \tag{8.23}$$

Equating $\partial \boldsymbol{\Omega}_o/\partial x^j$ (Eq. 8.23) to zero; substituting $j = 1, 2, 3$; and simplifying we get

$$
\begin{aligned}
\frac{\partial\varphi}{\partial x^1} &= \frac{\cos\psi}{h_2}\frac{\partial h_1}{\partial x^2} + \frac{\sin\psi}{h_3}\frac{\partial h_1}{\partial x^3} & \frac{\partial\psi}{\partial x^1} &= \cot\varphi\left[-\frac{\sin\psi}{h_2}\frac{\partial h_1}{\partial x^2} + \frac{\cos\psi}{h_3}\frac{\partial h_1}{\partial x^3}\right]\\
\frac{\partial\varphi}{\partial x^2} &= -\frac{\cos\psi}{h_1}\frac{\partial h_2}{\partial x^1} & \frac{\partial\psi}{\partial x^2} &= \frac{\cot\varphi\sin\psi}{h_1}\frac{\partial h_2}{\partial x^1} + \frac{1}{h_3}\frac{\partial h_2}{\partial x^3}\\
\frac{\partial\varphi}{\partial x^3} &= -\frac{\sin\psi}{h_1}\frac{\partial h_3}{\partial x^1} & \frac{\partial\psi}{\partial x^3} &= -\frac{\cot\varphi\cos\psi}{h_1}\frac{\partial h_3}{\partial x^1} - \frac{1}{h_2}\frac{\partial h_3}{\partial x^2}
\end{aligned}
\tag{8.24}
$$

Eq. (8.24) defines the generic form of the derivatives $\partial\varphi/\partial x^i$ and $\partial\psi/\partial x^i$ that are applicable in any orthogonal coordinate system. Note that this described form of the equation corresponds to the definition of the angles (φ, ψ) as illustrated in Fig. 8.2.

8.3.3 Equation of Phonon Radiative Transfer

In this section we state the EPRT in any orthogonal coordinate system in its final form. Substituting Eq. (8.24) into Eq. (8.17) and then Eq. (8.17) into Eq. (8.3), the EPRT in any orthogonal curvilinear coordinate system is obtained:

$$
\begin{aligned}
&\frac{1}{v_\omega}\frac{\partial I_\omega}{\partial t} + \frac{\cos\varphi}{h_1}\frac{\partial I_\omega}{\partial x^1} + \frac{\sin\varphi\cos\psi}{h_2}\frac{\partial I_\omega}{\partial x^2} + \frac{\sin\varphi\sin\psi}{h_3}\frac{\partial I_\omega}{\partial x^3}\\
&+\left[\frac{\cos\varphi}{h_1}\left(\frac{\cos\psi}{h_2}\frac{\partial h_1}{\partial x^2} + \frac{\sin\psi}{h_3}\frac{\partial h_1}{\partial x^3}\right) - \frac{\sin\varphi\cos^2\psi}{h_1 h_2}\frac{\partial h_2}{\partial x^1} - \frac{\sin\varphi\sin^2\psi}{h_1 h_3}\frac{\partial h_3}{\partial x^1}\right]\frac{\partial I_\omega}{\partial\varphi}\\
&+\left[\begin{aligned}&-\frac{\sin\psi\cos\varphi}{h_2}\left(\frac{\tan\varphi}{h_3}\frac{\partial h_3}{\partial x^2} + \frac{\cot\varphi}{h_1}\frac{\partial h_1}{\partial x^2}\right)\\ &+\frac{\cos\varphi\cos\psi}{h_3}\left(\frac{\cot\varphi}{h_1}\frac{\partial h_1}{\partial x^3} + \frac{\tan\varphi}{h_2}\frac{\partial h_2}{\partial x^3}\right) + h_3\frac{\cos\varphi\sin\psi\cos\psi}{h_1 h_2}\frac{\partial}{\partial x^1}\left(\frac{h_2}{h_3}\right)\end{aligned}\right]\frac{\partial I_\omega}{\partial\psi}\\
&= \frac{\frac{1}{4\pi}\iint_{4\pi} I_\omega \sin\varphi\, d\varphi\, d\psi - I_\omega}{\Lambda_\omega}
\end{aligned}
\tag{8.25}
$$

8.3.4 EPRT in the Cylindrical Coordinate System

As an example, the EPRT is now written explicitly in a cylindrical coordinate system. To do this the following assignments are made.

$$\rho \to x^1; \quad \theta \to x^2; \quad z \to x^3 \tag{8.26}$$

This particular labeling corresponds to the selection of the radial direction as the polar axis. Any other direction may also be selected as the polar axis. The scale factors are

$$h_1 = 1; \quad h_2 = \rho; \quad h_3 = 1 \tag{8.27}$$

The derivatives $\partial\varphi/\partial x^i$ and $\partial\psi/\partial x^i$ are calculated as

$$\frac{\partial\varphi}{\partial\rho} = 0; \quad \frac{\partial\varphi}{\partial\theta} = -\cos\psi; \quad \frac{\partial\varphi}{\partial z} = 0 \tag{8.28}$$

$$\frac{\partial\psi}{\partial\rho} = 0; \quad \frac{\partial\psi}{\partial\theta} = \cot\varphi\sin\psi; \quad \frac{\partial\psi}{\partial z} = 0 \tag{8.29}$$

The path-length derivative $\nabla\cdot(\mathbf{\Omega}I)$ as defined by Eq. (8.17) simplifies to

$$\nabla\cdot(\mathbf{\Omega}I_\omega) = \frac{1}{\rho}\begin{bmatrix} \cos\varphi\dfrac{\partial\rho I_\omega}{\partial\rho} + \sin\varphi\cos\psi\dfrac{\partial I_\omega}{\partial\theta} + \rho\sin\varphi\sin\psi\dfrac{\partial I_\omega}{\partial z} \\ -\cos^2\psi\dfrac{\partial\sin\varphi I_\omega}{\partial\varphi} + \cos\varphi\sin\psi\dfrac{\partial\cos\psi I_\omega}{\partial\psi} \end{bmatrix} \tag{8.30}$$

Finally, the EPRT can be written as

$$\begin{aligned} &\frac{1}{v_\omega}\frac{\partial I_\omega}{\partial t} + \frac{\cos\varphi}{\rho}\frac{\partial\rho I_\omega}{\partial\rho} + \frac{\sin\varphi\cos\psi}{\rho}\frac{\partial I_\omega}{\partial\theta} + \sin\varphi\sin\psi\frac{\partial I_\omega}{\partial z} \\ &-\frac{\cos^2\psi}{\rho}\frac{\partial\sin\varphi I_\omega}{\partial\varphi} + \frac{\cos\varphi\sin\psi}{\rho}\frac{\partial\cos\psi I_\omega}{\partial\psi} = \frac{\frac{1}{4\pi}\iint_{4\pi} I_\omega\sin\varphi d\varphi d\psi - I_\omega}{\Lambda_\omega} \end{aligned} \tag{8.31}$$

8.3.5 EPRT in the Spherical Coordinate System

As another example, the EPRT is now written explicitly in a spherical coordinate system. We make the following assignments.

$$\rho\to x^1; \quad \theta\to x^2; \quad \phi\to x^3 \tag{8.32}$$

This particular labeling corresponds to the selection of the radial direction as the polar axis. Any other direction may also be selected as the polar axis. The scale factors are

$$h_1 = 1; \quad h_2 = \rho; \quad h_3 = \rho\sin\theta \tag{8.33}$$

The derivatives $\partial\varphi/\partial x^i$ and $\partial\psi/\partial x^i$ are calculated as

$$\frac{\partial\varphi}{\partial x^1} = 0; \quad \frac{\partial\varphi}{\partial x^2} = -\cos\psi; \quad \frac{\partial\varphi}{\partial x^3} = -\sin\theta\sin\psi \tag{8.34}$$

$$\frac{\partial\psi}{\partial x^1} = 0; \quad \frac{\partial\psi}{\partial x^2} = \cot\varphi\sin\psi; \quad \frac{\partial\psi}{\partial x^3} = -(\cot\varphi\cos\psi\sin\theta + \cos\theta) \tag{8.35}$$

The path-length derivative $\nabla \cdot (\boldsymbol{\Omega} I)$ as defined in Eq. (8.17) simplifies to

$$\nabla \cdot (\boldsymbol{\Omega} I_\omega) = \cos\varphi \frac{\partial I_\omega}{\partial \rho} + \frac{\sin\varphi\cos\psi}{\rho}\frac{\partial I_\omega}{\partial \theta} + \frac{\sin\varphi\sin\psi}{\rho\sin\theta}\frac{\partial I_\omega}{\partial \phi} - \frac{\sin\varphi}{\rho}\frac{\partial I_\omega}{\partial \varphi} - \frac{\sin\varphi\sin\psi}{\rho\tan\theta}\frac{\partial I_\omega}{\partial \psi} \tag{8.36}$$

Finally, the EPRT can be written as

$$\frac{1}{v_\omega}\frac{\partial I_\omega}{\partial t} + \cos\varphi\frac{\partial I_\omega}{\partial \rho} + \frac{\sin\varphi\cos\psi}{\rho}\frac{\partial I_\omega}{\partial \theta} + \frac{\sin\varphi\sin\psi}{\rho\sin\theta}\frac{\partial I_\omega}{\partial \phi} - \frac{\sin\varphi}{\rho}\frac{\partial I_\omega}{\partial \varphi} - \frac{\sin\varphi\sin\psi}{\rho\tan\theta}\frac{\partial I_\omega}{\partial \psi} = \frac{\frac{1}{4\pi}\iint_{4\pi} I_\omega \sin\varphi d\varphi d\psi - I_\omega}{\Lambda_\omega} \tag{8.37}$$

8.4 EPRT IN NONORTHOGONAL COORDINATE SYSTEMS

A nonorthogonal coordinate system is again defined by the equations (given in Section 8.3.1, Eq. 8.5),

$$y^1 = y^1(x^1, x^2, x^3), \quad y^2 = y^2(x^1, x^2, x^3), \quad y^3 = y^3(x^1, x^2, x^3)$$

However, now, the covariant basis vectors $\mathbf{E}_i$ in general, are mutually nonorthogonal. Similarly, the contravariant basis vectors $\mathbf{E}^j$ are also mutually nonorthogonal. The two sets of basis vectors are again related by the equation (given in Section 8.3.1, Eq. 8.8)

$$\mathbf{E}_i \cdot \mathbf{E}^j = \delta_i^j$$

The problem now is to be able to define a direction in the nonorthogonal coordinate system. In other words we should find a way to define the unit vector $\boldsymbol{\Omega}$ at any generic point in the nonorthogonal curvilinear coordinate system x^i. The equation given in Section 8.3.1, Eq. (8.2),

$$\boldsymbol{\Omega} = \cos\varphi\,\hat{\mathbf{e}}_1 + \sin\varphi\cos\psi\,\hat{\mathbf{e}}_2 + \sin\varphi\sin\psi\,\hat{\mathbf{e}}_3$$

can be used to define the unit vector $\boldsymbol{\Omega}$, but it is first necessary to establish a set of mutually orthogonal unit vectors $(\hat{\mathbf{e}}_1, \hat{\mathbf{e}}_2, \hat{\mathbf{e}}_3)$ at each point in the nonorthogonal curvilinear coordinate system. These orthonormal vectors are to be represented as functions of the covariant $\mathbf{E}_i$ and contravariant $\mathbf{E}^j$ basis vectors. Once this representation is obtained, the unit vector $\boldsymbol{\Omega}$ can then be written as a function of $\mathbf{E}_i$ and $\mathbf{E}^j$. Finally, the EPRT can be written in terms of the nonorthogonal coordinates x^i.

In this section, the complete procedure to derive the EPRT in a general nonorthogonal curvilinear coordinate system is described.

8.4.1 Formulation of the Path-Length Derivative

The EPRT in any coordinate system whether Cartesian, orthogonal curvilinear, or nonorthogonal curvilinear is written in coordinate-independent form (as given in Section 8.3.1, Eq. 8.3), as

$$\frac{1}{v_\omega}\frac{\partial I_\omega}{\partial t} + \nabla\cdot(\mathbf{\Omega} I_\omega) = \frac{\frac{1}{4\pi}\iint_{4\pi} I_\omega \sin\varphi d\varphi d\psi - I_\omega}{\Lambda_\omega}$$

In this section we would like to focus our attention on the general case of a nonorthogonal curvilinear coordinate system. To use Eq. (8.3) in computations it is necessary to write the path-length derivative $\nabla\cdot(\mathbf{\Omega} I_\omega)$ in terms of the coordinates x^i. This has already been described in Section 8.3.1, Eq. (8.11); i.e.,

$$\nabla\cdot(\mathbf{\Omega} I) = (\Omega^i I)_{|i} = \frac{\partial}{\partial x^i}(\Omega^i I) + \Gamma^i_{ki}(\Omega^k I)$$

Furthermore, the following relations also exist:

$$\mathbf{\Omega} = \Omega^i \mathbf{E}_i = \Omega^1 \mathbf{E}_1 + \Omega^2 \mathbf{E}_2 + \Omega^3 \mathbf{E}_3 \tag{8.38}$$

$$|\mathbf{\Omega}| = 1 \tag{8.39}$$

The unit vector $\mathbf{\Omega}$ is used to specify the direction at any point in the space. This is usually done by introducing an orthogonal coordinate system at that point and then using the following expression for $\mathbf{\Omega}$:

$$\mathbf{\Omega} = \cos\varphi\, \hat{\mathbf{e}}_1 + \sin\varphi \cos\psi\, \hat{\mathbf{e}}_2 + \sin\varphi \sin\psi\, \hat{\mathbf{e}}_3 \tag{8.40}$$

In Eq. (8.40) $(\hat{\mathbf{e}}_1, \hat{\mathbf{e}}_2, \hat{\mathbf{e}}_3)$ are mutually orthogonal unit vectors. The polar angle φ is measured from the $\hat{\mathbf{e}}_1$ unit vector. The azimuthal angle ψ is measured from the $\hat{\mathbf{e}}_2$ unit vector. If the original coordinate system is orthogonal—i.e., if $(\mathbf{E}_1, \mathbf{E}_2, \mathbf{E}_3)$ are mutually orthogonal vectors, not necessarily of unit magnitude—then the orthonormal basis vectors are defined by means of the following prescription, which has already been discussed in Section 8.3.

$$\begin{aligned} \hat{\mathbf{e}}_1 &= \frac{\mathbf{E}_1}{|\mathbf{E}_1|}, \quad \hat{\mathbf{e}}_2 = \frac{\mathbf{E}_2}{|\mathbf{E}_2|}, \quad \hat{\mathbf{e}}_3 = \frac{\mathbf{E}_3}{|\mathbf{E}_3|} \\ \hat{\mathbf{e}}_1 &= \frac{\mathbf{E}_1}{h_1}, \quad \hat{\mathbf{e}}_2 = \frac{\mathbf{E}_2}{h_2}, \quad \hat{\mathbf{e}}_3 = \frac{\mathbf{E}_3}{h_3} \end{aligned} \tag{8.41}$$

In the general case of a nonorthogonal, curvilinear coordinate system, the covariant basis vectors $(\mathbf{E}_1, \mathbf{E}_2, \mathbf{E}_3)$ are not mutually orthogonal, and therefore we have to select a set of orthonormal basis vectors $(\hat{\mathbf{e}}_1, \hat{\mathbf{e}}_2, \hat{\mathbf{e}}_3)$ at any generic point (x^i) by some method. This is required so as to be able to specify a particular direction by means of Eq. (8.40). The important point to note here is that, whatever the choice of the orthonormal basis vectors $(\hat{\mathbf{e}}_1, \hat{\mathbf{e}}_2, \hat{\mathbf{e}}_3)$, to finally make use of Eq. (8.11) we have to express $(\hat{\mathbf{e}}_1, \hat{\mathbf{e}}_2, \hat{\mathbf{e}}_3)$ as functions of the covariant basis vectors $(\mathbf{E}_1, \mathbf{E}_2, \mathbf{E}_3)$. This means that ultimately the unit vector specifying the direction will be represented by

$$\boldsymbol{\Omega} = \Omega^i \mathbf{E}_i = \Omega^1 \mathbf{E}_1 + \Omega^2 \mathbf{E}_2 + \Omega^3 \mathbf{E}_3 \tag{8.42}$$

Consequently, the objective now is to select a set of appropriate, orthonormal, basis vectors $(\hat{\mathbf{e}}_1, \hat{\mathbf{e}}_2, \hat{\mathbf{e}}_3)$ at any point, and then represent them in terms of the covariant basis vectors $(\mathbf{E}_1, \mathbf{E}_2, \mathbf{E}_3)$ at that same point, by means of a prescription that is analogous to Eq. (8.41). One of perhaps many ways of doing this is described now. Let

$$\hat{\mathbf{e}}_1 = \frac{\mathbf{E}_1}{|\mathbf{E}_1|}; \quad \hat{\mathbf{e}}_2 = \frac{\mathbf{E}^2}{|\mathbf{E}^2|}; \quad \hat{\mathbf{e}}_3 = \frac{\mathbf{E}_1 \times \mathbf{E}^2}{|\mathbf{E}_1 \times \mathbf{E}^2|} \tag{8.43}$$

Then,

$$|\mathbf{E}_1| = \sqrt{g_{11}} \tag{8.44}$$

$$\mathbf{E}^2 = g^{21}\mathbf{E}_1 + g^{22}\mathbf{E}_2 + g^{23}\mathbf{E}_3 \tag{8.45}$$

$$|\mathbf{E}^2| = \sqrt{g^{22}} \tag{8.46}$$

$$\begin{aligned} \mathbf{E}_1 \times \mathbf{E}^2 &= (g_{11}\mathbf{E}^1 + g_{12}\mathbf{E}^2 + g_{13}\mathbf{E}^3) \times \mathbf{E}^2 \\ &= g_{11}\mathbf{E}^1 \times \mathbf{E}^2 + g_{13}\mathbf{E}^3 \times \mathbf{E}^2 \\ &= \frac{(g_{11}\mathbf{E}_3 - g_{13}\mathbf{E}_1)}{\mathrm{V}} \end{aligned} \tag{8.47}$$

where

$$\mathrm{V} = \frac{1}{\mathbf{E}^1 \cdot (\mathbf{E}^2 \times \mathbf{E}^3)} \tag{8.48}$$

$$|\mathbf{E}_1 \times \mathbf{E}^2| = \frac{\sqrt{g_{11}(g_{11}g_{33} - g_{13}^2)}}{\mathrm{V}} \tag{8.49}$$

Ultimately,

$$\mathbf{e}_1 = \frac{\mathbf{E}_1}{\sqrt{g_{11}}}; \quad \mathbf{e}_2 = \frac{g^{21}\mathbf{E}_1 + g^{22}\mathbf{E}_2 + g^{23}\mathbf{E}_3}{\sqrt{g^{22}}}; \quad \mathbf{e}_3 = \frac{g_{11}\mathbf{E}_3 - g_{13}\mathbf{E}_1}{\sqrt{g_{11}(g_{11}g_{33} - g_{13}^2)}} \tag{8.50}$$

Substituting Eq. (8.50) into Eq. (8.40) we get

$$\begin{aligned}\boldsymbol{\Omega} &= \cos\varphi\frac{\mathbf{E}_1}{\sqrt{g_{11}}} + \sin\varphi\cos\psi\left[\frac{g^{21}\mathbf{E}_1 + g^{22}\mathbf{E}_2 + g^{23}\mathbf{E}_3}{\sqrt{g^{22}}}\right] \\ &+ \sin\varphi\sin\psi\left[\frac{g_{11}\mathbf{E}_3 - g_{13}\mathbf{E}_1}{\sqrt{g_{11}(g_{11}g_{33} - g_{13}^2)}}\right]\end{aligned} \tag{8.51}$$

Rearranging this equation,

$$\begin{aligned}\boldsymbol{\Omega} = &\left[\frac{\cos\varphi}{\sqrt{g_{11}}} + \frac{g^{21}\sin\varphi\cos\psi}{\sqrt{g^{22}}} - \frac{g_{13}\sin\varphi\sin\psi}{\sqrt{g_{11}(g_{11}g_{33} - g_{13}^2)}}\right]\mathbf{E}_1 + \left[\sqrt{g^{22}}\sin\varphi\cos\psi\right]\mathbf{E}_2 \\ &+ \left[\frac{g^{23}\sin\varphi\cos\psi}{\sqrt{g^{22}}} + \frac{g_{11}\sin\varphi\sin\psi}{\sqrt{g_{11}(g_{11}g_{33} - g_{13}^2)}}\right]\mathbf{E}_3\end{aligned} \tag{8.52}$$

Eq. (8.52) presents the mathematical form of the unit vector $\boldsymbol{\Omega}$ in a direction specified by (φ, ψ) in terms of the basis vectors $(\mathbf{E}_1, \mathbf{E}_2, \mathbf{E}_3)$. The polar angle varies in the range $0 < \varphi < \pi$ and the azimuthal angle varies in the range $0 < \psi < 2\pi$. Comparing Eq. (8.52) with Eq. (8.38) we get

$$\Omega^1 = \frac{\cos\varphi}{\sqrt{g_{11}}} + \frac{g^{21}\sin\varphi\cos\psi}{\sqrt{g^{22}}} - \frac{g_{13}\sin\varphi\sin\psi}{\sqrt{g_{11}(g_{11}g_{33} - g_{13}^2)}} \tag{8.53}$$

$$\Omega^2 = \sqrt{g^{22}}\sin\varphi\cos\psi \tag{8.54}$$

$$\Omega^3 = \frac{g^{23}\sin\varphi\cos\psi}{\sqrt{g^{22}}} + \frac{g_{11}\sin\varphi\sin\psi}{\sqrt{g_{11}(g_{11}g_{33} - g_{13}^2)}} \tag{8.55}$$

Using Eq. (8.51) it can be shown conveniently that

$$|\boldsymbol{\Omega}| = 1 \tag{8.56}$$

8.4.2 Calculation of the Derivatives $\partial\varphi/\partial x^i$ and $\partial\psi/\partial x^i$

Let $\boldsymbol{\Omega}_o$ be a fixed global direction. Then as the coordinates x^i change, the angles (φ, ψ) also change so as to keep the direction $\boldsymbol{\Omega}_o$ fixed. This is because the basis vectors $(\mathbf{E}_1, \mathbf{E}_2, \mathbf{E}_3)$ change with x^i. However, since $\boldsymbol{\Omega}_o$ is fixed we may write

$$\frac{\partial\boldsymbol{\Omega}_o}{\partial x^j} = 0 \tag{8.57}$$

As described in Section 8.3, this equation is written as

$$\frac{\partial \boldsymbol{\Omega}_o}{\partial x^j} = \left(\frac{\partial \Omega^i}{\partial x^j} + \Omega^k \Gamma^i_{kj}\right)\mathbf{E}_i = 0 \tag{8.58}$$

or,

$$\frac{\partial \Omega^i}{\partial x^j} = -\Gamma^i_{kj}\Omega^k \tag{8.59}$$

Eq. (8.59) is actually nine equations corresponding to the various values of the indices i and j. In the following development, the different values of these indices are substituted in Eq. (8.59), in a systematic manner to derive the expressions for the derivatives of the polar and azimuthal angles $\partial\varphi/\partial x^i$ and $\partial\psi/\partial x^i$.

$$i = 1 \quad \text{and} \quad j = 1$$

$$\frac{\partial \Omega^1}{\partial x^1} = \frac{\partial}{\partial x^1}\left[\frac{\cos\varphi}{\sqrt{g_{11}}} + \frac{g^{21}\sin\varphi\cos\psi}{\sqrt{g^{22}}} - \frac{g_{13}\sin\varphi\sin\psi}{\sqrt{g_{11}(g_{11}g_{33}-g_{13}^2)}}\right] \tag{8.60}$$

$$\begin{aligned}\frac{\partial \Omega^1}{\partial x^1} = &\left[-\frac{\sin\varphi}{\sqrt{g_{11}}} + \frac{g^{21}\cos\varphi\cos\psi}{\sqrt{g^{22}}} - \frac{g_{13}\cos\varphi\sin\psi}{\sqrt{g_{11}(g_{11}g_{33}-g_{13}^2)}}\right]\frac{\partial\varphi}{\partial x^1}\\ &-\left[\frac{g^{21}\sin\varphi\sin\psi}{\sqrt{g^{22}}} + \frac{g_{13}\sin\varphi\cos\psi}{\sqrt{g_{11}(g_{11}g_{33}-g_{13}^2)}}\right]\frac{\partial\psi}{\partial x^1}\\ &+\cos\varphi\frac{\partial}{\partial x^1}\left(\frac{1}{\sqrt{g_{11}}}\right) + \sin\varphi\cos\psi\frac{\partial}{\partial x^1}\left(\frac{g^{21}}{\sqrt{g^{22}}}\right)\\ &-\sin\varphi\sin\psi\frac{\partial}{\partial x^1}\left(\frac{g_{13}}{\sqrt{g_{11}(g_{11}g_{33}-g_{13}^2)}}\right)\end{aligned} \tag{8.61}$$

Substituting Eq. (8.61) in the relation, $\frac{\partial \Omega^1}{\partial x^1} = -\Gamma^1_{k1}\Omega^k$ and rearranging,

$$\begin{aligned}&\left[-\frac{\sin\varphi}{\sqrt{g_{11}}} + \frac{g^{21}\cos\varphi\cos\psi}{\sqrt{g^{22}}} - \frac{g_{13}\cos\varphi\sin\psi}{\sqrt{g_{11}(g_{11}g_{33}-g_{13}^2)}}\right]\frac{\partial\varphi}{\partial x^1}\\ &-\left[\frac{g^{21}\sin\varphi\sin\psi}{\sqrt{g^{22}}} + \frac{g_{13}\sin\varphi\cos\psi}{\sqrt{g_{11}(g_{11}g_{33}-g_{13}^2)}}\right]\frac{\partial\psi}{\partial x^1} = -\Gamma^1_{k1}\Omega^k - \cos\varphi\frac{\partial}{\partial x^1}\left(\frac{1}{\sqrt{g_{11}}}\right)\\ &-\sin\varphi\cos\psi\frac{\partial}{\partial x^1}\left(\frac{g^{21}}{\sqrt{g^{22}}}\right) + \sin\varphi\sin\psi\frac{\partial}{\partial x^1}\left(\frac{g_{13}}{\sqrt{g_{11}(g_{11}g_{33}-g_{13}^2)}}\right)\end{aligned} \tag{8.62}$$

Let

$$
\begin{aligned}
A &= \left[-\frac{\sin\varphi}{\sqrt{g_{11}}} + \frac{g^{21}\cos\varphi\cos\psi}{\sqrt{g^{22}}} - \frac{g_{13}\cos\varphi\sin\psi}{\sqrt{g_{11}(g_{11}g_{33}-g_{13}^2)}} \right] \\
B &= \left[\frac{g^{21}\sin\varphi\sin\psi}{\sqrt{g^{22}}} + \frac{g_{13}\sin\varphi\cos\psi}{\sqrt{g_{11}(g_{11}g_{33}-g_{13}^2)}} \right] \\
G &= -\Gamma^1_{k1}\Omega^k - \cos\varphi\frac{\partial}{\partial x^1}\left(\frac{1}{\sqrt{g_{11}}}\right) - \sin\varphi\cos\psi\frac{\partial}{\partial x^1}\left(\frac{g^{21}}{\sqrt{g^{22}}}\right) \\
&\quad + \sin\varphi\sin\psi\frac{\partial}{\partial x^1}\left(\frac{g_{13}}{\sqrt{g_{11}(g_{11}g_{33}-g_{13}^2)}}\right)
\end{aligned}
\tag{8.63}
$$

Then Eq. (8.62) can be written as

$$
A\frac{\partial\varphi}{\partial x^1} - B\frac{\partial\psi}{\partial x^1} = G \tag{8.64}
$$

$i=1$ and $j=2$

We may observe that substituting $i=1$ and $j=2$ in Eq. (8.59) will result in a relation similar to Eqs. (8.62) and (8.64) except that the right-hand side will be different. Here we only state the final relation without going into the details,

$$
A\frac{\partial\varphi}{\partial x^2} - B\frac{\partial\psi}{\partial x^2} = H \tag{8.65}
$$

where

$$
\begin{aligned}
H &= -\Gamma^1_{k2}\Omega^k - \cos\varphi\frac{\partial}{\partial x^2}\left(\frac{1}{\sqrt{g_{11}}}\right) - \sin\varphi\cos\psi\frac{\partial}{\partial x^2}\left(\frac{g^{21}}{\sqrt{g^{22}}}\right) \\
&\quad + \sin\varphi\sin\psi\frac{\partial}{\partial x^2}\left(\frac{g_{13}}{\sqrt{g_{11}(g_{11}g_{33}-g_{13}^2)}}\right)
\end{aligned}
\tag{8.66}
$$

$i=1$ and $j=3$

Considerations similar to the previous case also apply to the current case and again only the final results are stated:

$$
A\frac{\partial\varphi}{\partial x^3} - B\frac{\partial\psi}{\partial x^3} = I \tag{8.67}
$$

where

$$
\begin{aligned}
I &= -\Gamma^1_{k3}\Omega^k - \cos\varphi\frac{\partial}{\partial x^3}\left(\frac{1}{\sqrt{g_{11}}}\right) - \sin\varphi\cos\psi\frac{\partial}{\partial x^3}\left(\frac{g^{21}}{\sqrt{g^{22}}}\right) \\
&\quad + \sin\varphi\sin\psi\frac{\partial}{\partial x^3}\left(\frac{g_{13}}{\sqrt{g_{11}(g_{11}g_{33}-g_{13}^2)}}\right)
\end{aligned}
\tag{8.68}
$$

Eqs. (8.64), (8.65), and (8.67) are three equations for six unknowns. These unknowns are

$$\frac{\partial \varphi}{\partial x^1}, \frac{\partial \varphi}{\partial x^2}, \frac{\partial \varphi}{\partial x^3} \quad \text{and} \quad \frac{\partial \psi}{\partial x^1}, \frac{\partial \psi}{\partial x^2}, \frac{\partial \psi}{\partial x^3} \tag{8.69}$$

To obtain the other three equations we now substitute $i = 2$ and $j = 1, 2, 3$ in Eq. (8.59):

$$i = 2 \text{ and } j = 1$$

$$\frac{\partial \Omega^2}{\partial x^1} = \frac{\partial}{\partial x^1}\left[\sqrt{g^{22}} \sin \varphi \cos \psi\right] \tag{8.70}$$

$$\frac{\partial \Omega^2}{\partial x^1} = \sqrt{g^{22}} \cos \varphi \cos \psi \frac{\partial \varphi}{\partial x^1} - \sqrt{g^{22}} \sin \varphi \sin \psi \frac{\partial \psi}{\partial x^1} + \sin \varphi \cos \psi \frac{\partial}{\partial x^1}\left(\sqrt{g^{22}}\right) \tag{8.71}$$

Substituting Eq. (8.71) in the relation $\frac{\partial \Omega^2}{\partial x^1} = -\Gamma^2_{k1}\Omega^k$ and rearranging,

$$\sqrt{g^{22}}\cos \varphi \cos \psi \frac{\partial \varphi}{\partial x^1} - \sqrt{g^{22}}\sin \varphi \sin \psi \frac{\partial \psi}{\partial x^1} = -\Gamma^2_{k1}\Omega^k - \sin \varphi \cos \psi \frac{\partial}{\partial x^1}\left(\sqrt{g^{22}}\right) \tag{8.72}$$

Let

$$\begin{aligned} C &= \sqrt{g^{22}}\cos \varphi \cos \psi \\ D &= \sqrt{g^{22}}\sin \varphi \sin \psi \\ J &= -\Gamma^2_{k1}\Omega^k - \sin \varphi \cos \psi \frac{\partial}{\partial x^1}\left(\sqrt{g^{22}}\right) \end{aligned} \tag{8.73}$$

Then Eq. (8.72) can be written as

$$C\frac{\partial \varphi}{\partial x^1} - D\frac{\partial \psi}{\partial x^1} = J \tag{8.74}$$

$$i = 2 \text{ and } j = 2$$

We again observe that substituting $i = 2$ and $j = 2$ in Eq. (8.59) will result in a relation similar to Eqs. (8.72) and (8.74) except that the right-hand side will be different. Here we only state the final relation without going into the details,

$$C\frac{\partial \varphi}{\partial x^2} - D\frac{\partial \psi}{\partial x^2} = K \tag{8.75}$$

where

$$K = -\Gamma^2_{k2}\Omega^k - \sin \varphi \cos \psi \frac{\partial}{\partial x^2}\left(\sqrt{g^{22}}\right) \tag{8.76}$$

$$i = 2 \text{ and } j = 3$$

Considerations similar to the previous case also apply to the current case and again only the final results are stated,

$$C\frac{\partial\varphi}{\partial x^3} - D\frac{\partial\psi}{\partial x^3} = L \tag{8.77}$$

where

$$L = -\Gamma^2_{k3}\Omega^k - \sin\varphi\cos\psi\frac{\partial}{\partial x^3}\left(\sqrt{g^{22}}\right) \tag{8.78}$$

Eqs. (8.74), (8.75), and (8.77) constitute the other set of three equations required to find the derivatives of the polar and azimuthal angles. The question now arises that whether by substituting $i = 3$ and $j = 1, 2, 3$ in Eq. (8.59) we get another set of three independent relations between the various derivatives of the polar and azimuthal angles. The answer is in the negative. To prove this assertion we now return to Eq. (8.39),

$$|\mathbf{\Omega}| = 1 \tag{8.39}$$

$$|\mathbf{\Omega}|^2 = (\Omega^m\mathbf{E}_m)\cdot(\Omega^n\mathbf{E}_n) = g_{mn}\Omega^m\Omega^n = 1 \tag{8.79}$$

Differentiating Eq. (8.79) with respect to the coordinates x^j,

$$\frac{\partial}{\partial x^j}(g_{mn}\Omega^m\Omega^n) = 0 \tag{8.80}$$

$$\Omega^m\Omega^n\frac{\partial g_{mn}}{\partial x^j} + 2g_{mn}\Omega^n\frac{\partial\Omega^m}{\partial x^j} = 0 \tag{8.81}$$

If we now let $j = 1$ in Eq. (8.81) and carry out the summations implied by the indices m and n, we get a relation between the derivatives $\partial\Omega^m/\partial x^1$, $m = 1, 2, 3$. Specifically, the derivative $\partial\Omega^3/\partial x^1$ can be represented in terms of the derivatives, $\partial\Omega^1/\partial x^1$ and $\partial\Omega^2/\partial x^1$. In a similar fashion by changing the value of j and carrying out the summations we end up representing $\partial\Omega^3/\partial x^2$ in terms of the derivatives $\partial\Omega^1/\partial x^2$ and $\partial\Omega^2/\partial x^2$, and the derivative $\partial\Omega^3/\partial x^3$ in terms of the derivatives $\partial\Omega^1/\partial x^3$ and $\partial\Omega^2/\partial x^3$. We thus conclude that the remaining three cases corresponding to the substitution of $i = 3$ and $j = 1, 2, 3$ in Eq. (8.59) are redundant and consequently not needed.

Consider now Eqs. (8.64) and (8.74). These equations can be written in matrix form as

$$\begin{bmatrix} A & -B \\ C & -D \end{bmatrix}\begin{pmatrix} \partial\varphi/\partial x^1 \\ \partial\psi/\partial x^1 \end{pmatrix} = \begin{pmatrix} G \\ J \end{pmatrix} \tag{8.82}$$

Solving the matrix in Eq. (8.82) we get

$$\frac{\partial\psi}{\partial x^1} = \frac{\begin{vmatrix} A & G \\ C & J \end{vmatrix}}{\begin{vmatrix} A & -B \\ C & -D \end{vmatrix}} = \frac{AJ-GC}{BC-AD} \quad \text{and} \quad \frac{\partial\varphi}{\partial x^1} = \frac{\begin{vmatrix} G & -B \\ J & -D \end{vmatrix}}{\begin{vmatrix} A & -B \\ C & -D \end{vmatrix}} = \frac{BJ-DG}{BC-AD} \tag{8.83}$$

Similarly, writing Eqs. (8.65) and (8.75) in matrix form and then solving the matrix equation we get

$$\begin{bmatrix} A & -B \\ C & -D \end{bmatrix} \begin{pmatrix} \partial\varphi/\partial x^2 \\ \partial\psi/\partial x^2 \end{pmatrix} = \begin{pmatrix} H \\ K \end{pmatrix} \tag{8.84}$$

$$\frac{\partial\psi}{\partial x^2} = \frac{\begin{vmatrix} A & H \\ C & K \end{vmatrix}}{\begin{vmatrix} A & -B \\ C & -D \end{vmatrix}} = \frac{AK-CH}{BC-AD} \quad \text{and} \quad \frac{\partial\varphi}{\partial x^2} = \frac{\begin{vmatrix} H & -B \\ K & -D \end{vmatrix}}{\begin{vmatrix} A & -B \\ C & -D \end{vmatrix}} = \frac{BK-DH}{BC-AD} \tag{8.85}$$

Similarly, writing Eqs. (8.67) and (8.77) in matrix form and then solving the matrix equation we get

$$\begin{bmatrix} A & -B \\ C & -D \end{bmatrix} \begin{pmatrix} \partial\varphi/\partial x^3 \\ \partial\psi/\partial x^3 \end{pmatrix} = \begin{pmatrix} I \\ L \end{pmatrix} \tag{8.86}$$

$$\frac{\partial\psi}{\partial x^3} = \frac{\begin{vmatrix} A & I \\ C & L \end{vmatrix}}{\begin{vmatrix} A & -B \\ C & -D \end{vmatrix}} = \frac{AL-CI}{BC-AD} \quad \text{and} \quad \frac{\partial\varphi}{\partial x^3} = \frac{\begin{vmatrix} I & -B \\ L & -D \end{vmatrix}}{\begin{vmatrix} A & -B \\ C & -D \end{vmatrix}} = \frac{BL-DI}{BC-AD} \tag{8.87}$$

For convenience we reproduce the final relations:

$$\frac{\partial\psi}{\partial x^1} = \frac{AJ-GC}{BC-AD}, \quad \frac{\partial\varphi}{\partial x^1} = \frac{BJ-DG}{BC-AD}$$
$$\frac{\partial\psi}{\partial x^2} = \frac{AK-CH}{BC-AD}, \quad \frac{\partial\varphi}{\partial x^2} = \frac{BK-DH}{BC-AD}$$
$$\frac{\partial\psi}{\partial x^3} = \frac{AL-CI}{BC-AD}, \quad \frac{\partial\varphi}{\partial x^3} = \frac{BL-DI}{BC-AD}$$

These relations define the derivatives of the polar and azimuthal angles with respect to the coordinate variables x^i.

8.4.3 Equation of Phonon Radiative Transfer

Returning back to the path-length derivative, Section 8.3.1, Eq. (8.11), we may now expand it and write it in terms of the derivatives of the polar and azimuthal angles.

$$\nabla.(\mathbf{\Omega} I_\omega) = \frac{\partial}{\partial x^i}\left(\Omega^i I_\omega\right) + \Gamma^i_{ki}\left(\Omega^k I_\omega\right) \tag{8.88}$$

$$\nabla\cdot(\boldsymbol{\Omega}I_\omega)=\Omega^i\frac{\partial I_\omega}{\partial x^i}+I_\omega\frac{\partial\Omega^i}{\partial x^i}+\Gamma^i_{ki}(\Omega^k I_\omega)$$

$$\nabla\cdot(\boldsymbol{\Omega}I_\omega)=\Omega^i\left(\frac{\partial I_\omega}{\partial x^i}+\frac{\partial I_\omega}{\partial\varphi}\frac{\partial\varphi}{\partial x^i}+\frac{\partial I_\omega}{\partial\psi}\frac{\partial\psi}{\partial x^i}\right)+\left(\frac{\partial\Omega^i}{\partial x^i}+\Gamma^i_{ki}\Omega^k\right)I_\omega$$

The last term on the R.H.S. of the above equation vanishes. This result is obtained by contracting the indices i and j in Eq. (8.59)

$$\frac{\partial\Omega^i}{\partial x^i}+\Gamma^i_{ki}\Omega^k=0 \tag{8.89}$$

$$\begin{aligned}\nabla\cdot(\boldsymbol{\Omega}I_\omega)&=\Omega^1\left(\frac{\partial I_\omega}{\partial x^1}+\frac{\partial I_\omega}{\partial\varphi}\frac{\partial\varphi}{\partial x^1}+\frac{\partial I_\omega}{\partial\psi}\frac{\partial\psi}{\partial x^1}\right)+\Omega^2\left(\frac{\partial I_\omega}{\partial x^2}+\frac{\partial I_\omega}{\partial\varphi}\frac{\partial\varphi}{\partial x^2}+\frac{\partial I_\omega}{\partial\psi}\frac{\partial\psi}{\partial x^2}\right)\\&+\Omega^3\left(\frac{\partial I_\omega}{\partial x^3}+\frac{\partial I_\omega}{\partial\varphi}\frac{\partial\varphi}{\partial x^3}+\frac{\partial I_\omega}{\partial\psi}\frac{\partial\psi}{\partial x^3}\right)\end{aligned} \tag{8.90}$$

$$\begin{aligned}\nabla\cdot(\boldsymbol{\Omega}I_\omega)&=\left(\Omega^1\frac{\partial I_\omega}{\partial x^1}+\Omega^2\frac{\partial I_\omega}{\partial x^2}+\Omega^3\frac{\partial I_\omega}{\partial x^3}\right)+\left(\Omega^1\frac{\partial\varphi}{\partial x^1}+\Omega^2\frac{\partial\varphi}{\partial x^2}+\Omega^3\frac{\partial\varphi}{\partial x^3}\right)\frac{\partial I_\omega}{\partial\varphi}\\&+\left(\Omega^1\frac{\partial\psi}{\partial x^1}+\Omega^2\frac{\partial\psi}{\partial x^2}+\Omega^3\frac{\partial\psi}{\partial x^3}\right)\frac{\partial I_\omega}{\partial\psi}\end{aligned} \tag{8.91}$$

Finally the EPRT is written as

$$\begin{aligned}&\frac{1}{v_\omega}\frac{\partial I_\omega}{\partial t}+\Omega^1\frac{\partial I_\omega}{\partial x^1}+\Omega^2\frac{\partial I_\omega}{\partial x^2}+\Omega^3\frac{\partial I_\omega}{\partial x^3}+\left(\Omega^1\frac{\partial\varphi}{\partial x^1}+\Omega^2\frac{\partial\varphi}{\partial x^2}+\Omega^3\frac{\partial\varphi}{\partial x^3}\right)\frac{\partial I_\omega}{\partial\varphi}\\&+\left(\Omega^1\frac{\partial\psi}{\partial x^1}+\Omega^2\frac{\partial\psi}{\partial x^2}+\Omega^3\frac{\partial\psi}{\partial x^3}\right)\frac{\partial I_\omega}{\partial\psi}=\frac{\frac{1}{4\pi}\iint_{4\pi}I_\omega\sin\varphi\,d\varphi\,d\psi-I_\omega}{\Lambda_\omega}\end{aligned} \tag{8.92}$$

where the expression for variables Ω^1, Ω^2, and $\Omega^3$are given in Section 8.4.1 (Eqs. (8.53)–(8.55)). We can find the expression for the terms $\partial\varphi/\partial x^i$ and $\partial\psi/\partial x^i$ in Section 8.4.2 (Eqs. (8.83), (8.85), and (8.87), respectively).

It can be verified that for an orthogonal coordinate system in which

$$g_{11}=\frac{1}{g^{11}}=h_1^2,\qquad g_{22}=\frac{1}{g^{22}}=h_2^2,$$
$$g_{33}=\frac{1}{g^{33}}=h_3^2,\qquad g_{ij}=g^{ij}=0,\quad i\neq j$$

Eqs. (8.83), (8.85), and (8.87) reduce to Eq. (8.24) and Eq. (8.92) reduces to Eq. (8.25).

8.5 APPLICATION TO TWO-DIMENSIONAL PHONON RADIATIVE TRANSFER

In domains that are not regular, the EPRT (Eq. 8.3) can be solved numerically. In this case the first step is to numerically generate a curvilinear coordinate system based on the domain boundaries. Mathematically, this means that the transformation equations in two dimensions,

$$y^1 = y^1(x^1, x^2), \quad y^2 = y^2(x^1, x^2) \tag{8.93}$$

are given numerically as in the form of a table or as an array stored in the computer. The algorithms for accomplishing this task are well documented and are known as numerical grid generation. In this section we shall consider the case of the solution of the EPRT in a generic quadrilateral. The quadrilateral is first covered by an orthogonal grid that is generated by means of a suitable algorithm. In the next step, the EPRT is solved numerically in the computational domain by means of the discrete-ordinates method. The solution is then mapped back to the physical domain. Additionally, the algorithm that is developed in this section is also used to solve the EPRT in a square domain. The solution for the square domain problem can be obtained by solving the well-established form of the EPRT in Cartesian coordinates. The two solutions are compared so as to validate the algorithm, but more importantly to validate the mathematical formalism developed in Section 8.3.

Consider a wedge-shaped, microscale, silicon flake as shown in Fig. 8.3. This flake is covered by an orthogonal grid by using an algorithm based on the numerical grid generation technique [8]. This algorithm in

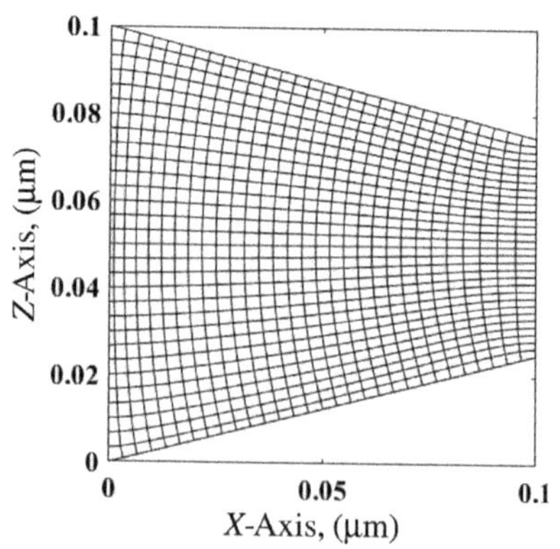

■ FIGURE 8.3 Orthogonal grid in a generic quadrilateral.

reality constructs the transformation Eq. (8.93), but in a way such that the resulting basis vectors are mutually orthogonal at each and every point in the physical domain. Since this is a numerical procedure, Eq. (8.93) are only given at discrete values of the variables x^i. Once these equations are available, the scale factors h_1, h_2, h_3 can then be calculated numerically. This is done by first formulating the position vector, Eq. (8.6), then differentiating it with respect to the coordinates x^i to obtain $\mathbf{E}_i$, and finally calculating the scale factors h_i according to the formula

$$h_i = \sqrt{\mathbf{E}_i \cdot \mathbf{E}_i} \quad i = 1, 2 \tag{8.94}$$

All the steps just described are performed numerically. Once the scale factors are known, the coefficients of the various derivatives occurring in Eq. (8.25) are calculated numerically at all values of the coordinates x^i as well as at all values of the polar and azimuthal angles. In the next step, the EPRT given by Eq. (8.25) is discretized by means of the finite-difference method. This is actually termed the discrete-ordinates method. The discretization of Eq. (8.25) results in a system of simultaneous, linear, algebraic equations. These equations are then solved numerically by a suitable algorithm. The details of the discrete-ordinates method applied to the EPRT in Cartesian coordinates can be obtained from Yilbas et al. [9]. For the solution of the EPRT in an orthogonal, curvilinear coordinate system see Mansoor et al. [10].

An important detail regarding the solution of Eq. (8.25) is concerned with the boundary conditions. The EPRT in the Cartesian coordinate system, Eq. (8.4), involves derivatives of the phonon intensity I_ω with respect to the position variables and not with respect to the polar and azimuthal angles. Hence we need boundary conditions related to position coordinates only. On the other hand, Eq. (8.25) includes derivatives of the phonon intensity I_ω with respect to the polar φ and azimuthal angles ψ also; that is, the derivatives $\partial I_\omega/\partial\varphi$ and $\partial I_\omega/\partial\psi$. Therefore two additional boundary conditions are also required that are related to the polar and azimuthal angles. These boundary conditions are described next.

8.5.1 Boundary Condition Related to the Polar Angle φ

$I_\omega(x^i, \varphi = 0, \psi)$ is unique for all values of ψ at the point x^i.

$I_\omega(x^i, \varphi = \pi, \psi)$ is unique for all values of ψ at the point x^i.

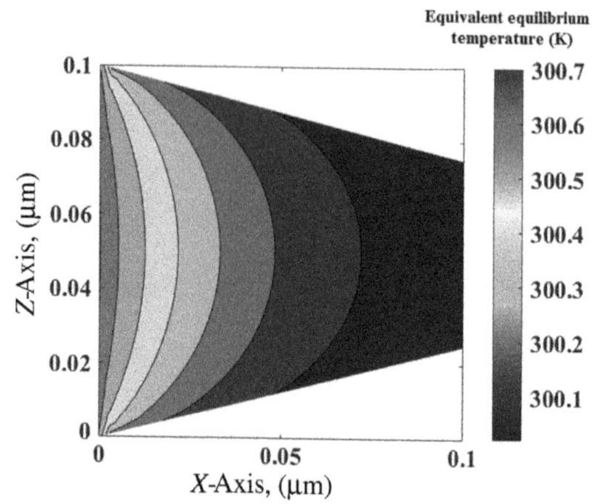

■ **FIGURE 8.4** Steady-state contours for the equivalent equilibrium temperature.

8.5.2 Boundary Condition Related to the Azimuthal Angle ψ

$$I_\omega(x^i, \varphi, \psi = 0) = I_\omega(x^i, \varphi, \psi = 2\pi)$$

The phonon intensity I_ω is continuous as the azimuthal angle takes on extreme values. This may also be termed the periodic boundary condition.

In words we may say that the variable I_ω remains single-valued and continuous as the polar φ and azimuthal ψ angles take on extreme values. These boundary conditions can easily be implemented numerically.

Returning back to the wedge problem, the previously described procedure is employed to solve the two-dimensional, frequency-independent, steady-state EPRT in the wedge-shaped flake. The boundary conditions are that of constant temperature. At the left boundary the temperature is fixed at 301K, whereas along the remaining boundaries the temperature is fixed at 300K. The material is selected as silicon with the following properties: $k = 157\ \mathrm{Wm^{-1}K^{-1}}$, $C = 1.653 \times 10^6\ \mathrm{Jm^{-3}K^{-1}}$, $v = 8430\ \mathrm{ms^{-1}}$, $\Lambda = 33.8\ \mathrm{nm}$. The physical domain along with the orthogonal grid as well as the temperature contours are shown in Figs. 8.3 and 8.4.

8.6 CONCLUDING REMARKS

This chapter was devoted to the derivation and development of the EPRT in general orthogonal and nonorthogonal curvilinear coordinate systems. This derivation closely follows the development of the ERT, which was initiated, among others, by Vaillon et al. [2] and completed and extended by J. Freimanis [3]. In this process, it becomes clear that tensor analysis is the required mathematical tool and this tool is becoming increasingly important for the engineer to learn. The EPRT thus obtained is useful in solving problems in domains that are not rectangular but that can still fit into one of the several available orthogonal coordinate systems. The general EPRT can also be used in irregular domains for which no existing coordinate system will suffice. In these situations the solution is obtained numerically. The domain is first covered by an orthogonal or nonorthogonal grid by means of one of several well-documented techniques. These techniques are referred to as numerical grid generation. Once the grid is generated, the relevant geometric properties such as the metric tensor and connection coefficients can then be determined. Finally, the most general form of the EPRT can be solved numerically by first discretizing it by means of the finite-difference method and then solving the resulting system of simultaneous, linear/nonlinear equations by some suitable algorithm. In the context of the radiative transfer problem, the finite-difference method is labeled as the discrete-ordinates method.

REFERENCES

[1] A. Majumdar, Microscale heat conduction in dielectric thin films, J. Heat Transfer 115 (1) (1993) 7.

[2] R. Vaillon, M. Lallemand, D. Lemonnier, Radiative heat transfer in orthogonal curvilinear coordinates using the discrete ordinates method, J. Quant. Spectrosc. Radiat. Transfer 55 (1) (1996) 7–17.

[3] J. Freimanis, On vector radiative transfer equation in curvilinear coordinate systems, J. Quant. Spectrosc. Radiat. Transfer 112 (2011) 2134–2148.

[4] J. Freimanis, Polarized radiative transfer equation in some curvilinear coordinate systems, J. Quant. Spectrosc. Radiat. Transfer 146 (2014) 250–269.

[5] J.H. Heinbockel, Introduction to Tensor Analysis and Continuum Mechanics, Trafford Publishing, Canada, 2001.

[6] J.K. Synge, A. Schild, Tensor Calculus, University of Toronto Press, Canada, 1969.

[7] R. Aris, Vectors, Tensors, and the Basic Equations of Fluid Mechanics, Dover, 1962.

[8] J.F. Thomas, Z.U.A Warsi, Numerical Grid Generation: Foundations and Applications. North-Holland, 1985.

[9] B.S. Yilbas, S.B. Mansoor, Phonon transport in two-dimensional silicon thin film: influence of film width and boundary conditions on temperature distribution, Eur. Phys. J. B 85 (7) (2012) 243.

[10] S.B. Mansoor, B.S. Yilbas, Phonon transport across nano-scale curved thin films, Phys. B 503 (2016) 130–140.

Chapter 9

Concluding Remarks

Some basics of crystal structures, relevant considerations in statistical thermodynamics, and phonon radiative transport are presented first to introduce the fundamental physical aspects in relation to the micro/nanoscale heat transfer. Both analytical and numerical approaches are included for the solution of the governing equations of the micro/nanoscale heat transfer. Although some simplifications are introduced for the analytical solution of the hyperbolic heat conduction equation and the steady phonon radiative transport equation, the findings are well in line with the numerical predictions. This is also true for the thermal stress formulations and the closed form results of the momentum (stress) equation. In the case of the metallic substrates, short duration of thermal disturbance gives rise to the nonequilibrium heating situation, which results in thermal separation of the electron and lattice subsystems. The analytical formulation of energy transport across the electron and lattice subsystems is presented, when metallic substrate is subjected to a local heating with a short duration, by using the electron kinetic theory approach. The findings of temperature distributions in electron and lattice subsystems agree well with the experimental data and the predictions of the two-equation heating model.

The formulation of the phonon radiative transport equation is presented for dielectric and metallic thin films. For the applications of the radiative phonon transport equation, several cases are introduced and the phonon intensity distribution within the films is quantified in terms of the equivalent equilibrium temperature. Thermal boundary resistance based on the diffusive mismatch and the cut-off mismatch models are introduced and the applications of the thermal boundary resistance across the thin films pair are demonstrated. The minute size vacuum gap is also introduced at the interface of the thin films pair and thermal energy transport due to blackbody and near field radiations is formulated. The predictions of the energy transport due to thermal radiation are compared with that of the energy transport via ballistic phonons.

The main conclusions derived from the solutions of the micro/nanoscale heat equations can be categorized into two main headings, which include

Heat Transport in Micro- and Nanoscale Thin Films.
DOI: http://dx.doi.org/10.1016/B978-0-323-42979-5.00009-8

results of the analytical approaches relevant to micro/nanoscale heat transfer and thermal stress field, and the solution of the phonon radiative transport equation pertinent to thin films and thin film pairs. The main concluding remarks are presented herein under the appropriate subheadings.

9.1 ANALYTICAL SOLUTIONS FOR HEAT EQUATION AND THERMAL STRESS FIELD

The closed form solution of the hyperbolic heat equation is presented for the case involved with laser short-pulse heating. The thermal stress field in the laser irradiated region is also formulated and the closed form solution for the thermal stress developed is obtained. The possible temperature and stress distributions in the irradiated region are presented in the light of the substrate material and laser pulse properties. The thermal stress does not follow the temperature distribution in the laser irradiated region. It demonstrates wave behavior inside the substrate material with progressing time. In addition, thermal stress is tensile in the surface vicinity and remains compressive in the region next to the surface vicinity. This behavior is attributed to the free expansion of the surface during the laser heating pulse. However, thermal strain remains high below the surface because of the constraints for free expansion in this region. This, in turn, results in compressive stress levels in this region. Since thermal separation of the electron and lattice subsystems gives rise to a nonequilibrium heating situation, the temperature field in both systems is formulated using the electron kinetic theory approach.

Although the resulting equations for energy transport in both subsystems are solved numerically, the formulation of those equations relies on heavily mathematical arguments. In the analysis, the external heating source is considered to be a laser irradiation pulse. This is because of the fact that laser short-pulse heating combines with the nonequilibrium heating situation, in which case, heating can be formulated by a hyperbolic heat equation. The findings revealed that electron temperature rise is rapid and temporal variation of the electron temperature does not exactly follow the laser intensity because of the collision processes taking place between the electrons and the lattice site. In this case, electrons gain excess energy from the laser irradiated field and transfer some fraction of their excess energy to the lattice site through electron–phonon collision, which is defined through the electron–phonon coupling parameter. Depending on the electron–phonon coupling, electron temperature reduces accordingly with progressing heating time. However, lattice temperature rise becomes gradual with progressing time because of the energy gain from the collisional process. Energy exchange between electron and lattice subsystems ceases once the temperature equilibrium is reached in between these subsystems.

On the other hand, an analytical solution for the radiative transport equation in thin films is presented through incorporating the Fredholm integral equation of the second kind. In order to simplify the energy transport in the film, it is assumed that the film is a gray body and the transport is independent of phonon frequencies. It is also assumed that the film edges are at constant temperature (high temperature edge is at 301K and low temperature edge is at 300K) and phonon transport takes place at a steady state. The influence of the film thickness and the film properties are incorporated during the computation of the temperature field from the analytical solution. Temperature jump is resulted at the edges of the film, which is more pronounced for the diamond film. In addition, the influence of the film thickness on the temperature distribution is significant for the diamond film. This is attributed to the mean free path of the diamond film, which is larger than that of the silicon film. In this case, for the film thicknesses comparable to the mean free path of the diamond film, the emitted phonons from the high temperature edge behave like ballistic phonons without going through the large-scale scattering in the film. This lowers the phonon resistance in the film. Since the mean free path of silicon is considerably smaller than the film thicknesses, the emitted phonons suffer from the large-scale scattering while generating the phonon resistance in the film. The influence of low phonon speed also contributes to the large-scale scattering of phonons in the silicon film prior to reaching the low temperature edge.

9.2 NUMERICAL SOLUTIONS TO EQUATION FOR PHONON RADIATIVE TRANSPORT

The findings related to the numerical solution of the equation for phonon radiative transport (EPRT) are given according to the applications introduced in the following subsections.

9.2.1 Frequency-Dependent Solution of EPRT for Silicon Thin Film (One-Dimensional Case)

Phonon transport in thin silicon film is examined and the influence of quasiballistic and ballistic phonons on the transport characteristics is assessed. The phonon intensity distribution ratio ($\frac{I}{I^o_{max}}$, where I^o_{max} is the maximum equilibrium intensity, I is the phonon intensity corresponding to any braches at any location and time in the film) is introduced, which provides information on normalized phonon intensity distribution in the film during different heating periods. The phonon transport characteristics are assessed by the wavelength ratio ($\eta_k = \frac{k_{max} - k_{min}}{k_{max} - k_{cut-off}}$, where k is the wave number, $k = 2\pi/\lambda$, and $k_{cut\text{-}off}$ is the cut-off wave number). In this case, the influence of quasiballistic and ballistic phonons on the thermal conductivity and specific heat of the film is determined. The phonon transport properties are examined due to the temperature disturbance at the film edges; in which case, the left edge of the film is kept at 301K and the right edge of the film is at 300K. The transient- and frequency-dependent phonon radiative transport equation is solved numerically, incorporating the discrete ordinate method. The phonon intensity ratio, because of the longitudinal acoustic branch, is considerably higher than that of the transverse acoustics branch. Consequently, energy transport mainly takes place by the longitudinal acoustics phonon branch. In addition, time required to reach quasisteady steady transport characteristics is shorter for the longitudinal acoustic branch than that corresponding to the transverse branch.

The effect of wave number ratio $\left(\eta_k = \frac{k_{max} - k_{min}}{k_{max} - k_{cut-off}}\right)$ is not considerably high on the intensity ratio due to longitudinal acoustics branch; in this case, the phonon ratio reduces slightly with increasing wave number ratio. In the case of the transverse acoustic branch, an increasing wave number ratio lowers the phonon intensity ratio significantly; in this case, limiting the maximum wavelength of the emitted phonons from the high temperature edge of the film alters the phonon scattering in the film. This, in turn, modifies thermal conductivity and specific heat of the film. Therefore, reducing the wave number ratio creates the size effect in the film while lowering thermal conductivity and specific heat of the film. This is more pronounced when the reduction in wave number ratio becomes more than 20%.

9.2.2 Frequency-Independent Solution of EPRT for Aluminum Thin Film (One-Dimensional Case)

The influence of film thickness on nonequilibrium energy transfer in the thin aluminum film is investigated. Electron and phonon temperatures are predicted using the Boltzmann equation and the modified two-equation model. The findings are compared for different film thicknesses. The contribution of ballistic phonons to phonon temperature rise is considerable for the small film thickness; in this case, phonon temperature reduces in the region of the high temperature edge of the film. This, in turn, results in a difference in phonon and electron temperature distributions obtained from the solution of the Boltzmann equation. As the film thickness increases, electron and phonon temperature distribution becomes almost identical in the film. This is associated with the emitted and reflected phonons from the low temperature, which contribute to the rise of phonon temperature in the film. In addition, increasing film thickness increases the number of electron–phonon collisions while attaining almost the same value of both temperatures. Temperature difference due to the solutions of the two-equation model and the Boltzmann equation attains high values for small film width, and as the film width increases this difference reduces. Phonon temperature obtained from the solution of the Boltzmann equation becomes higher than that corresponding to the two-equation model in the region of low temperature edge. This, in turn, causes a negative temperature difference in this region. The location of the maximum phonon temperature difference moves away from the high and low temperature edges as the film thickness increases. The similar behavior is also observed for electron temperature difference due to the solutions of the two-equation model and the Boltzmann equation. In general, the two-equation model predicts higher phonon and electron temperatures than that of the Boltzmann equation solution in the region of the high temperature edge. The opposite is true in the region of the low temperature edge.

9.2.3 Frequency-Dependent Solution of EPRT for Silicon Thin Film (Two-Dimensional Case)

Phonon transport in two-dimensional silicon film is considered. The frequency-dependent Boltzmann transport equation is solved numerically for constant temperature boundary conditions at the film edges. The transient effect of the phonon transport in the film is incorporated. The influence of film thickness on the phonon transport is analyzed through computing equivalent equilibrium temperature. The rate of phonon transport in the early heating period is low in the film resulting in less phonon scattering and dispersion in the film. This, in turn, gives rise to attainment of low equivalent equilibrium temperature in the film. The rate of equivalent equilibrium temperature rise is high in the early heating period ($t \leq 50$ ps) and it becomes gradual as the heating period progresses. This is attributed to the rate of phonon scattering and dispersion in the film, which increases in the early heating period. Equivalent equilibrium temperature reduces toward the low temperature edges of the film due to the fact that low temperature edges act like heat sinks, lowering the rate of phonon reflection from the cold edges toward the film. Increasing film thickness enhances equivalent equilibrium temperature rise in the film. In addition, the rate of rise of equivalent equilibrium temperature becomes high for the wider films. This is associated with the phonon scattering and dispersion, which increases with increasing film width.

9.2.4 Frequency-Independent Solution of EPRT for Aluminum Thin Film (Two-Dimensional Case)

Energy transfer in a two-dimensional aluminum thin film due to laser short-pulse irradiation is investigated. Since the heating duration is short and the film thickness is small, a nonequilibrium heating model is introduced. In this case, thermal separation of electron and lattice subsystems is considered and the modified Boltzmann equation is used to model energy transfer in the electron and lattice subsystems. The transient coupled energy equations are solved numerically via computer code that has been developed. Since the laser irradiated energy is absorbed by electrons, a source term is introduced in the energy equation for the electron subsystem while electron–phonon coupling is incorporated to account for electron excess energy transfer to the lattice subsystem.

To demonstrate the energy balance between the absorbed laser energy and the predicted energy gain in the electron and the lattice subsystems, an integral equation is developed. The integral equation satisfies the energy balance between the amount of absorbed irradiated energy by the film and the predicted energy gain in the electron and the lattice subsystems for all the heating periods. Equivalent equilibrium temperature increases sharply in the electron subsystem due to energy gain from the irradiated field in the film. As the heating period progresses, equivalent equilibrium temperature in the electron subsystem increases; however, further increase in the heating time results in reduction of equivalent equilibrium temperature because of the temporal behavior of laser pulse intensity, which decays exponentially with progressing time. The time occurrence of the peak equivalent equilibrium temperature changes at different locations along the x axis. This behavior is attributed to the spatial distribution of the laser peak intensity along the x axis, which is Gaussian. Equivalent equilibrium temperature in the lattice subsystem does follow its counterpart corresponding to the electron subsystem despite the presence of constant electron–phonon coupling factor. This is related to the phonon scattering in the irradiated region due to film resistance, which modifies phonon intensity distribution in this region. In addition, temporal variation of equivalent equilibrium temperature in the lattice subsystem does not follow that occurring in the electron subsystem. Equivalent equilibrium temperature remains higher in the electron subsystem than that of the lattice subsystem due to electron excess energy gain from the irradiated field.

9.2.5 Solution of EPRT for Cross-Plane Transport Across Dielectric and Metallic Thin Films Pair (One-Dimensional Case)

Laser short-pulse heating of silicon-diamond-aluminum films composite is considered and energy transport across the films is investigated. The Boltzmann equation is used to formulate energy transport in the films; however, modification in the Boltzmann equation is made to account for the thermal separation in the aluminum film during the laser short heating duration. In this case, electron–phonon coupling is introduced to account for the energy transfer across the electron and lattice subsystems. The transfer matrix method is incorporated to formulate transmission, reflectance, and absorption of the incident laser radiation in the films. The simulation conditions are altered to compare the predictions of the modified Boltzmann equation with the two-equation model. In this case, temperature disturbance is introduced across the aluminum film edges and equivalent equilibrium temperature variation across the film is predicted. The predictions of the modified Boltzmann equation differs slightly from those predicted from the two-equation model; however, this difference is small. Equivalent equilibrium temperature rise is high in the electron system during the early heating period, which is attributed to the excess energy gain of electrons from the irradiated field. As the heating period increases, laser pulse intensity reduces; in this case, electron temperature reduces significantly with progressing time.

Equivalent equilibrium temperature behavior in the lattice subsystem differs from that of electron subsystem despite that thermal coupling is present via electron–phonon coupling constant in between both subsystems. In this case, relaxation time for phonons is smaller than that of electrons in the electron subsystem. Consequently, temperature increase and decay rate differ for both subsystems. The presence of thermal boundary resistance in between the aluminum and diamond films suppresses the amount of phonon transport from the aluminum subsystem to the diamond film; nevertheless, temperature decay in the aluminum film is visible despite the presence of the thermal boundary resistance. In addition, phonon scattering at the interface boundary and thermal boundary resistance result in a temperature jump at the diamond interface. In the case of silicon and diamond films, the temperature attains lower values and a temperature jump also takes place at the interface due to phonon scattering at the boundary because of the thermal boundary resistance.

9.2.6 Solution of EPRT for Cross-Plane Transport Across Silicon and Diamond Thin Films Pair: Consideration of Diffusive Mismatch Model at Films Pair Interface (Two-Dimensional Case)

Two-dimensional transient phonon transport in the silicon–diamond thin film pair is investigated. The frequency-dependent Boltzmann transport equation is solved numerically with the appropriate boundary conditions; in this case, the thermal boundary resistance at the interface of the silicon and diamond films is incorporated. The diffusive and acoustic mismatch models are incorporated when formulating the thermal boundary resistance across the film pairs. Equivalent equilibrium temperature is predicted in the silicon–diamond films pair. The ballistic phonons make a significant contribution to energy transport in the silicon and diamond films, which is more pronounced with progressing heating duration. The presence of ballistic phonons suppresses the rise of equivalent equilibrium temperature in both films, and thus equivalent equilibrium temperature obtained from the frequency-dependent solution decays sharply as compared with its counterpart, corresponding to the frequency-independent solution. The low-temperature edges of the diamond film act like phonon sink, modifying equivalent equilibrium temperature across the film thickness. Because of the mismatch of phonon frequencies and properties of the films, thermal boundary resistance results in a temperature jump at the interface of the films pair. This is more pronounced for the frequency-independent solution of the phonon transport equation.

9.2.7 Solution of EPRT for Cross-Plane Transport Across Silicon and Diamond Thin Films Pair With Presence of Thermal Boundary Resistance: Consideration of Diffusive and Cut-Off Mismatch Models (Two-Dimensional Case)

A cross-plane phonon transport across the silicon–diamond films pair is considered and thermal boundary resistance at the interface of the films pair is formulated using the cut-off mismatch and diffusive mismatch models. The phonon intensity ratio due to both mismatch models is predicted for longitudinal acoustic and transverse acoustic phonon branches. A frequency-dependent transient solution of the Boltzmann transport equation across the films pair is obtained numerically incorporating the discrete ordinate method. Thermal disturbance across the edges of the films pair is introduced, initiating the cross-plane phonon transport. In the cut-off mismatch model, frequency matching of the phonons across the silicon–diamond interface is considered in accordance with the dispersion relations. Therefore, only phonons at different branches having the frequency matching between silicon and diamond can possibly transmit from the silicon film edge to the diamond film edge across the interface while satisfying the energy balance at the interface.

The temporal behavior of the thermal boundary resistance due to cut-off mismatch and diffusive mismatch models is compared. In addition, phonon equilibrium intensity ratios due to both models are obtained for the longitudinal and transverse acoustic branches. The study is extended to include the effect of heat source size, which is located at the high temperature (301K) edge of the silicon film, on the thermal boundary resistance at the films pair interface. Thermal boundary resistance predicted using the cut-off mismatch model is higher than its counterpart obtained from the diffusive mismatch model. However, the predictions of the thermal boundary resistance, due to cut-off mismatch model, are in the order of those reported in the previous study. Since the frequency mismatch is considered for each acoustic branch of phonons across the silicon-diamond interface, high frequency phonons do not possibly transmit from the silicon film edge to the diamond film edge due to frequency mismatch. This in turn results in a high equilibrium phonon intensity ratio at the silicon film edge. Increasing heat source size at the silicon film edge increases the phonon intensity ratio at the silicon film interface. This is associated with the quasiballistic behavior of the emitted phonons having the high frequencies, which do not contribute to the film resistance through scattering in the film. Phonon intensity ratio due to cut-off mismatch and diffusive mismatch models remains less than 1 in the diamond film.

9.2.8 Solution of EPRT for Cross-Plane Transport Across Silicon and Silicon Thin Films Pair With Presence of Minute Size Vacuum Gap at Interface (One-Dimensional Case)

Phonon transport across thin silicon films pair with the presence of a gap is examined. To assess the phonon transport characteristics across the gap within the films pair, equivalent equilibrium temperature is introduced. Equivalent equilibrium temperature is defined anywhere in the thin silicon films pair and it corresponds to the average energy of all phonons around a local point, which is equivalent to the equilibrium temperature of phonons when they redistribute adiabatically to an equilibrium state. The transient solution of the frequency-dependent Boltzmann equation is presented and the phonon intensity distribution in the films pair is predicted numerically in accordance with the relevant boundary conditions. Since the gap size between the films pair is small (Casimir limit), radiative energy transport in between the edges of the films pair across the gap is incorporated together with phonon transmission and reflection at the edges of the gap within the films while satisfying the energy conservation. The presence of the gap in between the films pair modifies the phonon transport characteristics significantly. In this case, the difference between equivalent equilibrium temperatures across the gap increases with increasing gap size. This behavior is attributed to the amount of phonons transmitted from the first film edge to the second film edge across the gap. The gap size enlargement results in a phonon intensity increase at the first film edge because only those phonons having wavelengths larger than the gap size can possibly transmit to the second film interface. This, in turn, increases the phonon intensity at the film edge while causing attainment of large values of equivalent equilibrium temperature at the interface. Although increase in temperature at the first film edge results in increased rate of radiation heat transfer from the first film edge to second film edge across the gap, temperature at the second film edge remains low. This is associated with the amount of phonon transmittance from the first film edge to the second film edge, which becomes low with increasing gap size. Consequently, the amount of phonon transmittance to the second film edge has more effect on the temperature difference across the gap than the net radiation heat transfer due to Casimir limit consideration. Thermal conductivity of the films pair (k_{eff}) reduces with increasing gap size because of small rate of thermal energy transfer to the second film. The thermal conductivity ratios predicted in this study may guide the researchers to obtain the maximum benefit by altering the gap size.

9.2.9 Solution of EPRT for Cross-Plane Transport Across Silicon and Silicon Thin Films Pair With Presence of Minute Size Vacuum Gap at Interface and Near Field Radiation (One-Dimensional Case)

Thermal energy transfer across silicon thin films pair with the presence of a vacuum gap in between the thin films is considered and the transport characteristics are examined. The Boltzmann transport equation is incorporated to formulate frequency-dependent transient phonon transport across the films pair. Since the gap size is comparable to the mean free path of the silicon, energy transfer due to near field radiation from the film edges, located at the gap interface, is accommodated in the analysis. Temperature disturbance is applied at the outer edges of the films pair and equivalent equilibrium temperature is introduced to assess the phonon intensity distribution in the film during the heating period. Equivalent equilibrium temperature at both edges, located across the gap interface, of the films pair varies with the first film thickness. Increasing the first film thickness reduces equivalent equilibrium temperature at the film edges. Although heat transfer due to near field radiation is incorporated across the gap, the effect of near field radiation heat transfer on temperature increase at the edges of the film, across the gap interface, is not considerable as compared to that corresponding to phonons transmitted across the gap. Increasing the gap size increases temperature at the first film edge. This is because of the reduction of the number of phonons transmitted to the second film interface across the gap. In this case, only phonons having larger wavelengths than the gap size can possibly be transmitted to the second film edge across the gap. Increasing the gap size lowers the effective thermal conductivity of the films pair. In addition, reducing the first film thickness lowers the effective thermal conductivity for all gap sizes incorporated in the simulations.

9.2.10 Solution of EPRT for Cross-Plane Transport Across Aluminum and Silicon Thin Films Pair With Presence of Minute Size Vacuum Gap at Interface and Laser Short-Pulse Irradiation (One-Dimensional Case)

Energy transport across the aluminum and silicon thin films with the presence of the minute size vacuum gap at the interface is examined. Laser short-pulse irradiation is considered to initiate the nonequilibrium heating in the aluminum film. Since aluminum has electrons in the conduction band, thermal separation of electron and lattice subsystems takes place during the short heating period. This in turn alters the phonon intensity in the aluminum film because of the thermal communication between the electron and lattice subsystems through electron–phonon coupling. In order to resemble the energy transport in the films pair, the frequency-dependent Boltzmann equation is incorporated in each subsystem and transport equations are coupled through the electron–phonon coupling. The absorption of incident laser radiation is incorporated in the electron subsystem as a volumetric heat source after the assumption of the Lamberts Beer law, which assumes the exponential decay of the laser intensity within the absorption depth of the substrate material. Since the vacuum gap size is considered to be less than the mean free path, Casimir limit is introduced to model the thermal radiation across the vacuum gap. Since the frequency-dependent Boltzmann equation is incorporated, only the ballistic phonons having wavelengths greater than the vacuum gap and satisfying the energy balance at the interface are allowed to jump across the vacuum gap. In order to quantify the phonon intensity distribution in the films pair, equivalent equilibrium temperature is introduced, which is defined anywhere in the films pair. It bases the analog of the thermodynamic temperature defined in the diffusive limit in any medium and represents the average energy of all phonons around a local point. Therefore, it is equivalent to the equilibrium temperature of phonons when they redistribute adiabatically to an equilibrium state. The gap size has significant effect on equivalent equilibrium temperature at the aluminum and silicon edges in the lattice subsystem; in this case, increasing vacuum gap size increases temperature difference across the vacuum gap. Energy transfer is mainly governed by the ballistic phonon transfer across the vacuum gap and contribution of thermal radiation, because of Casimir limit, to the energy transfer across the vacuum gap remains small. Equivalent equilibrium temperature in the electron subsystem does not follow the temporal variation of the laser pulse intensity.

Index

Note: Page numbers followed by "*f*" and "*t*" refer to figures and tables, respectively.

F

G

H

P

Q

R

S

T

U

V

W

X

Y